Innovative Conservation Cropping Systems and Practices

Innovative Conservation Cropping Systems and Practices

Editors

Chengfang Li
Lijin Guo

MDPI • Basel • Beijing • Wuhan • Barcelona • Belgrade • Manchester • Tokyo • Cluj • Tianjin

Editors
Chengfang Li
Huazhong Agricultural University
China

Lijin Guo
Fujian Agriculture and Forestry University
China

Editorial Office
MDPI
St. Alban-Anlage 66
4052 Basel, Switzerland

This is a reprint of articles from the Special Issue published online in the open access journal *Agriculture* (ISSN 2077-0472) (available at: https://www.mdpi.com/journal/agriculture/special_issues/Innovative_Conservation_Cropping_Systems).

For citation purposes, cite each article independently as indicated on the article page online and as indicated below:

LastName, A.A.; LastName, B.B.; LastName, C.C. Article Title. *Journal Name* **Year**, *Volume Number*, Page Range.

ISBN 978-3-0365-6776-1 (Hbk)
ISBN 978-3-0365-6777-8 (PDF)

Contents

About the Editors

Chengfang Li

Chengfang Li, Doctor, associate professor, College of Plant Science and Technology of Huazhong Agricultural University; major in Conservation tillage management and agricultural ecology, including conservation tillage, plant fertilization, water-saving irrigation, global warming potential, soil nutrient cycles, soil carbon sequestration, crop yield; soil microbial community, greenhouse gas emission, etc.

Lijin Guo

Lijin Guo, Doctor, Lecturer, College of Resources and Environment, Fujian Agriculture and Forestry University; major in conservation tillage management and soil organic carbon, including conservation tillage, soil organic carbon, carbon sequestration, soil microbial community, mycorrhizal fungi, abiotic stress, greenhouse gas emission, climate change impact and adaptation, etc.

Preface to "Innovative Conservation Cropping Systems and Practices"

Maintaining economically and environmentally sustainable cropping systems and practices is one of the most imperative challenges in innovative sustainable agriculture. In this view, it is essential to note that the sacrifice of crop yields in the farmland ecosystem may not appeal to farmers heavily focused on increasing economic outcomes. Therefore, a deeper understanding of how to innovate cropping systems and practices with the aim of maintaining sustainability in agriculture is of crucial importance. Innovative conservation cropping systems and practices can improve agroecosystem productivity, reduce energy input, increase synergies between food production and ecosystem conservation, and increase farmers' profits.

Chengfang Li and Lijin Guo
Editors

agriculture

MDPI

Article

Effects of Lake Sediment on Soil Properties, Crop Growth, and the phoD-Harboring Microbial Community

Feng Zhong [1,†], Naling Bai [2,†], Xiangqian Chu [3], Yu He [2], Hanlin Zhang [2,*] and Haibo Li [1,*]

1 Faculty of Resources and Environmental Science, Hubei University, Wuhan 430062, China
2 Institute of Eco-Environment and Plant Protection, Shanghai Academy of Agricultural Sciences, Shanghai 201403, China
3 College of Life Sciences, Shanghai Normal University, Shanghai 200234, China
* Correspondence: zhanghanlin@saas.sh.cn (H.Z.); lhb@hubu.edu.cn (H.L.)
† These authors contributed equally to this work.

Abstract: Removal of lake sediment has been shown to be an effective method for lake restoration. High phosphorus (P) content makes it possible for lake sediment to provide fertility for agricultural production. However, little research has focused on the responses of the soil-phosphorus-related microbial community to the sediment-derived fertilizer enriched in phosphorus content. The phoD-harboring gene, important to the global phosphorus cycle, encodes alkaline phosphatase hydrolyzing organic P in soil. Accordingly, a plot experiment was performed to compare the effects of four different fertilization treatments—no-fertilizer control (CK), 50% chemical fertilization with compressed sediment (CS), 50% chemical fertilization with original lake sediment (S), and conventional chemical fertilization treatment (CT)—on the phoD gene community using QPCR and high-throughput sequencing analysis. Relationships among soil physicochemical properties, phoD-harboring microbial community abundance and composition were also evaluated. Results showed that compared to CT, CS significantly increased soil organic matter (SOM) content by 20.29%, and S enhanced the humus content by 20.75% ($p < 0.05$). There was no significant influence on phoD gene microbial community richness (Chao1 and Sobs indexes) and diversity (Shannon index) between all treatments. The CS treatment significantly altered the phoD community structure and enhanced the Chinese cabbage yield by 40.19% ($p < 0.05$). Pearson analysis showed that phoD gene abundance (copy number) had significant and negative relationships with SOM, total nitrogen (TN), total phosphorus (TP), available nitrogen (AN), available phosphorus (AP), and the Chao1 index. Redundancy analysis showed that shifts in the phoD community structure were related to soil physicochemical properties (SOM, TN, TP, AN, AP, and humus) rather than soil pH. In conclusion, the compressed sediment can be used in farmland since it optimizes the phoD-harboring microbial community abundance, composition, and structure, and thus significantly increases the Chinese cabbage yield.

Keywords: lake sediment; phosphorus; Chinese cabbage; phoD gene; high-throughput sequencing

Citation: Zhong, F.; Bai, N.; Chu, X.; He, Y.; Zhang, H.; Li, H. Effects of Lake Sediment on Soil Properties, Crop Growth, and the phoD-Harboring Microbial Community. *Agriculture* **2022**, *12*, 2065. https://doi.org/10.3390/agriculture12122065

Academic Editors: Chengfang Li and Lijin Guo

Received: 28 September 2022
Accepted: 28 November 2022
Published: 1 December 2022

1. Introduction

Phosphorus (P) is one of the major nutrients necessary for plants and also a crucial nutrient element in the farmland ecosystem [1]. However, the P utilization rate is only 5–25% in Chinese agricultural production [2]. Soil organic P accounts for 30–80% of the total P but cannot be directly used by plants [3]. It can only be converted into inorganic P through phosphatase and then used for growth and metabolism by plants [4]. Alkaline phosphatase (ALP) is an enzyme that hydrolyzes soil organic P to orthophosphate available for plants [5], mainly including phoA, phoX, and phoD [6]. Among them, phoA and phoX genes are mainly distributed in the aquatic environment, and phoD is usually found in terrestrial ecosystems [7]. Among ALP homologous genes, phoD is the most common gene in 16s rRNA metagenomic datasets and has become an important indicator of the soil P cycle [6,8]. Under the condition of low P, the phoD gene expression is induced to increase the phosphate absorption and transport enzymes, thus accelerating the transformation process of soil organic P. In contrast, under the condition of sufficient P, the expression of the phoD gene is inhibited, and the ALP activity is decreased, which is conducive to the stable accumulation of organic P [6]. Luo et al. [9] found that long-term organic and inorganic fertilizer applications significantly increased the phoD gene abundance, while a single application of inorganic fertilizer showed the opposite behavior. They also noticed that the content of available phosphorus (AP) was increased in long-term fertilizer treatment, which significantly reduced the abundance and diversity index of the phoD microbial community. Several reports have indicated that ALP activity is related to phoD gene abundance and community structure [5,6]. Factors such as soil pH [10,11], land use [12], and vegetation [13] might significantly affect the phoD gene abundance and microbial community structure in the soil.

In recent years, with the excessive application of chemical fertilizers, nutrients in the soil have become more easily lost with water [14]. These extra nutrients tend to accumulate at the lake bottom and are at a risk of recycling to the above water body (i.e., internal nutrient loading), resulting in eutrophication [15]. There is in total around 200 million m^3 of sediment removed from water bodies in European nations each year [16]. Removal of sediment from lakes is effective for lake control and restoration [17]; it also can provide fertility for agricultural production because of the relatively high contents of nutrients such as nitrogen (N) and P [18]. Canet et al. [19] found that silt from the lake bottom enhanced the lettuce yield but did not affect the output of tomatoes. Kazberuk et al. [20] suggested that the yield of mustard was significantly increased by adding 5% of the dam sediment, but the contents of heavy metals in soil also increased. Previous studies have mainly focused on the impact of sediment on plant yield, soil nutrients, and heavy metal content, but there are few reports on the impact of sediment application on the phoD community.

In this study, high-throughput pyrosequencing and QPCR were used to assess the soil phoD community abundance and composition based on a plot trial with lake sediment addition. The aim of this research was to clarify the impact of lake sediment addition on soil biochemical properties, crop growth, and phoD gene microbial communities, and to provide scientific support for the efficient and ecological utilization of lake sediment with respect to a stable or increasing crop yield.

2. Materials and Methods

2.1. Site Description

The experiment was performed in the greenhouse test plot of Maqiao Youran Agricultural Base, China (30°59′44″ N, 121°20′39″ E). In this area, the average annual temperature and precipitation were 15.8 °C and 1178.0 mm, respectively. The soil texture was sandy loam, and the planting crop was *Brassica chinensis* L. The initial soil physicochemical characteristics in the test plots were: pH value of 8.47, total nitrogen (TN) value of 1.35 $g{\cdot}kg^{-1}$, total phosphorus (TP) value of 1.25 $g{\cdot}kg^{-1}$, soil organic matter (SOM) value of 31.26 $g{\cdot}kg^{-1}$.

2.2. Experimental Design

Brassica chinensis L. was planted and harvested on 1 March 2021 and 2 May 2021, respectively. Four treatments were set for the test: no fertilizer (CK), 50% chemical fertilization with compressed sediment (CS), 50% chemical fertilization with original lake sediment (S), and conventional inorganic fertilizer treatment (CT). Each treatment was performed in four replicates in a random block design. The planting area was 20 m^2. Based on the local regime, the pure amounts of N, P, and potassium (K) were the same in every fertilized treatment, which were N value of 375 kg/ha, P (P_2O_5) value of 225 kg/ha, and K (K_2O) value of 225 kg/ha, respectively. The original lake sediment was derived from Xuanmiaoguan lake in Yichang, Hubei Province, China and underwent sedimentation for 10~20 min and mechanical compression to form the compressed sediment (water content = 40%). The properties of the original lake sediment and the compressed sediment used in the experiment are shown in Table S1, and the content of heavy metals were all below the control standards for pollutants in sludges from agricultural use in China (GB 4284-84). In the trial test, the original lake sediment and compressed sediment accounted for half of the P supply in the S and CS treatments, respectively. The remaining nutrients in the experiment were supplemented by urea, $Ca(H_2PO_4)_2$, and K_2SO_4. All the fertilizers were used as base fertilizer before planting.

2.3. Sampling and Measurements

After crop harvest, the 0–20 cm layer of soil samples were collected using a five-point sampling method with a soil sampler of each plot and mixed together. After removing the dead leaves, stones, and roots, the soil samples were kept in sterilized bags and then brought back to the laboratory for study. One portion of the samples was ground to 0.25 mm in a sieve after natural air drying to determine the physicochemical properties, and the other portion was kept at −80 °C for QPCR analysis and high-throughput sequencing of the phoD-harboring microbial community.

The SOM was analyzed by potassium dichromate oxidation [21]. The soil pH was determined by potentiometry (water: soil = 2.5:1) [22]. The TN content was evaluated by the Kjeldahl method [21]. The soil AN was determined by the alkali diffusion method [23]. The soil TP was determined by the sulfuric acid–perchloric acid digestion method [24]. The soil AP was determined by the $NaHCO_3$–Mo-Sb anti spectrophotometric technique [25]. The measurement of humus content was according to Wu et al. (2020) [26].

2.4. Microbial Analysis

Soil DNA was extracted using an Omega E.Z.N.A.® Soil DNA Kit (D5625-02) following the manufacturer's instructions. The concentration and purity of the isolated DNA were measured using a spectrophotometer (RS232G, Eppendorf, Germany). The primer pairs for the phoD gene were ALPS-F73 (5′-CAGTGGGACGACCACGAGGT-3′) and ALPS-1101 (5′-GAGGCCGATCGGC-ATGTCG-3′) [27]. The QPCR quantitative tests were run in duplicates, and the amplifications were performed with a total volume of 20 μL, which contained 4 μM of respective primer, 1 μL DNA template (approximately 20 ng), and SYBR real-time PCR premix (Takara, Dalian, China). The thermal conditions were: 95 °C for 5 min, followed by 40 cycles of 95 °C for 5 s, and 60 °C for 40 s [28]. The standard curve was established using a serial dilution of purified plasmid DNA harboring phoD genes.

The PCR reactions were carried out in 25 μL mixtures containing 12.5 μL of Phusion Master Mix (New England Biolabs, Ipswich, MA, USA) (2×), 0.5 μL of each primer, 1 μL of DNA template, and 10.5 μL of H_2O. Samples were subjected to the following amplification program: 95 °C denaturation for 30 s, 40 cycles of 95 °C for 5 s and at 60 °C for 34 s, and a final extension at 72 °C for 7 min [28]. After quantification, the PCR amplicons were pooled in equal amounts ($ng \cdot \mu L^{-1}$), and paired-end 2 × 300 bp sequencing was performed using the Illumina NovaSeq platform at Allwegene Tech. (Beijing, China).

2.5. Statistical Analysis

To identify the significant differences between the average values of different treatments, one-way analyses of variance (ANOVA) with least significant difference (LSD) tests ($p < 0.05$) were performed using SPSS 26.0 software. Venn diagrams were generated by the Venn Diagram package (Adrian Dusa, University of Bucharest, Bucharest, Romina) of R (version 3.5.1) software. The phoD gene richness and diversity indexes (Chao1 and observed species; PD_whole_tree and Shannon) were calculated using Mothur (version v.1.30.1) (Patrick Schloss, University of Michigan, Ann Arbor, MI, USA). Partial least squares discriminant analysis (PLS-DA) was performed to investigate the differences of phoD-harboring microbial community structure among the treatments using R software. Redundancy analysis (RDA) was employed to determine the relationships between the phoD-harboring microbial community composition and the soil properties using R software. Using the AMOS (IBM SPSS AMOS 25) (IBM, Armonk, NY, USA) software, structural equation modelling (SEM) was established to evaluate the direct and indirect relationships among the diversity of the phoD-harboring microbial community, input of fertilizer and sediment, and soil physicochemical characteristics.

3. Results and Discussion

3.1. Soil Properties

Fertilization significantly affected the soil pH with an increase and decrease in the CT and CS treatments, respectively (Table 1, $p < 0.05$). The SOM, TN, and AP contents were highest in the CS treatment and lowest in the CK treatment ($p < 0.05$), while similar and intermediate values were observed in the CT and S treatments. The AN content in the CT treatment (98.03 mg·kg^{-1}) was 1.70, 1.31, and 1.26 times more than those of the CK, S, and CS treatments, respectively ($p < 0.05$). Fertilizer addition (CT, S, and CS treatments) notably increased the TN, TP, and AP contents compared to the CK treatment ($p < 0.05$). The humus content increased significantly only in the S treatment (4.48 g·kg^{-1}) in comparison with CK and CT ($p < 0.05$).

Table 1. Soil properties under different fertilization treatments.

Items	pH	SOM (g·kg^{-1})	TN (g·kg^{-1})	TP (g·kg^{-1})	Humus (g·kg^{-1})	AN (mg·kg^{-1})	AP (mg·kg^{-1})
CK	8.44 ± 0.06ab	30.36 ± 0.88c	1.33 ± 0.04b	1.19 ± 0.21b	3.41 ± 0.25b	57.6 ± 2.92c	24.64 ± 1.97b
CS	7.77 ± 0.04c	49.26 ± 2.89a	1.84 ± 0.19a	1.87 ± 0.22a	4.06 ± 0.16ab	77.89 ± 6.64b	46.30 ± 2.45a
S	8.23 ± 0.22b	37.68 ± 3.25b	1.64 ± 0.06a	1.90 ± 0.03a	4.48 ± 0.46a	74.55 ± 6.13b	44.66 ± 2.16a
CT	8.52 ± 0.04a	40.95 ± 5.18b	1.73 ± 0.08a	1.83 ± 0.06a	3.71 ± 0.18b	98.03 ± 1.6a	45.11 ± 1.34a

Note: The values present the average ± standard deviation ($n = 4$). CK: no fertilizer application; CS: 50% chemical fertilization with compressed sediment; S: 50% chemical fertilization with original lake sediment; CT: conventional treatment; SOM: soil organic matter; TN, total nitrogen; TP: total phosphorus; AN: available nitrogen; AP: available phosphorus. Different lowercase letters indicate significant differences between treatments analyzed by ANOVA using LSD test ($p < 0.05$).

3.2. Crop Yield and Growth-Related Traits

Fertilization significantly improved the yield and quality of Chinese cabbage (Table 2). Fertilizer application (CT, S, and CS treatments) resulted in 2.25~3.15 times the vegetable yield of that in the CK treatment, among which the production in the CS treatment was the highest (3992.67 kg·667 m^{-2}) ($p < 0.05$), and no significant difference was observed between the CT and S treatments. The S and CS treatments exhibited more obvious enhancement in plant height than the CK and CT treatments (Table 2, $p < 0.05$), but there were no differences between them. Similarly, lake-sediment-derived substances (S and CS treatments) increased the chlorophyll contents to 34.93 mg·kg^{-1} and 36.27 mg·kg^{-1}, respectively. Different fertilizers did not change the p-TN content (1.98~2.06 g·kg^{-1}); the highest p-TP contents were in the S and CS vegetables, followed by that of the CT sample (0.30 g·kg^{-1}) ($p < 0.05$). The CK vegetable had the lowest p-TP content (0.20 g·kg^{-1}), which was almost 1.53~1.76-fold lower than that in the other treatments.

The application of chemical fertilizer or organic matter is a widely used agricultural practice to improve the soil P supply [29]. The S and CS sediments have a relatively neutral pH (7.30~7.45) (Table S1), which might influence the soil pH to a certain degree. The CS possesses nutrients to an extent almost 10-fold higher than S, so they had similar contents of TN, TP, AN, and AP in the soil, but not SOM content. Adding fertilizers significantly increased the storage (TP) and availability (AP) of P (Table 1). There was no significant difference of TP content in CT, S, and CS treatments; this may correspond to the same concentration of total P applied in the three treatments. However, the p-TP contents in the S and CS vegetables were 13.33~16.67% higher than in the CT treatment ($p < 0.05$, Table 2). Regarding the crop yield, the CS treatment had the highest P apparent utilization efficiency.

Table 2. The yield and quality of Chinese cabbage under different treatments.

Items	Yield (kg·667m^{-2})	Plant Height (cm)	Chlorophyll (mg·kg^{-1})	p-TN (g·kg^{-1})	p-TP (g·kg^{-1})
CK	1266.55 ± 45.13c	11.3 ± 0.88c	28.85 ± 0.91c	1.19 ± 0.08b	0.20 ± 0.01c
CS	3992.67 ± 85.27a	23.38 ± 0.56a	34.93 ± 0.47ab	2.04 ± 0.06a	0.34 ± 0.01a
S	2835.2 ± 108.69b	23.74 ± 1.04a	36.27 ± 1.32a	1.98 ± 0.16a	0.35 ± 0.00a
CT	2847.95 ± 105.48b	19.46 ± 1.38b	33.18 ± 1.57b	2.06 ± 0.12a	0.30 ± 0.01b

Note: p-TN: TN content in Chinese cabbage; p-TP: TP content in Chinese cabbage. Dates in the table are Mean ± SE; Different letters in the same column indicate a significant difference ($p < 0.05$).

3.3. phoD-Harboring Microbial Abundance Analysis

In the case of phosphorus-derived organic materials as an alternative to chemical fertilizers, it is important to understand how these organic materials affect soil phoD microbes, since they could accelerate the mineralization of organophosphates. CK soil had the highest phoD gene abundance (5.85×10^7 copy·g^{-1} soil) (Figure 1a). Fertilization led to a substantial 2.02~2.95-fold reduction in the phoD gene abundance compared to CK ($p < 0.05$), with no significant differences between the CT, S, and CS treatments.

In this study, a total of 1,198,288 clean data were obtained, and subsequently 9761 OTUs were generated. After rarefication, 9280 OTUs remained in the soil samples. The level of Good's coverage per sample was >97%, suggesting that the majority of phoD gene diversity in the sample soils was captured. The total OTUs in the CK, CT, S, and CS soils were 3893, 4966, 5320, and 4523, respectively (Figure 1b). The shared OTUs of the four treatments were 1158; the unique OTUs were 1641, 450, 1051, and 914, respectively. The shared OTUs mainly belonged to p_unidentified, Actinobacteria, and Proteobacteria. Although the CK soil had the lowest total OTUs (3893), it exhibited the highest number of unique OTUs (1641). The S and CS treatments alleviated the decrease in unique OTUs compared to the CT treatment. Therefore, fertilization regimes reshaped the phoD gene microbial community.

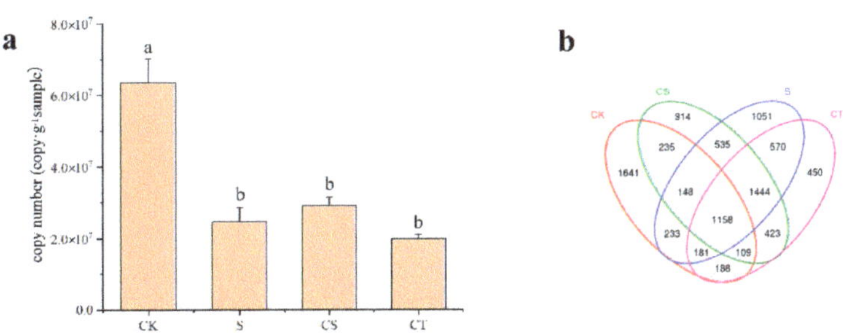

Figure 1. (**a**) Abundance of microbial phoD gene (copy number) quantified by QPCR. (**b**) Unique and shared OTUs among the different treatments by Venn analysis. Different lowercase letters indicate a significant difference between treatments ($p < 0.05$).

The high phoD gene abundance was thought to be caused by the available P deficiency, so microbes need to increase the high expression of genes related to P transportation and absorption in order to facilitate plant uptake. In addition, it has been well proved that low available P favors the synthesis of phosphatases [30], which was also identified by the Pearson analysis (Table 3). However, in the CT treatment, the P input comprised mainly small molecules which could be directly absorbed and utilized by plants, so the expression of phoD was definitely inhibited. Organic fertilizer, rich in C substrate but low in available P, probably favored the proliferation of some phoD-harboring species to mineralize organic phosphonate, thus increasing the phoD gene abundance (Figure 1a). Luo et al. [9] reported that chemical-only fertilization had the lowest phoD gene abundance, and organic–inorganic mixed fertilization would significantly increase it when compared with the control. However, Chen et al. reported that phoD gene abundance was highest in organic–inorganic mixed soil samples, and lowest in inorganic chemical fertilization [29]. Such conflicting results regarding the exogenous organic substance may be due to fertilization regime, crop type, and intrinsic soil property. Long-term soil with no P had the lowest phoD microbial diversity and total bacterial diversity [31]. Long-term utilization of chemical fertilizers would inhibit the growth of the phoD bacterial community [9,32]. Such a phenomenon was not noticeable in this short-term trial assay, and the temporal variation needs further analysis.

Table 3. Pearson correlations between phoD gene abundance, soil properties, and α diversity under different treatments.

	phoD	pH	SOM	TN	AN	TP	AP	Humus	Chao1	Shannon
PhoD	1									
pH	0.193	1								
SOM	−0.534 *	−0.662 **	1							
TN	−0.683 *	−0.500 *	0.772 **	1						
AN	−0.712 *	0.045	0.496	0.670 **	1					
TP	−0.888 **	−0.369	0.630 **	0.700 **	0.599 *	1				
AP	−0.885 **	−0.398	0.718 **	0.835 **	0.689 **	0.869 **	1			
Humus	−0.471	−0.493	0.375	0.371	0.110	0.599 *	0.592 *	1		
Chao1	−0.628 **	−0.379	0.431	0.399	0.371	0.561 *	0.682 **	0.422	1	
Shannon	−0.084	−0.221	0.073	−0.083	−0.056	−0.057	0.033	−0.146	0.628 **	1

Note: *: correlation is significant at the 0.05 level; **: correlation is significant at the 0.01 level.

3.4. *phoD Microbial Community α and β Diversity Analysis*

The Chao1, Observed_species (Sobs), PD_whole_tree, and Shannon indexes were calculated to analyze the phoD microbial α diversity (Figure 2). For Chao1 and Sobs indexes, S and CS treatments showed significant increases compared with the unfertilized soil (CK) ($p < 0.05$). The CT treatment had lower microbial richness, but there was no significant difference between the CT, S, and CS treatments. The microbial community diversity index, PD_whole_tree, was higher under fertilizer application treatments than the CK; however, it is noteworthy that there was no notable difference between the CT, S, and CS treatments (Figure 2c). Fertilization, or not, had no effect on the Shannon index (Figure 2d).

The PLS-DA revealed a good model to differentiate the CK from other fertilization treatments at the OTU level, with a total explanatory degree of 26.79% (Figure 3). Unfertilized soil (CK) and fertilized soil (CT, S, and CS) were separated by the PC1 axis (17.75%). The PC2 axis distinguished the CS treatment from the CT and S treatments, with an interpretation of 9.04%. Samples of the CT and S treatments gathered together, indicating a similar phoD microbial community structure in their soil samples. Furthermore, PERMANOVA showed a notable structural difference between the treatments with $p = 0.001$ (data not shown). In the present study, the compressed sediment application significantly affected the phoD gene community structure in the vegetable soil.

Chen et al. [33] reported that long-term P fertilizer input enhanced phoD gene diversity. However, in the present study, only the S treatment increased the PD_whole_tree compared with the CK, and fertilization showed no influence on the Shannon index (Figure 2). A contradictory finding concluded that no detectable effects were observed on soil microbial abundance and diversity after repeated applications of sediments for two seasons; at least, such effects on soil microbial ecology variation seemed to be more remarkable in long-term fertilization experiments [34]. Sapp et al. [35] in a short-term greenhouse trial concluded that digestate application decreased the bacterial community diversity. Similarly, in the present study, the original lake/compressed sediment attenuated the decrease in phoD gene microbial richness and diversity.

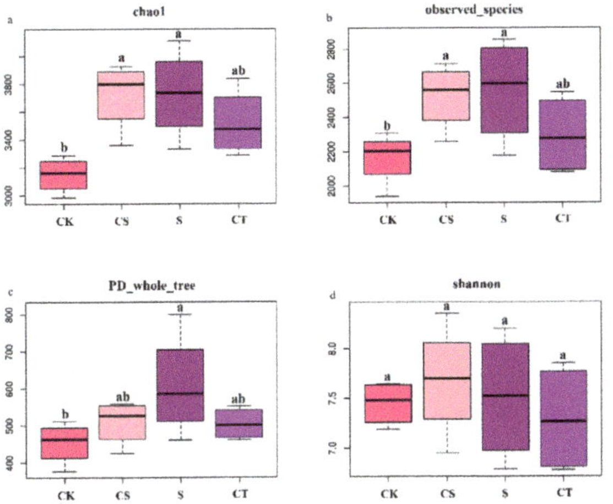

Figure 2. Comparison of the estimated OTU α-diversity indexes ((**a**): chao1, (**b**): observed_species, (**c**): PD_whole_tree, (**d**): shannon) of the phoD microbial community in different fertilizer treatments. Different lowercase letters indicate a significant difference between treatments ($p < 0.05$).

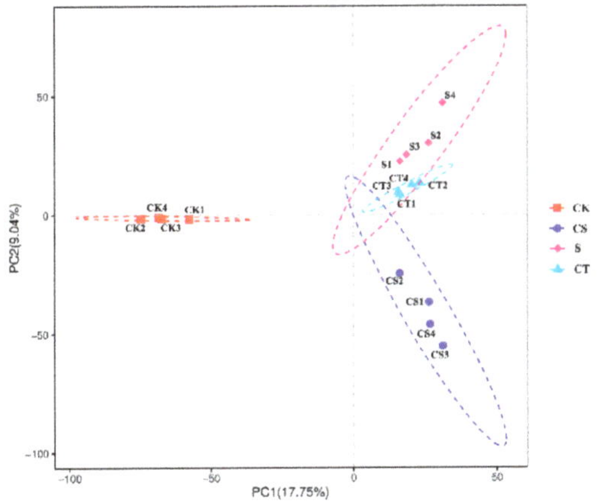

Figure 3. Analysis of the phoD microbial community structure in different fertilizer treatments by PLS-DA.

3.5. Comparison of phoD Microbial Community Composition among the Different Treatments

Fertilization did not change the composition of the phoD microbial community, but significantly affected its relative abundances (Figure 4). The species across all soil samples were classified into 17 phyla, with the predominated phyla p_unclassified, Actinobacteria, and Proteobacteria occupying 66.86~78.62%, 7.36~19.99%, and 1.12~5.85%, respectively. The minimal phyla were Nitrospirae, Lentisphaerae, Ascomycota, Euryarchaeota, and Fornicata. One-way ANOVA of the 17 phyla taxa showed that the relative abundances of nine bacterial phyla were markedly influenced by the fertilization regime compared with CK ($p < 0.05$). Compared to the CT treatment, the relative abundances of Lentisphaerae, Nitrospiare, and Fornicata were significantly decreased in the CS treatment; on the contrary, Ascomycota and Chloroflexi contents were increased ($p < 0.05$). Ascomycota, Bacteroidetes, and Fornicata were distinctly affected in the S soil compared with the CT treatment ($p < 0.05$) (Figure 4c). The relative abundances of dominant genera are shown in Figure 4b,d, and the genera with an abundance of less than 1% was classified as "Others". The top five dominant genera were g_unidentified, Planctomyces, Streptomyces, Bradyrhizobium, and Phenylobacterium. In comparison with the CK treatment, fertilization treatments (CT, S, and CS) increased the relative abundances of Plantactinospora, Bosea, Phenylobacterium, Scytonema, Auraticoccus, Planctomyces, and g_unidentified ($p < 0.05$). Vriovorax, Pseudomonas, Ramlibacter, Luteipulveratus, Deinococcus, Bradyrhizobium, Thermobispora, Streptomyces, and Saccharopolyspora contents were decreased by fertilizer application ($p < 0.05$). The S and CS treatments significantly decreased the contents of Phenylobacterium (47.04% and 68.45%, respectively) and Bradyrhizobium (18.48% and 55.26%, respectively), compared with the CT treatment ($p < 0.05$).

a b

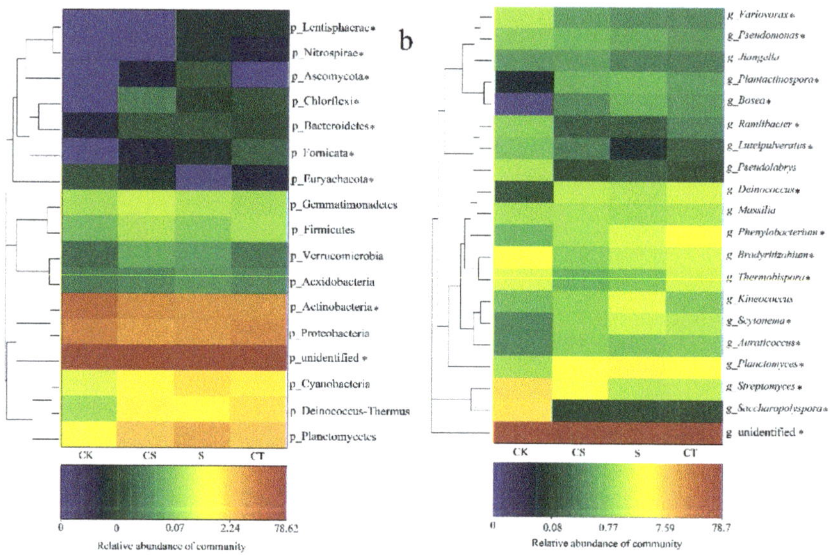

c

	CK	CS	S	CT
p__Lentisphaerae	b	b	a	a
p__Nitrospirae	b	b	a	a
p__Ascomycota	c	b	a	c
p__Chloroflexi	b	a	b	b
p__Bacteroidetes	b	b	a	b
p__Fornicata	d	c	b	a
p__Euryarchaeota	a	ab	c	bc
p__Gemmatimonadetes	a	a	a	a
p__Firmicutes	a	a	a	a
p__Verrucomicrobia	a	a	a	a
p__Acidobacteria	a	a	a	a
p__Actinobacteria	a	b	b	b
p__Proteobacteria	a	a	a	a
p__unidentified	b	a	ab	ab
p__Cyanobacteria	a	a	a	a
p__Deinococcus-Thermus	a	a	a	a
p__Planctomycetes	a	a	a	a

d

	CK	CS	S	CT
g__Variovorax	a	b	b	b
g__Pseudomonas	a	ab	ab	b
g__Jiangella	a	a	a	a
g__Plantactinospora	b	a	a	a
g__Bosea	b	a	a	a
g__Ramlibacter	a	b	b	b
g__Luteipulveratus	a	ab	b	ab
g__Pseudolabrys	a	a	a	a
g__Deinococcus	a	b	b	b
g__Massilia	a	a	a	a
g__Phenylobacterium	c	b	b	a
g__Bradyrhizobium	a	c	bc	ab
g__Thermobispora	a	b	ab	ab
g__Kineococcus	a	a	a	a
g__Scytonema	b	ab	a	a
g__Auraticoccus	b	a	ab	ab
g__Planctomyces	b	ab	a	ab
g__Streptomyces	a	b	b	b
g__Saccharopolyspora	a	b	b	b
g__unidentified	b	a	ab	ab

Figure 4. Comparison of the phoD microbial community composition in different treatments at the phylum (**a**) and genus levels (**b**). Different letters in the same column indicate a significant difference ($p < 0.05$) ((**c**): at phylum level, (**d**): at genus level).

Many micoorganisms can degrade and transform nutrients from soil organic matter, improve soil quality, and increase crop yield [36]. The dominant phoD-harboring microbes include Actinobacteria, Proteobacteria, Planctomycetes, and Cyanobacteria, which was in accordance with some other studies [9,37]. Generally, the variation in relative abundances of the phoD gene was not notable between the CT, S, and CS treatments, suggesting that the phoD-containing bacteria were not as sensitive to phosphate fertilizer as the 16S rRNA bacteria. The P fertilizer, soil types, and experimental conditions may explain the divergent results reported in the various studies [38].

Ragot et al. [10] emphasized that Bradyrhizobium and Streptomyces were the dominant genera of phoD-harboring microbes and are not affected by changes in environmental conditions. Bradyrhizobium and Streptomyces were both monitored in this study, but their relative abundances were significantly decreased by fertilization (Figure 4d), which was in accordance with the relatively low copy number of the phoD gene in the CT, S, and CS treatments (Figure 1a). Bradyrhizobium is a symbiotic N_2 fixer and may play important roles in coupling the soil N and P cycles. Some α-Proteobacteia (e.g., Bradyrhizobium) increased ALP activity and P transport rates as a response to P stress, so CK had the highest relative abundance of Bradyrhizobium (3.26%). Furthermore, Bradyrhizobium may fix N in the air to avoid N limitation in the CK and CT treatments. Streptomyces is closely related with P transformation, particularly mineralizing organic P and phosphate solubilization [39]. Their relative abundance seemed to sensitively respond to P limitation, at least in the present study (Figure 4d). Pseudomonas content was significantly decreased by inorganic fertilizer (CT treatment) (Figure 4d); bacteria belonging to this genus usually contribute to mineralizing organophosphorus and thus play an important role in promoting plant growth. Phenylobacterium was also known as a N-fixer bacterium, belonging to the Proteobacteria phylum; CT significantly increased its relative abundance compared to CK and the other fertilizer addition treatments (S and CS). Therefore, future study should focus on the investigation of N cycling and P turnover under various conditions.

In relation to the phoD microbial community composition tested in the fertilization application and their agricultural importance, most dominant groups were found to have unknown or not well-reported roles in the soil P cycle and/or crop growth, such as the "unidentified" species shown in Figure 4a,b. This might embody the need for a deeper sequencing depth between metagenomics and phoD alleles with more comprehensive primers. The ALP activity and phoD expression should be considered in future research.

3.6. Comparison of phoD Microbial Community Composition among the Different Treatments

Correlations were investigated between soil properties and the phoD gene communities (Table 3, Figure 5). Results showed that phoD gene copy number was significantly and negatively correlated with SOM*, TN*, AN*, TP**, AP**, and phoD gene microbial community richness (Chao1**) (Table 3, * $p < 0.05$, ** $p < 0.01$). On the other hand, phoD gene richness also positively correlated with soil TP ($p < 0.05$) and AP ($p < 0.01$), while phoD gene community diversity (Shannon index) only positively correlated with the Chao1 index ($p < 0.01$) and had no correlations with soil properties and phoD gene abundance. Soil pH showed no obvious correlations with phoD gene α diversity (Chao1 and Shannon), which was in contrast with Chen et al. reports [29,33].

The first two axes of RDA explained 49.54% and 9.50% of the total variation (Figure 5). The phoD gene community of the CK soil was separated from the other fertilized soils (CT, S, and CS) along the RDA1 axis. However, the CS samples could not differentiate from the CT and S samples by the RDA2 axis. The structure of the phoD gene community was strongly correlated with TP (r2 = 0.702, $p = 0.001$) and AP (r2 = 0.827, $p = 0.001$), and to a lesser extent with SOM (r2 = 0.523, $p = 0.007$), TN (r2 = 0.544, $p = 0.012$), AN (r2 = 0.487, $p = 0.009$), and humus (r2 = 0.458, $p = 0.019$) except soil pH (r2 = 0.123, $p = 0.431$) (Table S2). Soil AP was strongly correlated with the RDA2 axis, indicating that the phoD bacterial community structure was mainly related to available nutrient changes caused by P inputs. Environmental factors played more roles in the phoD microbial community structure of fertilized treatments (CT, S, and CS) than in the CK treatment.

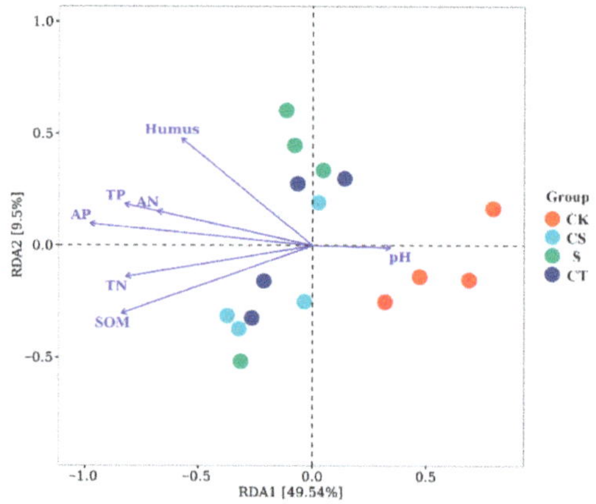

Figure 5. Ordination plots by RDA at the phylum level to explore the relationships between the phoD community and soil properties. Samples from different fertilization treatments were marked with different colors.

3.7. *Structural Equation Model Analysis of the phoD Microbial Communities*

The SEM analysis quantified the impact of each environmental factor and were expressed as path coefficients (Figure 6). Fertilizers had significant effects on all the indicators (SOM, TN, AN, TP, and AP) with the path coefficients ranging from 0.770 to 0.851, except in pH and humus. Original lake sediment could indirectly improve the phoD gene diversity by increasing the humus content (path coefficient = 0.678). Compressed sediment had negative correlations with pH and AN with path coefficients of −0.554 and −0.387, respectively, and was positively correlated with the SOM content (path coefficient = 0.409). The original lake sediment and compressed sediment both had no notable influence on the AP and TP contents, though they were intended to be used to reduce the inorganic P input. The pH and humus were significantly and positively correlated with phoD gene diversity, with path coefficients of 0.585 and 0.296, respectively. The AP and TP were negatively correlated with phoD gene diversity, with the path coefficients of −0.832 and −0.462, respectively. Different fertilization treatments regulated the variation in phoD gene diversity mainly by changing the soil pH, humus, TP, and AP.

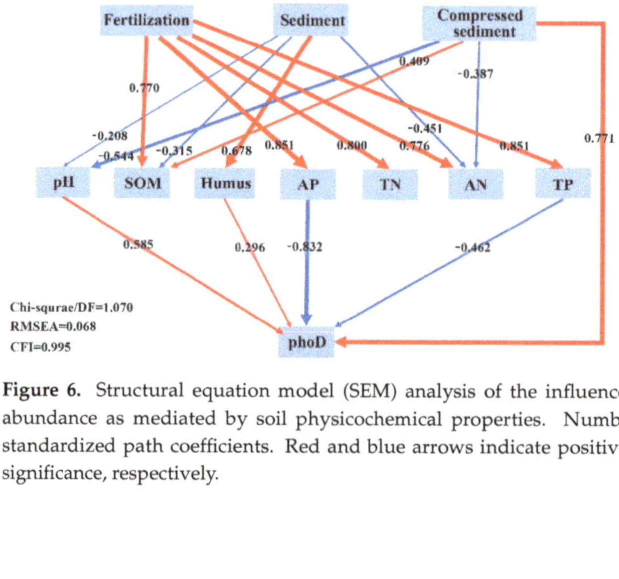

Figure 6. Structural equation model (SEM) analysis of the influences of fertilization on phoD abundance as mediated by soil physicochemical properties. Numbers adjacent to arrows are standardized path coefficients. Red and blue arrows indicate positive and negative effects with significance, respectively.

Many studies have emphasized that pH is associated with microbial and phoD communities and could be reflected as a strong indicator [10,11,29]. In our present study, pH exerted insignificant effects on phoD gene abundance, microbial community diversity, and structure. In addition, whether the dominant microbial community from the original lake/compressed sediment survived and thus affected the native microbial population of the soil was not analyzed in the present study. Bacteria and soil properties always altered by temporal and spatial variations [9].

4. Conclusions

The compressed sediment addition could enhance Chinese cabbage yield and increase the P cycling through the promotion of the soil phoD-harboring microbial community. Compressed sediment addition is a feasible technical means to improve the soil P supply and promote crop growth. The optimal addition quantity of compressed sediment in various agricultural fields will be tested in a further study.

Supplementary Materials: The following supporting information can be downloaded at: https://www.mdpi.com/article/10.3390/agriculture12122065/s1, Table S1: Basic properties of the original lake sediment and compressed sediment used in the experiment; Table S2: Monte Carlo permutation test of soil phoD microbial communities and soil properties.

Author Contributions: Conceptualization, H.Z. and N.B.; formal analysis, N.B.; investigation, F.Z. and Y.H.; data curation, H.L.; writing—original draft preparation, F.Z.; writing—review and editing, H.Z.; visualization, X.C.; project administration, H.L. and H.Z. All authors have read and agreed to the published version of the manuscript.

Funding: This work was supported by the Hubei Province Yangtze River Water Ecological Environmental Protection Research Project [Grant No. 2022-LHYJ-02-0506-04], the Shanghai Natural Science Foundation of the Shanghai Science and Technology Commission [Grant No. 21ZR1464500], and the Outstanding Team Program of the Shanghai Academy of Agricultural Science [Grant No. Hu-Nong-Ke-Zhuo 2022 (008)].

Institutional Review Board Statement: Not applicable.

Data Availability Statement: The datasets presented in this study can be found in online repositories. The names of the repository/repositories and accession number(s) can be found below: NCBI, PRJNA882984, https://www.ncbi.nlm.nih.gov/bioproject/PRJNA882984 (21 September 2022).

Acknowledgments: We thank Maqiao Youran Agricultural Base for providing the long-term experimental fields and the agricultural management.

Conflicts of Interest: The authors declare no conflict of interest.

References

1. Tiessen, H. Phosphorus in the global environment. In *The Ecophysiology of Plant-Phosphorus Interactions*; Springer: Berlin/Heidelberg, Germany; Dordrecht, The Netherlands, 2008; pp. 1–7.
2. Guang-Can, T.A.O.; Shu-Jun, T.I.A.N.; Miao-Ying, C.A.I.; Guang-Hui, X.I.E. Phosphate-solubilizing and-mineralizing abilities of bacteria isolated from soils. *Pedosphere* **2008**, *18*, 515–523.
3. Turner, B.L.; Frossard, E.; Baldwin, D.S. (Eds.) *Organic Phosphorus in the Environment*; CABI Pub.: Wallingford, UK, 2005; pp. 269–294.
4. Rawat, P.; Das, S.; Shankhdhar, D.; Shankhdhar, S.C. Phosphate-solubilizing microorganisms: Mechanism and their role in phosphate solubilization and uptake. *J. Soil Sci. Plant Nutr.* **2021**, *21*, 49–68. [CrossRef]
5. Ahmed, S.; Iqbal, N.; Tang, X.; Ahmad, R.; Irshad, M.; Irshad, U. Organic amendment plus inoculum drivers: Who drives more P nutrition for wheat plant fitness in small duration soil experiment. *PLoS ONE* **2022**, *17*, e0266279. [CrossRef] [PubMed]
6. Fraser, T.; Lynch, D.H.; Entz, M.H.; Dunfield, K.E. Linking alkaline phosphatase activity with bacterial phoD gene abundance in soil from a long-term management trial. *Geoderma* **2015**, *257*, 115–122. [CrossRef]
7. Neal, A.L.; Blackwell, M.; Akkari, E.; Guyomar, C.; Clark, I.; Hirsch, P.R. Phylogenetic distribution, biogeography and the effects of land management upon bacterial non-specific Acid phosphatase Gene diversity and abundance. *Plant Soil* **2018**, *427*, 175–189. [CrossRef]

8. Sarr, P.S.; Tibiri, E.B.; Fukuda, M.; Zongo, A.N.; Compaore, E.; Nakamura, S. Phosphate-solubilizing fungi and alkaline phosphatase trigger the P solubilization during the co-composting of sorghum straw residues with Burkina Faso phosphate rock. *Front. Environ. Sci.* **2020**, *8*, 559195. [CrossRef]
9. Luo, G.; Ling, N.; Nannipieri, P.; Chen, H.; Raza, W.; Wang, M.; Guo, S.; Shen, Q. Long-term fertilisation regimes affect the composition of the alkaline phosphomonoesterase encoding microbial community of a vertisol and its derivative soil fractions. *Biol. Fertil. Soils* **2017**, *53*, 375–388. [CrossRef]
10. Ragot, S.A.; Huguenin-Elie, O.; Kertesz, M.A.; Frossard, E.; Bünemann, E.K. Total and active microbial communities and phoD as affected by phosphate depletion and pH in soil. *Plant Soil* **2016**, *408*, 15–30. [CrossRef]
11. Hu, M.; Peñuelas, J.; Sardans, J.; Tong, C.; Chang, C.T.; Cao, W. Dynamics of phosphorus speciation and the phoD phosphatase gene community in the rhizosphere and bulk soil along an estuarine freshwater-oligohaline gradient. *Geoderma* **2020**, *365*, 114236. [CrossRef]
12. Neal, A.L.; Rossmann, M.; Brearley, C.; Akkari, E.; Guyomar, C.; Clark, I.M.; Allen, E.; Hirsch, P.R. Land-use influences phosphatase gene microdiversity in soils. *Environ. Microbiol.* **2017**, *19*, 2740–2753. [CrossRef]
13. Wang, F.; Zhang, Y.; Xia, Y.; Cui, Z.; Cao, C. Soil Microbial Community Succession Based on phoD and Gcd Genes along a Chronosequence of Sand-Fixation Forest. *Forests* **2021**, *12*, 1707. [CrossRef]
14. Itelima, J.U.; Bang, W.J.; Onyimba, I.A.; Sila, M.D.; Egbere, O.J. Bio-fertilizers as key player in enhancing soil fertility and crop productivity: A review. *Direct Res. J. Agric. Food Sci.* **2018**, *6*, 73–83.
15. Qin, B.; Gao, G.; Zhu, G.; Zhang, Y.; Song, Y.; Tang, X.; Xu, H.; Deng, J. Lake eutrophication and its ecosystem response. *Chin. Sci. Bull.* **2013**, *58*, 961–970. [CrossRef]
16. Kiani, M.; Raave, H.; Simojoki, A.; Tammeorg, O.; Tammeorg, P. Recycling lake sediment to agriculture: Effects on plant growth, nutrient availability, and leaching. *Sci. Total Environ.* **2021**, *753*, 141984. [CrossRef]
17. Huser, B.J.; Futter, M.; Lee, J.T.; Perniel, M. In-lake measures for phosphorus control: The most feasible and cost-effective solution for long-term management of water quality in urban lakes. *Water Res.* **2016**, *97*, 142–152. [CrossRef] [PubMed]
18. Nixon, S.W. Replacing the Nile: Are anthropogenic nutrients providing the fertility once brought to the Mediterranean by a great river? *AMBIO A J. Hum. Environ.* **2003**, *32*, 30–39. [CrossRef]
19. Canet, R.; Chaves, C.; Pomares, F.; Albiach, R. Agricultural use of sediments from the Albufera Lake (eastern Spain). *Agric. Ecosyst. Environ.* **2003**, *95*, 29–36. [CrossRef]
20. Kazberuk, W.; Szulc, W.; Rutkowska, B. Use bottom sediment to agriculture—Effect on plant and heavy metal content in soil. *Agronomy* **2021**, *11*, 1077. [CrossRef]
21. Zhang, H.; Sun, H.; Zhou, S.; Bai, N.; Zheng, X.; Li, S.; Zhang, J.; Lv, W. Effect of Straw and Straw Biochar on the Community Structure and Diversity of Ammonia-oxidizing Bacteria and Archaea in Rice-wheat Rotation Ecosystems. *Sci. Rep.* **2019**, *9*, 9367. [CrossRef]
22. Dupont, A.Ö.C.; Griffiths, R.I.; Bell, T.; Bass, D. Differences in soil micro-eukaryotic communities over soil pH gradients are strongly driven by parasites and saprotrophs. *Environ. Microbiol.* **2016**, *18*, 2010–2024. [CrossRef]
23. Khan, S.A.; Mulvaney, R.L.; Mulvaney, C.S. Accelerated diffusion methods for inorganic-nitrogen analysis of soil extracts and water. *Soil Sci. Soc. Am. J.* **1997**, *61*, 936–942. [CrossRef]
24. Song, H.; Li, Z.; Du, B.; Wang, G.; Ding, Y. Bacterial communities in sediments of the shallow Lake Dongping in China. *J. Appl. Microbiol.* **2012**, *112*, 79–89. [CrossRef] [PubMed]
25. Tan, Y.; Cui, Y.; Li, H.; Kuang, A.; Li, X.; Wei, Y.; Ji, X. Rhizospheric soil and root endogenous fungal diversity and composition in response to continuous Panax notoginseng cropping practices. *Microbiol. Res.* **2017**, *194*, 10–19. [CrossRef] [PubMed]
26. Wu, J.; Wei, Z.; Zhu, Z.; Zhao, Y.; Jia, L.; Lv, P. Humus formation driven by ammonia-oxidizing bacteria during mixed materials composting. *Bioresour. Technol.* **2020**, *311*, 123500. [CrossRef] [PubMed]
27. Huang, B.; Yan, D.; Ouyang, C.; Zhang, D.; Zhu, J.; Liu, J.; Li, Y.; Wang, Q.; Han, Q.; Cao, A. Chloropicrin fumigation alters the soil phosphorus and the composition of the encoding alkaline phosphatase PhoD gene microbial community. *Sci. Total Environ.* **2020**, *711*, 135080. [CrossRef] [PubMed]
28. Lagos, L.M.; Acuña, J.J.; Maruyama, F.; Ogram, A.; de la Luz Mora, M.; Jorquera, M.A. Effect of phosphorus addition on total and alkaline phosphomonoesterase-harboring bacterial populations in ryegrass rhizosphere microsites. *Biol. Fertil. Soils* **2016**, *52*, 1007–1019. [CrossRef]
29. Chen, X.; Jiang, N.; Condron, L.M.; Dunfield, K.E.; Chen, Z.; Wang, J.; Chen, L. Impact of long-term phosphorus fertilizer inputs on bacterial phoD gene community in a maize field, Northeast China. *Sci. Total Environ.* **2019**, *669*, 1011–1018. [CrossRef]
30. Nannipieri, P.; Giagnoni, L.; Renella, G.; Puglisi, E.; Ceccanti, B.; Masciandaro, G.; Fornasier, F.L.A.V.I.O.; Moscatelli, M.C.; Marinari, S.A.R.A. Soil enzymology: Classical and molecular approaches. *Biol. Fertil. Soils* **2012**, *48*, 743–762. [CrossRef]
31. Tan, H.; Barret, M.; Rice, O.; Dowling, D.N.; Burke, J.; Morrissey, J.P.; O'Gara, F. Long-term agrichemical use leads to alterations in bacterial community diversity. *Plant Soil Environ.* **2012**, *58*, 452–458. [CrossRef]
32. Liu, W.; Ling, N.; Luo, G.; Guo, J.; Zhu, C.; Xu, Q.; Liu, M.; Shen, Q.; Guo, S. Active phoD-harboring bacteria are enriched by long-term organic fertilization. *Soil Biol. Biochem.* **2021**, *152*, 108071. [CrossRef]
33. Chen, X.; Jiang, N.; Chen, Z.; Tian, J.; Sun, N.; Xu, M.; Chen, L. Response of soil phoD phosphatase gene to long-term combined applications of chemical fertilizers and organic materials. *Appl. Soil Ecol.* **2017**, *119*, 197–204. [CrossRef]

34. Coelho, J.J.; Hennessy, A.; Casey, I.; Bragança, C.R.S.; Woodcock, T.; Kennedy, N. Biofertilisation with anaerobic digestates: A field study of effects on soil microbial abundance and diversity. *Appl. Soil Ecol.* **2020**, *147*, 103403. [CrossRef]
35. Sapp, M.; Harrison, M.; Hany, U.; Charlton, A.; Thwaites, R. Comparing the effect of digestate and chemical fertiliser on soil bacteria. *Appl. Soil Ecol.* **2015**, *86*, 1–9. [CrossRef]
36. Huang, B.; Yan, D.; Wang, Q.; Fang, W.; Song, Z.; Cheng, H.; Li, Y.; Ouyang, C.; Han, Q.; Jin, X.; et al. Effects of dazomet fumigation on soil phosphorus and the composition of phoD-harboring microbial communities. *J. Agric. Food Chem.* **2020**, *68*, 5049–5058. [CrossRef] [PubMed]
37. Cheng, J.; Zhang, Y.; Wang, H.; Cui, Z.; Cao, C. Sand-fixation plantation type affects soil phosphorus transformation microbial community in a revegetation area of Horqin Sandy Land, Northeast China. *Ecol. Eng.* **2022**, *180*, 106644. [CrossRef]
38. Liu, J.; Ma, Q.; Hui, X.; Ran, J.; Ma, Q.; Wang, X.; Wang, Z. Long-term high-P fertilizer input decreased the total bacterial diversity but not phoD-harboring bacteria in wheat rhizosphere soil with available-P deficiency. *Soil Biol. Biochem.* **2020**, *149*, 107918. [CrossRef]
39. Wang, J.; Liu, S.; Li, S. Effect of long-term plastic film mulching and fertilization on inorganic N distribution and organic N mineralization in brown earth. *J. Soil Water Conserv.* **2006**, *20*, 107–110.

agriculture

Article

In Arid Regions, Forage Mulching between Fruit Trees Rows Enhances Fruit Tree Light and Lowers Soil Salinity

Shuai Zhang [1], Tingting Liu [1], Wenwen Wei [1], Lei Shen [1], Xiuyuan Wang [1,2], Tayir Tuertia [1], Luhua Li [1] and Wei Zhang [1,*]

[1] College of Agriculture, Shihezi University, Shihezi 832003, China
[2] College of Horticulture and Forestry Sciences, Huazhong Agricultural University, Wuhan 430070, China
* Correspondence: bluesky2002040@shzu.edu.cn

Abstract: Agroforestry is considered a means to provide sustainable and productive agriculture. This work aims to study the effect of fruit-grass agroforestry patterns on the soil moisture, salinity, growth, and yield of fruit trees, as well as to provide a reference for the development of agroforestry complex systems in Northwest China. The study has been designed with two cropping patterns: monocropped apple and apple-ryegrass intercropping. The results showed that compared to monocropped apples, intercropped apples have increased soil moisture content by 33.38–39.02%, net photosynthetic rate by 35.33–42.26%, transpiration rate by 29.62–29.76%, and stomatal conductance by 15.65–16.55% in the 0–60 cm soil layer. Intercrop reduced the total soil salt content by 36.41–38.58%, and the intercellular CO_2 concentration decreased by 5.96–6.61%. In addition, intercropping improves fruit yield and quality by improving the orchard environment and increasing tree height, breast height, north-south crown spread, and east-west crown spread. Therefore, increased yield and quality can be achieved by changing the fruit tree and ryegrass planting method, which is beneficial to the sustainable development of agriculture in Northwest China.

Keywords: apple-ryegrass intercropping; photosynthetic characteristics; soil salinity; soil moisture content; fruit quality

Citation: Zhang, S.; Liu, T.; Wei, W.; Shen, L.; Wang, X.; Tuertia, T.; Li, L.; Zhang, W. In Arid Regions, Forage Mulching between Fruit Trees Rows Enhances Fruit Tree Light and Lowers Soil Salinity. *Agriculture* **2022**, *12*, 1895. https://doi.org/10.3390/agriculture12111895

Academic Editor: Daniela Farinelli

Received: 9 September 2022
Accepted: 8 November 2022
Published: 10 November 2022

1. Introduction

Soil salinization is a major constraint to sustainable agricultural development in arid and semi-arid regions [1]. Xinjiang, which is located in the hinterland of Asia and Europe, where precipitation is scarce, and evaporation is high, making it a typical arid and semi-arid region, is a region in China where wind-sand disasters are serious, with the widest distribution of saline soil areas and the greatest degree of salinization. In particular, in the south of Xinjiang, the high level of sand and wind damage and saline soils greatly impair the growth and development of crops, resulting in low yields and economic returns.

This area is located at the edge of the desert and is often affected by wind-sand disasters and increased saline–alkali soils, which further exacerbates the quality of the land. As such, how to use reasonable measures to control and govern soil salinization is an issue that must be addressed. At present, the main measures used in Xinjiang to amend soil salinization are engineering, chemical, and biological measures [2]. Among them, engineering measures, although fast and effective, are costly and constrained by water shortages, making it difficult to promote them in arid and semi-arid areas. These measures can wash away water-soluble nutrients in the soil, resulting in the loss of soil nutrients [3]. Chemical measures are quick to work in the short term, but there are restrictions to their use on a large scale due to the limitations of the improvers and costs [4]. Biological measures refer to planting saline-tolerant plants in salinized areas to gradually reduce the harm of salinity to plants through the physiological function of plants and the influence of the root system on the physical and chemical properties of soil, which achieves the purpose of improving saline soil.

At present, most fruit trees are planted in saline soils in South Xinjiang, which seriously affects the yield and quality of these trees. The tree-based intercropping system is largely favored by the local population to achieve better yield and benefits. Intercropping of trees and crops has been traditionally practiced in China for centuries [5]. Scientists and growers understand that adding trees to an agricultural system offers various additional benefits, also known as ecosystem services, to the farmer and society [6]. Meanwhile, planting salt-tolerant forage grasses on saline soils can increase the ground cover, replace soil evaporation with plant transpiration and reduce soil evaporation, thus reducing the rate of soil salt accumulation and reducing the accumulation of salt in the surface layer, thus achieving desalination [7]. Therefore, the tree-based agroforestry system has been established on the salt-affected soils in the south of Xinjiang, and fruit tree/crop intercropping has become the dominant land-use type for ameliorating the saline conditions of soils in the area.

Different land use patterns determine the differences in the soil water content and salinity distribution. The factors affecting soil water content are variable at different spatial and temporal scales [8,9]. Planting alfalfa between tree rows allows the open space between the rows to be covered, allowing the system to effectively block soil moisture loss due to seasonal drought [10]. However, in arid and semi-arid regions, strong plant transpiration leads to large consumption of soil water, negatively affects the natural distribution characteristics of soil water, reduces the differences in soil water distribution, and changes the soil structure [11]. Forest-grass intercropping has a greater leaf area index and consumes soil water too quickly [12].

Intercropping is generally considered to have a greater increase in tree height and crown growth rate than clear-cutting. However, some studies [13] have shown that whole orchard grassing can significantly reduce tree growth, with newly planted trees being particularly sensitive. Intercropping not only provides favorable conditions for the growth of fruit trees but also improves fruit quality. Orchard grassing increased the soluble solids, hardness, and Vc content of apples, but the effect on yield per plant and fruit quality per fruit was not significantly different from that of clear-cut areas [14]. At the same time, a good orchard environment can also create conditions for high fruit quality and yield. Grass in orchards can reduce air temperature during the high-temperature period, moderate air temperature fluctuations, increase the relative humidity of the orchard air, maintain a high air humidity in the orchard, and form an air temperature and humidity environment that is conducive to the growth and development of fruit trees and various physiological activities [15].

Halophytes can absorb salt ions in the soil and improve saline soil [16]. At the same time, halophytes have desalination and restoration potential with respect to saline soils and can be used in phytoremediation [17]. Some research has shown that alfalfa can absorb salts from the soil, resulting in a significant decrease in the soluble salt content of the soil [18,19]. Planting alfalfa in woodland can play a role in reducing soil salinity and can effectively reduce soil salinity beneath the woodlands. Meanwhile, canopy shading has a promotional effect on the growth of pastures in the presence of some water stress [20].

South Xinjiang is located near the Taklamakan Desert, where sandy and dust weather is frequent. This can have a negative impact on the growing environment of native plants. The effects of dust on plant growth and development are manifold, especially on photosynthesis [21]. It was speculated that dust decreased the photosynthetic rate by shading the leaf surface. Meanwhile, the dust increased the transpiration rate by increasing the leaf temperature [22]. In addition, dust may cause physical injury to tree leaves and bark, reducing fruit setting and resulting in a general reduction in growth [23]. Dust can also cause important changes to leaf physiology, affecting tree productivity [24].

In conclusion, soil salinity, sand, and dust have certain harmful effects on plant growth. At present, there are few studies on the effects of intercropping on soil salinization and sand and dust [7,23]. In this study, we select South Xinjiang, where soil salinization and sand and dust are serious, as the study area and use apple and ryegrass as experimental materials. The purpose was to understand the following.

1. Which cropping pattern has a beneficial effect on managing soil salinization?
2. Which planting pattern can weaken sand and dust hazards and improve the photosynthetic capacity and fruit quality of fruit trees?

We hypothesize that (1) this type of intercropping system can have a beneficial effect on salinity management by changing the soil salt ion concentrations and (2) this type of intercropping system has a positive impact on weakening dust damage and improving fruit yield and quality.

2. Materials and Methods

2.1. Experimental Site

The field experiments were carried out in 2020 and 2021 at the Fruit Core Demonstration Base of the 44th Regiment of the 3rd Division of the Xinjiang Production and Construction in crops (79°18′ E, 39°84′ N). This area has a temperate, extremely arid desert climate with long sunshine hours and large temperature differences between day and night. The average annual temperature is 11.6 °C, and in the hottest month (July), the average temperature is 25–26.7 °C, the coldest month (January) has an average temperature of −6.6–−7.3 °C, the annual average frost-free period is 221–225 days, and the annual rainfall is 38.3 mm.

2.2. Experimental Design

The experimental design employed two treatments: sole-cropped apple (MA) and apple/ryegrass intercropping (IR). The experimental plot area is 24 × 20 m^2. Each treatment plot is 24 × 8 m^2 in size and has 72 apple trees. The apple tree + ryegrass intercropping system was established in 2019 in arable cropland fields. Apple trees were transplanted in 2019. Apple trees were three years old. The average height of apple trees was 1.90 m, the diameter at breast height was 2.11 cm, the north-south crown amplitude was 1.25 m, and the east-west crown amplitude was 1.42 m.

The planting layout of each test plot is shown in Figure 1. The ARI system plot, had a 100 cm wide bare area between apples and ryegrass. The spacing of tree rows is 4.0 m, and the spacing between trees in each row is 1.0 m. Apple trees were planted 2.5×10^3 plants/hm^{-2}. Ryegrass was seeded between the apple rows at a seeding density of 20 kg/hm^2, a row spacing of 0.25 m, and a sowing depth of 2.0 cm. The sole-cropped apple system used the same pattern as intercropping, bare land between trees and no other crops. The sole-cropping fields in each experimental site were used as the control plots that included the same cultivars, the same planting and cultivation modes, the same previous agricultural system, and the same agricultural management system in intercropping and sole-cropping systems. The samples were collected in mid-July (flowering and fruiting period), late August (fruit enlarging period), and late September (fruit maturing period).

2.3. Soil Sample Collection

2.3.1. Soil Sample Collection

According to the sampling plan described above, soil samples were collected from a randomly selected apple tree of good growth in the middle row of the monocrop and intercrop systems to study the distribution of soil moisture and salinity. Sampling was carried out by the auger sampling method, using the tree as the origin and sampling in a horizontal radial direction at 50 cm, 100 cm, 150 cm, and 200 cm from the tree at depths of 0–20 cm, 20–40 cm and 40–60 cm, replicated three times.

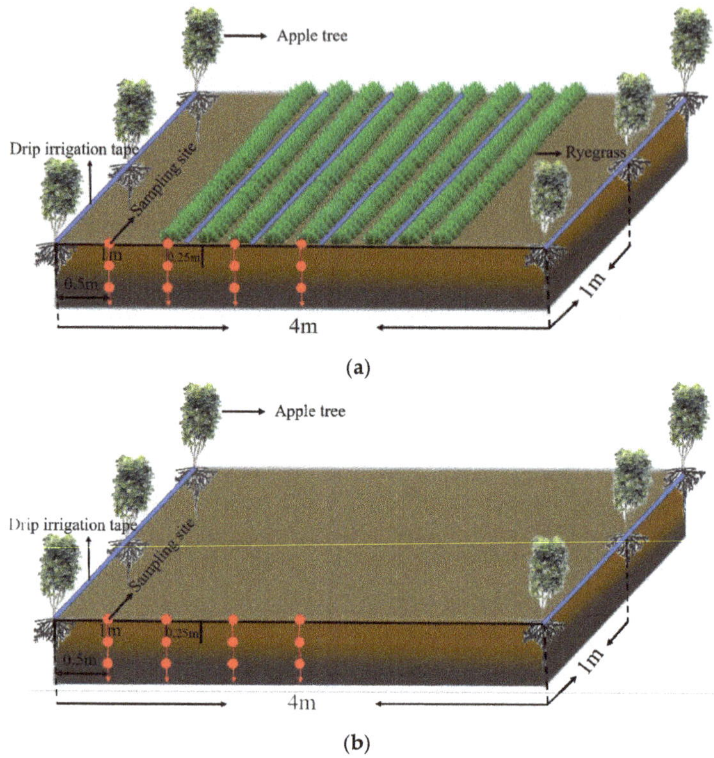

Figure 1. (**a**): diagram of apple and ryegrass planting in an apple/ryegrass intercropping system; (**b**): diagram of apple growing in an apple monoculture system.

2.3.2. Measurement of Soil Moisture Content and Soil Salinity

In accordance with the sampling time, the completed soil samples are placed partly in aluminum boxes for the determination of the soil water content and partly for the determination of the soil K^+ and total salt content. Soil moisture content is determined by the drying method, whereby the soil samples are dried in an oven in an aluminum box, and the soil moisture content is calculated by comparing the soil weight before and after drying. The other part of the soil sample is air dried and passed through a 2 mm sieve, mixed at a 5:1 ratio of water to soil, oscillated, and the supernatant extracted to determine the soil K^+ and total salt content. The soil K^+ content was determined using Plasma Emitted Atomic Spectrometer (ICAP6300), while the soil's total salt content was determined by conductivity.

2.4. Measurement of Microclimate on Farmland

According to the sampling period, air temperature, relative air humidity, and wind speed were measured in the apple monoculture and the apple-ryegrass intercropping systems. A day of continuous good weather was selected, and measurements were taken once in the morning and once in the afternoon using a Kestrel 4000 handheld weather station and repeated three times.

2.5. Measurement of Photosynthetic Characteristics

Depending on the above measurement times, photosynthetic parameters were selected from 10:00 to 12:00 on a sunny day. Within the apple monocropping and apple-ryegrass intercropping systems, three well-grown apple trees were randomly selected, and the photosynthetic parameters, including net photosynthetic rate (Pn), stomatal conductance (Gs), intercellular CO_2 concentration (Ci) and translocation rate (Tr) of the middle functional leaves of apple trees were measured using an LI-6400 photosynthetic instrument and replicated three times.

2.6. Measurement of Fruit Tree Growth Indicators, Yield and Fruit Quality

In October 2020 and 2021, we randomly selected three good growth apple trees and measured the standard tree height and crown width with a steel tape measure and the diameter at breast height with a vernier caliper, replicated three times. At the same time, the number of fruits from standard apple trees was counted, and 60 apples were collected randomly in each treatment plot and preserved in ice boxes. Individual fruit weights were determined using a 1/100 electronic balance, and individual plant yields were calculated. Soluble solids were determined using a WYT-4 handheld sugar meter; hardness was determined using a GY-1 fruit hardness tester; soluble sugar content was determined using the anthrone colorimetric method, and total acid content was determined using the NaOH neutralizing titration method.

2.7. Statistical Analysis

Analysis of variance was performed on the data using SPSS ver. 22.0. Among them, the least significant difference method was used for multiple comparison analysis, and the significance level was set as a = 0.05. Origin 2022 and Sufer15 were used for graphing.

3. Results

3.1. Effect of Planting Patterns on the Soil Moisture Content

The two-dimensional distribution characteristics of the soil water content (Figure 2) showed that the soil water content increased with increasing soil depth and decreased with increasing distance from the tree row in both monocropped and intercropped apple systems. A comparison of the soil water content between monocropped and intercropped apples revealed that intercropping ryegrass with apples significantly increased the soil water content at the flowering and fruiting periods (50, 40), (150, 20), (150, 60) and (200, 20) by 33.66%, 53.90%, 34.62%, and 45.66%, respectively (Figure 2a,d,g,j). During the fruit enlarging period, intercropping significantly increased the soil water content of (50, 20), (100, 20), (150, 20), (200, 20), and (200, 40) by 46.95%, 42.92%, 33.80%, 41.53%, and 35.03%, respectively (Figure 2b,e,h,k). At the fruit maturing period, intercropping significantly increased the soil water content in (50, 20), (50, 40), (100, 20), (100, 40), (150, 20), (150, 40), (200, 20), (200, 40), and (200, 60) by 46.38%, 45.73%, 64.86%, 52.95%, 92.55%, 53.72%, 62.16%, 53.29%, and 42.84%, respectively (Figure 2c,f,i,l). The variation in the soil water content was higher in the zero to 20 cm soil layer and lower in the 40–60 cm soil layer, showing that apples intercropped with ryegrass can effectively increase the soil water content.

3.2. Effect of Planting Patterns on Soil Salinity

The two−dimensional distribution characteristics of the soil Na^+ content are shown in Figure 3. In both monocropped and intercropped apple systems, the soil Na^+ content increased with increasing soil depth and did not vary significantly with increasing distance from the tree, and the variation in the soil Na^+ content between different fertility periods was not significant. A comparison of the soil Na^+ content between monocropping and intercropping showed that apple intercropping with ryegrass significantly reduced (50, 20), (50, 40), (100, 20), (100, 40), (150, 20), (150, 40), (200, 20), (200, 40) the soil Na^+ content by 52.33%, 41.96%, 50.09%, 33.89%, 52.13%, 34.98%, 51.31%, and 34.31%, respectively. Intercropping significantly reduced the Na^+ content in the zero to 40 cm soil layer. This is because the addition of mulch between fruit rows changed the soil structure, increased the soil water content, and contributed to the downward transport of soil Na^+ with water.

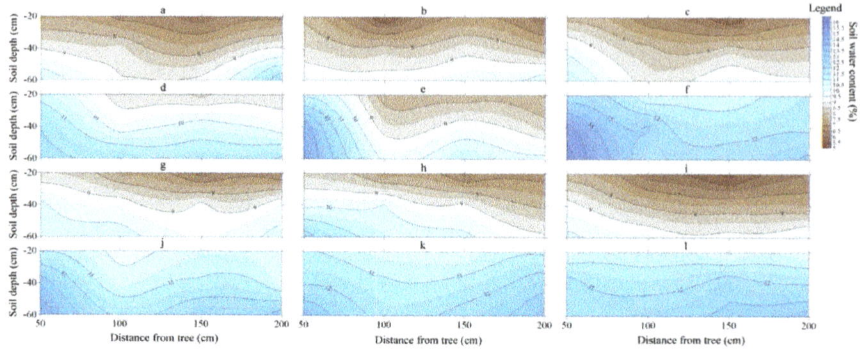

Figure 2. Two−dimensional soil moisture content distribution in different stages of monocropping apple and ryegrass apple intercropping system ((**a**,**d**,**g**,**j**): flowering and fruiting period; (**b**,**e**,**h**,**k**): fruit enlarging period; (**c**,**f**,**i**,**l**): fruit maturing period; monocropping apple (first and third rows) and ryegrass apple intercropping system (second and fourth rows); in the 2020 (top two rows) and 2021 (bottom two rows) seasons).

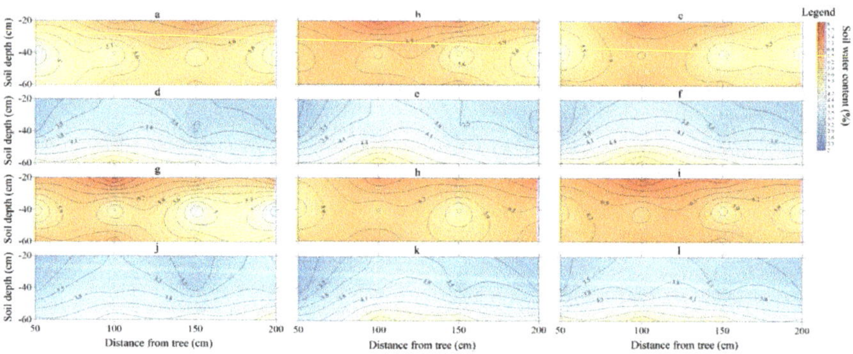

Figure 3. Two−dimensional soil Na^+ content distribution in different stages of monocropping apple and ryegrass apple intercropping system ((**a**,**d**,**g**,**j**): flowering and fruiting period; (**b**,**e**,**h**,**k**): fruit enlarging period; (**c**,**f**,**i**,**l**): fruit maturing period; monocropping apple (first and third rows) and ryegrass apple intercropping system (second and fourth rows); in the 2020 (top two rows) and 2021 (bottom two rows) seasons).

The two-dimensional distribution characteristics of the soil's total salt content are shown in Figure 4. In both monocropped and intercropped apple systems, the soil total salt content increased with increasing soil depth and did not change significantly with increasing distance from the tree. The comparison between the monocrop and intercrop revealed that intercropping significantly reduced (50, 20), (100, 20), (150, 20), and (200, 20) the soil total salt content by 54.29%, 51.79%, 54.95%, and 48.8%, respectively, during the flowering and fruiting period (Figure 4a,d,g,j). At the fruit enlarging period, intercropping reduced (50, 20), (50, 40), (100, 20), (150, 20), (200, 20), and (200, 40) the soil total salt content by 56.39%, 52.33%, 50.63%, 51.38%, 56.12%, and 44.27%, respectively (Figure 4b,e,h,k). During the fruit maturing period, intercropping significantly reduced (50, 20), (50, 40), (100, 20), (150, 20), and (200, 20) the soil total salt content by 49.1%, 41.12%, 50.14%, 53.22%, and 52.43%, respectively (Figure 4c,f,i,l). Intercropping significantly reduced the total salt content in the zero to 20 cm soil layer, and the reduction was greater at the fruit expansion stage. This is because, during the fruit expansion period, plants need more water, salts are transported downward with water, and the total salt content in the soil surface layer is significantly reduced. Therefore, intercropping can effectively improve the distribution of total soil salt in the 0–20 cm soil layer and reduce the soil salt content in the 0–40 cm soil layer, thus reducing the toxic effect on plants.

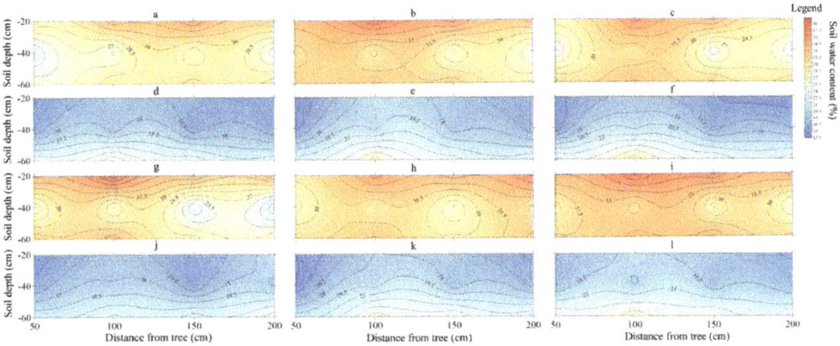

Figure 4. Two−dimensional soil total salt content distribution in different stages of monocropping apple and ryegrass apple intercropping system ((**a**,**d**,**g**,**j**): flowering and fruiting period; (**b**,**e**,**h**,**k**): fruit enlarging period; (**c**,**f**,**i**,**l**): fruit maturing period; monocropping apple (first and third rows) and ryegrass apple intercropping system (second and fourth rows); in the 2020 (top two rows) and 2021 (bottom two rows) seasons).

3.3. *Effect of Planting Pattern on the Orchard Microclimate*

Observations showed that temperature, humidity, and wind speed of intercropped ryegrass orchards differed significantly from those of monocropped orchards (Figure 5). The air temperature and wind speed in intercropped ryegrass orchards were lower than those in monocropped orchards at different fertility periods, with decreases of 5.93%, 6.26%, 6.03% (Figure 5A), 12.33%, 11.21%, and 11.91% (Figure 5C). The relative humidity of air in intercropped ryegrass orchards was greater than in monocropped orchards, increasing by 10.76%, 16.11%, and 14.37% (Figure 5B). It can be seen that the planting of ryegrass reduced the wind speed, weakened the translational velocity of the airflow, reduced the turbulent exchange effect, increased the soil moisture between apple rows, and increased the relative humidity of air in the orchard. Ryegrass also transpires during the growth process, absorbing some of the heat and lowering the orchard temperature. Therefore, intercropping can significantly increase the relative humidity of orchard air, significantly reduce the temperature and wind speed in the orchard, and improve the wind protection effectiveness of the orchard. Intercropping had no significant interannual effect on the orchard microclimate.

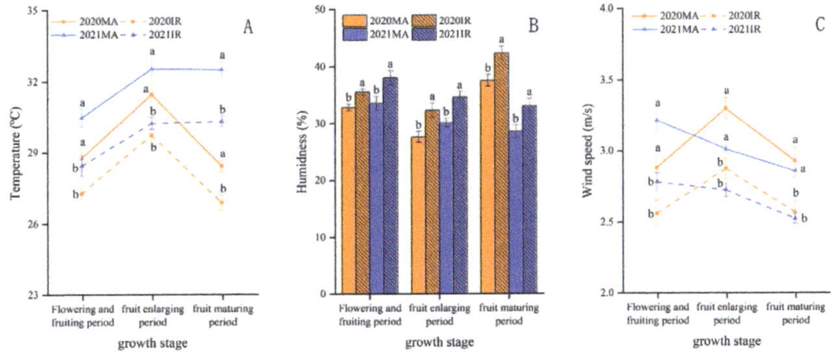

Figure 5. Schematic diagram of temperature, humidness, and wind speed changes in different stages of monocropping apple and ryegrass apple intercropping system. ((**A**): temperature; (**B**): humidness; (**C**): wind speed, Different lowercase letters in the graph indicates significant differences across treatments ($p < 0.05$).).

3.4. Effect of Planting Pattern on the Photosynthesis of Fruit Trees

Plant photosynthesis is a complex physiological and biochemical reaction influenced by several environmental factors, such as light intensity, temperature, and water. As shown in Figure 6, "the measuring parameters" Pn, Tr, Ci, and Gs of intercropped apples were significantly different from those of monocropped apples. As apple fertility progressed, the Pn, Tr, and Gs of intercropped apples were greater than those of monocropped apples, reaching the maximum at fruit maturity, which increased by 13.18%, 40.04%, and 58.7% (Figure 6A); 19.62%, 32.59%, and 36.86% (Figure 6B); 6.51%, 16.01%, and 25.78% (Figure 6D). The Ci of intercropped apples was less than that under monocropping and reached the minimum value at fruit ripening, which decreased by 3.43%, 6.17%, and 9.26% (Figure 6C). Intercropping increased Pn, Tr, and Gs and decreased Ci because intercropping reduced the orchard temperature, increased humidity, and reduced the "lunch break" phenomenon of apple leaves. In addition, due to the local dusty weather, the lack of forage mulching between the rows of monocropped apples, and the high amount of bare land blown by the wind, there was more sand and dust on the surface of the apple leaves, which blocked the stomata and reduced the transpiration rate of the apple leaves. Intercropping had no significant interannual effect on the photosynthesis of the apple leaves.

3.5. Effect of the Planting Pattern on Fruit Tree Yield and Fruit Quality

Different cropping patterns can change the growth traits of apple trees (Table 1). In 2020, the height, breast height, north-south crown amplitude, and east-west crown amplitude of intercropped apples increased by 0.09 m, 0.08 cm, 0.12 m, and 0.12 m, respectively, compared to monocropped apples, with increases of 3.37%, 2.69%, 7.89%, and 7.64%. In 2021, intercropping increased the height, breast height, north-south crown amplitude and east-west crown amplitude, by 0.07 m, 0.15 cm, 0.16 m, and 0.12 m, or 2.03%, 3.92%, 9.04%, and 6.98%, respectively. However, although intercropping increased the height, breast height, north-south crown amplitude, and east-west crown amplitude of apples, the magnitude of the increase did not vary significantly. At the same time, the inter-annual variation was not significant. Therefore, intercropping improves the growth of apple trees by increasing the soil water content and improving the microclimate of the orchard.

As shown in Table 2, intercropping affected the external quality of apples, and there were significant differences between monocropping and intercropping. The single fruit weight, fruit hardness, and single plant yield of intercropped apples were significantly greater than those of monocropped apples, which increased by 10.8%, 8.2%, and 11.02%, respectively. The soluble sugar and sugar-acid ratios of intercropped apples were significantly greater than those of monocropped apples, with increases of 5.07% and 10.9%; titratable acid was significantly lower than that of monocropped apples, with a decrease of 5.5%; soluble solids and Vc were not significantly different from those of monocropped apples. This showed that intercropping significantly increased the single fruit weight, fruit hardness, yield per plant, and soluble sugar and sugar-acid ratio and reduced the titratable acid content of the apples. Intercropping had no significant effect on apple quality from year to year.

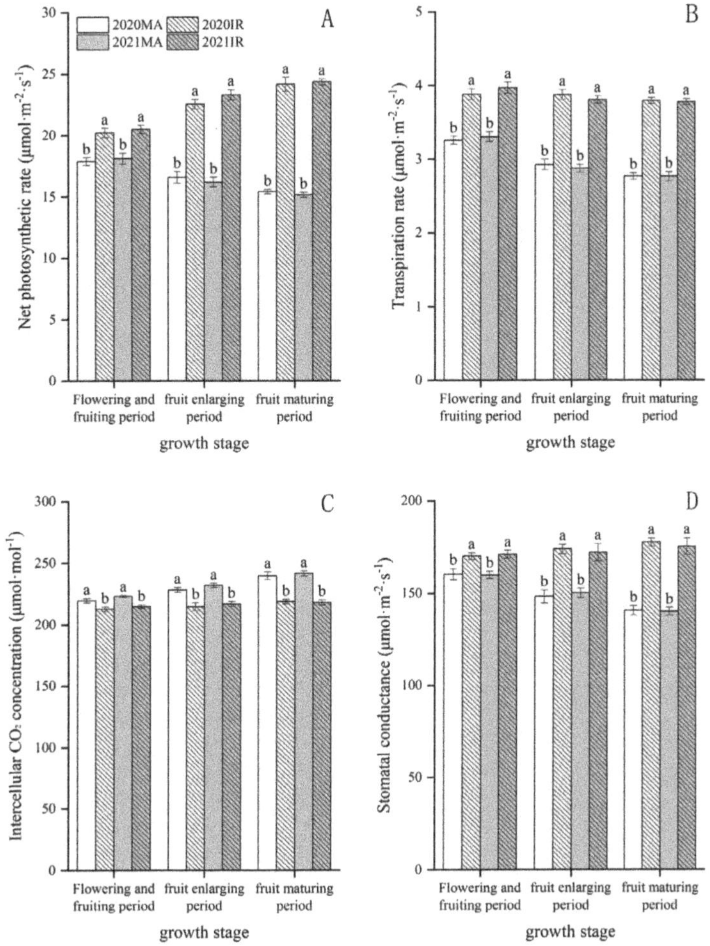

Figure 6. Schematic diagram of changes in net photosynthetic rate, transpiration rate, intercellular CO_2 concentration, and stomatal conductance in different stages of monocropping apple and ryegrass apple intercropping system. ((**A**): net photosynthetic rate; (**B**): transpiration rate; (**C**): intercellular CO_2 concentration; (**D**): stomatal conductance, Different lowercase letters in the graph indicates significant differences across treatments ($p < 0.05$)).

Table 1. The growth indicators of fruit trees in different planting patterns.

Year	Treatment	Tree Height (m)	Breast Height (cm)	The North-South Crown Amplitude (m)	The East-West Crown Amplitude (m)
2020	MA	2.67 ± 0.14 a	2.97 ± 0.16 a	1.52 ± 0.05 a	1.57 ± 0.02 a
	IR	2.76 ± 0.17 a	3.05 ± 0.08 a	1.64 ± 0.06 a	1.65 ± 0.05 a
2021	MA	3.44 ± 0.11 a	3.83 ± 0.12 a	1.77 ± 0.04 a	1.72 ± 0.05 a
	IR	3.51 ± 0.09 a	3.98 ± 0.19 a	1.93 ± 0.06 a	1.84 ± 0.06 a

MA- monocropped apples; IR- intercropped apples. The same letters in the same column indicated no significant difference at 0.05 level in Duncan's analysis.

Table 2. The appearance and internal quality of the fruit in different planting patterns.

Year	Treatment	Single Fruit Weight (g)	Fruit Firmness (kg/cm^2)	Soluble Solids (%)	Soluble Sugar (%)	Titratable Acid (%)	Sugar to Acid Ratio	Vc (mg/100g)	Single Average Yield (kg)
2020	MA	206.06 ± 10.76 b	9.89 ± 0.44 a	14.82 ± 0.32 a	10.14 ± 0.33 a	0.42 ± 0.05 a	24.59 ± 3.50 a	5.57 ± 0.13 a	1.03 ± 0.05 a
	IR	227.54 ± 7.78 a	10.56 ± 0.39 a	15.07 ± 0.29 a	10.67 ± 0.28 a	0.39 ± 0.03 a	27.66 ± 1.45 a	5.83 ± 0.19 a	1.14 ± 0.04 a
2021	MA	206.70 ± 8.00 b	9.75 ± 0.41 b	14.71 ± 0.35 a	10.18 ± 0.22 b	0.41 ± 0.07 a	25.02 ± 3.85 a	5.49 ± 0.19 a	4.75 ± 0.18 b
	IR	229.79 ± 8.28 a	10.69 ± 0.35 a	15.01 ± 0.18 a	10.68 ± 0.12 a	0.39 ± 0.04 a	27.35 ± 2.94 a	5.92 ± 0.08 a	5.29 ± 0.19 a

MA- monocropped apples; IR- intercropped apples. The same letters in the same column indicated no significant difference at 0.05 level in Duncan's analysis.

4. Discussion

4.1. Soil Moisture Content

Soil moisture is a key factor limiting plant growth in semi-arid regions [25]. It has been suggested that grass planting can improve soil water retention and increase soil moisture content [26], and it has also been suggested that grass planting can promote water depletion in orchards, which is detrimental to fruit tree growth [27]. Intercropping is a means of providing sustainable and productive agriculture [28]. Intercropping ryegrass with apples is mainly influenced by the combined effects of transpiration, soil evaporation, and irrigation of apple trees and cover crops. Ryegrass increases ground cover, which can reduce ineffective evaporation of soil surface water and has a moderating effect on the orchard soil water content [29]. In this study, the soil moisture content was measured continuously in a zero to 60 cm soil layer to study the effect of the different cropping patterns on the soil moisture dynamics in the orchard. It was found that when apple was intercropped with ryegrass, the soil water content increased with the increase in soil depth; the change in the soil water content in the zero to 40 cm layer was the most prominent, mainly because inter-row grass planting can reduce soil evaporation and plays a role in water retention, which is consistent with the results of Sun et al. [10].

In general, the degree of water uptake by different soil layers depends on the amount of available water in the soil profile [30]. Spatial variability of soil water in orchards exists mainly because of the differences in the degree of soil water utilization by apple trees caused by differences in the spatial distribution of their root systems. Planting ryegrass between apple tree rows can effectively reduce soil water evaporation, increase soil water content between apple tree rows, and increase the use of soil water between rows by apple trees. At the same time, ryegrass transpiration and surface evaporation in grassy orchards were greater than those in clear-cut orchards, producing a water competition effect [31].

4.2. Soil Salinity

Soil salinization is significantly harmful to the ecological environment, agricultural production, engineering construction, and soil and water resources, especially in the northern arid regions of China and the western irrigated agricultural areas; this has become an important factor limiting the sustainable development of the region [32]. It was found that intercropping salt-tolerant plants in cotton fields with sub-membrane drip irrigation can reduce soil salinity, increase water content, increase cotton yield, and improve water use efficiency [33]. Li et al. planted alfalfa on saline land in the Chidamu Basin and found that the total salt content of zero to 30 cm soil decreased from 1.52% before planting to 0.13%, and the salt phenology was no longer obvious [34].

In this study, the Na^+ and total salt contents in apples intercropped with ryegrass and monocropped apples were determined. Apple intercropped with ryegrass had a significant desalination effect on the zero to 40 cm soil layer, while there was salt enrichment in the 40–60 cm soil layer. This is because the root system of the crop disrupts the original structure of the soil and increases the permeability of the soil so that the salts are transported downward; the result is consistent with the first hypothesis that apple-ryegrass intercropping systems can have a beneficial effect on salinity management by changing the concentrations of salt ions in the soil. Different plants impact soil salt differently, which can also affect the salt ion content [35]. Alfalfa grass was able to significantly reduce the total soluble salts in the 0–40 cm soil layer of coastal saline soils, with a desalination rate of 65.5%, while the bottom layer changed less [36], which is consistent with the results of this study. In this study, we have only investigated the desalination effect of various patterns on 0–60cm soils. In the future, we should focus on the desalination effect of various patterns on deeper soils in order to facilitate other soil improvement measures to reduce salt and improve soil fertility.

4.3. Photosynthesis

Daily changes in plant photosynthesis show different trends depending on the growing environment. A date/wheat and cotton complex system improved the canopy structure of intercropped crops by reasonably optimizing the crop layout, which in turn improved the photosynthetic rate of the crops and increased their light interception and light energy utilization [37]. The apple/wheat complex system increased light intensity and improved light energy utilization [38]. The present study showed that the inhibition of photosynthesis in apple leaves by high temperature and strong light was significantly alleviated, and the utilization of strong light was improved, the same as the findings of Yang et al. [39]. The net photosynthetic rate and stomatal conductance of all intercropped apple leaves were higher than those of monocropped apples, which is generally consistent with the findings of Yang et al. [40]. The transpiration rate is an important physiological indicator of plant water metabolism. The present study showed that intercropping ryegrass with apple significantly increased the transpiration rate of the apple leaves, which may be that sand and dust blocked the stomata of apple leaves and thus reduced the transpiration rate of the apple leaves. The effect of sand and dust on monocropped apples was greater than that on intercropped apples. The result is consistent with the second hypothesis that apple-ryegrass intercropping has a positive effect on the reduction of dust damage in apple leaves. Intercropping significantly reduced the intercellular CO_2 concentration of apple leaves, which is consistent with the findings of Hu et al. [41].

4.4. Microclimate

The Microclimate is the main environmental factor affecting the growth and development of crops, not only photosynthesis and the growth and development of crops but the change in the soil structure, nitrogen migration and transformation, and microbial diversity. Intercropping changes the natural conditions such as the field temperature, humidity, and light, forming a microclimate environment favorable to crop growth and development and improving the crop yield. Furthermore, orchard grass cultivation can regulate orchard temperature and humidity, improve the orchard microclimate, improve the soil organic matter, water retention, water storage capacity, and enhance fruit quality and yield, thus forming balanced orchard ecosystems [42].

Intercropping also reduced wind speed and had a significant wind protection effect. Date palm intercropped with cotton and alfalfa resulted in significantly lower wind speed due to the planting of the date palm trees, and the wind protection effect of date palm intercropped with alfalfa was lower than that of intercropped cotton in the case of high winds, but both were greater than that of the monocrop system [43]. This is due to the relatively small row spacing of date palm trees, which increases the wind friction and thus reduces the wind braking force and improves the wind protection effect. In addition, intercropping reduces temperature and increases humidity. Apple intercropping with white clover produced a suitable climate for fruit tree growth, improved resource utilization, and increased fruit tree yield due to the reduction of the lower surface and subsurface temperatures and increased humidity [44]. The present study showed that intercropping significantly reduced the temperature and wind speed, increased the relative air humidity, and enhanced the wind protection effectiveness at different fertility stages, which is generally consistent with the results of Mao et al. [45] and Moreira et al. [46].

4.5. Fruit Tree Growth Indicators, Yield and Fruit Quality

During apple growth and development, the orchard microclimate environment and photosynthesis are closely related to fruit tree growth indicators and quality, and intercropping with ryegrass in orchards significantly improved the orchard microclimate environment and fruit tree growth indicators, thus improving the apple fruit quality. Grassing improved the ecological environment of the orchard and increased leaf photosynthesis, thereby promoting tree growth [15], which is consistent with the results of this study. In addition, low concentrations of salinity can promote plant growth, while high concentrations of saline stress inhibit plant growth and can lead to death in severe cases. The growth of fruit trees under saline stress can be used as indicators to judge the salinity tolerance of fruit trees [47]. Both apple plant height and stem thickness decreased gradually with increasing stress concentration. Some researchers have suggested that fruit tree and grass intercropping can reduce the occurrence of fruit diseases and thus improve the quality of their processed products [48,49]. The present study showed that intercropping significantly increased the single fruit weight, fruit hardness, and yield of apples, which is consistent with the results of Jia et al. [50]. The significant increase in the single fruit weight of intercropped apples may be related to the increase in soil nutrients and the water content; the increase in fruit hardness may be related to the increased uptake of soil mineral elements by fruit trees; the increase in the sugar-acid ratio may be related to the decrease in the titratable acid content and the increase in soluble solids. It can be seen that the improvement in apple quality is directly related to intercropping. The result is consistent with the second hypothesis that apple-ryegrass intercropping has a positive effect on yield and quality.

5. Conclusions

In this study, it was found that apple-ryegrass intercropping affected the soil water, soil salinity, photosynthetic index, fruit tree growth indicators, yield, and quality. The soil water content, net photosynthetic rate, transpiration rate, stomatal conductance, and fruit tree growth indicators in the intercropping system were higher than those under monocropping. The intercellular CO_2 concentration, soil Na^+, and total soil content in the intercropping system were lower than those under monocropping. Intercropping significantly improved the orchard microclimate, increasing fruit yield and improving fruit quality. Thus, intercropping should be properly promoted to improve soil salinization, crop yield and quality.

Author Contributions: Conceptualization, S.Z.; Data curation, S.Z. and W.W.; Formal analysis, S.Z.; Funding acquisition, W.Z.; Methodology, S.Z.; Project administration, W.Z.; Resources, W.Z., T.T. and L.L.; Supervision, W.Z. and L.L.; Visualization, S.Z. and T.L.; Writing—original draft, S.Z.; Writing—review and editing, T.T., T.L., L.S. and X.W. All authors have read and agreed to the published version of the manuscript.

Funding: This work was financially supported by the Innovation and Development Project of Shihezi University (CXFZ202008), the National Natural Science Foundation of China (32260723), the special funds of agricultural science and technology of Huyanghe City (2022C18).

Institutional Review Board Statement: Not applicable.

Informed Consent Statement: Informed consent was obtained from all subjects involved in the study.

Data Availability Statement: The datasets used and/or analyzed during the current study are available from the first author upon reasonable request.

Conflicts of Interest: The authors declare that they have no known competing financial interests or personal relationships that could have appeared to influence the work reported in this paper.

References

1. Yang, J.S. Development and Prospect of the research on salt-affected soils in China. *Acta Pedol. Sin.* **2008**, *9*, 837–845. (In Chinese) [CrossRef]
2. Zhang, Y.F.; Li, W.Y.; Hu, H.; Chen, W.Z.; Wang, X.L. Current status and outlook of saline land improvement research. *Jiangsu Agric. Sci.* **2017**, *45*, 7–10. (In Chinese) [CrossRef]
3. Qadir, M.; Schubert, S.; Ghafoor, A.; Murtaza, G. Amelioration strategies for sodic soils: A review. *Land Degrad. Dev.* **2001**, *12*, 357–386. [CrossRef]
4. Qadir, M.; Noble, A.D.; Schubert, S.; Thomas, R.J.; Arslan, A. Sodicity-induced land degradation and its sustainable management: Problems and prospects. *Land Degrad. Dev.* **2006**, *17*, 661–676. [CrossRef]
5. Yang, L.; Ding, X.; Liu, X.; Li, P.; Eneji, A.E. Impacts of long-term jujube tree/winter wheat–summer maize intercropping on soil fertility and economic efficiency—A case study in the lower North China Plain. *Eur. J. Agron.* **2016**, *75*, 105–117. [CrossRef]
6. Tsonkova, P.; Böhm, C.; Quinkenstein, A.; Freese, D. Ecological benefits provided by alley cropping systems for production of woody biomass in the temperate region: A review. *Agrofor. Syst.* **2012**, *85*, 133–152. [CrossRef]
7. Wei, Z.P.; Pan, W.L.; Fan, J.G. Effects of soil amelioration of saline-alkali soils on forest-grass ecological patterns. *J. Cent. South Univ. For. Technology.* **2012**, *32*, 100–104. (In Chinese) [CrossRef]
8. Gómez-Plaza, A.; Martínez-Mena, M.; Albaladejo, J.; Castillo, V. Factors regulating spatial distribution of soil water content in small semiarid catchments. *J. Hydrol.* **2001**, *253*, 211–226. [CrossRef]
9. Wang, J.; Fu, B.; Qiu, Y.; Chen, L.; Wang, Z. Geostatistical analysis of soil moisture variability on Da Nangou catchment of the loess plateau, China. *Environ. Earth Sci.* **2001**, *41*, 113–120. [CrossRef]
10. Sun, R.J.; Chen, W.F.; Song, X.L.; Luo, Y.F.; Liu, L.J. Distribution characteristics and edge effect of soil water and salt in silvopastoral system of the Yellow River Delta , China. *Chin. J. Appl. Ecology.* **2019**, *30*, 2549–2557. (In Chinese) [CrossRef]
11. Xu, B.C.; Shan, L.; Chen, Y.M. Review and discuss on the effect and influence factors of vegetation construction on soil water in semi-arid area on Loess Plateau. *Sci. Soil Water Conserv.* **2003**, *1*, 32–35. (In Chinese) [CrossRef]
12. Grace, J. Plant response to wind. *Agric. Ecosyst. Environ.* **1988**, *22–23*, 71–88. [CrossRef]
13. Welker, W.V.; Gleen, D.M. The relationship of sod proximity to the growth and nutrient composition of newly planted peach trees. *J. Am. Soc. Hortic. Sci.* **1985**, *20*, 417–418. [CrossRef]
14. Pang, Q.H.; Wang, J.; Sun, Q.; Xu, X.; Wang, R. Effects of grass covering on soil nutrients and berry quality in winegrowers. *Southwest China J. Agric. Sciences.* **2019**, *32*, 1833–1838. [CrossRef]
15. Bai, G.S.; Guo, J.P.; Du, J.H. Effects of natural grass on microclimate of apple orchard, fruit sunburn and leaf early deciduous diesease in Werbei Dry Plateau. *Acta Agrestia Sinica* **2021**, *29*, 324–332. (In Chinese) [CrossRef]
16. Fountoulakis, M.; Sabathianakis, G.; Kritsotakis, I.; Kabourakis, E.; Manios, T. Halophytes as vertical-flow constructed wetland vegetation for domestic wastewater treatment. *Sci. Total Environ.* **2017**, *583*, 432–439. [CrossRef]
17. Hasanuzzaman, M.; Nahar, K.; Alam, M.M.; Bhowmik, P.C.; Hossain, M.A.; Rahman, M.M.; Prasad, M.N.; Ozturk, M.; Fujita, M. Potential Use of Halophytes to Remediate Saline Soils. *BioMed Res. Int.* **2014**, *2014*, 1–12. [CrossRef]
18. Qiu, Q.H.; Deng, S.Y. Influence of *Veliveria zizanioides*, Alfalfa and Clover on Soil Saliniazation. *J. Anhui Agric.* **2012**, *40*, 15204–15206. (In Chinese) [CrossRef]
19. Zhang, Y.; Luo, S.W.; Wang, B.-L. Effect of alfalfa on saline-alkali soil improvement. *Mod. Agric. Sci. Technol.* **2009**, *20*, 121–125. (In Chinese)
20. Callaway, R.M.; Brooker, R.W.; Choler, P.; Kikvidze, Z.; Lortie, C.J.; Michalet, R.; Paolini, L.; Pugnaire, F.I.; Newingham, B.; Aschehoug, E.T.; et al. Positive interactions among alpine plants increase with stress. *Nature* **2002**, *417*, 844–848. [CrossRef]
21. Zhou, R.L.; Jia, Y.Y.; Hou, Y.L.; Shi, L.L. Relationship between changes in photosynthetic characteristics and plant compensatory growth under different sand burial depths. *Acta Ecol. Sin.* **2016**, *36*, 8111–8119. (In Chinese) [CrossRef]
22. Hirano, T.; Kiyota, M.; Aiga, I. Physical effects of dust on leaf physiology of cucumber and kidney bean plants. *Environ. Pollut.* **1995**, *89*, 255–261. [CrossRef]
23. Farmer, A.M. The effects of dust on vegetation—A review. *Environ. Pollut.* **1993**, *79*, 63–75. [CrossRef]
24. Nanos, G.D.; Ilias, I.F. Effects of inert dust on olive (*Olea europaea* L.) leaf physiological parameters. *Environ. Sci. Pollut. Res.* **2007**, *14*, 212–214. [CrossRef]
25. Li, L.L.; Zhang, P.F.; Song, Y.Q.; Li, J.; Li, J. Research Progress of Water Absorption in Fruit Tree Roots. *J. Shanxi Agric. Univ. (Nat. Sci. Ed.)* **2015**, *35*, 331–336. (In Chinese) [CrossRef]
26. Palese, A.M.; Vignozzi, N.; Celano, G.; Agnelli, A.; Pagliai, M.; Xiloyannis, C. Influence of soil management on soil physical characteristics and water storage in a mature rainfed olive orchard. *Soil Tillage Res.* **2014**, *144*, 96–109. [CrossRef]
27. Ramos, M.E.; Benítez, E.; García, P.A.; Robles, A.B. Cover crops under different managements vs. frequent tillage in almond orchards in semiarid conditions: Effects on soil quality. *Appl. Soil Ecol.* **2010**, *44*, 6–14. [CrossRef]
28. Brooker, R.W.; Bennett, A.E.; Cong, W.-F.; Daniell, T.J.; George, T.S.; Hallett, P.D.; Hawes, C.; Iannetta, P.P.M.; Jones, H.G.; Karley, A.J.; et al. Improving intercropping: A synthesis of research in agronomy, plant physiology and ecology. *New Phytol.* **2015**, *206*, 107–117. [CrossRef]
29. Li, H.F.; Wei, W.; Chen, L.D.; Guo, E.H.; Huang, Y. Progress in the Study of Soil Water Balance under Forest and Grassland Covers on the loess Plateau. *Res. Soil Water Conserv.* **2013**, *20*, 287–293. (In Chinese)

30. Song, X.; Gao, X.; Dyck, M.; Zhang, W.; Wu, P.; Yao, J.; Zhao, X. Soil water and root distribution of apple tree (*Malus pumila Mill*) stands in relation to stand age and rainwater collection and infiltration system (RWCI) in a hilly region of the Loess Plateau, China. *CATENA* **2018**, *170*, 324–334. [CrossRef]
31. Monteiro, A.; Lopes, C.M. Influence of cover crop on water use and performance of vineyard in Mediterranean Portugal. *Agric. Ecosyst. Environ.* **2007**, *121*, 336–342. [CrossRef]
32. Liu, C.M.; Zhao, K.; Luo, D.H.; Guo, C.; Liu, E.Z. Spatial Distribution Characteristics of Soil Salinization in Xinjiang Tumushuke area. *Arid. Environ. Monit.* **2020**, *34*, 9–15.
33. He, Z.J.; Shi, W.J.; Yang, J.Q. Water and salt transport and desalination effect of halophytes intercropped cotton field with drip irrigation under film. *Trans. Chin. Soc. Agric. Eng.* **2017**, *33*, 129–138. (In Chinese) [CrossRef]
34. Li, H.Y.; Peng, H.C.; Niu, D.L.; Wang, Q.J. Analysis of the Effect of Biomeasures on Discarded Salkaline Land in Chidamu Basin. *Acta Agrestia Sin.* **2002**, *10*, 63–68. (In Chinese)
35. Dong, L.P.; Cao, J.; Li, X.T.; Dai, L.L.; Su, Y.B. Dynamic change of salt contents in rhizosphere soil of salt-tolerant plants. *Acta Ecol. Sin.* **2011**, *31*, 2813–2821. (In Chinese)
36. Dong, X.X.; Guo, H.H.; Kong, L.A. Effect of alfalfa planting on soil salinity characteristics and fertility in coastal saline soils. *Shandong Agric. Sci.* **2001**, *1*, 24–25. (In Chinese) [CrossRef]
37. Xue, J.; Huang, C.; Chang, J.; Sun, H.; Zeng, F.; Lei, J.; Liu, G. Water use efficiencies, economic tradeoffs, and portfolio optimizations of diversification farm systems in a desert oasis of Northwest China. *Agrofor. Syst.* **2021**, *95*, 1703–1718. [CrossRef]
38. Bai, P.-H.; Liu, Q.-Z.; Li, X.-Y.; Liu, Y.-B.; Zhang, L.-L. Response of the wheat rhizosphere soil nematode community in wheat/walnut intercropping system in Xinjiang, Northwest China. *Appl. Èntomol. Zoöl.* **2018**, *53*, 297–306. [CrossRef]
39. Hui, Z.M.; Li, H.; Wang, M.L.; Zhao, G.F.; Wang, N. The effect of green cover on photosynthetic characteristics of grapevine (*vitis vinifera* L. cv. cabernet sauvignon) leaves. *Acta Agric. Boreali-Occident. Sin.* **2005**, *14*, 83–86.
40. Yang, W.Q.; Kou, J.C.; Han, M.Y. Effects of the orchard-planted grasses on the 1-year-old apple photosynthesis characteristics. *Acta Agrestia Sin.* **2011**, *19*, 20–25. (In Chinese)
41. Yang, W.Q.; Kou, J.C.; Han, M.Y.; Wang, A.Z. Effects of inter-planted trifolium repens on the photosynthetic physiology of peach tree. *Acta Agrestia Sin.* **2011**, *19*, 771–775.
42. Wu, J.J.; Li, Q.S.; Yan, L.J. Soil ecological effects of intercropping forages in young Platycodon grandiflorum and their effects on the growth of Platycodon grandiflorum. *Chin. J. Ecol.* **1996**, *15*, 10–14. (In Chinese)
43. Wang, X.-Y.; Yang, T.; Shen, L.; Zhang, W.-L.; Wan, S.-M.; Li, L.-H. Formation of factors influencing cotton yield in jujube–cotton intercropping systems in Xinjiang, China. *Agrofor. Syst.* **2021**, *95*, 177–189. [CrossRef]
44. Zhang, W.; Xie, H.; Zhang, P.; Zhong, H.-X.; Zhang, F.C.; Zhuang, H.M.; Yang, L.; Xu, Y.T.; Gong, P.; Lu, C.S. Effect of tree canopy structure on light condition in almond-winter wheat intercropping systems. *Chin. J. Eco-Agric.* **2016**, *24*, 753–761. (In Chinese) [CrossRef]
45. Mao, P.C.; Meng, L.; Zhang, G.F.; Hou, F.Q.; Zhang, Z.G.; Cui, T.-H. Effect of planting white clover as a cover crop on the microclimate of a peach orchard and the peach quality. *Acta Agrestia Sin.* **2006**, *14*, 360–364. (In Chinese)
46. Moreira, S.L.S.; Pires, C.V.; Marcatti, G.E.; Santos, R.H.; Imbuzeiro, H.M.; Fernandes, R.B. Intercropping of coffee with the palm tree, macauba, can mitigate climate change effects. *Agric. For. Meteorol.* **2018**, *256–257*, 379–390. [CrossRef]
47. Munns, R. Comparative physiology of salt and water stress. *Plant Cell Environ.* **2022**, *25*, 239–250. [CrossRef]
48. Li, G.H.; Yi, H.L. Influences of sod culture on the soil water content, effect of soil nutrients, fruit yield and quality in citrus orchard. *Chin. J. Eco-Agric.* **2005**, *13*, 161–163. (In Chinese)
49. Zeng, M.; Li, D.G.; Xiong, B.Q.; Lu, Z.M.; Luo, Y. Effects of sod culture management in citrus orchards on am infection in citrus roots and fruit quality. *J. Southwest Agric. Univ. (Nat. Sci.)* **2004**, *26*, 105–107. (In Chinese)
50. Jia, X.H.; Du, Y.M.; Wang, W.H.; Li, Z. Hang, B. Effect of natural grass on quality and storage tolerance of Huahong apple. *China Fruits* **2015**, *6*, 30–32. (In Chinese) [CrossRef]

agriculture

Article

Controlled-Release Fertilizer Improves Rice Matter Accumulation Characteristics and Yield in Rice–Crayfish Coculture

Qiangsheng Wang *, Kunlong Yu and Hui Zhang

Lab of Modern Farming System and Ecological Circular Agriculture, Nanjing Agricultural University, Nanjing 210095, China
* Correspondence: qswang@njau.edu.cn

Abstract: In recent years, rice–fish coculture has gained more popularity at a growing pace in China. Controlled-release fertilizer can provide nutrients in a timely manner and increase nutrient efficiency. A 2-year field experiment, which adopted both conventional japonica and two indica hybrid rice varieties, was performed to evaluate the effects of controlled-release fertilizer and inorganic compound fertilizer on rice matter accumulation and yield in rice–crayfish coculture and conventional rice farming. The results showed that compared to conventional rice farming, rice–crayfish coculture decreased dry matter accumulation at mature stage and yield by 4.02–8.15% and 4.13–9.34%, respectively. This was mainly due to a decrease in the crop growth rate, net assimilation rate, leaf area index, and light accumulation duration before elongation stage. Compared to inorganic compound fertilizer, controlled-release fertilizer increased dry matter accumulation at the mature stage and yield by 5.02–6.95% and 3.29–6.21%, respectively. Compared to conventional rice farming, rice–crayfish coculture decreased N partial factor productivity and N agronomic use efficiency by 4.13–9.34% and 3.96–8.98%, respectively. Compared to inorganic compound fertilizer, controlled-release fertilizer increased those by 3.29–6.15% and 7.36–14.01%. There was a positive linear correlation between the N partial factor productivity, N agronomic use efficiency, and yield.

Keywords: rice–crayfish coculture; different rice varieties; controlled-release fertilizer; dry matter accumulation; rice growth characteristics parameters; yield

Citation: Wang, Q.; Yu, K.; Zhang, H. Controlled-Release Fertilizer Improves Rice Matter Accumulation Characteristics and Yield in Rice–Crayfish Coculture. *Agriculture* **2022**, *12*, 1674. https://doi.org/10.3390/agriculture12101674

Academic Editors: Chengfang Li and Lijin Guo

Received: 22 August 2022
Accepted: 8 October 2022
Published: 12 October 2022

1. Introduction

Increased and stable rice production is essential to ensure food security around the world since rice feeds more than half of the world's population [1]. However, in order to pursue higher rice production, the excessive application of N fertilizer, one of the main factors in ensuring rice yield, leads to lower N use efficiency, growing environmental pollution, and more diseases, pests and weeds, thus disturbing rice yield [2–4].

As an alternative to the traditional chemical fertilizer, controlled-release fertilizer can make sure nutrient release (release rate and time) correspond to the S-shaped curve of the N requirement at the rice growth stage by coating and adding various biochemical inhibitors [5,6]. This saves fertilizer, increases efficiency, reduces nutrient volatilization, and provides nutrients in a timely manner. Some studies have indicated that controlled-release fertilizer could significantly improve rice yield, the photosynthesis rate, dry matter accumulation, N accumulation, and N use efficiency [7,8], but its effects are impacted by various factors such as rice varieties, fertilizer application, planting methods, and soil types. Some studies, however, have shown that controlled-release fertilizer reduces rice yield. For example, Ye et al. [9] proved that rice yield reduced by 11.5–12.0% with 70% controlled-release N and 30% inorganic N fertilizer compared to an equal amount of inorganic N fertilizer. Mi et al. [10] determined that compared to the step-by-step application of urea, yield increased by 5.2% with the application of controlled-release fertilizer during the manual trans-plantation of rice seedlings and decreased by 3% when rice seedlings were sown directly. Furthermore, controlled-release fertilizer could also reduce N loss in paddy fields [11–13].

Rice–crayfish coculture, a typical eco-circular agricultural mode, has experienced widespread adoption in recent years. China has saw the fastest adoption of rice–fish coculture in the world, with the area surpassing 1.26 million hectares [14]. In this system, rice leaves provide a shady environment for crayfish, which feed on the pathogens and pests that are living at the lower part of stem, reducing the incidence of pests in paddy fields. Therefore, a beneficial symbiotic relationship between rice and crayfish formed. Many studies have concluded that the rice yield of rice–crayfish coculture is not lower than that of conventional rice farming. Using the food-equivalent unit method and arable-land-equivalent unit method, Jin et al. [15] evaluated the overall sustainability of rice systems and showed that the rice yield of RC increased by 4.48% compared to rice monoculture on the basis of 10% arable land being occupied by excavated ring trenches. Hou et al. [16] pointed out the rice yield of rice–crayfish coculture with deep groundwater (50–100 cm below the soil surface) decreased by 30–55% compared to typical rice–rapeseed rotations, while the rice yield of rice–crayfish coculture with shallow groundwater (40–60 cm below the soil surface) was similar to that of conventional rice farming. Moreover, rice–crayfish coculture helps to promote soil nutrient cycling and improve soil quality [17,18].

Currently, there has been some achievements in rice yield, soil fertility, N loss in rice–crayfish coculture, and N use efficiency of controlled-release fertilizer in conventional rice farming [19–21]. In contrast, studies on production management of large-scale rice–crayfish coculture with controlled-release fertilizer have not been reported yet, and in addition whether it can improve rice yield and N use efficiency under deep water irrigation at the middle and later stage of rice–crayfish coculture needs to be further exploration. Therefore, this study aims to examine the effects of controlled-release fertilizer on rice yield formation and dry matter accumulation as well as the correlation of dry matter accumulation, dry matter utilization characteristics, and yield through different types of rice varieties. The results can provide a reference for controlled-release fertilizer application in rice–crayfish coculture.

2. Materials and Methods

2.1. Experimental Site and Materials

The field experiment was carried out in 2019 and 2020 at the experimental farm of Nanjing Agricultural University (31°47′ N, 118°55′ E), which is located at the middle and lower reaches of the Yangtze River and is characterized by a subtropical monsoon climate, with an annual average air temperature of 16.8 °C, precipitation of 1017.4 mm, and a frost-free period of 237 d. The properties of the 0–20 cm soil in the study area are as follows: pH 6.5, organic matter, 24.4 $g \cdot kg^{-1}$; total N, 1.4 $g \cdot kg^{-1}$; available N, 86.6 $mg \cdot kg^{-1}$; available P, 4.8 $mg \cdot kg^{-1}$; and available K, 72.3 $mg \cdot kg^{-1}$.

The conventional japonica rice varieties that were used in the experiment were Wuyunjing23 (WYJ23) and Nanjingjinggu (NJJG), and the indica hybrid rice varieties were Quanyou0861 (QY0861) and Shenliangyou600 (SLY600). The rice growth period of the two japonica rice varieties totaled 157 d. The number of days from transplanting stage to effective tillering termination stage, from effective tillering termination stage to elongation stage, from elongation stage to heading stage, and from heading stage to mature stage were 28 d, 17 d, 29 d, and 63 d, respectively. The rice growth period of the two indica hybrid rice varieties totaled 137 d, and the number of days from transplanting stage to effective tillering termination stage, from effective tillering termination stage to elongation stage, from elongation stage to heading stage, and from heading stage to mature stage was 26 d, 15 d, 26 d, and 47 d, respectively. Crayfish (Procambarus clarkii) were used as the tested fish in this experiment, and were fed with drones. No chemical pesticides were used in rice–crayfish coculture.

The nutrient ratios of the tested fertilizer were as follows: resin-coated controlled-release fertilizer ($N-P_2O_5-K_2O$ = 26%-11%-11%), inorganic compound fertilizer ($N-P_2O_5-K_2O$ = 17%-17%-17%). The fertilizer used in the auxiliary experiment contained the following: urea (N = 46.2%), calcium superphosphate (P_2O_5 = 12.4%), and potassium chloride (K_2O = 60%).

2.2. Experimental Design

The experiment took fertilizer type as the main plot, set conventional rice farming with inorganic compound fertilizer (CR-IF), conventional rice farming with controlled-release fertilizer (CR-CF), rice–crayfish coculture with inorganic compound fertilizer (RC-IF), rice-crayfish coculture with controlled-release fertilizer (RC-CF) as four treatments. Each treatment was conducted with three replicates. Each replicate covered an area of 4 ha and was separated by film-wrapped ridges. The ditches, which were shaped like homocentric squares around the rice-crayfish coculture, were 3 m wide and 1.5 m deep. The rice varieties were sown on 20 May, and seedlings were mechanically transplanted into a hole with a row spacing of 30 cm and a plant spacing of 12 cm on 12 June. Four seedlings of the japonica rice varieties were placed in each hole, while two seedlings of the indica rice varieties were placed in each hole. The indica rice varieties were harvested on 3 October, and the japonica rice varieties were harvested on 26 October.

The experiment was conducted with an equal amount of nutrients, including 240 kg N ha^{-1}, 102 kg P_2O_5 ha^{-1}, and 102 kg K_2O ha^{-1}. The specific fertilization methods are shown in Table 1. At transplanting stage, the machine insertion mode and deep fertilization were adopted, where a trench was dug at a depth of 5 cm and at a distance of 4 cm away from one side of the seedlings, and then fertilizer was spread evenly over the trench. Conventional rice farming without N fertilizer and rice–crayfish coculture without N fertilizer were used as auxiliary experiments.

Table 1. Fertilizer management in rice–crayfish coculture and conventional rice farming (kg·ha^{-1}).

Treatments	Basic Fertilizer	The First Tiller Fertilizer at 7th Day after Transplanting Stage	The Second Tiller Fertilizer at 15th Day after Transplanting Stage	Spikelet Promoting Fertilizer	Spikelet Developing Fertilizer
		Urea			Urea
Inorganic compound fertilizer	373.0	93	94	224.0	112.0
Controlled-release fertilizer	577.0	-	-	346.0	-

During rice growth, the following water management of conventional rice farming were maintained: a 3–5 cm water layer was maintained from transplanting stage until initial tillering occurrence stage, with the wet and dry irrigation modes being alternated until effective tillering termination stage. From effective tillering termination stage to elongation stage, the water was drained. After the repeated application of alternating wet and dry irrigation until the 7th day prior to mature stage, water was cut off. Water management in the rice–crayfish coculture was similar to that in conventional rice farming before effective tillering termination stage, with the exception of a deeper water layer. After elongation stage, a 20–40 cm water layer was irrigated but was drained on the 10th day before mature stage.

2.3. Plant Sampling and Analysis

2.3.1. Tiller Dynamics

The number of tillers was investigated every 7 d after transplanting stage through the selection of twenty contiguous holes in each plot until elongation stage in order to calculate the panicle rate according to the proportion of effective panicles to the number of peak seedlings.

2.3.2. Dry Matter Weight

Plants with the average number of tillers ± 1 were chosen from five holes in each plot at effective tillering termination stage, elongation stage, heading stage, and mature stage according to the average number of tillers surveyed at each stage.

The stem and sheath as well as the leaves and panicles (heading stage and mature stage) were separated and dried at 105 °C for 30 min, oven-dried at 80 °C until a constant weight was reached, and the dry matter weight was then determined.

2.3.3. Yield and Its Components

Representative plants from five holes per plot were selected to measure the yield components, including the number of grains per panicle, the seed-setting rate, and the grain weight. In each plot, 1/3 ha of rice was harvested to determine the yield at mature stage. The yields of the japonica and indica rice varieties were conversed according to the standard water contents of 14.5% and 13.5%, respectively.

2.3.4. Data Calculation

The rice growth characteristics and dry matter utilization evaluation index used in this experiment were carried out according to equations of Ye et al. [9], Liu et al. [22], Dou et al. [23], and Chen et al. [24]. The relevant evaluation methods were as follows:

$$\text{Leaf area index} = \frac{\text{plant green leaf area}}{\text{per unit land area}} \tag{1}$$

$$\text{Panicle rate } (\%) = \frac{\text{effective panicle number at mature stage}}{\text{peak seedling number at tillering stage}} \times 100 \tag{2}$$

$$\text{Crop growth rate } \left(\text{g}\cdot\text{m}^{-2}\cdot\text{d}^{-1}\right) = \frac{(\text{W}_2 - \text{W}_1)}{(\text{T}_2 - \text{T}_1)} \tag{3}$$

where W_1 and W_2 represent the dry matter weight per unit land area at the beginning and end of a time interval, respectively, and T_1 and T_2 are the corresponding days.

$$\text{Light accumulation duration } \left(\text{m}^2\cdot\text{d}\cdot\text{m}^{-2}\right) = \frac{(\text{L}_1 + \text{L}_2) \times (\text{T}_2 - \text{T}_1)}{2} \tag{4}$$

where L_1 and L_2 represent the leaf area per unit land area at the beginning and end of a time interval, respectively, and T_1 and T_2 are the corresponding days.

$$\text{Net assimilation rate } \left(\text{g}\cdot\text{m}^{-2}\cdot\text{d}^{-1}\right) = \frac{\left[\frac{(\text{lnL}_2 - \text{lnL}_1)}{(\text{L}_2 - \text{L}_1)}\right] \times (\text{W}_2 - \text{W}_1)}{(\text{T}_2 - \text{T}_1)}$$

where L_1 and L_2 are the leaf area index at the beginning and end of a time interval, respectively; W_1 and W_2 are the dry matter weight per unit land area at the beginning and end of a time interval; and T_1 and T_2 are the corresponding days.

$$\text{N partial factor productivity } \left(\text{kg}\cdot\text{kg}^{-1}\right) = \frac{\text{yield}}{\text{the amount of applied N}} \tag{5}$$

$$\text{N agronomic use efficiency } \left(\text{kg}\cdot\text{kg}^{-1}\right) = \frac{(\text{rice yield} - \text{rice yield without N fertilizer})}{\text{the amount of applied N}} \tag{6}$$

$$\text{Harvest index} = \frac{\text{yield}}{\text{dry matter weight at mature stage}} \tag{7}$$

2.3.5. Data Analysis

SPSS (IBM SPSS 24, USA) was adopted to statistically analyze experimental data, which were expressed as the mean value ± standard error. One-way ANOVA was used to analyze the significance of variation among treatments and the least significant difference (LSD) tests were used for pairwise comparisons. The correlation between N partial factor productivity, N agronomic use efficiency, harvest index, and rice yield were analyzed by Pearson's correlation analysis. Excel 2019 and Origin (version 8.1, USA) was employed for figure preparation.

3. Results

3.1. Rice Yield and Its Components

As shown in Table 2, the rice variety has a significant effect on rice yield. The yield of the japonica rice varieties was 0.13–1.60 t·ha^{-1} higher than that of the indica rice varieties. Among the four rice varieties, NJJG obtained the highest actual yield, while obtained QY0861 lowest.

The yields of the japonica and indica rice varieties in rice–crayfish coculture were significantly reduced by 6.28–9.34% and 4.13–5.91%, respectively, compared to the yield in conventional rice farming, which was mainly due to a steep decline of 3.61–6.03% and 4.45–6.45% in the 1000-grain weight of the japonica and indica rice varieties, respectively, in rice–crayfish coculture. Nonetheless, the panicle rate of the japonica and indica rice varieties in rice–crayfish coculture grew by 1.42–3.31% and 1.41–6.23% compared to in conventional rice farming, respectively.

The yields of japonica and indica rice varieties in controlled-release fertilizer treatments significantly increased by 3.29–4.70% and 4.73–6.21%, respectively, compared to that of inorganic fertilizer. In addition, effective panicles, peak seedlings, panicle rate, the number of grains per panicle of japonica and indica rice varieties in controlled-release fertilizer treatments increased by 4.51–6.71% and 5.94–8.29%, 1.51–3.46% and 1.02–4.41%, 1.80–5.07% and 2.97–7.15%, and 2.54–3.39% and 2.07–3.86%, respectively. In addition, the yield of CR-CF was 3.29–6.21% higher than that of CR-IF; the yield of RC-CF was 4.22–6.15% higher than that of RC-IF.

3.2. Dry Matter Accumulation

Compared to the indica rice varieties, there was a decrease in the dry matter accumulation of the japonica rice varieties before elongation stage but an increase after elongation stage compared to indica rice varieties (Figure 1). There were significant differences in the dry matter accumulation of difference rice varieties at mature stage among the different treatments, which was expressed by the lowercase letters in Figure 1. The dry matter accumulation of the japonica rice varieties at the mature stage increased by 0.06–3.48 t·ha^{-1} compared to the indica rice varieties. Among the four rice varieties, NJJG had the highest dry matter accumulation at the mature stage, while QY0861 had the lowest.

Compared to conventional rice farming, rice–crayfish coculture significantly decreased the dry matter accumulation of the japonica and indica rice varieties by 5.48–7.83% and 3.00–7.06% from elongation stage to heading stage, by 8.21–10.80% and 2.69–8.80% from heading stage to mature stage, and by 5.94–8.15% and 4.02–5.69% at mature stage, respectively.

The dry matter accumulation of the japonica and indica rice varieties grew significantly by 4.26–6.78% and 5.15–8.76% from elongation stage to heading stage and by 6.84–8.88 and 4.81–10.42% from heading stage to mature stage, respectively. The dry matter accumulation of the japonica and indica rice varieties at mature stage increased by 5.02–6.76% and 5.54–6.95% in rice–crayfish coculture, respectively, compared to in conventional rice farming. In addition, the dry matter accumulation of CR-CF at mature stage was 5.22–6.86% higher than that of CR-IF, while the dry matter accumulation of RC-CF was 5.02–6.95% higher than that of RC-IF.

Table 2. Yield and its components in different rice varieties under two types of fertilizer treatments.

Year	Varieties	Treatments	Effective Panicle ($\times 10^4$ ha^{-1}))	Peak Seedling ($\times 10^4$ ha^{-1})	Panicle Rate (%)	No. of Grains Per Panicle	Seed-Settingrate (%)	1000-Grain Weight (%)	Actual Yield (t·ha^{-1})
2019	WYJ23	CR-IF	336.06 ± 4.93 e	409.05 ± 10.76 d	82.16 ± 0.62 e	138.01 ± 2.45 fg	91.08 ± 0.57 ab	26.71 ± 0.16 ab	10.18 ± 0.20 c
		CR-CF	358.41 ± 4.54 c	415.22 ± 9.28 a	86.32 ± 0.46 c	141.51 ± 3.08 f	90.93 ± 0.56 ab	26.12 ± 0.24 bc	10.55 ± 0.21 bc
		RC-IF	330.77 ± 4.76 e	389.73 ± 6.61 k	84.87 ± 0.02 d	133.91 ± 3.77 g	90.89 ± 0.64 b	25.48 ± 0.44 cd	9.25 ± 0.15 f
		RC-CF	352.96 ± 6.26 c	403.20 ± 8.51 g	87.54 ± 0.28 b	137.38 ± 2.89 fg	90.73 ± 0.82 c	24.75 ± 0.13 de	9.64 ± 0.20 e
	NJJG	CR-IF	349.72 ± 7.05 c	414.75 ± 10.59 b	84.32 ± 0.17 d	134.60 ± 2.27 g	92.02 ± 0.59 ab	27.12 ± 0.18 a	10.72 ± 0.21 bc
		CR-CF	368.73 ± 6.69 a	425.1 ± 10.37 a	86.74 ± 0.21 c	138.60 ± 2.47 g	91.88 ± 0.58 ab	26.33 ± 0.36 b	11.08 ± 0.16 a
		RC-IF	343.32 ± 6.92 d	397.80 ± 8.45 i	86.30 ± 0.13 c	131.35 ± 3.86 g	91.82 ± 0.55 ab	26.14 ± 0.21 bc	9.92 ± 0.07 d
		RC-CF	361.45 ± 4.39 b	407.25 ± 13.67 e	88.75 ± 1.28 a	134.82 ± 3.53 g	91.74 ± 0.73 ab	25.35 ± 0.45 cd	10.38 ± 0.15 c
	QY0861	CR-IF	238.04 ± 4.12 f	385.82 ± 6.41 l	61.70 ± 0.11 k	193.44 ± 2.17 b	84.88 ± 0.40 ef	25.76 ± 0.25 c	9.13 ± 0.13 f
		CR-CF	252.85 ± 3.56 f	395.41 ± 6.92 j	63.95 ± 0.09 j	198.58 ± 2.31 a	84.67 ± 0.39 f	25.13 ± 0.45 d	9.69 ± 0.08 de
		RC-IF	233.95 ± 4.23 f	369.90 ± 8.87 n	63.25 ± 0.23 j	188.18 ± 4.0 6c	84.79 ± 0.59 ef	24.56 ± 0.34 e	8.70 ± 0.04 g
		RC-CF	247.84 ± 4.60 f	380.44 ± 12.41 m	65.15 ± 0.62 i	194.47 ± 3.08 ab	84.61 ± 0.70 f	23.82 ± 0.32 f	9.18 ± 0.17 f
	SLY600	CR-IF	262.05 ± 6.10 f	405.15 ± 10.55 f	64.68 ± 0.01 i	182.03 ± 3.88 d	85.84 ± 0.43 e	25.45 ± 0.20 cd	9.47 ± 0.18 ef
		CR-CF	281.70 ± 5.89 f	409.28 ± 7.42 c	68.83 ± 0.33 g	186.32 ± 2.65 cd	85.66 ± 0.40 ef	24.58 ± 0.41 e	9.92 ± 0.09 d
		RC-IF	258.34 ± 5.35 f	381.30 ± 9.37 m	67.75 ± 0.03 h	176.87 ± 1.62 e	85.75 ± 0.70 ef	23.89 ± 0.10 f	9.00 ± 0.21 f
		RC-CF	278.42 ± 5.53 f	398.11 ± 7.49 k	69.94 ± 0.26 f	183.70 ± 3.06 cd	85.59 ± 1.29 ef	23.06 ± 0.38 g	9.51 ± 0.08 e
2020	WYJ23	CR-IF	331.40 ± 7.16 e	406.05 ± 8.32 c	81.62 ± 0.25 f	134.24 ± 3.84 ef	90.83 ± 1.24 a	26.58 ± 0.25 ab	9.86 ± 0.12 cd
		CR-CF	346.92 ± 7.34 c	414.44 ± 7.80 a	83.71 ± 0.36 d	137.74 ± 2.26 ef	90.78 ± 0.63 a	25.56 ± 0.13 bc	10.24 ± 0.16 bc
		RC-IF	325.36 ± 7.52 e	387.91 ± 8.50 i	83.88 ± 0.31 d	130.56 ± 3.74 f	90.64 ± 0.53 b	25.12 ± 0.34 c	8.94 ± 0.13 f
		RC-CF	342.78 ± 7.01 c	401.12 ± 9.82 f	85.46 ± 0.05 b	134.35 ± 2.41 ef	90.62 ± 0.42 b	24.16 ± 0.13 d	9.36 ± 0.17 e
	NJJG	CR-IF	342.30 ± 4.91 c	412.20 ± 8.71 a	83.04 ± 0.33 e	130.86 ± 2.06 f	91.87 ± 0.89 a	26.88 ± 0.67 a	10.33 ± 0.17 b
		CR-CF	357.82 ± 4.74 a	423.25 ± 6.14 a	84.54 ± 0.04 c	135.29 ± 3.69 ef	91.79 ± 0.88 a	26.08 ± 0.16 b	10.75 ± 0.09 a
		RC-IF	338.53 ± 4.72 d	395.25 ± 6.90 f	85.65 ± 0.09 b	127.42 ± 2.54 f	91.72 ± 0.45 a	25.26 ± 0.46 c	9.62 ± 0.19 d
		RC-CF	353.81 ± 5.34 b	405.45 ± 7.21 d	87.26 ± 0.25 a	131.47 ± 2.21 f	91.61 ± 0.83 a	24.73 ± 0.34 cd	10.03 ± 0.14 c
	QY0861	CR-IF	231.49 ± 5.17 g	381.75 ± 6.77 j	60.64 ± 0.36 m	187.66 ± 1.76 b	84.68 ± 0.38 cd	25.62 ± 0.23 bc	8.82 ± 0.07 fg
		CR-CF	249.70 ± 5.06 fg	392.34 ± 11.85 g	63.64 ± 0.37 k	192.51 ± 3.21 a	84.45 ± 0.68 e	24.92 ± 0.42 cd	9.25 ± 0.19 e
		RC-IF	228.34 ± 5.67 g	364.28 ± 12.25 l	62.68 ± 0.33 l	182.30 ± 2.81 c	84.53 ± 0.38 d	24.48 ± 0.23 d	8.31 ± 0.10 h
		RC-CF	243.20 ± 5.04 fg	376.80 ± 8.03 j	64.54 ± 0.13 j	188.69 ± 3.62 ab	84.35 ± 0.33 e	23.65 ± 0.31 e	8.72 ± 0.14 f
	SLY600	CR-IF	255.43 ± 5.66 fg	403.35 ± 13.28 e	63.33 ± 0.47 k	179.63 ± 1.43 c	85.66 ± 0.71 c	25.28 ± 0.10 c	9.16 ± 0.20 ef
		CR-CF	276.61 ± 6.39 f	407.65 ± 11.30 b	67.85 ± 0.05 h	183.34 ± 1.67 bc	85.43 ± 0.50 cd	24.43 ± 0.68 d	9.72 ± 0.10 d
		RC-IF	251.87 ± 4.53 fg	374.40 ± 10.70 k	67.27 ± 0.42 i	174.27 ± 2.51 d	85.52 ± 0.39 cd	23.65 ± 0.28 e	8.62 ± 0.13 g
		RC-CF	272.19 ± 4.74 fg	389.25 ± 7.84 h	69.93 ± 0.04 g	179.24 ± 3.78 c	85.38 ± 0.58 cd	22.88 ± 0.37 f	9.15 ± 0.14 ef
	V		**	**	**	**	**	**	**
	R		ns	**	**	**	ns	**	**
	F		**	**	**	**	ns	ns	**
	V*R		*	*	**	ns	ns	ns	**
	V*F		**	**	**	ns	ns	ns	ns
	R*F		ns	*	ns	ns	ns	ns	ns
	V*R*F		ns	ns	**	ns	ns	ns	**

All the data presented are the means of three replicates ± standard errors. Different lowercase letters indicate the significant differences among all treatments in the four varieties at the 5% level in the same year. CR, conventional rice cultivation; RC, rice–crayfish coculture system; IF, inorganic compound fertilizer; CF, controlled-release fertilizer; V, rice variety; R, rice cropping mode; F, fertilizer type. * and ** indicate significant differences at the $p < 0.05$ and 0.01 levels, respectively. "ns" means not significant.

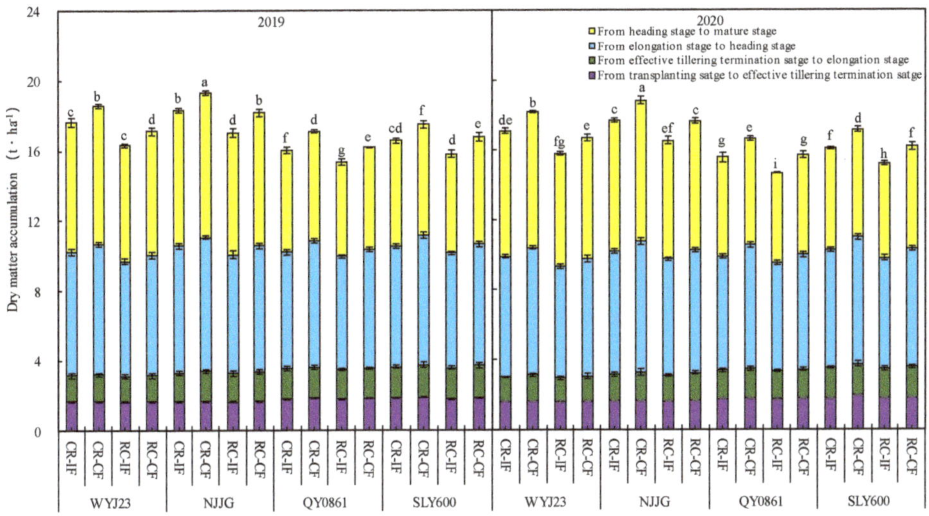

Figure 1. Dry matter accumulation of different rice varieties under two types of fertilizer treatments. The error bars were presented as the standard errors of the dry matter accumulation of transplanting stage to effective tillering termination stage, effective tillering termination stage to elongation stage, elongation stage to heading stage, heading stage to mature stage. Different lowercase letters indicate significant differences of dry matter accumulation at mature stage among all treatments of four varieties in the same year at the 5% level according to LSD tests (0.05).

3.3. Leaf Area Index

The leaf area index represents the ability of rice leaves to capture and utilize light energy during rice growth. During rice growth, leaf area indexes of the different rice varieties increased gradually from transplanting stage to heading stage, peaked at heading stage, and then decreased (Figure 2). Compared to the indica rice varieties, the leaf area indexes of the japonica rice varieties decreased by 0.21–0.42, 0.92–1.07, and 0.35–0.02 at effective tillering termination stage, elongation stage, and heading stage, respectively, but increased by 0.13–0.37 at mature stage.

There was a significant difference in the leaf area index at heading stage and mature stage in rice–crayfish coculture compared to conventional rice farming, and rice–crayfish coculture significantly decreased the leaf area index of the japonica and indica rice varieties by 5.05–6.20% and 5.12–5.98% at heading stage, respectively. Compared to inorganic compound fertilizer, controlled-release fertilizer increased the leaf area index of the japonica and indica rice varieties by 3.42–4.79% and 3.49–5.12% at the heading stage and by 3.12–4.97% and 4.21–5.39% at the mature stage.

3.4. Light Accumulation Duration and Net Assimilation Rate

Light accumulation duration, a measure of leaf area population and photosynthetic sustainability, marks the accumulation of crop leaf area during the whole or a certain growth stage in rice and plays an important role in evaluating dry matter accumulation. The results showed light accumulation duration of japonica rice varieties decreased before elongation stage but increased after elongation stage compared to the indica rice varieties (Table 3). Meanwhile, SLY600 had the highest light accumulation duration and WYJ23 had the lowest before elongation stage, but after elongation stage, NJJG had the highest light accumulation duration and QY0861 had the lowest.

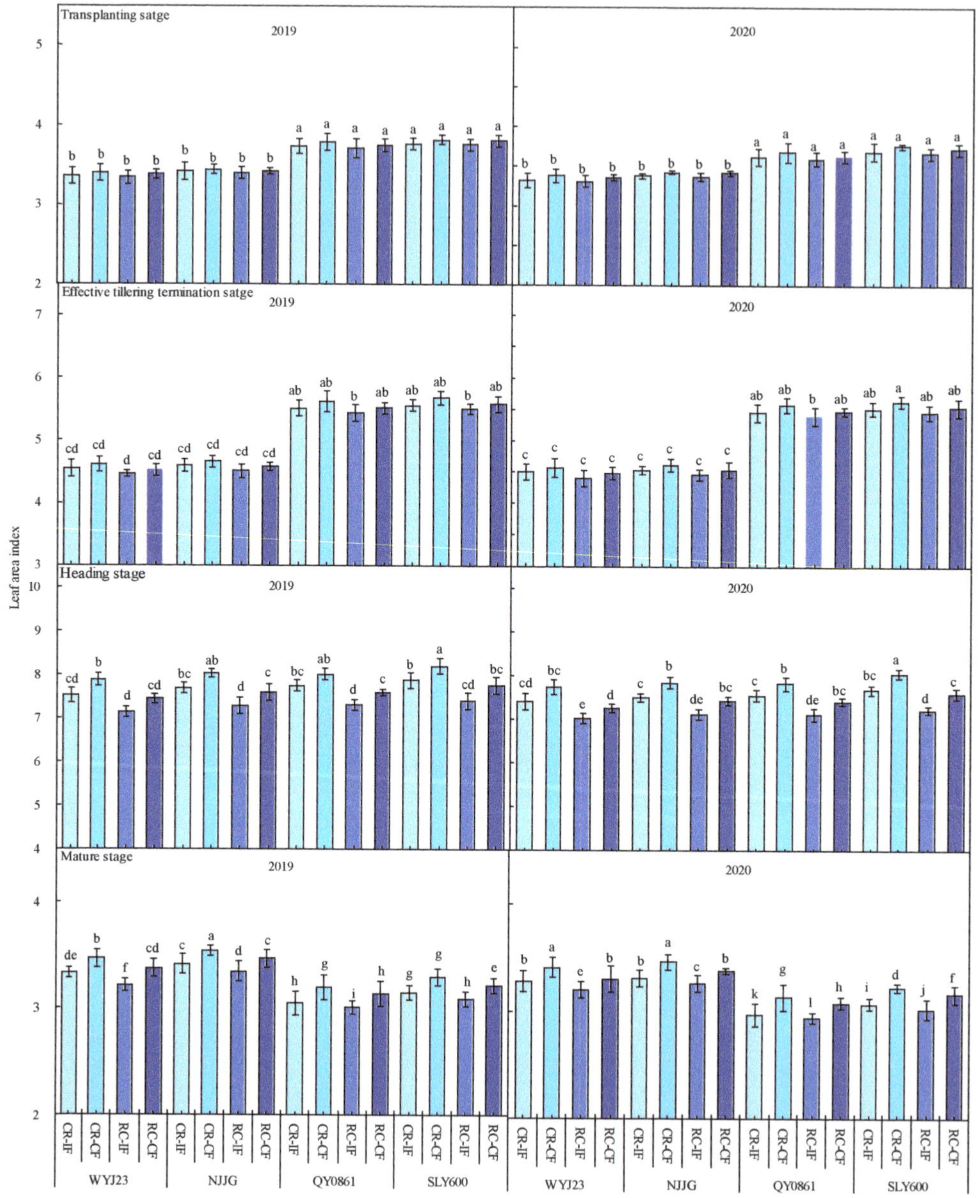

Figure 2. Leaf area indexes of different rice varieties under two types of fertilizer treatments. Different lowercase letters indicate significant differences among all treatments of four varieties in the same year at the 5% level according to LSD tests (0.05).

Table 3. Light accumulation duration and net assimilation rate of different rice varieties under two types of fertilizer treatments.

Year	Varieties	Treatments	Light Accumulation Duration($m^2 \cdot d \cdot m^{-2}$)				Net Assimilation Rate ($g \cdot m^{-2} \cdot d^{-1}$)			
			Transplanting Stage-Effective Tillering Termination Stage	Effective Tillering Termination Stage-Elongation Stage	Elongation Stage-Heading Stage	Heading Stage-Mature Stage	Transplanting Stage-Effective Tillering Termination Stage	Effective Tillering Termination Stage-Elongation Stage	Elongation Stage-Heading Stage	Heading Stage-Mature Stage
2019	WYJ23	CR-IF	48.86 ± 1.19 b	67.15 ± 1.45 b	174.87 ± 3.10 bc	342.09 ± 7.09 d	5.94 ± 0.19 ab	2.25 ± 0.05 c	4.13 ± 0.04 bc	2.29 ± 0.02 d
		CR-CF	49.42 ± 1.25 b	68.09 ± 1.78 b	181.11 ± 3.17 ab	357.53 ± 7.15 cd	6.00 ± 0.13 ab	2.25 ± 0.07 c	4.20 ± 0.04 ab	2.35 ± 0.03 cd
		RC-IF	48.58 ± 1.12 b	66.22 ± 1.32 b	173.85 ± 2.47 bc	367.78 ± 6.75 bc	5.93 ± 0.08 ab	2.22 ± 0.11 c	3.86 ± 0.07 d	1.90 ± 0.04 h
		RC-CF	49.14 ± 0.72 b	67.15 ± 1.53 b	179.40 ± 2.63 ab	384.11 ± 7.98 ab	5.95 ± 0.14 ab	2.25 ± 0.08 c	3.90 ± 0.09 d	1.95 ± 0.02 h
	NJJG	CR-IF	49.70 ± 0.76 b	68.09 ± 1.34 b	177.92 ± 2.46 b	349.65 ± 6.84 d	5.90 ± 0.12 ab	2.45 ± 0.03 b	4.17 ± 0.08 ab	2.34 ± 0.04 d
		CR-CF	49.98 ± 0.70 b	68.77 ± 1.25 ab	183.72 ± 2.55 a	364.14 ± 6.77 c	5.99 ± 0.15 ab	2.48 ± 0.04 b	4.27 ± 0.06 a	2.4 ± 0.03 cd
		RC-IF	49.42 ± 0.84 b	67.15 ± 1.84 b	176.70 ± 2.84 bc	377.37 ± 6.61 b	5.83 ± 0.06 b	2.44 ± 0.05 b	3.91 ± 0.03 d	1.94 ± 0.02 h
		RC-CF	49.84 ± 0.56 b	68.00 ± 1.52 b	182.55 ± 2.97 ab	392.99 ± 6.78 a	5.90 ± 0.24 ab	2.47 ± 0.08 b	4.05 ± 0.05 c	2.03 ± 0.03 g
	QY0861	CR-IF	50.96 ± 0.88 ab	69.38 ± 1.71 ab	172.25 ± 3.16 c	253.57 ± 8.31 h	5.91 ± 0.25 ab	2.52 ± 0.11 ab	3.90 ± 0.05 d	2.47 ± 0.04 b
		CR-CF	51.61 ± 1.30 a	70.58 ± 1.52 ab	177.19 ± 3.44 b	263.44 ± 6.67 gh	5.97 ± 0.02 ab	2.53 ± 0.08 ab	4.12 ± 0.06 bc	2.55 ± 0.05 a
		RC-IF	50.57 ± 0.81 ab	68.63 ± 1.35 b	171.99 ± 2.43 c	278.37 ± 7.12 fg	5.90 ± 0.16 ab	2.51 ± 0.06 b	3.78 ± 0.09 d	2.06 ± 0.04 fg
		RC-CF	51.09 ± 0.72 ab	69.53 ± 1.42 ab	177.12 ± 2.86 b	289.98 ± 6.68 ef	5.95 ± 0.19 ab	2.52 ± 0.04 ab	3.87 ± 0.07 d	2.15 ± 0.02 e
	SLY600	CR-IF	51.35 ± 0.85 a	69.98 ± 1.58 ab	174.59 ± 4.11 bc	258.97 ± 6.94 gh	6.03 ± 0.01 ab	2.56 ± 0.09 ab	4.00 ± 0.04 d	2.49 ± 0.03 b
		CR-CF	52.00 ± 0.67 a	71.25 ± 1.41 a	180.44 ± 2.65 ab	270.25 ± 6.71 g	6.11 ± 0.16 a	2.65 ± 0.08 a	4.16 ± 0.08 bc	2.50 ± 0.02 ab
		RC-IF	51.22 ± 0.30 ab	69.45 ± 1.29 ab	174.15 ± 2.57 bc	283.23 ± 6.91 f	5.94 ± 0.12 ab	2.54 ± 0.12 ab	3.81 ± 0.03 d	2.12 ± 0.04 f
		RC-CF	51.87 ± 0.58 a	70.43 ± 1.90 ab	180.09 ± 2.65 ab	296.46 ± 7.62 e	6.03 ± 0.18 ab	2.65 ± 0.13 a	3.88 ± 0.07 d	2.21 ± 0.05 e
2020	WYJ23	CR-IF	48.30 ± 0.98 c	66.47 ± 1.83 b	172.84 ± 2.72 bc	337.05 ± 6.75 d	5.84 ± 0.09 ab	2.10 ± 0.02 d	4.08 ± 0.05 c	2.25 ± 0.02 f
		CR-CF	49.14 ± 0.36 c	67.58 ± 1.16 b	178.50 ± 2.47 ab	351.23 ± 6.84 cd	5.87 ± 0.11 ab	2.21 ± 0.04 cd	4.17 ± 0.06 a	2.34 ± 0.03 de
		RC-IF	48.16 ± 0.32 c	65.54 ± 1.02 b	171.30 ± 4.33 bc	362.81 ± 6.35 bc	5.75 ± 0.11 ab	2.09 ± 0.05 d	3.80 ± 0.03 de	1.87 ± 0.04 m
		RC-CF	48.86 ± 0.84 c	66.64 ± 1.42 b	176.10 ± 3.73 ab	374.88 ± 6.78 ab	5.79 ± 0.02 ab	2.17 ± 0.08 cd	3.88 ± 0.02 d	1.95 ± 0.03 k
	NJJG	CR-IF	49.14 ± 0.45 c	67.24 ± 1.22 b	174.44 ± 3.44 b	340.52 ± 6.54 d	5.80 ± 0.16 ab	2.27 ± 0.09 c	4.13 ± 0.05 b	2.33 ± 0.05 e
		CR-CF	49.84 ± 0.76 bc	68.43 ± 1.90 a	180.82 ± 2.46 a	356.27 ± 7.11 c	5.84 ± 0.06 ab	2.37 ± 0.04 bc	4.23 ± 0.07 a	2.39 ± 0.02 c
		RC-IF	49.00 ± 0.45 c	66.56 ± 1.57 b	173.70 ± 2.74 b	368.49 ± 7.29 b	5.74 ± 0.08 b	2.26 ± 0.03 c	3.90 ± 0.08 d	1.93 ± 0.04 l
		RC-CF	49.70 ± 0.30 bc	67.66 ± 1.75 b	179.55 ± 3.58 ab	383.76 ± 6.82 a	5.81 ± 0.11 ab	2.35 ± 0.04 bc	4.00 ± 0.04 c	2.02 ± 0.03 j
	QY0861	CR-IF	49.40 ± 0.50 bc	68.10 ± 2.26 ab	169.26 ± 3.21 c	247.46 ± 6.46 g	5.85 ± 0.08 ab	2.42 ± 0.10 b	3.88 ± 0.04 d	2.46 ± 0.03 b
		CR-CF	50.18 ± 0.56 b	69.45 ± 1.73 a	174.33 ± 2.69 b	257.56 ± 6.19 g	5.87 ± 0.13 a	2.46 ± 0.01 ab	4.10 ± 0.06 b	2.53 ± 0.02 a
		RC-IF	49.14 ± 0.80 c	67.50 ± 1.06 b	169.16 ± 2.80 c	271.89 ± 6.55 f	5.77 ± 0.12 ab	2.40 ± 0.09 b	3.69 ± 0.08 e	2.03 ± 0.04 ij
		RC-CF	49.53 ± 0.38 bc	68.33 ± 1.44 ab	174.15 ± 3.37 b	283.5 ± 6.63 ef	5.86 ± 0.08 ab	2.46 ± 0.12 ab	3.81 ± 0.06 de	2.15 ± 0.03 h
	SLY600	CR-IF	50.31 ± 0.41 ab	69.08 ± 1.47 a	171.73 ± 2.63 bc	252.86 ± 6.54 g	5.89 ± 0.18 ab	2.53 ± 0.11 ab	3.96 ± 0.08 cd	2.47 ± 0.05 b
		CR-CF	51.22 ± 0.34 ab	70.43 ± 1.07 a	177.97 ± 2.46 ab	265.08 ± 7.00 fg	6.33 ± 0.09 a	2.56 ± 0.05 a	4.12 ± 0.07 b	2.50 ± 0.03 ab
		RC-IF	50.05 ± 0.77 bc	68.48 ± 1.90 a	171.32 ± 3.03 bc	276.75 ± 6.97 f	5.85 ± 0.10 ab	2.50 ± 0.09 ab	3.73 ± 0.03 e	2.08 ± 0.04 i
		RC-CF	50.70 ± 0.83 ab	69.45 ± 2.31 a	177.39 ± 3.23 ab	290.52 ± 6.65 e	5.89 ± 0.14 ab	2.55 ± 0.11 a	3.84 ± 0.05 d	2.15 ± 0.02 gh
	V		**	**	**	**	**	**	**	**
	R		ns	*	ns	**	**	ns	**	**
	F		**	**	**	**	ns	**	**	**
	V*R		ns	ns	ns	ns	ns	ns	*	*
	V*F		ns	ns	ns	ns	ns	ns	*	ns
	R*F		ns	ns	**	ns	*	**	**	*
	V*R*F		ns	ns	**	*	ns	ns	*	*

All the data presented are the means of three replicates ± standard errors. Different lowercase letters indicate the significant differences among all treatments in the four varieties at the 5% level in the same year. CR, conventional rice cultivation; RC, rice–crayfish coculture system; IF, inorganic compound fertilizer; CF, controlled-release fertilizer; V, rice variety; R, rice cropping mode; F, fertilizer type. * and ** indicate significant differences at the $p < 0.05$ and 0.01 levels, respectively. "ns" means not significant.

Light accumulation duration of the japonica and indica rice varieties in rice–crayfish coculture decreased before elongation stage but showed significant increases of 3.76–8.82% and 9.37–10.08%, respectively, from heading stage to mature stage compared to conventional rice farming. However, light accumulation duration of different rice varieties of controlled-release fertilizer significantly increased after elongation stage compared to that of inorganic compound fertilizer. The light accumulation duration of RC-CF increased by 2.80–3.66% from elongation stage to heading stage and by 2.87–3.63% from heading stage to mature stage compared to that of RC-IF, while the light accumulation duration of the CR-CF increased by 3.33–4.63% from elongation stage to heading stage and by 3.89–4.98% from heading stage to mature stage compared to those of CR-IF.

Net assimilation rate represents the ability of rice to synthesize dry matter per unit leaf area per unit time. According to the results, SLY600 had the highest net assimilation rate, and NJJG had the lowest from transplanting stage to effective tillering termination stage, from effective tillering termination stage to elongation stage, SLY600 had the highest net accumulation rate and WYJ23 had the lowest (Table 3). NJJG had the highest net accumulation rate and QY0861 had the lowest from elongation stage to heading stage, and SLY600 had the highest net accumulation rate and WYJ23 had the lowest from heading stage to mature stage.

Net assimilation rate of the japonica and indica rice varieties of rice–crayfish coculture significantly decreased by 5.17–7.06% and 3.09–7.14% from elongation stage to heading stage and by 15.31–17.22% and by 11.66–17.55% from heading stage to mature stage, respectively, compared to those of conventional rice farming. Controlled-release fertilizer increased net assimilation rate of the different rice varieties after elongation stage. Net assimilation rate of CR-CF increased by 1.70–5.67% from elongation stage to heading stage and by 0.36–3.74% from heading stage to mature stage compared to CR-IF, while the net assimilation rate of RC-CF increased by 1.23–3.59% from elongation stage to heading stage and by 2.58–5.81% from heading stage to mature stage compared to RC-IF.

3.5. Crop Growth Rate

Crop growth rate represents the dry matter accumulation per unit land area per unit time and is an important index describing the rate of population dry matter production and photosynthetic accumulation effect. During the rice growth period, the crop growth rate of different rice varieties increased before heading stage but decreased after heading stage (Table 4). Among the four rice varieties, crop growth rate of SLY600 had the highest crop growth rate, and WYJ23 had the lowest at each growth stage.

Crop growth rate of the japonica and indica rice varieties grown under rice–crayfish coculture decreased significantly by 8.63–10.90% and 6.60–10.50% from elongation stage to heading stage and by 18.55–20.85% and 15.30–20.62% from heading stage to mature stage, respectively, compared to those under conventional rice farming. Controlled-release fertilizer increased the crop growth rate of different rice varieties after elongation stage compared to inorganic compound fertilizer. Crop growth rate of CR-CF, increased by 5.1–8.8% from elongation stage to heading stage and by 4.8–8.1% from heading stage to mature stage compared to CR-IF, respectively, while crop growth rate of RC-CF increased by 4.26–6.78% from elongation stage to heading stage and by 7.23–10.42% from heading stage to mature stage compared to those of RC-IF.

3.6. Dry Matter Utilization Characteristics

N partial factor productivity of the japonica rice varieties was 0.54–6.66 kg·kg^{-1} higher than that of the indica rice varieties (Table 5). Among the four rice varieties, N partial factor productivity of NJJG was the highest, but that of QY0861 was the lowest. Rice–crayfish coculture significantly reduced N partial factor productivity of the japonica and indica rice varieties by 6.28–9.34% and 4.13–5.91%, respectively, compared to that of conventional rice farming. In rice–crayfish coculture, the decrease in the japonica rice varieties was greater than that in the indica rice varieties. Compared to inorganic compound fertilizer, controlled-release fertilizer significantly improved N partial factor productivity. CR-CF increased N partial factor productivity by 3.29–6.21% compared to CR-IF, while RC-CF increased N partial factor productivity by 4.22–6.15% compared to RC-IF.

Table 4. Crop growth rate of different rice varieties under two types of fertilizer treatments (Unit: $g{\cdot}m^{-2}{\cdot}d^{-1}$).

Year	Varieties	Treatments	Transplanting Stage-Effective Tillering Termination Stage	Effective Tillering Termination Stage-Elongation Stage	Elongation Stage-Heading Stage	Heading Stage-Mature Stage
2019	WYJ23	CR-IF	5.90 ± 0.10 de	8.81 ± 0.41 d	24.38 ± 0.44 f	11.78 ± 0.31 c
		CR-CF	6.01 ± 0.18 de	8.95 ± 0.19 d	25.61 ± 0.37 de	12.63 ± 0.25 b
		RC-IF	5.86 ± 0.08 e	8.57 ± 0.14 d	21.93 ± 0.44 h	9.35 ± 0.26 g
		RC-CF	5.93 ± 0.11 de	8.82 ± 0.28 d	22.87 ± 0.35 g	10.03 ± 0.30 f
	NJJG	CR-IF	5.94 ± 0.06 de	9.73 ± 0.33 c	25.03 ± 0.33 e	12.30 ± 0.28 b
		CR-CF	6.05 ± 0.10 d	9.97 ± 0.30 c	26.41 ± 0.32 c	13.14 ± 0.22 ab
		RC-IF	5.84 ± 0.10 e	9.56 ± 0.21 c	22.60 ± 0.44 g	9.83 ± 0.33 f
		RC-CF	5.95 ± 0.17 de	9.83 ± 0.39 c	24.13 ± 0.46 f	10.70 ± 0.26 e
	QY0861	CR-IF	6.93 ± 0.13 bc	11.51 ± 0.26 b	25.62 ± 0.42 de	12.45 ± 0.29 b
		CR-CF	7.08 ± 0.10 bc	11.73 ± 0.22 b	27.77 ± 0.48 b	13.36 ± 0.24 a
		RC-IF	6.88 ± 0.07 c	11.35 ± 0.14 b	23.93 ± 0.27 f	9.96 ± 0.18 f
		RC-CF	7.00 ± 0.11 bc	11.55 ± 0.14 b	25.16 ± 0.29 de	10.85 ± 0.19 e
	SLY600	CR-IF	7.11 ± 0.10 ab	11.81 ± 0.48 b	26.62 ± 0.32 c	12.83 ± 0.24 ab
		CR-CF	7.28 ± 0.06 a	12.40 ± 0.44 a	28.54 ± 0.38 a	13.45 ± 0.31 a
		RC-IF	7.00 ± 0.07 bc	11.61 ± 0.21 b	24.38 ± 0.31 f	10.48 ± 0.19 e
		RC-CF	7.17 ± 0.13 ab	12.27 ± 0.32 ab	25.67 ± 0.27 d	11.39 ± 0.28 d
2020	WYJ23	CR-IF	5.75 ± 0.21 c	8.14 ± 0.84 f	23.82 ± 0.48 ef	11.43 ± 0.33 c
		CR-CF	5.85 ± 0.05 c	8.71 ± 0.49 ef	25.10 ± 0.5 cd	12.35 ± 0.30 b
		RC-IF	5.64 ± 0.10 c	8.02 ± 0.33 f	21.33 ± 0.32 h	9.07 ± 0.31 g
		RC-CF	5.75 ± 0.22 c	8.43 ± 0.17 ef	22.37 ± 0.48 g	9.77 ± 0.38 fg
	NJJG	CR-IF	5.78 ± 0.23c	8.92 ± 0.27 de	24.31 ± 0.38 e	11.92 ± 0.33 bc
		CR-CF	5.89 ± 0.13c	9.48 ± 0.22 d	25.79 ± 0.52 bc	12.83 ± 0.46 ab
		RC-IF	5.71 ± 0.10c	8.78 ± 0.25 e	22.17 ± 0.37 g	9.55 ± 0.26 g
		RC-CF	5.85 ± 0.15c	9.31 ± 0.38 de	23.47 ± 0.44 f	10.39 ± 0.32 ef
	QY0861	CR-IF	6.71 ± 0.1b	10.86 ± 0.16 bc	25.04 ± 0.41 cd	12.09 ± 0.32 b
		CR-CF	6.81 ± 0.33 ab	11.24 ± 0.24 b	27.23 ± 0.46 a	12.98 ± 0.40 ab
		RC-IF	6.59 ± 0.10 b	10.64 ± 0.16 c	22.94 ± 0.46 fg	9.59 ± 0.26 fg
		RC-CF	6.73 ± 0.21 b	11.06 ± 0.27 bc	24.37 ± 0.48 d	10.59 ± 0.33 d
	SLY600	CR-IF	6.85 ± 0.15 ab	11.51 ± 0.31 ab	25.92 ± 0.34 b	12.43 ± 0.30 b
		CR-CF	7.04 ± 0.10 a	11.85 ± 0.20 a	27.88 ± 0.34 a	13.17 ± 0.42 a
		RC-IF	6.77 ± 0.10 ab	11.26 ± 0.16 ab	23.49 ± 0.37 f	10.02 ± 0.34 fg
		RC-CF	6.88 ± 0.19 ab	11.65 ± 0.55 ab	25.04 ± 0.46 d	10.89 ± 0.29 cd
	V		**	**	**	**
	R		**	**	**	**
	F		**	**	**	**
	V*R		ns	ns	ns	ns
	V*F		ns	ns	**	ns
	R*F		*	**	*	ns
	V*R*F		ns	ns	**	*

All the data presented are the means of three replicates ± standard errors. Different lowercase letters indicate the significant differences among all treatments in the four varieties at the 5% level in the same year. CR, conventional rice cultivation; RC, rice–crayfish coculture system; IF, inorganic compound fertilizer; CF, controlled-release fertilizer; V, rice variety; R, rice cropping mode; F, fertilizer type. * and ** indicate significant differences at the $p < 0.05$ and 0.01 levels, respectively. "ns" means not significant.

N agronomic use efficiency of the japonica rice varieties was 0.34–3.17 $kg{\cdot}kg^{-1}$ higher than that of the indica rice varieties. Among the four rice varieties, N agronomic use efficiency of NJJG was the highest and that of QY0861 was the lowest. Rice–crayfish coculture reduced significantly N agronomic use efficiency of the japonica and indica rice varieties by 3.96–8.95% and 4.26–8.09%, respectively, compared to conventional rice farming. Compared to inorganic compound fertilizer, controlled-release fertilizer significantly increased N agronomic use efficiency. CR-CF increased N agronomic use efficiency by 7.36–13.88% compared to CR–IF, while RC-CF increased N agronomic use efficiency by 9.15–14.01% compared to RC-IF.

No obvious differences in the harvest index can be found between rice–crayfish coculture and conventional rice farming or controlled-release fertilizer and inorganic compound fertilizer. There was a slight decrease of 0.30–1.83% and 0.07–0.46%, respectively, in the harvest index of the japonica and indica rice varieties in rice–crayfish coculture compared to that in conventional rice farming. In contrast to inorganic fertilizer, controlled-release fertilizer decreased the harvest index by 0.76–2.32% in two years.

Table 5. Evaluation indexes of the dry matter utilization of different rice varieties under two types of fertilizer treatments.

Year	Varieties	Treatments	N Partial Factor Productivity ($kg \cdot kg^{-1}$)	N Agronomic Use Efficiency ($kg \cdot kg^{-1}$)	Harvest Index
2019	WYJ23	CR-IF	42.43 ± 0.77 cd	18.67 ± 0.25 d	49.16 ± 0.86 ab
		CR-CF	43.94 ± 0.82 bc	20.18 ± 0.23 b	48.32 ± 0.85 b
		RC-IF	38.54 ± 0.32 fg	17.76 ± 0.16 e	48.26 ± 0.78 b
		RC-CF	40.17 ± 0.77 e	19.38 ± 0.55 c	47.90 ± 0.85 b
	NJJG	CR-IF	44.68 ± 0.44 b	19.94 ± 0.22 bc	49.85 ± 0.84 a
		CR-CF	46.15 ± 0.75 a	21.40 ± 0.39 a	48.82 ± 0.81 ab
		RC-IF	41.33 ± 0.62 d	18.48 ± 0.26 d	49.63 ± 0.75 ab
		RC-CF	43.25 ± 0.79 c	20.40 ± 0.43 b	48.64 ± 0.79 b
	QY0861	CR-IF	38.02 ± 0.08 f	17.00 ± 0.52 f	48.41 ± 0.88 b
		CR-CF	40.38 ± 0.96 de	19.36 ± 0.60 c	48.24 ± 0.60 b
		RC-IF	36.25 ± 0.75 g	16.28 ± 0.39 g	48.27 ± 0.57 b
		RC-CF	38.25 ± 0.37 fg	18.28 ± 0.21 de	48.20 ± 0.58 b
	SLY600	CR-IF	39.47 ± 0.71 ef	17.66 ± 0.35 e	48.64 ± 0.56 b
		CR-CF	41.33 ± 0.79 d	19.53 ± 0.43 c	48.27 ± 0.50 b
		RC-IF	37.48 ± 0.46 fg	16.31 ± 0.10 g	48.42 ± 0.48 b
		RC-CF	39.63 ± 0.87 ef	18.45 ± 0.51 d	48.21 ± 0.53 b
2020	WYJ23	CR-IF	41.09 ± 0.62 cd	18.25 ± 0.26 cd	49.12 ± 0.84 ab
		CR-CF	42.65 ± 0.41 bc	19.81 ± 0.32 b	47.98 ± 0.81 b
		RC-IF	37.25 ± 0.25 f	16.61 ± 0.33 e	48.23 ± 0.76 b
		RC-CF	39.00 ± 0.79 e	18.36 ± 0.43 cd	47.76 ± 0.78 b
	NJJG	CR-IF	43.05 ± 0.37 b	19.13 ± 0.18 b	49.74 ± 0.78 a
		CR-CF	44.79 ± 0.87 a	20.87 ± 0.51 a	48.67 ± 0.78 ab
		RC-IF	40.08 ± 0.50 d	18.00 ± 0.14 cd	49.59 ± 0.64 ab
		RC-CF	41.79 ± 0.66 cd	19.71 ± 0.30 b	48.44 ± 0.66 b
	QY0861	CR-IF	36.76 ± 0.08 fgh	15.96 ± 0.41 f	48.22 ± 0.84 b
		CR-CF	38.54 ± 0.83 e	17.74 ± 0.47 d	47.31 ± 0.85 b
		RC-IF	34.63 ± 0.46 h	15.11 ± 0.10 g	48.13 ± 0.71 b
		RC-CF	36.33 ± 0.96 fgh	16.82 ± 0.60 e	47.22 ± 0.73 b
	SLY600	CR-IF	38.17 ± 0.41 ef	17.15 ± 0.33 de	48.46 ± 0.72 ab
		CR-CF	40.50 ± 0.91 d	19.48 ± 0.55 b	47.87 ± 0.85 b
		RC-IF	35.92 ± 0.50 g	15.76 ± 0.14 f	48.23 ± 0.75 b
		RC-CF	38.13 ± 0.62 ef	17.97 ± 0.26 cd	48.07 ± 0.76 b
	V		**	**	**
	R		**	**	ns
	F		**	**	**
	V*R		**	*	ns
	V*F		ns	*	ns
	R*F		**	*	ns
	V*R*F		ns	ns	ns

The yield of conventional rice farming and rice–crayfish coculture without N fertilizer in 2019 and 2020 were as follows: WYJ23:380.13 $kg \cdot kg^{-1}$, 332.52 $kg \cdot kg^{-1}$ and 365.45 $kg \cdot kg^{-1}$, 328.18 $kg \cdot kg^{-1}$; NJJG: 395.90 $kg \cdot kg^{-1}$, 365.65 $kg \cdot kg^{-1}$ and 382.68 $kg \cdot kg^{-1}$, 353.35 $kg \cdot kg^{-1}$; QY0861: 336.37 $kg \cdot kg^{-1}$, 319.57 $kg \cdot kg^{-1}$ and 332.81 $kg \cdot kg^{-1}$, 312.24 $kg \cdot kg^{-1}$; SLY600: 348.88 $kg \cdot kg^{-1}$, 338.78 $kg \cdot kg^{-1}$, and 336.38 $kg \cdot kg^{-1}$, 322.48 $kg \cdot kg^{-1}$, respectively. All the data presented are the means of three replicates ± standard errors. Different lowercase letters indicate the significant differences among all treatments in the four varieties at the 5% level in the same year. CR, conventional rice cultivation; RC, rice–crayfish coculture system; IF, inorganic compound fertilizer; CF, controlled-release fertilizer; V, rice variety; R, rice cropping mode; F, fertilizer type. * and ** indicate significant differences at the $p < 0.05$ and 0.01 levels, respectively. "ns" means not significant.

The determination coefficient R^2 can reflect the fitting of the model and data. Results showed that there was a positive correlation between N partial productivity, N agronomic use efficiency and rice yield; the fitting (R^2) of N partial productivity and yield was 0.998 **, N agronomic use efficiency and yield was 0.483 **, while harvest index and yield was not significantly correlated (Figure 3).

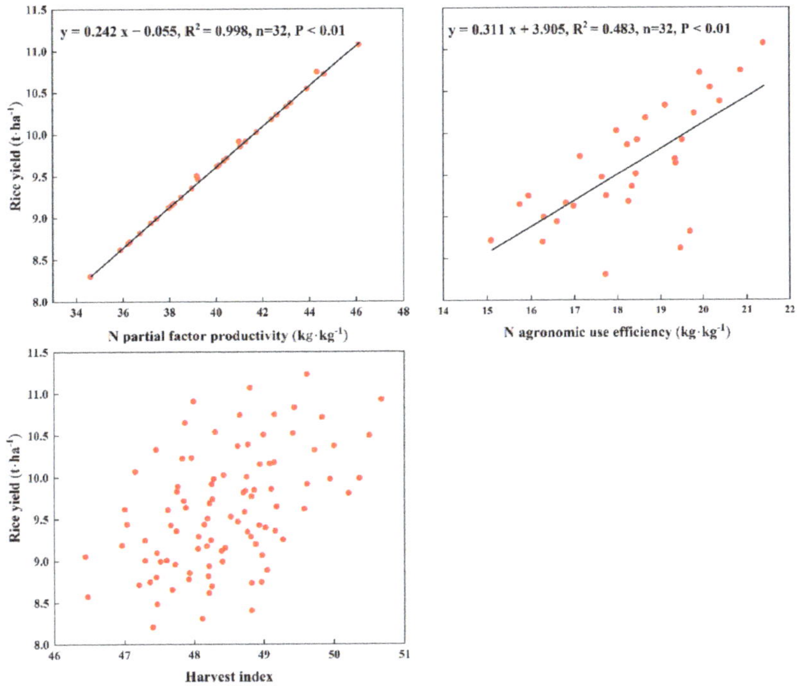

Figure 3. Correlation between rice yield and the dry matter utilization characteristics.

4. Discussion and Conclusions

4.1. Dry Matter Accumulation and Rice Yield

This study suggests that the yield and the dry matter accumulation of japonica rice varieties at mature stage were higher than those of the indica rice varieties. Compared to the indica rice varieties, the number of grains per panicle of the japonica rice varieties decreased, while the effective panicle number and the 1000-grain weight increased (Table 2 and Figure 1). This was mainly due to the roots of the indica rice varieties featuring premature senescence [25], leading to a reduction in the nutrient uptake and dry matter accumulation from heading to mature stage. Wei et al. [26] showed that compared to indica rice varieties, the dry matter accumulation of japonica rice from elongation stage to heading stage and from heading stage to mature stage increased by 16.33–32.00% and 14.06–23.00%, respectively, and the yield increased by 3.81–6.73%. Xu et al. [27] showed that japonica rice varieties have a higher yield and better quality compared to indica rice varieties. Uyeh et al. [28] concluded that japonica rice varieties feature a higher yield than indica hybrid rice varieties. However, Sun et al. [29] showed that compared to japonica rice, indica hybrid rice has a higher yield, dry matter, and N uptake throughout all growth stages. Generally speaking, the higher yield observed in the japonica and indica rice varieties was a result of genetic characteristics as well as cultivation measures, including the rice growth period, the number of panicle types, the seed setting rate, dry matter accumulation at the mature stage, and climatic influence. Additionally, this was also due to higher plant density of japonica rice varieties, compared to indica rice varieties.

Whether the rice yield of rice–crayfish coculture is higher than that of conventional rice farming, the results are different at present. Sun et al. [30] stated the rice yield resulting from rice–crayfish coculture had declined by 30.8–30.9% compared to rice monocultures. In contrast, Sun et al. [31] pointed out that rice–crayfish coculture reduced rice yield by 28.5% compared to rice monocultures. However, Wu et al. [32] believed that compared to conventional rice production, rice–crayfish coculture with extremely high nutrients status was accompanied by 14% rice yields reduction, while rice–crayfish coculture with appropriately improved soil quality created favorable nutrient status accompanied by 15% rice yield increase. This study showed that rice–crayfish coculture decreased rice yield by 4.14–9.34% compared to conventional rice farming, mainly because the deep-water irrigation from elongation stage to mature stage in rice–crayfish coculture made it difficult for the available nutrients in the soil to fully meet the uptake and utilization requirements of rice. Moreover, the poor root activity of rice led to a decline in its photosynthetic production capacity in deep water conditions [33], as well as in the amount of dry matter accumulation after elongation stage, disrupting rice yield.

Controlled-release fertilizer was able to control the release rate of nutrients and promote their uptake, meeting the needs of N demands for rice, reducing fertilization times, and boosting N use efficiency [34,35]. Mohammad et al. [36] showed that compared to urea, different types of controlled-release fertilizer contributed to a better uptake of N in rice plants, especially in rice grains. In addition, compared to inorganic fertilizer, controlled-release fertilizer decreased the amylose content by 3.05%, improved peak viscosity and breakdown by 9.62% and 8.53%, and decreased setback by 19.39% [37]. In this study, compared to inorganic compound fertilizer, controlled-release fertilizer improved the dry matter accumulation of rice from effective tillering termination stage to mature stage, increased the effective panicles and the rice yield, and reduced the seed setting rate and 1000-grain weight. This might because that controlled-release fertilizer prolonged N release and provided synchronous N supply to rice. At the same time, controlled-release fertilizer also promoted the growth of rice roots and their distribution in deep soil, which was conducive to maintaining root activity [38,39].

4.2. Photosynthetic Production Characteristics

As the characteristics of photosynthetic production serves as the main determinant of rice yield, a better photosynthetic capacity after heading stage is a key factor that is responsible for high rice yield. The present results are different regarding the photosynthetic production characteristics of different rice varieties. Li et al. [40] reported rice yield decreased as rice seedlings aged, because younger seedlings promote rice growth after the heading stage, achieving higher photosynthetic production and higher yield. Yamori et al. [41] indicated that photosynthesis was an important biochemical process supporting plant growth and grain yield. Gong et al. [42] indicated that dry matter accumulation, light accumulation duration, the crop growth rate, and the net assimilation rate of conventional japonica rice varieties were lower than those of indica rice varieties from transplanting stage to elongation stage but higher from elongation stage to mature stage. This study showed that compared to the indica rice varieties, the japonica rice varieties decreased the crop growth rate, net assimilation rate, and leaf area index before heading stage and decreased the light accumulation duration before elongation stage but increased crop growth rate, net assimilation rate, and leaf area index after heading stage and increased the light accumulation duration after elongation stage. Therefore, the japonica rice varieties were better than the indica rice varieties in terms of the photosynthetic production characteristics at the later rice growth stages, laying a sound foundation for yield formation.

In addition to the genetic factors of rice varieties, the rice cropping mode also had a significant effect on the photosynthetic production characteristics of rice. Related studies have compared conventional rice production to six coculture modes (rice–crayfish, rice–turtle, rice–loach, rice–catfish, rice–carp, and rice–duck), showing a decrease in leaf area index, light accumulation duration, crop growth rate, and net assimilation rate at main growth stages [22,43]. Yao et al. [44] thought that compared to conventional rice cultivation modes, rice–crayfish coculture demonstrated a slight decline in the leaf area index at the later growth stages, a higher population growth rate, and better leaf photosynthetic characteristics. This study showed that compared to conventional rice farming, rice–crayfish coculture decreased the crop growth rate, net assimilation rate, and leaf area index from transplanting stage to mature stage and the light accumulation duration before elongation stage but increased the light accumulation duration after heading stage.

The proper application of controlled-release fertilizer is an important measure to improve the light environment and to optimize the characteristics of photosynthetic production. Xu et al. [19] reported that the controlled release fertilizer met N demands at each rice growth stage and had higher N accumulation, photosynthetic activity, and above-ground matter accumulation as well as a higher leaf area index at heading stage and mature stage. Yang et al. [6] indicated that controlled-release fertilizer enhanced the activities of glutamine synthetase, glutamine 2-oxoglutarate aminotransferase, and nitrate reductase in leaves, thus improving the photosynthetic production of rice. In this study, controlled-release fertilizer increased the leaf area index, light accumulation duration, crop growth rate, and net assimilation rate, thus obviously promoting the photosynthetic production characteristics, because the slow nutrient release of controlled-release fertilizer in the early growth stages provides a sufficient nutrient supply for rice in the middle and later stages. Therefore, the separate application of controlled-release fertilizer can also improve the light environment that is necessary for crop growth.

4.3. Dry Matter Utilization Characteristics and the Relationship with Rice Yield

The dry matter utilization characteristics of rice involves multiple physiological and biochemical processes such as carbohydrate metabolism, the transmission of nutrient signals, N metabolism, plant protein synthesis, and degradation, which are important for the exploration of rice growth and yield formation. In this study, N partial factor productivity and N agronomic use efficiency of the japonica rice varieties were higher than those of the indica rice varieties. Compared to conventional rice farming, rice–crayfish coculture decreased N partial factor productivity, N agronomic use efficiency, harvest index. N partial factor productivity and N agronomic use efficiency in controlled-release fertilizer were higher than those in inorganic compound fertilizer, but harvest index was opposite. In this study, there was a positive correlation between N partial factor productivity, N agronomic use efficiency, and the rice yield, but there was no significant correlation between the harvest index and the rice yield, which was mainly due to similar fertility characteristics and the same climate conditions in this experiment.

5. Conclusions

Rice–crayfish coculture has gained widespread popularity in recent years, features high ecological and economic benefits. Compared to conventional rice farming, a decline could be seen in rice–crayfish coculture in terms of rice dry matter accumulation, N partial factor productivity, N agronomic use efficiency, and yield. This was mainly due to the fact that rice–crayfish coculture was not beneficial to the leaf area index, crop growth rate, net assimilation rate, and light accumulation duration of different rice varieties at different growth stages. However, the application of controlled-release fertilizer in rice–crayfish coculture increased the dry matter accumulation at the mature stages as well as N partial factor productivity and N agronomic use efficiency, and yield, mainly because controlled-release fertilizer improved the leaf area index, crop growth rate, and net assimilation rate of different rice varieties at different growth stages as well as the light accumulation duration before the elongation stage. In addition, whether in indica rice or japonica rice varieties, controlled release fertilizer can increase rice yield in rice–crayfish coculture, but in this study, japonica rice varieties have more yield advantages. The application of controlled-release fertilizer in rice–crayfish coculture can not only enhance rice dry matter accumulation, N partial productivity, and N agronomic use efficiency, but also increase rice yield, so as to boost the benefit of large-scale rice–crayfish coculture.

Author Contributions: Q.W. conceived and designed the research, revised the manuscript; K.Y. provided experimental data and wrote the manuscript; H.Z. tested and experimental analysis. All authors have read and agreed to the published version of the manuscript.

Funding: This study was financially supported by the Earmarked Fund for Jiangsu Agricultural Industry Technology System (JATS[2022]482, JATS[2021]482, JATS[2020]431); Jiangsu Agricultural Science and Technology Innovation Fund (CX(22)3158), Belt and Road Initiative Cooperation Project (DL2022145004L).

Institutional Review Board Statement: Not applicable for studies not involving humans or animals.

Informed Consent Statement: This paper does not contain any studies with human participants and/or animals. Informed consent was obtained from all individual participants included in this study.

Data Availability Statement: The data presented in this study are available on request from the corresponding author.

Conflicts of Interest: The authors declare that they have no conflict of interest.

References

1. Hu, L.; Ren, W.; Tang, J.; Li, N.; Zhang, J.; Chen, X. The productivity of traditional rice-fish co-culture can be increased without increasing nitrogen loss to the environment. *Agr. Ecosyst. Environ.* **2013**, *177*, 28–34. [CrossRef]
2. Zhu, J.; Peng, H.; Ji, X.; Li, C.; Li, S. Effects of reduced inorganic fertilization and rice straw recovery on soil enzyme activities and bacterial community in double-rice paddy soils. *Eur. J. Soil. Biol.* **2019**, *94*, 103116. [CrossRef]
3. Rahman, K.M.A.; Zhang, D. Effects of fertilizer broadcasting on the excessive use of inorganic fertilizers and environmental sustainability. *Sustainability* **2018**, *10*, 759. [CrossRef]
4. Hu, R.; Cao, J.; Huang, J.; Peng, S.; Huang, J.; Zhong, X.; Zou, Y.; Yang, J.; Buresh, R.J. Farmer participatory testing of standard and modified site-specific nitrogen management for irrigated rice in China. *Agr. Syst.* **2007**, *94*, 331–340. [CrossRef]
5. Azeem, B.; KuShaari, K.; Man, Z.B.; Basit, A.; Thanh, T.H. Review on materials & methods to produce controlled release coated urea fertilizer. *J. Control. Release* **2014**, *181*, 11–21.
6. Yang, Y.; Zhang, M.; Li, Y.; Fan, X.; Geng, Y. Controlled release urea improved nitrogen use efficiency, activities of leaf enzymes, and rice yield. *Soil Sci. Soc. Am. J.* **2012**, *76*, 2307–2317. [CrossRef]
7. Cao, X.; Yuan, L.; Liu, X.; Zhu, L.; Zhu, C.; Kong, Y.; Wu, L.; Tian, C.; Lu, R.; Zhang, J.; et al. Benefits of controlled-release/stable fertilizers plus biochar for rice grain yield and nitrogen utilization under alternating wet and dry irrigation. *Eur. J. Agron.* **2021**, *129*, 126338. [CrossRef]
8. Deng, F.; Li, W.; Wang, L.; Hu, H.; Liao, S.; Pu, S.L.; Tao, Y.F.; Li, G.H.; Ren, W.J. Effect of controlled-release fertilizers on leaf characteristics, grain yield, and nitrogen use efficiency of machine-transplanted rice in southwest China. *Arch. Agron. Soil Sci.* **2021**, *67*, 1739–1753. [CrossRef]
9. Ye, Y.; Liang, X.; Chen, Y.; Liu, J.; Gu, J.; Guo, R.; Li, L. Alternate wetting and drying irrigation and controlled-release nitrogen fertilizer in late-season rice. Effects on dry matter accumulation, yield, water and nitrogen use. *Field Crop. Res.* **2013**, *144*, 212–224. [CrossRef]
10. Mi, W.; Yang, X.; Wu, L.; Ma, Q.; Liu, Y.; Zhang, X. Evaluation of nitrogen fertilizer and cultivation methods for agronomic performance of rice. *Agron. J.* **2016**, *108*, 1907–1916. [CrossRef]
11. Li, P.; Lu, J.; Hou, W.; Pan, Y.; Wang, Y.; Khan, M.R.; Ren, T.; Cong, R.; Li, X. Reducing nitrogen losses through ammonia volatilization and surface runoff to improve apparent nitrogen recovery of double cropping of late rice using controlled release urea. *Environ. Sci. Pollut. Res.* **2017**, *24*, 11722–11733. [CrossRef] [PubMed]
12. Yu, K.; Fang, X.; Zhang, Y.; Miao, Y.; Liu, S.; Zou, J. Low greenhouse gases emissions associated with high nitrogen use efficiency under optimized fertilization regimes in double-rice cropping systems. *Appl. Soil Ecol.* **2021**, *160*, 103846. [CrossRef]
13. Sun, H.; Zhang, H.; Min, J.; Feng, Y.; Shi, W. Controlled-release fertilizer, floating duckweed, and biochar affect ammonia volatilization and nitrous oxide emission from rice paddy fields irrigated with nitrogen-rich wastewater. *Paddy Water Environ.* **2016**, *14*, 105–111. [CrossRef]
14. Xu, Q.; Peng, X.; Guo, H.; Che, Y.; Dou, Z.; Xing, Z.; Xing, Z.; Hou, J.; Styles, D.; Gao, H.; et al. Rice-crayfish coculture delivers more nutrition at a lower environmental cost. *Sustain. Prod. Consum.* **2022**, *29*, 14–24. [CrossRef]
15. Jin, T.; Ge, C.; Gao, H.; Zhang, H.; Sun, X. Evaluation and screening of co-culture farming models in rice field based on food productivity. *Sustainability* **2020**, *12*, 2173. [CrossRef]
16. Hou, J.; Styles, D.; Cao, Y.; Ye, X. The sustainability of rice-crayfish coculture systems: A mini review of evidence from Jianghan plain in China. *J. Sci. Food Agr.* **2021**, *101*, 3843–3853. [CrossRef] [PubMed]
17. Yuan, P.; Wang, J.; Li, C.; Xiao, Q.; Liu, Q.; Sun, Z.; Wang, J.; Cao, C. Soil quality indicators of integrated rice-crayfish farming in the Jianghan Plain, China using a minimum data set. *Soil Till. Res.* **2020**, *204*, 104732. [CrossRef]
18. Si, G.; Peng, C.; Yuan, J.; Xu, X.; Zhao, S.; Xu, D.; Wu, J. Changes in soil microbial community composition and organic carbon fractions in an integrated rice-crayfish farming system in subtropical China. *Sci. Rep.* **2017**, *7*, 2856. [CrossRef] [PubMed]
19. Xu, D.; Zhu, Y.; Zhu, H.; Hu, Q.; Liu, G.; Wei, H.; Zhang, H. Effects of a one-time application of controlled-release nitrogen fertilizer on yield and nitrogen accumulation and utilization of late japonica rice in China. *Agriculture* **2021**, *11*, 1041. [CrossRef]
20. Zhang, X.; Ding, J.; Liu, Y.; Gu, Y.; Han, K.; Wu, L. Effects of mechanical transplanting of rice with controlled release bulk blending fertilizer on rice yield and soil fertility. *Chin. J. Appl. Ecol.* **2014**, *25*, 78.

21. Liu, T.; Huang, J.; Chai, K.; Cao, C.; Li, C. Effects of N fertilizer sources and tillage practices on NH_3 volatilization, grain yield, and N use efficiency of rice fields in Central China. *Front. Plant. Sci.* **2018**, *9*, 385. [CrossRef] [PubMed]
22. Liu, X.; Xu, G.; Wang, Q.; Hang, Y. Effects of insect-proof net cultivation, rice-duck farming, and organic matter return on rice dry matter accumulation and nitrogen utilization. *Front. Plant. Sci.* **2017**, *8*, 47. [CrossRef] [PubMed]
23. Dou, Z.; Li, Y.; Guo, H.; Chen, L.; Jiang, J.; Zhou, Y.; Zou, Y.; Xu, Q.; Xin, Z.; Gao, H.; et al. Effects of mechanically transplanting methods and planting densities on yield and quality of Nanjing 2728 under rice-crayfish continuous production system. *Agronomy* **2021**, *11*, 488. [CrossRef]
24. Chen, Y.; Tu, P.; Yang, Y.; Xue, X.; Feng, Z.; Dan, C.; Cheng, F.; Yang, Y.; Deng, L. Diversity of rice rhizosphere microorganisms under diferent fertilization modes of slow-release fertilizer. *Sci. Rep.* **2022**, *12*, 2694. [CrossRef]
25. Gong, J.; Xing, Z.; Hu, Y.; Zhang, H.; Dai, Q.; Huo, Z.; Xu, K.; Wei, H.; Gao, H. Difference of characteristics of photosynthesis, matter production and translocation between indica and japonica super rice. *Acta Agron. Sin.* **2014**, *40*, 497–510. [CrossRef]
26. Wei, H.; Meng, T.; Li, C.; Xu, K.; Huo, Z.; Wei, H.; Guo, B.; Zhang, H.; Dai, Q. Comparisons of grain yield and nutrient accumulation and translocation in high-yielding japonica/indica hybrids, indica hybrids, and japonica conventional varieties. *Field Crop. Res.* **2017**, *204*, 101–109. [CrossRef]
27. Xu, Q.; Chen, W.; Xu, Z. Relationship between grain yield and quality in rice germplasms grown across different growing areas. *Breed. Sci.* **2015**, *65*, 226–232. [CrossRef] [PubMed]
28. Uyeh, D.D.; Asem-Hiablie, S.; Park, T.; Kim, K.M.; Mikhaylov, A.; Woo, S.; Ha, Y. Could japonica rice be an alternative variety for increased global food security and climate change mitigation? *Foods* **2021**, *10*, 1869. [CrossRef]
29. Sun, T.; Yang, X.; Tan, X.; Han, K.; Tang, S.; Tong, W.; Zhu, S.; Hu, Z.; Wu, L. Comparison of agronomic performance between japonica/indica hybrid and japonica cultivars of rice based on different nitrogen rates. *Agronomy* **2020**, *10*, 171. [CrossRef]
30. Sun, Z.; Guo, Y.; Li, C.; Cao, C.; Yuan, P.; Zou, F.; Wang, J.; Jia, P.; Wang, J. Effects of straw returning and feeding on greenhouse gas emissions from integrated rice-crayfish farming in Jianghan Plain, China. *Environ. Sci. Pollut. Res.* **2019**, *26*, 11710–11718. [CrossRef]
31. Sun, G.; Sun, M.; Du, L.; Zhang, Z.; Wang, Z.; Zhang, G.; Nie, S.-A.; Xu, H.; Wang, H. Ecological rice-cropping systems mitigate global warming—A meta-analysis. *Sci. Total Environ.* **2021**, *789*, 147900. [CrossRef] [PubMed]
32. Wu, Y.; Li, Y.; Niu, L.; Zhang, W.; Wang, L.; Zhang, H. Nutrient status of integrated rice-crayfish system impacts the microbial nitrogen-transformation processes in paddy fields and rice yields. *Sci. Total Environ.* **2022**, *836*, 155706. [CrossRef]
33. Thakuk, A.K.; Mandal, K.G.; Mohanty, R.K.; Ambast, S.K. Rice root growth, photosynthesis, yield and water productivity improvements through modifying cultivation practices and water management. *Agric. Water Manag.* **2018**, *206*, 67–77. [CrossRef]
34. Kenawy, E.R.; Hosny, A.; Saad-Allah, K. Reducing nitrogen leaching while enhancing growth, yield performance and physiological traits of rice by the application of controlled-release urea fertilizer. *Paddy Water Environ.* **2021**, *19*, 173–188. [CrossRef]
35. Lyu, Y.; Yang, X.; Pan, H.; Zhang, X.; Cao, H.; Ulgiati, S.; Wu, J.; Zhang, Y.; Wang, G.; Xiao, Y. Impact of fertilization schemes with different ratios of urea to controlled release nitrogen fertilizer on environmental sustainability, nitrogen use efficiency and economic benefit of rice production: A study case from Southwest China. *J. Clean. Prod.* **2021**, *293*, 126198. [CrossRef]
36. Mohammad, M.A.H.; Mohd, K.Y.; Radziah, O.; Samsuri, A. Field evaluation of newly-developed controlled release fertilizer on rice production and nitrogen uptake. *Sains Malays.* **2017**, *46*, 925–932.
37. Liu, Q.; Ma, H.; Lin, X.; Zhou, X.; Zhao, Q. Effects of different types of fertilizers application on rice grain quality. *Chil. J. Agr. Res.* **2019**, *79*, 202–209. [CrossRef]
38. Rahman, M.H.; Haque, K.M.S.; Khan, M.Z.H. A review on application of controlled released fertilizers influencing the sustainable agricultural production: A cleaner production process. *Environ. Technol. Inno.* **2021**, *23*, 101697. [CrossRef]
39. Aasmi, A.A.; Li, J.; Hamoud, Y.A.; Lan, Y.B.; Alordzinu, K.E.; Appiah, S.A.; Hiba, S.; Mohamed, S.; Wang, H.; Qiao, S.; et al. Impacts of slow-release nitrogen fertilizer rates on the morpho-physiological traits, yield, and nitrogen use efficiency of rice under different water regimes. *Agriculture* **2022**, *12*, 86. [CrossRef]
40. Li, Y.X.; Liu, Y.; Wang, Y.H.; Ding, Y.F.; Wang, S.H.; Liu, Z.H.; Li, G.H. Effects of seedling age on the growth stage and yield formation of hydroponically grown long-mat rice seedlings. *J. Integr. Agr.* **2020**, *19*, 1755–1767. [CrossRef]
41. Yamori, W.; Kondo, E.; Sugiura, D.; Terashima, I.; Suzuki, Y.; Makino, A. Enhanced leaf photosynthesis as a target to increase grain yield: Insights from transgenic rice lines with variable Rieske FeS protein content in the cytochrome b6/f complex. *Plant. Cell Environ.* **2016**, *39*, 80–87. [CrossRef] [PubMed]
42. Gong, G.; Xing, Z.; Hu, Y.; Zhang, H.; Dai, Q.; Huo, Z.; Xu, K.; Wei, H.; Gao, H.; Guo, B. Difference of root morphological and several physiological characteristics between indica and japonica super rice varieties. *Acta Agron. Sin.* **2014**, *40*, 1066–1080. [CrossRef]
43. Che, Y.; Cheng, S.; Tian, J.; Tao, Y.; Liu, Q.; Xing, Z.; Dou, Z.; Xu, Q.; Hu, Y.; Guo, B. Characteristics and differences of rice yield, quality and economic benefits under different modes of comprehensive planting-breeding in paddy fields. *Acta Agron. Sin.* **2021**, *47*, 1953–1965.
44. Yao, Y.; Tang, J.; Chen, J.; Xie, C.; Zhang, M.; Lu, P.; Min, S. Effects of rice-shrimp co-cultivation and density on material production and yield of good eating quality japonica rice. *J. Yangzhou Univ.* **2021**, *42*, 98–104.

agriculture

Article

Soil Bacteria Mediate Soil Organic Carbon Sequestration under Different Tillage and Straw Management in Rice-Wheat Cropping Systems

Lijin Guo [1,2,†], Jie Shi [3,†], Wei Lin [4], Jincheng Liang [2], Zhenhua Lu [5], Xuexiao Tang [2], Yue Liu [2], Purui Wu [2] and Chengfang Li [6,*]

1 International Magnesium Institute, College of Resources and Environment, Fujian Agriculture and Forestry University, Fuzhou 350002, China
2 Key Laboratory of Genetics and Germplasm Innovation of Tropical Special Forest Trees and Ornamental Plants, College of Forestry, Hainan University, Ministry of Education, Haikou 570228, China
3 State Key Laboratory of Hydraulic Engineering Simulation and Safety, School of Civil Engineering, Tianjin University, 135 Yaguan Road, Jinnan District, Tianjin 300350, China
4 College of Forestry, Fujian Agriculture and Forestry University, Fuzhou 350002, China
5 Kaifeng Academy of Agriculture and Forestry, Kaifeng 475004, China
6 MOA Key Laboratory of Crop Ecophysiology and Farming System in the Middle Reaches of the Yangtze River, College of Plant Science & Technology, Huazhong Agricultural University, Wuhan 430070, China
* Correspondence: lichengfang@mail.hzau.edu.cn
† These authors contributed equally to this work.

Citation: Guo, L.; Shi, J.; Lin, W.; Liang, J.; Lu, Z.; Tang, X.; Liu, Y.; Wu, P.; Li, C. Soil Bacteria Mediate Soil Organic Carbon Sequestration under Different Tillage and Straw Management in Rice-Wheat Cropping Systems. *Agriculture* **2022**, *12*, 1552. https://doi.org/10.3390/agriculture12101552

Academic Editor: Pavel Krasilnikov

Received: 11 August 2022
Accepted: 22 September 2022
Published: 26 September 2022

Publisher's Note: MDPI stays neutral with regard to jurisdictional claims in published maps and institutional affiliations.

Abstract: Soil organic carbon (SOC) largely influences soil quality and sustainability. The effects of no-till (NT) and crop straw return practices (SR) on soil organic carbon sequestration have been well documented. However, the mechanism of soil bacterial community in regulating soil organic carbon under NT and SR remains unclear. In this study, we investigated the impacts of tillage (conventional tillage (CT) and NT) and crop straw return practices (crop straw removal (NS) and SR) on topsoil layer (0–5 cm) bacterial community, CH_4 and CO_2 emissions and SOC fractions in rice-wheat cropping system. Overall, in the wheat season following the annual rice-wheat rotation in two cycles, NT significantly increased SOC by 4.4% for 1–2 mm aggregates in the 0–5 cm soil layer, but decreased CO_2 emissions by 7.4%. Compared with NS, SR notably increased the contents of SOC in the topsoil layer by 6.5% and in macro-aggregate by 17.4% in 0–5 cm soil layer, and promoted CH_4 emissions (by 22.3%) and CO_2 emissions (by 22.4%). The combination of NT and NS resulted in relatively high SOC and low CH_4 emissions along with high bacterial community abundance. The most abundant genus under different treatments was *Gp6*, which significant impacted SOC and MBC. Bacterial communities like Subdivision3 had the most impact on CH_4 emissions. Structural equation modeling further suggested that the soil bacterial community indirectly mediated the SOC through balancing SOC in 1–2 mm aggregates and CH_4 emissions. This study provides a new idea to reveal the mechanism of short-term tillage and straw return on SOC.

Keywords: no-till; straw return; soil organic carbon fractions; soil aggregate; bacterial diversity

1. Introduction

Enhancing soil organic carbon storage is vital to achieving sustainable agriculture and alleviating the negative impacts of climate change [1–5]. Tillage and straw return practices greatly affect the storage of organic carbon in the soil [6–8]. Conventional intensive tillage (CT), which accompanies removing crop straw from the field, results in soil organic carbon decline, soil structural degradation, and greenhouse gas (GHG) emission increase [7,9,10]. On the contrary, no-till (NT) and crop straw return (SR) are regarded as effective ways to increase soil organic carbon (SOC) sequestration [11–13].

Compared with CT, NT can increase SOC content by improving soil aggregation and decreasing CH_4 and CO_2 emissions in the topsoil layer [14,15]. NT reduces soil disturbance, prevents the soil macro-aggregate from being destroyed, provides better physical protection for SOC, and slows down SOC decomposition [2,16]. Moreover, NT can increase the input of organic residues in soil surface, which can be broken down by soil microbes and thus provide microbial binding agents for macroscopic aggregates to form [17,18]. NT can also reduce soil temperature and enhance soil humidity, which leads to the decline in microbial activity and the emissions of CH_4 and CO_2, reducing the loss of SOC [19,20]. However, the effect of NT on SOC content was regulated by complex biochemical processes, such as the GHG emissions and the formation of soil aggregate, and thus consistent conclusions were not obtained [6,21]. The soil microbial regulation mechanism of NT affecting SOC content is still unclear.

Crop straw is a source of organic carbon, and crop straw return is shown to result in enhancing SOC sequestration. Compared to no straw (NS), SR may have positively influenced the SOC content in the topsoil layer by elevating organic carbon input and improving soil microbial community and aggregate stability [9,12,22]. However, straw return also stimulates GHG emissions [17,23], which may offset the positive effects of SR on SOC sequestration [12]. Studies reported that soil microbial community [21], soil aggregate size [7,18] and GHG emissions [24] are closely related to SOC sequestration [25]. Yet, no experiment has so far been conducted to reveal the relationships among soil microbial community, the SOC in soil aggregates, GHG emissions and SOC sequestration as influenced by SR. Further study is needed to reveal the soil microbial regulation mechanism of SR affecting SOC sequestration.

The objectives of this study were to evaluate the effects of NT and SR on SOC content and soil microbial communities in a rice-wheat cropping system and to reveal the mechanisms that enable soil bacterial communities to regulate SOC under NT and SR. Therefore, we studied the effects of tillage (CT and NT) and crop straw return practices (NS and SR) on soil bacterial communities, CH_4 and CO_2 emissions, crop yields, and SOC aggregates in rice-wheat cropping systems. During the experiment, we found that tillage and straw return management had significant effects on SOC content in the 0–5 cm soil layer after two cycles of the rice-wheat rotation, but no significant effect on soil SOC in the 5–10 cm and 10–20 cm soil layers. Therefore, we focused on the 0–5 cm soil layer in this study. We hypothesized that soil bacteria could mediate SOC content through affecting soil aggregate SOC and CH_4 emissions under tillage and straw return management.

2. Materials and Methods

2.1. Experimental Site

The experiment site lies at Dafashi Town (30°01′ N, 115°34′ E), Hubei province, China, and was established in June 2012. The area has a humid mid-subtropical monsoon climate, in which the annual mean air temperature is 16.8 °C, and the average annual precipitation from 2012 to 2014 is 1408.7 mm (Figure 1). The experimental soil (0–20 cm) is a silty clay loam (containing clay 40%, sandy 25%, and silt 40%), which is defined as Gleysol (FAO classification). Besides, the total organic carbon content is 1.64%, total nitrogen content is 0.24%, the pH is 5.9, and the bulk density is 1.20 g cm^{-3}. This site has been dominated by a cropping system of rice (HHZ, *Oryza sativa* L.) and wheat (ZM9023, *Triticum aestivum* L.).

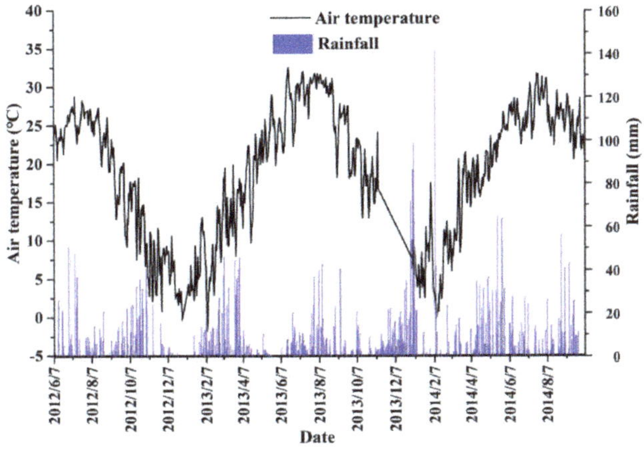

Figure 1. Average daily temperature and rainfall at the experimental site from 2012 to 2014 [2].

2.2. Experimental Design

This experiment was set up in a split-plot design, where tillage (CT and NT) and straw treatment (NS and SR) were set as the main plot and subplots, respectively (Figure 2). Four treatments including: (i) CTNS; (ii) CTSR; (iii) NTNS; and (iv) NTSR were arranged, and each treatment was conducted in triplicates. The area of each plot was 90 m^2 (9 m × 10 m). Under CT treatment, the soil was moldboard plowed twice in one year at a depth of 20 cm before planting rice and wheat. Moldboard plowing was omitted in NT treatment. Crop straw was removed from the field for both CTNS and NTNS treatments. A 6 cm length of crop straw harvested from each plot, was covered on the soil surface under NTSR treatment and incorporated into the soil under CTSR treatment. For all treatments, crop stubbles were kept in the fields. Rice was thrown manually at the rate of 190,000 seedlings per hectare in June and reaped in October. Wheat was directly sown at 150 kg ha^{-1} in October and harvested in May the following year.

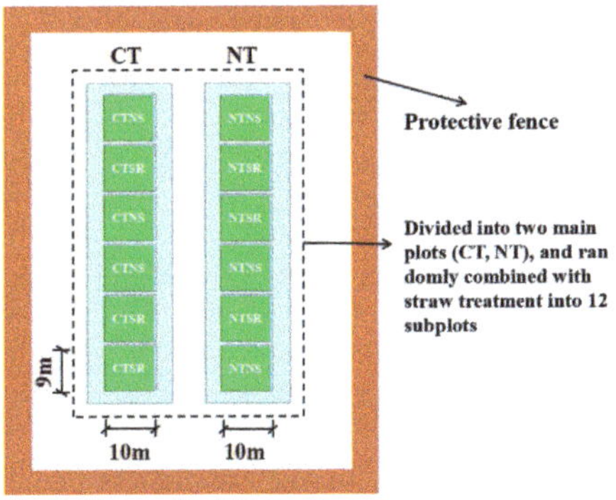

Figure 2. Design drawing of field experiment. Note: CT, conventional tillage; NT, no tillage; CTNS, conventional intensive tillage with straw removal; CTSR, conventional intensive tillage with straw return; NTNS, no tillage with straw removal; NTSR, no-tillage with straw return.

For all treatments, weeds were controlled by spraying 30% chlorpromazine emulsifiable oil containing 10% fenorim. The application rate of chemical fertilizer was 180 kg N ha^{-1}, 90 kg P_2O_5 ha^{-1} and 180 kg K_2O ha^{-1} in rice season and was 144 kg N ha^{-1}, 72 kg P_2O_5 ha^{-1}, and 144 kg K_2O ha^{-1} in wheat season. Commercial compound fertilizer (N:P_2O_5:K_2O = 15%:15%:15%) were used in rice and wheat seasons. P and K fertilizers were applied immediately as basal fertilizers after throwing or sowing. N fertilizers were applied with four splits (seedling stages: tillering stages: jointing stages: booting stages = 25:10:6:9) for the rice season, and three splits (seedling stages: tillering stages: boosting stages = 5:3:2) for the wheat season. During the rice growing season, the depth of waterlogging was kept at 8 cm, except for during the tillering and maturing stages. The wheat season was not irrigated, except after sowing.

2.3. Soil Sampling and Physicochemical Analysis

Soil samples were collected in May and October after the rice and wheat harvests from 2012 to 2014. The soil samples were taken from eight locations in each plot at a depth of 0–20 cm using a soil sampler with a diameter of 5 cm, before being divided into three categories (0–5 cm, 5–10 cm and 10–20 cm depths). Then part of soil samples in 0–5 cm soil layer were separated into the 1–2 mm, 0.25–1 mm, 0.053–0.25 mm, and <0.053 mm aggregate with a nest of sieves mounted (including 1 mm, 0.25 mm, and 0.053 mm). These soil samples were used to measure SOC, soil microbial biomass carbon (MBC) and dissolved organic carbon (DOC). The remaining soil samples were placed at −20 °C for DNA extraction.

The dry-sieving method was used to separate soil aggregates following the descriptions of Garzia-Bengoetxea et al. [26]. In the present study, the dry sieving method was used as the mechanical pressure exerted from outside is the main cause of soil aggregate destruction, and compared with the wet sieving method, the dry sieving method is less destructive to the soil. Furthermore, drying at a low temperature of 4 °C minimizes the effect on the soil microbial community and activities. Retsch AS200 control (Retsch Technology, Düsseldorf, Germany) was used to separate soil aggregates. Air-dried soil fragments (5 mm) were prepared for separation, and soil samples were separated into 1–2 mm, 0.25–1 mm, 0.053–0.25 mm and <0.053 mm soil aggregates by mechanical shaking (amplitude 1.5 mm) for 2 min.

The SOC content was measured with a FlashEA 1112 elemental analyzer (Thermo Finnigan, Milan, Italy). Fumigation-extraction method was used to measure MBC. MBC was calculated as the ratio of differences in organic carbon extracted from fumigated and non-fumigated soil and the conversion coefficient was 0.38. DOC could be measured by the methods of Jiang et al. [27].

2.4. Phospholipid Fatty Acid Pattern

Phospholipid fatty acids were extracted from a 3 g freeze-dried soil sample using the methods of Frostegård et al. [28]. Briefly, lipids were extracted in a single-phase chloroform–methanol–phosphate buffer system in a ratio of 1:2:0.8 (*v/v/v*). A stream of N_2 was used for drying the different phases. Separation of extracts was performed on solid phase extraction columns (Supelco Inc., Bellefonte, PA, USA). The phospholipid fractions were saponified and methylated to fatty acid methyl esters. Internal standard 19:0 fatty acid methyl esters were added to calculate the absolute amount of fatty acid methyl esters before measurement. We employed a gas chromatograph/mass spectrometry system (6890–5973N series GC/MS Agilent Technologies, Palo Alto, CA, USA) outfitted with a Flame Ionization Detector and HP-5 capillary column (30 m × 0.25 mm × 0.25 μm) with ultra-purified helium as carrier gas for the extraction. The quantification of fatty acid methyl esters was performed.

2.5. Measurement of Crop Grain Yields

Crop grains harvested from the 2012 rice season to the 2014 rice season were measured at the central position in each plot using a 5 m^2 frame. The rice and wheat grains were air dried, weighed, and adjusted to 14.0% and 12.5% moisture content, respectively.

2.6. Measurement of CH_4 and CO_2 Emissions

Static closed steel chamber method was used to monitor soil CH_4 emission [14]. After the crop straw was returned to fields, continuous gas sampling was conducted until the crop harvest. The inner diameter of the chamber was 34 cm and the height of the chamber was 50 or 120 cm (depending on the rice height). To mix the air in the chambers well, four fans were installed on the top of the chamber. During sampling, two rings were placed in each plot, the chambers were placed temporarily in the groove of rings, and water was added to create a sealed environment. After 0, 5, 10 and 15 min of chamber closure (according to the IGAC recommendations [29]), gas samples were gathered from each chamber. The gas samples were collected using a syringe (20 mL) at the chamber's headspace, and then transferred to 20 mL vacuum glass bottles. The gas samples were collected between 9:00 and 11:00 am once every 7–10 and 10–15 days (recommended by Buendia et al. [30]) in rice and wheat seasons, respectively. A chromatograph equipped with a flame ionization detector (Shimadzu GC-14B) was used to measure CH_4 fluxes [31].

The measurement of soil CO_2 flux was conducted according to the method proposed by Li et al. [31]. The CO_2 fluxes were measured three times for each plot with an 8100–103 short-term chamber connected to a LI-8100A soil CO_2 flux system (Li-Cor Inc., Lincoln, NE, USA). The final value of soil CO_2 flux was obtained by averaging the values of three separate measurements. The calculation of CH_4 and CO_2 fluxes was based on the linear variation in CH_4 and CO_2 fluxes [32]. The cumulative seasonal CH_4 and CO_2 emissions were derived by sequentially accumulating emissions from each of the two adjacent measurement intervals [31].

2.7. High-Throughput Sequencing

According to the instructions, the FastDNA kit for soil (MP Bio-medicals, Santa Ana, CA, USA) was used to extract DNA from the soil samples and then stored at −20 °C. The bacterial hypervariable regions, including V3, V4 and V5 of 16S rDNA, were amplified by PCR using primers 357F (5′-CCTACGGGAGGCAGCAG-3′) and 926R (5′-CCGTCAATTCMTTTRAGT-3′) [33]. The forward primer was modified to include the FLX-titanium adaptor "B" sequence (5′-CCTATCCCCTGTGTGCCTTGGCAGTCTCAG-3′), and the reverse primer was linked with the 454 FLX-titanium adaptor "A" (5′-CCATCTCATCCC TGCGTGTCTCCGACTCAG-3′) [21].

DNA samples (10 ng) were applied as templates in the polymerase chain reaction. Polymerase chain reaction was conducted at 95 °C for 3 min, 94 °C for 30 s, 55 °C for 45 s, 72 °C for 1 min with 25 cycles and a final extension step of 72 °C for 7 min. Polymerase Chain Reaction Purification Kit (Axygen Bio, Union City, California, CA, USA) was used for purification of polymerase chain reaction products. The amplitudes from each sample were then combined in equimolar concentrations into one tube prior to 454 pyrophosphate sequencing. Pyrophosphate sequencing was performed by Shanghai Personal Biotechnology Co., Ltd. using the 454 GS-FLX Titanium System (Roche, Switzerland). To ensure analytical accuracy, the Quantitative Insights into Microbial Ecology (QIIME) pipeline [34] was employed to fetch high quality sequences, following the descriptions of Fierer et al. [35]. The unique sequence set was classified into operational taxonomic units OTUs (a threshold of 97% pairwise identity) by the QIIME implementation. Extraction of the longest sequences of the most abundant OTUs was used as a proxy for taxonomic identification for comparison with the Green Gene Database (release 13.8 http://greengenes.secondgenome.com/ (accessed on 10 January 2012)).

2.8. Statistical Analysis

All data were expressed as means and standard deviations of three replicates. The main effects and interactions of tillage and straw returning were conducted using general linear model analysis of variance with SAS 9.0 (SAS Institute 1999) designed for split plot with tillage practice and straw returning methods as fixed factors and replicates as random factors. The least significant difference test was conducted to examine whether the influence of tillage practices, straw return practices, or their interactions were significant at the level of 0.05. To test the effect of experimental treatments on bacterial composition, redundancy analysis was performed using the "vegan" package in R v. 3.1.2 (R Development Core Team, 2014) [36].

Structural equation modeling was performed to reveal the influence paths of tillage and straw return practices on SOC content from the perspectives of the soil bacterial community, DOC, microbial biomass carbon, CH_4 and SOC in 1–2 mm aggregates. The use of structural equation modeling allowed the testing of complex path-relation networks. It should be noted that only the data for the 2013 rice season and 2014 wheat seasons were selected [37]. In the model, tillage (0 = no-till and 1 = tillage) and straw return (0 = straw removal and 1 = straw return) were considered as categorical variables. This approach allowed us to compare the effect of tillage and straw return practices on SOC content. Redundancy analysis results for bacterial communities in order level (relative abundance > 0.5%), were used as 'bacteria community' in the model. A 'robust' maximum likelihood estimation procedure of AMOS 20.0 (IBM SPSS, Chicago, IL, USA) software was conducted for the analysis. The chi-square test, goodness of fit index, comparative fit index, and root square mean error of approximation were used for testing the overall goodness of the fit of the model.

3. Results

3.1. Soil Organic Carbon

Tillage and straw return management significantly changed the SOC content in the 0–5 cm soil layer ($p < 0.05$, Table S1). Compared with CT treatment, NT treatment significantly increased SOC content in the 2013 wheat season (5.7%), 2013 rice season (15.3%) and 2014 rice season (4.4%). In comparison with NS treatment, SR treatment markedly enhanced SOC content in the 2012 rice season (6.6%), 2013 wheat season (8.3%), 2013 rice season (9.1%), 2014 wheat season (6.5%) and 2014 rice season (8.3%). Compared with CTNS, NTNS markedly increased SOC in the 2013 rice season, 2014 wheat season and 2014 rice season by 15.6%, 2.9% and 4.4%, respectively. In comparison with CTSR, NTSR significantly enhanced SOC in the 2013 wheat season, 2013 rice season and 2014 rice season by 7.1%, 15.1% and 4.5%, respectively. Interaction of tillage and straw return practices showed no remarkable effects on SOC content.

3.2. Distribution of Soil Aggregates

Tillage and straw returning methods had a significant effect on the distribution of soil aggregates in the soil layer within the topsoil layer (Table S2). Compared with CT treatment, NT treatment significantly increased the percentage of 1–2 mm soil aggregates in the 2014 wheat season (4.1%) ($p < 0.05$), whereas, there was a markedly reduced percentage of soil aggregates < 0.053 mm in the 2013 rice season (18.9%). Compared with NS treatment, SR treatment resulted in an increased the proportion of 1–2 mm soil aggregates in both wheat and rice seasons of 2013 (5.4%, 4.4%), and in the wheat season of 2014 (5.6%) ($p < 0.05$), but decreased the proportion of soil aggregates <0.053 mm in the 2014 wheat season (21%) ($p < 0.05$). NTNS significantly increased the proportion of 1–2 mm soil aggregates by 6.3% in the 2014 wheat season, and markedly reduced the percentage of soil aggregates < 0.053 mm in the 2013 rice season (7.6%) and 2014 wheat season (20.2%) compared to CTNS. NTSR, respectively, increased the proportion of 1–2 mm soil aggregates by 2.8%, 4.7%, 2.1% in the 2013 wheat season, 2013 rice season, 2014 wheat season, and markedly reduced the percentage of soil aggregates < 0.053 mm in 2013 rice season (29.9%) compared to CTSR. Interaction of tillage practices and straw returning methods showed no significant difference.

3.3. Soil Organic Carbon Content within Aggregates

Tillage and straw return practices greatly influenced the SOC content of aggregates in the 0–5 cm soil layer (Table 1). Compared to CT treatment, NT treatment increased the SOC content in 1–2 mm aggregates in the 2013 wheat season (17%), 2013 rice season (19.9%) ($p < 0.05$, Table 1). Higher SOC content in 0.25–1 mm aggregates was also observed under NT treatment than under CT treatment in the 2013 wheat (14.6%) and rice seasons (13.4%) ($p < 0.05$). Compared with NS treatment, SR treatment led to higher SOC content in

1–2 mm aggregates in the 2013 rice season (17.2%), and 2014 wheat season (17.4%) ($p < 0.05$). Moreover, higher SOC content in 0.25–1 mm aggregates was found in the 2013 rice season (7%) and 2014 wheat season (6.2%) under SR treatment than under NS treatment ($p < 0.05$). Compared to CTNS, NTNS significantly increased the SOC content in 1–2 mm aggregates in the 2013 rice season by 17.0%. Moreover, there were also significant differences in the SOC content of 0.25–1 mm aggregates in the 2013 rice season (16.5%), and in the 2014 wheat season (3.6%) between CTNS and NTNS. NTSR, respectively, increased the SOC content in 1–2 mm aggregates by 17.9%, 22.4%, 8.3% in the 2013 wheat season, 2013 rice season and 2014 wheat season, and enhanced SOC content in 0.25–1 mm aggregates in the 2013 rice season (10.6%) and 2014 wheat season (4.1%) relative to CTSR. In other soil layers, there was no remarkable difference between treatments. Interaction of tillage and straw return practices remarkably influenced the SOC content in 1–2 mm aggregates in the 2014 wheat season ($p < 0.05$).

Table 1. SOC contents (g kg^{-1}) of aggregate fractions under different tillage and straw return practices (2012–2014).

Crop Season	Soil Aggregate Fraction	CTNS	CTSR	NTNS	NTSR	T	SR	T × SR
2012 rice season	1–2 mm	17.03 ± 1.05 ^{a}	18.97 ± 0.60 ^{a}	18.61 ± 2.39 ^{a}	21.14 ± 1.70 ^{a}	ns	ns	ns
	0.25–1 mm	15.60 ± 0.32 ^{a}	17.07 ± 0.76 ^{a}	14.16 ± 0.03 ^{a}	16.94 ± 1.66 ^{a}	ns	ns	ns
	0.053–0.25 mm	19.67 ± 1.08 ^{a}	16.49 ± 1.72 ^{a}	20.62 ± 0.3 ^{a}	16.85 ± 2.00 ^{a}	ns	ns	ns
	<0.053 mm	17.97 ± 1.09 ^{a}	19.49 ± 1.04 ^{a}	18.72 ± 0.43 ^{a}	18.00 ± 0.38 ^{a}	ns	ns	ns
2013 wheat season	1–2 mm	16.32 ± 1.00 ^{c}	18.20 ± 0.29 ^{bc}	18.92 ± 0.91 b	21.45 ± 0.63 ^{a}	*	ns	ns
	0.25–1 mm	17.52 ± 0.55 ^{a}	18.52 ± 0.65 ^{a}	19.77 ± 0.12 ^{a}	21.53 ± 1.76 ^{a}	*	ns	ns
	0.053–0.25 mm	20.79 ± 1.62 ^{a}	21.59 ± 0.98 ^{a}	23.42 ± 1.91 ^{a}	26.14 ± 3.05 ^{a}	ns	ns	ns
	<0.053 mm	15.02 ± 1.34 ^{a}	16.59 ± 1.54 ^{a}	16.97 ± 1.75 ^{a}	19.79 ± 1.45 ^{a}	ns	ns	ns
2013 rice season	1–2 mm	16.14 ± 0.67 ^{c}	18.46 ± 0.28 ^{bc}	18.88 ± 0.43 ^{b}	22.60 ± 1.36 ^{a}	*	*	ns
	0.25–1 mm	18.06 ± 0.43 ^{c}	19.87 ± 0.18 b	21.04 ± 0.46 ^{ab}	21.98 ± 0.36 ^{a}	*	*	ns
	0.053–0.25 mm	16.90 ± 0.84 ^{a}	16.93 ± 1.82 ^{a}	18.23 ± 0.50 ^{a}	19.65 ± 1.00 ^{a}	ns	ns	ns
	<0.053 mm	15.51 ± 1.19 ^{a}	16.92 ± 1.95 ^{a}	20.18 ± 0.81 ^{a}	18.81 ± 0.77 ^{a}	ns	ns	ns
2014 wheat season	1–2 mm	16.25 ± 0.12 ^{c}	18.34 ± 0.18 ^{b}	16.30 ± 0.07 ^{c}	19.87 ± 0.34 ^{a}	ns	*	*
	0.25–1 mm	20.66 ± 0.33 ^{c}	21.89 ± 0.19 ^{ab}	21.40 ± 0.55 ^{bc}	22.78 ± 0.14 ^{a}	ns	*	ns
	0.053–0.25 mm	17.49 ± 0.88 ^{a}	18.22 ± 0.82 ^{a}	18.75 ± 0.85 ^{a}	19.63 ± 1.37 ^{a}	ns	ns	ns
	<0.053 mm	17.72 ± 4.40 ^{a}	17.83 ± 2.02 ^{a}	16.99 ± 3.11 ^{a}	17.67 ± 1.63 ^{a}	ns	ns	ns

Different letters in the columns denote statistical differences in the means of the variables between treatments by the least significant difference test ($p < 0.05$). * $p < 0.05$; ns, not significant. CTNS, conventional intensive tillage with straw removal; CTSR, conventional intensive tillage with straw return; NTNS, no tillage with straw removal; NTSR, no-tillage with straw return. T, tillage; SR, straw return practices. T × SR, the interactions between tillage and straw return. Values are mean ± standard deviation (n = 3).

3.4. Soil Dissolved Organic Carbon and Microbial Biomass Carbon

Tillage and straw returning methods had significant effects on DOC contents in the 0–5 cm soil layer (Table S3). Compared to CT treatment, NT treatment markedly increased the DOC contents in both wheat seasons and rice seasons in 2014 (12.3%, 8.8%) (Table S3, $p < 0.05$). Similarly, higher DOC contents were found in both wheat and rice seasons in 2013 (23.7%, 23.8%) and 2014 (18.5%, 13%) ($p < 0.05$) under the SR treatment than under the NS treatment. Compared with CTNS, NTNS showed a significant increase in the DOC contents in the 2013 rice season (3.3%) as well as in the 2014 wheat (13.4%) and rice seasons (10.4%). NTSR showed significantly higher DOC contents in both wheat and rice seasons in 2013 (16.5%, 13.8%) and 2014 (11.4%, 7.4%) relative to CTSR. The interaction between NT and SR remarkably influenced the DOC during the whole 2013 season.

Relative to CT treatment, NT treatment significantly increased the MBC contents in the 0–5 cm soil layer in both wheat and rice seasons in 2013 (15.1%, 14.3%) and 2014 (21.5%, 39.8%) (Table S4, $p < 0.05$). SR treatment also resulted in higher MBC contents in both wheat and rice seasons in 2013 (27.8%, 18.1%) and 2014 (26.6%, 20.1%) ($p < 0.05$) than NS treatment. Compared to CTNS, NTNS displayed a statistically significant improvement in MBC contents for both the 2013 (19.2%, 6.9%) and 2014 wheat and rice seasons (5.7%, 33.2%). NTSR also showed an improvement in the 2013 wheat season (12.0%), 2013 rice season (21.0%), 2014 wheat season (35.8%) and 2014 rice season (45.5%) against CTSR. Interactions of tillage and straw returning practices had significant effects on MBC contents with the exception of the 2012 and 2013 wheat seasons ($p < 0.05$).

3.5. Greenhouse Gas Emissions

Relative to CT treatment, NT treatment remarkably decreased CH_4 emissions during the 2014 rice seasons (15.6%) ($p < 0.05$) (Table 2). SR treatment resulted in higher CH_4 emissions during the rice (by 34.0–91.2%) and wheat (by 22.1–22.3%) seasons throughout all three experimental years ($p < 0.05$). Compared with CTSR, NTSR showed a significant reduction in CH_4 emissions in the 2013 wheat season (18.0%), 2014 wheat season (10.8%), and 2014 rice season (16.6%). However, the NTNS showed the lowest CH_4 emissions among all treatments in whole seasons. Interaction of tillage and straw return practices showed no significant effects on CH_4 emissions.

Table 2. Seasonal CH_4 emissions (kg hm^{-2}) under different tillage and straw return practices (2012–2014) (has been published by Guo et al. [2]).

Treatment	2012	2013		2014	
	Rice Season	Wheat Season	Rice Season	Wheat Season	Rice Season
CTNS	400 ± 7.51 ^{b}	4.86 ± 0.98 ^{c}	475 ± 21.7 ^{b}	5.39 ± 0.54 ^{c}	167 ± 11.37 ^{b}
CTSR	560 ± 30.73 ^{a}	16.91 ± 0.37 ^{a}	645 ± 12.0 ^{a}	15.95 ± 0.99 ^{a}	202 ± 13.68 ^{a}
NTNS	391 ± 21.16 ^{b}	3.99 ± 0.42 ^{c}	445 ± 7.7 ^{b}	4.81 ± 0.46 ^{c}	140 ± 10.60 ^{b}
NTSR	632 ± 27.09 ^{a}	12.53 ± 2.23 ^{b}	610 ± 9.7 ^{a}	12.33 ± 0.60 ^{b}	162 ± 2.35 ^{b}
T	ns	ns	ns	ns	*
SR	*	*	*	*	*
T × SR	ns	ns	ns	ns	ns

Different letters in the columns denote statistical differences in the means of the variables between treatments by the least sign difference test ($p < 0.05$). * $p < 0.05$; ns, not significant. CTNS, conventional intensive tillage with straw removal; CTSR, conventional intensive tillage with straw return; NTNS, no tillage with straw removal; NTSR, no-tillage with straw return. T, tillage; SR, straw return practices. T × SR, the interactions between tillage and straw return. Values are mean ± standard deviation (n = 3).

Compared with CT treatment, NT treatment lowered CO_2 emissions in the 2014 wheat (7.2%) and rice reasons (21.4%) ($p < 0.05$) (Table 3). SR treatment induced more CO_2 emissions ($p < 0.05$) than NS treatment in the 2012 rice season (91.2%), 2013 wheat (22.1%) and rice seasons (40.8%), and 2014 wheat (22.3%) and rice seasons (34.0%). Compared with CTNS, NTNS markedly elevated CO_2 emissions in the 2012 rice season by 19.3%, whereas, it reduced CO_2 emissions in the 2013 rice season by 22.7% and 2014 rice season by 10.9%. NTSR had lower CO_2 emissions in the 2012 rice season, 2013 rice season, 2014 rice season (by 9.7%, 14.6% and 28.5%, respectively) against CTSR. Interaction of tillage and straw return practices showed a significant effect on CO_2 emissions only in the 2012 rice seasons and 2014 rice seasons ($p < 0.05$). Meanwhile, the combination of NT and NS can reduce CO_2 emissions compared to other treatments.

Table 3. Seasonal CO_2 emissions (kg hm^{-2}) under different tillage and straw return practices (2012–2014) (has been published by Guo et al. [2]).

Treatment	2012	2013		2014	
	Rice Season	Wheat Season	Rice Season	Wheat Season	Rice Season
CTNS	2230 ± 92 ^{d}	5282 ± 123 ^{b}	4833 ± 397 ^{b}	3982 ± 84 ^{bc}	4283 ± 50 ^{c}
CTSR	4914 ± 27 ^{a}	6695 ± 408 ^{ab}	6503 ± 308 ^{a}	4989 ± 123 ^{a}	6332 ± 154 ^{a}
NTNS	2660 ± 86 ^{c}	6090 ± 298 ^{ab}	3734 ± 116 ^{c}	3799 ± 124 ^{c}	3817 ± 88 ^{d}
NTSR	4438 ± 47 ^{b}	7193 ± 939 ^{a}	5557 ± 265 ^{b}	4525 ± 452 ^{ab}	4525 ± 162 ^{b}
T	ns	ns	ns	*	*
SR	*	*	*	*	*
T × SR	*	ns	ns	ns	*

Different letters in the columns denote statistical differences in the means of the variables between treatments by the least sign difference test ($p < 0.05$). *, $p < 0.05$; ns, not significant. CTNS, conventional intensive tillage with straw removal; CTSR, conventional intensive tillage with straw return; NTNS, no tillage with straw removal; NTSR, no-tillage with straw return. T, tillage; SR, straw return practices. T × SR, the interactions between tillage and straw return. Values are mean ± standard deviation (n = 3).

3.6. Soil Microbial Community

Tillage and straw returning methods significantly influenced bacterial biomass in 0–5 cm soil layer (Table S5). Compared with CT treatment, NT treatment increased the bacterial biomass in the 2013 rice seasons (35.6%) and 2014 wheat seasons (56.2%) ($p < 0.05$). SR treatment led to higher bacterial biomass than NS treatment in the 2013 wheat season (76.5%), 2013 rice season (54.9%), 2014 wheat season (75.7%), 2014 rice season (59.7%) ($p < 0.05$). Compared with CTNS, NTNS had a greater impact on bacterial biomass in the 2013 rice season (33.1%), and in the 2014 wheat season (38.3%), while for fungi, NTNS had a significant reduction in the 2013 wheat season (25.2%). Compared to CTSR, NTSR had a greater effect on microorganisms in both wheat and rice seasons in 2013 (70.1%, 37.1%) and 2014 (67%,75%), while for fungi, NTSR showed a significant improvement in the 2013 wheat season (23.8%). The interaction of tillage and straw return practices had a significant effect on bacterial biomass in the 2014 wheat season ($p < 0.05$). Tillage and straw returning practices had no significant effects on fungal biomass in the 2013 wheat seasons (Table S5). The fungal PLFAs were not detected in the 2013–2014 rice seasons.

3.7. Soil Bacterial Community

Soil bacterial community was mainly composed of phylum Acidobacteria, Verrucomicrobia and Proteobacteria in the 2013 rice season (Table S6), while it was mainly composed of phylum Acidobacteria, Chloroflexi and Proteobacteria in the 2014 wheat season (Table S7).

In the 2013 rice season, compared with CT treatment, NT treatment affected the abundance of *Gp1* (−17.7%), *Gp18* (−19.4%), *Gp4* (65.2%), *Gp16* (21.6%), *Dehalogenimonas* (29.1%), *Caulobacterales* (27.6%), *Desulfuromonadales* (33.0%), *Myxococcales* (42.8%), and *Legionellale* (−9.7%) (Table S6, $p < 0.05$). Compared with NS treatment, SR treatment significantly affected the abundance of *Gp18* (−16.3%), *Gp4* (48.1%), *Gp17* (16.4%), *Chlamydiales* (−26.0%), *Caulobacterales* (19.7%), *Burkholderiales* (−37.6%) ($p < 0.05$). Compared with CTNS, NTNS significantly increased the abundance of *Rhizobiales* (7.0%), *Burkholderiales* (13.4%), and *Spartobacteria_genera_incertae_sedis* (46.2%), whereas, it markedly decreased the abundance of order *Gp1* (1.5%), *Gp18* (25.2%), *Gp4* (38.3%), and *Dehalogenimonas* (17.2%). Compared with CTSR, NTSR significantly elevated the abundance of *Gp4* (267%), *Holophagales* (85.4%), *Dehalogenimonas* (103.4%), *Chlamydiales* (33.9%), *Caulobacterales* (50.9%), *Burkholderiales* (27.8%), and *Spartobacteria_genera_incertae_sedis* (30.5%), whereas it significantly reduced the abundance of order *Gp1* (32.0%) and *Rhizobiales* (17.2%). Interaction of tillage and straw return practices significantly affected the abundance of *Gp1*, *Gp18*, *Gp4*, *Gp17*, *Holophagales*, *Dehalogenimonas*, *Chlamydiales*, *Caulobacterales*, *Syntrophobacterales*, and *Spartobacteria_genera_incertae_sedis* ($p < 0.05$). NTSR showed the highest or the lowest bacterial community richness compared with the NTNS and CTSR.

In the 2014 wheat season, NT treatment led to a higher abundance of *Myxococcales* (206.3%) than CT treatment (Table S7, $p < 0.05$). In comparison with NS treatment, SR treatment brought out higher enrichment of *Gp4* (573.9%), *Gp10* (502.4%), *Gp18* (97.5%), *Sphingobacteriales* (164.7%) *Gp7* (303.8%), *Sphingobacteriales* (164.7%), *Flavobacteriales* (89.4%), and *Myxococcales* (173.2%), while it decreased the abundance of *Gemmatimonadales* (47.8%) and *Rhodospirillales* (42.0%) ($p < 0.05$). Compared with CTNS, NTNS significantly increased the abundance of order *Myxococcales* (176.0%), whereas it markedly decreased the abundance of order *Xanthomonadales* (5.5%). Compared with CTSR, NTSR significantly increased the abundance of *Gp4* (161.9%), *Gp10* (429.1%), *Gp18* (106.2%), *Gp7* (145.9%), *Xanthomonadales* (17.7%), and *Myxococcales* (218.1%). The interplay of tillage and straw return practices significantly influenced the abundance of *Gp18* and *Flavobacteriales* ($p < 0.05$).

3.8. Crop Grain Yields and Their Relationship with Soil Properties

Crop grain yields in this study were reported in our previous study (Table S8) [2]. NT treatment significantly reduced crop yields by 8.8% in the 2014 wheat season compared to CT treatment ($p < 0.05$). SR treatment showed no significant difference relative to NS. There was no significant difference in grain yields between NTNS and CTNS. NTSR had a remarkable increase in crop yields in the 2014 wheat season compared to CTSR (19.1%, $p < 0.05$). Interaction of tillage and straw return practices showed a significant effect on crop yields in the 2014 wheat season ($p < 0.05$). A significant correlation was observed between DOC and crop yields (Table S9).

3.9. Relationship of Bacterial Community with Yield, Soil Aggregates and Soil Organic Carbon Fractions

Redundancy analysis (RDA) showed that soil bacterial community was considerably influenced by SOC content in 1–2 mm aggregates, MBC and CH_4 emissions (Figure 3, $p < 0.05$). MBC and SOC in 1–2 mm aggregates were closely related to *Gp6*, *Burkholderiales*, *Gp10*, *Sphingobacteriales*, *Myxococcales*, *Gp16*, *Flavobacteriales*, *Gp2*, *Gp3*, and *Xanthomonadales*. CH_4 emissions were closely related to *Subdivision3_genera_incertae_sedis*, *Gp18*, *Caulobacterales*, *Gp16*, and *Chlamydiales*. Besides, no significant correlation were found between crop yield and microbial community.

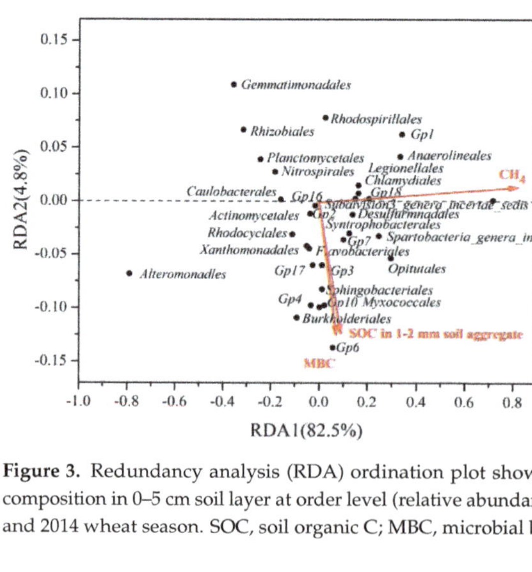

Figure 3. Redundancy analysis (RDA) ordination plot showing changes in bacterial community composition in 0–5 cm soil layer at order level (relative abundance > 0.5%) during the 2013 rice season and 2014 wheat season. SOC, soil organic C; MBC, microbial biomass C.

The structural equation modeling revealed that the predictors could explain 85.0% of the variances in SOC content (Figure 4). Soil bacterial community mediated SOC under tillage and straw systems through affecting SOC in 1–2 mm aggregates and CH_4 emissions.

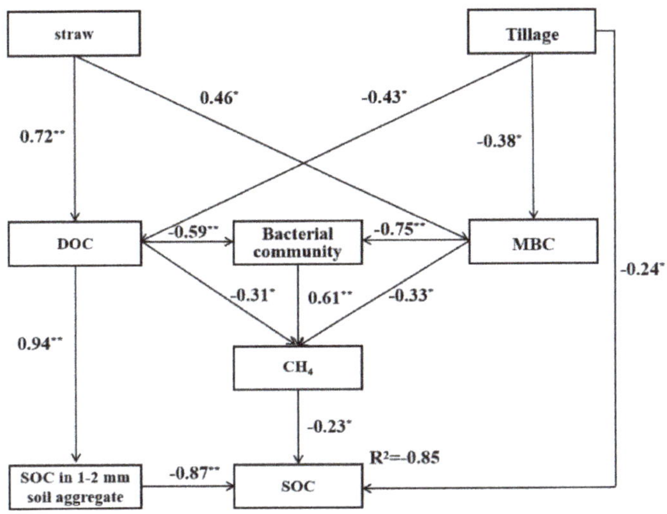

Figure 4. Selected structural equation modeling (data from 2013 rice season and 2014 wheat season were selected) for SOC in 0–5 cm soil layer (The chi-square test = 9.91; Goodness of fit index = 1.00; Comparative fit index = 0.91; Root square mean error of approximation = 0.00), based on the impact of tillage and straw return practices and SOC fractions. Values related to the solid arrows stand for the path coefficients. R^2 indicates the proportion of variance explained. Significance levels are as follows: * $p < 0.05$; ** $p < 0.01$. Straw indicates straw systems; DOC indicates dissolved organic carbon; MBC indicates microbial biomass carbon; SOC indicates soil organic carbon.

4. Discussion

4.1. Impact of NT and Straw Return on SOC Content in Aggregates

NT can enhance SOC content by promoting the SOC sequestration in macro- aggregate [7,38,39]. In this study, a higher proportion of 1–2 mm soil aggregates (Table S2), and more SOC content in 1–2 mm aggregates, were found under NT than under CT (Table 1). Tillage operations breaks soil macro-aggregates and results in SOC losses [40]. Conversely, NT keeps soil undisturbed, which is conducive for accelerating the formation of macro-aggregate, and reduces the degradation rate of SOC [41]. Moreover, NT can provide more physical protection for soil aggregates and promote the longevity of newly-formed macro-aggregates, leading to stabilization of SOC in the micro-aggregates formed within stable macro-aggregates [42,43].

As an essential organic matter source, SR can promote the formation of soil macro-aggregates and increase SOC content. Previous studies have well reported that SR can increase SOC content by increasing the input of organic carbon input [12,43]. In this work, higher SOC content in the topsoil layer (0–5 cm) was observed under SR than under NS (Table S1), which may be due to higher SOC sequestrated in 1–2 mm aggregates (Table S2). Straw degradation generates a large number of organic matter particles, contributing to the formation of macro-aggregates and the accumulation of SOC in macro-aggregate [7,17,44,45].

Some studies reported that interaction of tillage and straw return practices significantly affected SOC, possibly as NT can promote the accumulation of straw on the soil surface, thus enhancing SOC sequestration in the topsoil [21]. Similarly, we also found that both under NS or SR conditions, NT caused higher SOC in 1–2 mm and 0.25–1 mm aggregates than CT (Table 1). Moreover, the interaction of tillage and straw return practices significantly affected SOC in 1–2 mm aggregates in the 2014 wheat season (Table 1), suggesting that under straw return condition, NT can further promote SOC sequestration in 1–2 mm aggregates [7,21]. However, the interaction of tillage and straw return practices had no significant effect on SOC in other aggregates sizes (Table 1), possibly as 1–2 mm soil aggregates are more sensitive than soil aggregates of other smaller sizes [2,21].

4.2. Effect of NT and Straw Return on Greenhouse Gas Emissions

Emissions of CH_4 and CO_2 are important pathways of carbon loss from agricultural soil [24,31,46]. In this study, NT treatment reduced CH_4 and CO_2 emissions compared with CT treatment (Tables 2 and 3). CH_4 emissions are primarily affected by the availability of organic matter and oxygen [12,17,46]. NT decreases soil disturbance and improves gas diffusion, inhibiting the growth of methanogenic bacteria and reducing the production of CH_4 [23]. NT also can enhance soil moisture, reduce soil temperature, slow the organic residue degradation, and reduce the activity of soil microorganisms, thus reducing CH_4 and CO_2 emissions [12,19,20].

In contrast, straw return promoted CH_4 and CO_2 emissions (Tables 2 and 3) mainly by providing a large number of organic carbon for soil microorganisms [12,46]. Moreover, anaerobic degradation of crop residues can reduce soil Eh, thus increasing methanogenic populations and enhancing CH_4 emissions [17,47,48]. Nevertheless, a large number of straw-derived carbon can be sequestrated in soil by the formation of resistant organic matter, which may offset the losses of SOC caused by CH_4 and CO_2 emissions.

In this study, the interaction of tillage and straw return practices had no effect on CH_4 and CO_2 emissions in the experiment (Tables 2 and 3), which may be due to SR and NT having the opposite effect on CH_4 emissions. SR significantly enhanced CH_4 and CO_2 emissions, whereas, NT was found to have reduced CH_4 and CO_2 (Table 2). We also found that CTNS had no significant effect on CH_4 emissions relative to NTNS, while NTSR had lower CH_4 emissions than NTNS (Table 2). The reason may be the fact that NT leads to more straw being accumulated in the soil surface, which has better oxygen available than topsoil layers, thus inhibiting the production of CH_4 from the soil [17,48]. Besides, CTNS had higher CO_2 emissions than NTNS, and CTSR also had more CO_2 emissions than NTSR. This can be attributed to there being a lower soil temperature under NT than under CT, thus leading to a decrease in the activity of soil microorganisms, and subsequently to a decrease in CO_2 emissions [19,20].

4.3. Effects of NT and Straw Return on Bacterial Community

Microorganisms play a key role in regulating SOC turnover and sequestration [49]. The bacterial community accounts for the majority of soil microorganisms in the rice-wheat cropping system (Table S5) [50], which is probably due to the long-term flooding of the field during the rice season resulting in the formation of an anaerobic environment, inhibiting the growth of the soil fungal community [21]. The bacterial community is sensitive to tillage and straw management [51]. Common dominant bacteria such as Phylum Actinobacteria, Proteobacteria and Actinobacteria are recognized to be remarkable plant biomass decomposers [52–54]. In this study, Phylum Acidobacteria, Verrucomicrobia and Proteobacteria phylum were dominated in the 2013 rice season (Table S5) and phylum Acidobacteria, Chloroflexi and Proteobacteria phylum were dominated in the 2014 wheat season (Table S6).

We found that NT significantly affected the bacterial community in the 2013 rice season and 2014 wheat season (Tables S5 and S6). NT can enhance some bacterial abundance related to the decomposition of crop residue [52,54,55], for example, phylum Actinobacteria (including order *Gp4*, *Gp16*), phylum Chloroflex (including order *Dehalococcoidetes*), phylum Proteobacteria (including order *Myxococcales*), phylum Alphaproteobacteria (including order *Caulobacterales*), phylum Chloroflexi (including order Dehalococcoidates), and phylum (including *Desulfuromonadales* and *Myxococcales*) (Tables S5 and S6). This is probably due to the fact that NT can provide more available substrates and nutrients for soil microorganisms [56]. However, we also found NT decreased the abundance of bacteria such as order *GP1* and *Gp18* in 2013 rice season (Table S5), which may be due to *Gp1* and *Gp18* benefitting from a poor nutrition condition [57].

Straw return can input a large quantity of straw-derived carbon into soil, and thus affect the bacterial community [56,58,59]. In this study, straw return observably affected the bacterial community in the 2013 rice season and 2014 wheat season (Tables S6 and S7). Generally, SR can tend to increase the abundance of the bacterial community as SR can provide more metabolic substrates for bacteria [12]. SR can improve soil properties, such as soil permeability and water holding capacity, and provide comfortable habitat conditions for bacteria, thus improving the bacterial community [12,59]. However, in this study, SR decreased the abundance of some within the bacterial community, such as order *Caulobacterales*, which may be due to SR increasing the availability of oxygen and thus inhibiting the growth of *Caulobacterales* in the 2013 rice season [60]. SR also decreased the abundance of order *Gp18* in the 2013 rice season, which is probably due to the fact that order *Gp1* and *Gp18* could benefit from a poor nutrition condition [57].

In this study, the interaction of tillage and straw return practices significantly influenced the abundance of the bacterial community in 0–5 cm soil layer, such as order *Gp1, Gp18, Gp4, Gp17, Holophagales, Dehalogenimonas, Chlamydiales, Caulobacterales, Syntrophobacterales,* and *Spartobacteria_genera_incertae_sedis* (Tables S6 and S7). SR and NT tended to increase the abundance of the soil bacterial community, and the combination of SR and NT can provide better habitat conditions, such as higher availability of oxygen and greater organic carbon for the soil bacterial community [12,59]. However, some within the bacterial community, such as order *G16* and *Rhodospirillales*, were not significantly affected by the interaction (Tables S6 and S7). This can be attributed to high diversity of soil bacterial community, and the difference in the preference of soil microorganisms regarding habitat conditions, such as oxygen availability and carbon and nitrogen sources. Moreover, crop rotation can also reduce the interaction effect of tillage and straw return practices on the soil bacterial community [21,26].

4.4. Effect of NT and Straw Return on Crop Yields

The effect of NT and SR on crop yields was discussed in our previous study [2]. In this study, NT had no significant effect on grain yields during 2012–2014, except that NT significantly reduced crop yields in the 2014 wheat season (Table S8). In general, less than five years of continuous NT is not enough to change crop yields [2,61]. Lower yields under NT than under CT in the 2014 wheat season can be attributed to high rainfall during the growth season of wheat (Figure 1). NT can promote the accumulation of straw residue on the soil surface, and enhance the soil anaerobic condition in the case of high rainfall, inhabiting the growth of wheat under NT [2,62–64]. Moreover, NT can decrease crop yield and may be due to decreased productive tillers and increased weed growth [2].

On the contrary, straw return often increases crop yields, as SR can enhance the input level of organic matter, thus improving soil nutrient conditions [61]. In this study, SR had no effect on crop yields (Table S8). The reason may be the fact that a long time is required, usually, for straw to be degraded and then change soil physical-chemical properties. Therefore, short-term straw return may have little effect on crop yields [2].

In this study, the interaction of tillage and straw return practices had no effect on crop yields, except in the 2014 wheat season (Table S8). Generally, the interaction of long-term tillage and straw return can increase crop yield, as long-term NT or SR can promote straw residue input into the soil, thus providing more nutrition for crops [3]. However, short-term NT and SR cannot significantly change crop yields [11].

4.5. Relationships between Soil Organic Carbon and Bacterial Community under Different Tillage and Straw Return Practices

The soil bacterial community largely contributes to aggregates stabilization and SOC sequestration [25,41]. In this study, the bacterial community (such as *Gp6, Gp10, Gp16,*

Gp18), Planctomycetes (including *Burkholderiales* and *Subdivision3_genera_incertae_sedis*) and Actinobacteria (such as *Sphingobacteriales*) were significantly affected by MBC and SOC in 1–2 mm macro-aggregates (Figure 3), which may be due to the fact that bacteria can metabolize organic matter and be stabilized as microbial residues in organic mineral complexes [53,54]. During the process of the decomposition of organic matter, a large quantity of broken organic carbon is released, contributing to the formation of soil macro-aggregates [64].

We found that the bacterial community, such as *Subdivision3_genera_incertae_sedis*, *Gp18*, and *Caulobacterales*, observably affected CH_4 emissions (Figure 3). It was reported that the bacterial community could affect CH_4 emissions through changing the availability of oxygen and organic carbon for methanogens [65–67]. The bacterial community can provide organic carbon for methanogens by degrading crop residue and thus enhance CH_4 emissions [67,68]. Besides, methanotrophic bacteria are important mediators for CH_4 consumption, which plays a significant role in controlling CH_4 emissions [69]. Therefore, the bacterial community may contribute to the shift in SOC content in macro-aggregates and CH_4 emissions, thus affecting the dynamics of SOC [17,41].

In this study, structural equation modeling analysis showed that SOC in 1–2 mm aggregates and CH_4 emissions jointly affected SOC sequestration under tillage and straw return systems (Figure 4), suggesting that SOC content was regulated by the balance between the SOC sequestration in 1–2 mm aggregates and the SOC losses induced by CH_4 emissions. Compared with CT, NT enhanced the formation of macro-aggregates (Table S2) and the accumulation of SOC in 1–2 mm aggregates (Table 1), while it reduced CH_4 emissions (Table 2), resulting in an increase in SOC content in the topsoil layer [32,41,42]. Compared with NS, SR promoted the losses of SOC induced by CH_4 emissions compared with NS (Table 2), and accelerated an increase in SOC sequestration in 1–2 mm aggregates (Table 1). Moreover, part of the straw could be sequestrated in soil by forming recalcitrant organic matter [7], which leads to increase in SOC content. Therefore, it can be concluded that both NT and SR increased SOC content, which may be the results of the balance between SOC accumulation in 1–2 mm aggregates and CH_4 emissions.

5. Conclusions

Both NT and SR increased SOC content in 0–5 cm topsoil layers in a rice-wheat cropping system. Our study indicates that NT and SR increased SOC content in 1–2 mm soil aggregates. NT resulted in lower CO_2 and CH_4 emissions compared with CT. However, SR increased CO_2 and CH_4 emissions compared to NS. Bacterial communities (such as Gp6, Gp10, Gp16 and Gp18), had significant relationships with SOC in 1–2 mm aggregates and MBC. Bacterial communities like *Subdivision3_genera_incertae_sedis*, *Gp18*, and *Caulobacterales* had *the most effect on* CH_4 emissions. Our study highlights that 4.4–15.3% of increase in SOC contents under NT and straw return were mainly due to the balance between SOC accumulation in 1–2 mm soil aggregates and CH_4 emissions in rice and wheat cropping systems.

Supplementary Materials: The following supporting information can be downloaded at: https://www.mdpi.com/article/10.3390/agriculture12101552/s1, Table S1: Changes of soil organic carbon contents (g kg^{-1}) under different tillage and straw return practices from (2012–2014); Table S2: Changes in aggregate composition (%) under different tillage and straw return practices in 0–5 cm soil layer (2012–2014); Table S3: Changes of dissolved organic carbon (g kg^{-1}) contents in 0–5 cm soil layer under different tillage and straw return practices during 2012–2014; Table S4: Changes of soil microbial biomass carbon (mg kg^{-1}) contents in 0–5 cm soil layer under different tillage and straw return practices during 2012–2014; Table S5: Soil bacterial and fungal PLFA under different tillage practices and residue returning methods in 0–5 cm soil layer (2013–2014); Table S6: The change in bacterial community at order level (relative abundance > 0.5%) under different tillage and straw return practice in 2013 rice season; Table S7: The change in bacterial community at order level (relative abundance > 0.5%) under different tillage and straw return practice in 2014 wheat season. Bacterial community; Table S8: The change in bacterial community at order level (relative abundance > 0.5%)

under different tillage and straw return practice in 2014 wheat season. Bacterial community; Table S9: The relationship between crop yield and soil properties.

Author Contributions: Data curation and writing-original draft preparation, L.G.; methodology, W.L. and Z.L.; review and editing, J.S. and J.L.; investigation, P.W.; formal analysis, X.T. and Y.L.; validation, C.L.; supervision, C.L. All authors have read and agreed to the published version of the manuscript.

Funding: This research is funded by the Key Research and Development Project of Hubei Province (2021BBA224), the Hainan Provincial Natural Science Foundation of China (320RC470, 2019RC111), and Priming Scientific Research Foundation of Hainan University (KYQD(ZR)1982).

Institutional Review Board Statement: Not applicable.

Data Availability Statement: Not applicable.

Conflicts of Interest: The authors declare no conflict of interest in this manuscript.

References

1. Fiorini, A.; Boselli, R.; Maris, S.C.; Santelli, S.; Perego, A.; Acutis, M.; Brenna, S.; Tabaglio, V. Soil type and cropping system as drivers of soil quality indicators response to no-till: A 7-year field study. *Appl. Soil Ecol.* **2020**, *155*, 103646. [CrossRef]
2. Guo, L.; Zhang, L.; Liu, L.; Sheng, F.; Cao, C.; Li, C. Effects of long-term no tillage and straw return on greenhouse gas emissions and crop yields from a rice-wheat system in central China. *Agric. Ecosyst. Environ.* **2021**, *322*, 107650. [CrossRef]
3. Berhane, M.; Xu, M.; Liang, Z.; Shi, J.; Wei, G.; Tian, X. Effects of long-term straw return on soil organic carbon storage and sequestration rate in North China upland crops: A meta-analysis. *Glob. Chang. Biol.* **2020**, *26*, 2686–2701. [CrossRef] [PubMed]
4. Zeraatpisheh, M.; Ayoubi, S.; Mirbagheri, Z.; Mosaddeghi, M.R.; Xu, M. Spatial prediction of soil aggregate stability and soil organic carbon in aggregate fractions using machine learning algorithms and environmental variables. *Geoderma Reg.* **2021**, *27*, e00440. [CrossRef]
5. Ajami, M.; Heidari, A.; Khormali, F.; Gorji, M.; Ayoubi, S. Environmental factors controlling soil organic carbon storage in loess soils of a subhumid region, northern Iran. *Geoderma* **2016**, *281*, 1–10. [CrossRef]
6. Christopher, S.F.; Lal, R.; Mishra, U. Regional Study of No-Till Effects on Carbon Sequestration in the Midwestern United States. *Soil Sci. Soc. Am. J.* **2009**, *73*, 207–216. [CrossRef]
7. Six, J.; Paustian, K. Aggregate-associated soil organic matter as an ecosystem property and a measurement tool. *Soil Biol. Biochem.* **2014**, *68*, A4–A9. [CrossRef]
8. Wu, Q.-X.; Du, B.; Jiang, S.-C.; Zhang, H.-W.; Zhu, J.-Q. Side Deep Fertilizing of Machine-Transplanted Rice to Guarantee Rice Yield in Conservation Tillage. *Agriculture* **2022**, *12*, 528. [CrossRef]
9. Akhtar, K.; Wang, W.; Ren, G.; Khan, A.; Feng, Y.; Yang, G. Changes in soil enzymes, soil properties, and maize crop productivity under wheat straw mulching in Guanzhong, China. *Soil Tillage Res.* **2018**, *182*, 94–102. [CrossRef]
10. Modak, K.; Biswas, D.R.; Ghosh, A.; Pramanik, P.; Das, T.K.; Das, S.; Kumar, S.; Krishnan, P.; Bhattacharyya, R. Zero tillage and residue retention impact on soil aggregation and carbon stabilization within aggregates in subtropical India. *Soil Tillage Res.* **2020**, *202*, 104649. [CrossRef]
11. Mathew, R.P.; Feng, Y.; Githinji, L.; Ankumah, R.; Balkcom, K.S. Impact of No-Tillage and Conventional Tillage Systems on Soil Microbial Communities. *Appl. Environ. Soil Sci.* **2012**, *2012*, 1–10. [CrossRef]
12. Jin, V.L.; Schmer, M.R.; Stewart, C.E.; Sindelar, A.J.; Varvel, G.E.; Wienhold, B. Long-term no-till and stover retention each decrease the global warming potential of irrigated continuous corn. *Glob. Chang. Biol.* **2017**, *23*, 2848–2862. [CrossRef] [PubMed]
13. Shakoor, A.; Shahbaz, M.; Farooq, T.H.; Sahar, N.E.; Shahzad, S.M.; Altaf, M.M.; Ashraf, M. A global meta-analysis of greenhouse gases emission and crop yield under no-tillage as compared to conventional tillage. *Sci. Total Environ.* **2021**, *750*, 142299. [CrossRef]
14. Li, G.; Bian, S.; Lu, X.; Yang, T.; Lei, G.H. Field test on extruding soil caused of PHC pipe pile driving by static pressure for improving soft foundation of widened embankment. *Rock Soil Mech.* **2013**, *34*, 1089–1096.
15. Kan, Z.-R.; Liu, Q.-Y.; Virk, A.L.; He, C.; Qi, J.-Y.; Dang, Y.P.; Zhao, X.; Zhang, H.-L. Effects of experiment duration on carbon mineralization and accumulation under no-till. *Soil Tillage Res.* **2021**, *209*, 104939. [CrossRef]
16. Bandyopadhyay, P.K.; Saha, S.; Mani, P.K.; Mandal, B. Effect of organic inputs on aggregate associated organic carbon concentration under long-term rice–wheat cropping system. *Geoderma* **2010**, *154*, 379–386. [CrossRef]
17. Liu, C.; Lu, M.; Cui, J.; Li, B.; Fang, C. Effects of Straw Carbon Input on Carbon Dynamics in Agricultural Soils: A Meta-Analysis. *Glob. Chang. Biol.* **2014**, *20*, 1366–1381. [CrossRef] [PubMed]
18. Hidalgo, C.; Merino, A.; Osorio-Hernández, V.; Etchevers, J.D.; Figueroa, B.; Limon-Ortega, A.; Aguirre, E. Physical and chemical processes determining soil organic matter dynamics in a managed vertisol in a tropical dryland area. *Soil Tillage Res.* **2019**, *194*, 104348. [CrossRef]
19. Regina, K.; Alakukku, L. Greenhouse gas fluxes in varying soils types under conventional and no-tillage practices. *Soil Tillage Res.* **2010**, *109*, 144–152. [CrossRef]

20. Ruan, L.; Philip Robertson, G. Initial nitrous oxide, carbon dioxide, and methane costs of converting conservation reserve program grassland to row crops under no-till vs. conventional tillage. *Glob. Chang. Biol.* **2013**, *19*, 2478–2489. [CrossRef]
21. Guo, L.-J.; Lin, S.; Liu, T.-Q.; Cao, C.-G.; Li, C.-F. Effects of Conservation Tillage on Topsoil Microbial Metabolic Characteristics and Organic Carbon within Aggregates under a Rice (*Oryza sativa* L.)–Wheat (*Triticum aestivum* L.) Cropping System in Central China. *PLoS ONE* **2016**, *11*, e0146145. [CrossRef]
22. Zhao, X.; Virk, A.L.; Ma, S.-T.; Kan, Z.-R.; Qi, J.-Y.; Pu, C.; Yang, X.-G.; Zhang, H.-L. Dynamics in soil organic carbon of wheat-maize dominant cropping system in the North China Plain under tillage and residue management. *J. Environ. Manag.* **2020**, *265*, 110549. [CrossRef] [PubMed]
23. Ussiri, D.A.; Lal, R.; Jarecki, M.K. Nitrous oxide and methane emissions from long-term tillage under a continuous corn cropping system in Ohio. *Soil Tillage Res.* **2009**, *104*, 247–255. [CrossRef]
24. Pareja-Sánchez, E.; Cantero-Martínez, C.; Álvaro-Fuentes, J.; Plaza-Bonilla, D. Tillage and nitrogen fertilization in irrigated maize: Key practices to reduce soil CO_2 and CH_4 emissions. *Soil Tillage Res.* **2019**, *191*, 29–36. [CrossRef]
25. Duan, M.; Long, Y.; Fan, H.; Ma, L.; Han, S.; Li, S.; Wei, B.; Wang, L. Fenlong-Ridging Promotes Microbial Activity in Sugarcane: A Soil and Root Metabarcoding Survey. *Agriculture* **2022**, *12*, 244. [CrossRef]
26. Gartzia-Bengoetxea, N.; González-Arias, A.; Merino, A.; de Arano, I.M. Soil organic matter in soil physical fractions in adjacent semi-natural and cultivated stands in temperate Atlantic forests. *Soil Biol. Biochem.* **2009**, *41*, 1674–1683. [CrossRef]
27. Jiang, P.; Xu, Q.; Xu, Z.; Cao, Z. Seasonal changes in soil labile organic carbon pools within a Phyllostachys praecox stand under high rate fertilization and winter mulch in subtropical China. *For. Ecol. Manag.* **2006**, *236*, 30–36. [CrossRef]
28. Frostegård, Å.; Tunlid, A.; Bååth, E. Phospholipid Fatty Acid Composition, Biomass, and Activity of Microbial Communities from Two Soil Types Experimentally Exposed to Different Heavy Metals. *Appl. Environ. Microbiol.* **1993**, *59*, 3605–3617. [CrossRef]
29. Neue, H.U.; Sass, R.L. *Global Measurement Standardization of Methane Emissions from Irrigated Rice Cultivation: A Report of the Rice Cultivation and Trace Gas Exchange Activity (RICE) of the International Global Atmospheric Chemistry (IGAC) Project*; Massachusetts Institute of Technology: Cambridge, MA, USA, 1994; pp. 1–6. Available online: http://www.worldcat.org/title/197230180?__cf_chl_tk=DkYxejo5hvH01DTFE3XL_W_Aq7RcOSe3nCs2nFFzE9E-1664118918-0-gaNycGzNBqU (accessed on 10 August 2022).
30. Buendia, L.; Neue, H.; Wassmann, R.; Lantin, R.; Javellana, A.; Arah, J.; Wang, Z.; Wanfang, L.; Makarim, A.; Corton, T.; et al. An efficient sampling strategy for estimating methane emission from rice field. *Chemosphere* **1998**, *36*, 395–407. [CrossRef]
31. Li, C.; Zhang, Z.; Guo, L.; Cai, M.; Cao, C. Emissions of CH_4 and CO_2 from double rice cropping systems under varying tillage and seeding methods. *Atmos. Environ.* **2013**, *80*, 438–444. [CrossRef]
32. Zhang, Z.; Guo, L.; Liu, T.; Li, C.; Cao, C. Effects of tillage practices and straw returning methods on greenhouse gas emissions and net ecosystem economic budget in rice–wheat cropping systems in central China. *Atmos. Environ.* **2015**, *122*, 636–644. [CrossRef]
33. Zhao, J.; Carmody, L.A.; Kalikin, L.M.; Li, J.; Petrosino, J.F.; Schloss, P.; Young, V.; Lipuma, J.J. Impact of Enhanced Staphylococcus DNA Extraction on Microbial Community Measures in Cystic Fibrosis Sputum. *PLoS ONE* **2012**, *7*, e33127. [CrossRef] [PubMed]
34. Caporaso, J.; Kuczynski, J.; Stombaugh, J.; Bittinger, K.; Bushman, F.; Costello, E.K.; Fierer, N.; Peña, A.G.; Goodrich, J.K.; Gordon, J.I.; et al. QIIME allows integration and analysis of high-7 throughput community sequencing data. *Nat. Methods* **2010**, *7*, 335–336. [CrossRef] [PubMed]
35. Fierer, N.; Hamady, M.; Lauber, C.L.; Knight, R. The influence of sex, handedness, and washing on the diversity of hand surface bacteria. *Proc. Natl. Acad. Sci. USA* **2008**, *105*, 17994–17999. [CrossRef] [PubMed]
36. Oksanen, J.; Kindt, R.; Legendre, P.; O'Hara, B.; Stevens, M.H.H.; Oksanen, M.J.; Suggests, M.A.S.S. The vegan package. *Community Ecol. Package* **2007**, *10*, 719.
37. Grace, J.B.; Michael Anderson, T.; Smith, M.D.; Seabloom, E.; Andelman, S.J.; Meche, G.; Weiher, E.; Allain, L.K.; Jutila, H.; Sankaran, M.; et al. Does species diversity limit productivity in natural grassland communities? *Ecol. Lett.* **2007**, *10*, 680–689. [CrossRef]
38. Borie, F.; Rubio, R.; Rouanet, J.; Morales, A.; Borie, G.; Rojas, C. Effects of tillage systems on soil characteristics, glomalin and mycorrhizal propagules in a Chilean Ultisol. *Soil Tillage Res.* **2006**, *88*, 253–261. [CrossRef]
39. Biswakarma, N.; Pooniya, V.; Zhiipao, R.; Kumar, D.; Verma, A.; Shivay, Y.; Lama, A.; Choudhary, A.; Meena, M.; Bana, R.J.A.; et al. Five years integrated crop management in direct seeded rice–zero till wheat rotation of north-western India: Effects on soil carbon dynamics, crop yields, water productivity and economic profitability. *Agric. Ecosyst. Environ.* **2021**, *318*, 107492. [CrossRef]
40. Balota, E.L.; Colozzi Filho, A.; Andrade, D.S.; Dick, R.P. Long-term tillage and crop rotation effects on microbial biomass and C and N mineralization in a Brazilian Oxisol. *Soil Tillage Res.* **2004**, *77*, 137–145. [CrossRef]
41. Trivedi, P.; Delgado-Baquerizo, M.; Jeffries, T.C.; Trivedi, C.; Anderson, I.C.; Lai, K.; McNee, M.; Flower, K.; Pal Singh, B.; Minkey, D.; et al. Soil aggregation and associated microbial communities modify the impact of agricultural management on carbon content. *Environ. Microbiol.* **2017**, *19*, 3070–3086. [CrossRef]
42. Jia, S.; Liang, A.; Zhang, S.; Chen, X.; McLaughlin, N.B.; Sun, B.; Zhang, X.; Wu, D. Effect of tillage system on soil CO2 flux, soil microbial community and maize (*Zea mays* L.) yield. *Geoderma* **2021**, *384*, 114813. [CrossRef]
43. Kan, Z.-R.; He, C.; Liu, Q.-Y.; Liu, B.-Y.; Virk, A.L.; Qi, J.-Y.; Zhao, X.; Zhang, H.-L. Carbon mineralization and its temperature sensitivity under no-till and straw returning in a wheat-maize cropping system. *Geoderma* **2020**, *377*, 114610. [CrossRef]
44. Kan, Z.-R.; Ma, S.-T.; Liu, Q.-Y.; Liu, B.-Y.; Virk, A.L.; Qi, J.-Y.; Zhao, X.; Lal, R.; Zhang, H.-L. Carbon sequestration and mineralization in soil aggregates under long-term conservation tillage in the North China Plain. *CATENA* **2020**, *188*, 104428. [CrossRef]

45. Emde, D.; Hannam, K.D.; Most, I.; Nelson, L.M.; Jones, M.D. Soil organic carbon in irrigated agricultural systems: A meta-analysis. *Glob. Chang. Biol.* **2021**, *27*, 3898–3910. [CrossRef] [PubMed]
46. Tan, Y.; Wu, D.; Bol, R.; Wu, W.; Meng, F. Conservation farming practices in winter wheat–summer maize cropping reduce GHG emissions and maintain high yields. *Agric. Ecosyst. Environ.* **2019**, *272*, 266–275. [CrossRef]
47. Ma, J.; Xu, H.; Yagi, K.; Cai, Z. Methane emission from paddy soils as affected by wheat straw returning mode. *Plant Soil* **2008**, *313*, 167–174. [CrossRef]
48. Akter, M.; Deroo, H.; Kamal, A.M.; Kader, M.A.; Verhoeven, E.; Decock, C.; Boeckx, P.; Sleutel, S. Impact of irrigation management on paddy soil N supply and depth distribution of abiotic drivers. *Agric. Ecosyst. Environ.* **2018**, *261*, 12–24. [CrossRef]
49. Lin, Q.; Wang, S.; Li, Y.; Riaz, L.; Yu, F.; Yang, Q.; Han, S.; Ma, J. Effects and mechanisms of land-types conversion on greenhouse gas emissions in the Yellow River floodplain wetland. *Sci. Total Environ.* **2022**, *813*, 152406. [CrossRef]
50. Acosta-Martínez, V.; Dowd, S.; Sun, Y.; Allen, V. Tag-encoded pyrosequencing analysis of bacterial diversity in a single soil type as affected by management and land use. *Soil Biol. Biochem.* **2008**, *40*, 2762–2770. [CrossRef]
51. Ondreičková, K.; Piliarová, M.; Bušo, R.; Hašana, R.; Schreiber, L.; Gubiš, J.; Kraic, J. The Structure and Diversity of Bacterial Communities in Differently Managed Soils Studied by Molecular Fingerprinting Methods. *Sustainability* **2018**, *10*, 1095. [CrossRef]
52. Lewin, G.R.; Carlos, C.; Chevrette, M.G.; Horn, H.A.; McDonald, B.R.; Stankey, R.J.; Fox, B.G.; Currie, C.R. Evolution and ecology of Actinobacteria and their bioenergy applications. *Annu. Rev. Microbiol.* **2016**, *70*, 235. [CrossRef]
53. Francioli, D.; Schulz, E.; Lentendu, G.; Wubet, T.; Buscot, F.; Reitz, T. Mineral vs. Organic Amendments: Microbial Community Structure, Activity and Abundance of Agriculturally Relevant Microbes Are Driven by Long-Term Fertilization Strategies. *Front. Microbiol.* **2016**, *7*, 1446. [CrossRef] [PubMed]
54. Zhu, J.; Peng, H.; Ji, X.; Li, C.; Li, S. Effects of reduced inorganic fertilization and rice straw recovery on soil enzyme activities and bacterial community in double-rice paddy soils. *Eur. J. Soil Biol.* **2019**, *94*, 103116. [CrossRef]
55. Luo, Y.; Iqbal, A.; He, L.; Zhao, Q.; Wei, S.; Ali, I.; Ullah, S.; Yan, B.; Jiang, L. Long-Term No-Tillage and Straw Retention Management Enhances Soil Bacterial Community Diversity and Soil Properties in Southern China. *Agronomy* **2020**, *10*, 1233. [CrossRef]
56. Miura, T.; Niswati, A.; Swibawa, I.G.; Haryani, S.; Gunito, H.; Arai, M.; Yamada, K.; Shimano, S.; Kaneko, N.; Fujie, K. Shifts in the composition and potential functions of soil microbial communities responding to a no-tillage practice and bagasse mulching on a sugarcane plantation. *Biol. Fertil. Soils* **2016**, *52*, 307–322. [CrossRef]
57. Yu, C.; Hu, X.; Deng, W.; Li, Y.; Han, G.; Xiong, C. Response of Bacteria Community to Long-Term Inorganic Nitrogen Application in Mulberry Field Soil. *PLoS ONE* **2016**, *11*, e0168152. [CrossRef]
58. Dong, W.; Liu, E.; Yan, C.; Tian, J.; Zhang, H.; Zhang, Y. Impact of no tillage vs. conventional tillage on the soil bacterial community structure in a winter wheat cropping succession in northern China. *Eur. J. Soil Biol.* **2017**, *80*, 35–42. [CrossRef]
59. Sun, M.; Chen, B.; Wang, H.; Wang, N.; Ma, T.; Cui, Y.; Luan, T.; Chun, S.; Liu, C.; Wang, L. Microbial Interactions and Roles in Soil Fertility in Seasonal Freeze-Thaw Periods under Different Straw Returning Strategies. *Agriculture* **2021**, *11*, 779. [CrossRef]
60. Yamada, T.; Imachi, H.; Ohashi, A.; Harada, H.; Hanada, S.; Kamagata, Y.; Sekiguchi, Y. Bellilinea caldifistulae gen. nov., sp. nov. and Longilinea arvoryzae gen. nov., sp. nov., strictly anaerobic, filamentous bacteria of the phylum Chloroflexi isolated from methanogenic propionate-degrading consortia. *Int. J. Syst. Evol. Microbiol.* **2007**, *57*, 2299–2306. [CrossRef]
61. Xia, L.; Lam, S.K.; Wolf, B.; Kiese, R.; Chen, D.; Butterbach-Bahl, K. Trade-offs between soil carbon sequestration and reactive nitrogen losses under straw return in global agroecosystems. *Glob. Chang. Biol.* **2018**, *24*, 5919–5932. [CrossRef]
62. Arunrat, N.; Sansupa, C.; Kongsurakan, P.; Sereenonchai, S.; Hatano, R. Soil Microbial Diversity and Community Composition in Rice–Fish Co-Culture and Rice Monoculture Farming System. *Biology* **2022**, *11*, 1242. [CrossRef]
63. Rezgui, C.; Trinsoutrot-Gattin, I.; Benoit, M.; Laval, K.; Wassila, R.A. Linking changes in the soil microbial community to C and N dynamics during crop residue decomposition. *J. Integr. Agric.* **2021**, *20*, 3039–3059. [CrossRef]
64. Guo, L.-J.; Zhang, Z.-S.; Wang, D.-D.; Li, C.-F.; Cao, C.-G. Effects of short-term conservation management practices on soil organic carbon fractions and microbial community composition under a rice-wheat rotation system. *Biol. Fertil. Soils* **2015**, *51*, 65–75. [CrossRef]
65. Dubey, S.K.; Singh, A.; Singh, R.; Upadhyay, S. Changes in methanogenic population size and CH_4 production potential in response to crop phenology in tropical rice field. *Soil Biol. Biochem.* **2013**, *57*, 972–978. [CrossRef]
66. Fan, X.; Yu, H.; Wu, Q.; Ma, J.; Xu, H.; Yang, J.; Zhuang, Y. Effects of fertilization on microbial abundance and emissions of greenhouse gases (CH4 and N2O) in rice paddy fields. *Ecol. Evol.* **2016**, *6*, 1054–1063. [CrossRef]
67. Glissmann, K.; Weber, S.; Conrad, R. Localization of processes involved in methanogenic degradation of rice straw in anoxic paddy soil. *Environ. Microbiol.* **2001**, *3*, 502–511. [CrossRef]
68. Ortiz-Cornejo, N.L.; Romero-Salas, E.A.; Navarro-Noya, Y.E.; González-Zúñiga, J.C.; Ramirez-Villanueva, D.A.; Vásquez-Murrieta, M.S.; Verhulst, N.; Govaerts, B.; Dendooven, L.; Luna-Guido, M. Incorporation of bean plant residue in soil with different agricultural practices and its effect on the soil bacteria. *Appl. Soil Ecol.* **2017**, *119*, 417–427. [CrossRef]
69. Conrad, R. Microbial ecology of methanogens and methanotrophs. *Adv. Agron.* **2007**, *96*, 1–63.

agriculture

MDPI

Article

Optimizing Nitrogen Application for Chinese Ratoon Rice Based on Yield and Reactive Nitrogen Loss

Ren Hu [1,2], Zijuan Ding [1,2], Tingyu Li [3], Dingyue Zhang [4], Yingbing Tian [1,2], Yuxian Cao [5] and Jun Hou [1,2,*]

1 College of Agriculture, Yangtze University, Jingzhou 434025, China; 202072796@yangtzeu.edu.cn (R.H.); 201971382@yangtzeu.edu.cn (Z.D.); tyb918@yangtzeu.edu.cn (Y.T.)
2 Engineering Research Center of Ecology and Agricultural Use of Wet Land, Ministry of Education, Jingzhou 434025, China
3 College of Tropical Crops, Hainan University, Haikou 570228, China; lty@hainanu.edu.cn
4 Jiangsu Institute of Agricultural Sciences in Coastal Region, Yancheng 224002, China; 201872443@yangtzeu.edu.cn
5 College of Life Science, Yangtze University, Jingzhou 434025, China; caoyx2018@yangtzeu.edu.cn
* Correspondence: houjun@yangtzeu.edu.cn; Tel.: +86-18310707325

Abstract: Ratoon rice (RR) has been regarded as a labor-saving and beneficial production system. Nitrogen (N) surplus and reactive N losses (Nr losses) are effective environmental indicators used to evaluate the performance of N management. Few studies have assessed N surplus and Nr losses for Chinese RR. In this study, Chinese RR planting areas were divided into South China (SC), the southern part of East China (SEC), Central China (CC), the northern part of East China (NEC), and Southwest China (SW). N surplus and Nr losses were also calculated based on 782 studies using a quadratic model under optimized N management for the highest yield (OPT-yield), the highest N-use efficiency (NUE) (OPT-NUE), and the highest grain N uptake (OPT-N uptake). The RR yields in the five regions ranged from 9.98 to 13.59 t ha^{-1}. The high-yield record was observed in SEC, while the low-yield record was observed in NEC. The highest and the lowest Nr losses were found in NEC and SC, respectively. N surplus was reduced, while the yield was maintained in SEC, CC, NEC, and SW under OPT-yield and OPT-N uptake, and N surplus and Nr losses were reduced in the five regions when targeting the highest NUE. Farmers should be encouraged to plant RR in SEC and CC. RR was also a good choice when N management measures were conducted in three other regions. To achieve a win–win situation for both yield and the environment, OPT-yield could serve to improve the N management of current conventional practices.

Keywords: ratoon rice; nitrogen balance; reactive nitrogen losses; nitrogen surplus; nitrogen-use efficiency

Citation: Hu, R.; Ding, Z.; Li, T.; Zhang, D.; Tian, Y.; Cao, Y.; Hou, J. Optimizing Nitrogen Application for Chinese Ratoon Rice Based on Yield and Reactive Nitrogen Loss. *Agriculture* **2022**, *12*, 1064. https://doi.org/10.3390/agriculture12071064

Academic Editor: Antonio Carlo Barbera

Received: 15 June 2022
Accepted: 13 July 2022
Published: 20 July 2022

1. Introduction

With the world population increasing, rice production needs to reach 519.50 million tonnes in order to meet the world population's demand for rice in 2022 [1], and China is not exempt from this. Rice is a staple food for more than 65% of Chinese people, and it is a subsistence crop for rice farmers and consumers in Chinese rural areas lacking resources. About 20% more rice needs to be produced by 2030 to meet domestic demands if rice consumption per capita is to be kept at the present level in China [2]. Therefore, it is imperative to increase rice yield per hectare in the limited planting area. Ratoon rice (RR) is a kind of rice that can be harvested twice in one crop; dormant sprouts that survive on rice stubble germinate into ears and can then be harvested for another season (ratoon crop) after the harvest of the first crop (main crop). Two harvests and a higher multiple cropping index can be realized using this rice farming system [3]. Grain yield in the RR system is higher than that in middle-season rice, and the net energy ratio and the economic profit in the RR system are higher than those in double-season rice [4].

Nitrogen (N) is the main nutrient used to boost the growth and development of crops. The use of N fertilizer is necessary to obtain high crop yields [5]. Farmers usually input ample N fertilizer to ensure higher grain yields. The data released by the National Bureau of Statics in 2021 showed that the consumption of N fertilizer applied to crops in China was 51.91 Tg (1 Tg = 10^{12} g) [6], accounting for about 45.66% of the world's total N fertilizer consumption (113.70 Tg, International Fertilizer Industry) [7]. However, too much N does not increase the yield [8] but rather increases serious environmental pollution, including CH_4 and N_2O emissions in the RR system [9]. Therefore, increasing or maintaining the yield with a low input of N fertilizers has become a critical consideration for ensuring sustainability in RR production. To minimize environmental pollution, China achieved zero growth in using chemical fertilizers by 2020 [10]. For this purpose, N management practices that could sustain high yield and minimize Nr losses needed to be established. Many optimized N management (OPT) strategies can increase the yield and reduce Nr losses in RR systems, for example, special fertilizer for bud promotion [11] and optimal N application using a quadratic equation on yield and N application [12]. Besides N management, water-saving irrigation, such as alternative wetting and drying irrigation, has been found to be a promising option to mitigate environmental Nr losses while reducing irrigation water input in RR fields [9]. However, few studies have been conducted to assess environmental effects under different N applications in the RR cropping system.

Of the many indicators used to assess N management, N-use efficiency (NUE) and N surplus may be helpful in policymaking. The efficiency of all the N inputs transferring to harvested crop N is defined as NUE, and it is consistent with the definition used by Zhang et al. [13]. The difference between N input and harvested N output is defined as N surplus [14,15]. This helps provide guidelines for improvements in nutrient management within a specified boundary [16]. N surplus has been widely used as an indicator for N management by various countries and organizations [13], for example, the mineral accounting system in the Netherlands [17] and intensive farming in Denmark [18]. Several case studies have considered the effects of N surplus analyses in different systems, for example, understanding seasonal N dynamics in the maize–wheat double-cropping system [19] and determining the appropriate N rate and topdressing N ratio in rice–wheat rotation [20]. These studies have contributed to the efficient agricultural N management and helped in reducing Nr losses while maintaining or improving crop yields.

However, few studies have been conducted to assess the rice yield and the environmental load in different Chinese RR planting areas after optimal N application [8,13]. This study collected data from 782 studies on the RR system covering 16 provinces in China to quantify N surplus and Nr losses in the RR system. The aim was (1) to answer which region should be encouraged to develop RR by comparing yields, Nr losses, and NUE in five Chinese RR regions, and (2) to establish a model to simulate the N surplus and Nr losses under OPT for the highest yield, the highest NUE, and the highest grain N uptake.

2. Materials and Methods

2.1. Main Cropping Regions

According to the requirements of temperature, light, and water for RR growth, the critical meteorological indexes of suitable and unsuitable planting areas of RR were determined using the principal component analysis [21]. Then, the suitable RR planting zones in China were divided into 5 climatic ecological zones and 13 regions (Figure 1 and Table S1 in Supplementary Information). They were named South China (SC), the southern part of East China (SEC), Central China (CC), the northern part of East China (NEC), Southwest China (SW), and Others (Figure 1). For "Others", no data were available; therefore, these areas, which included Beijing, Tibet, Qinghai, Hong Kong, and Macao, were not included in this study [21].

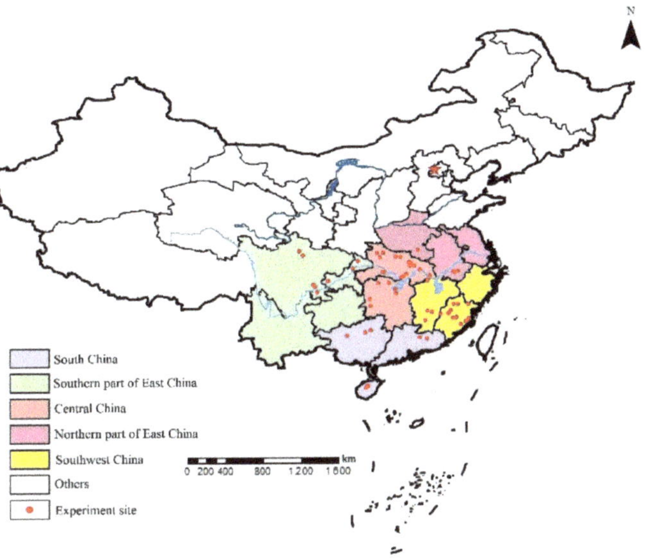

Figure 1. Experiment sites and main regions for RR cultivation in China. ★ indicates Beijing.

2.2. Data Source

We searched for peer-reviewed publications published between 2005 and 2020 on RR via Science Direct, Springer Journals, the Web of Science, and the China National Knowledge Infrastructure using the search terms ratoon rice, nitrogen fertilizers, and yield. All studies that met the following criteria were included: (1) the crops in all studies were RR; (2) the start and end years of the experiment were available; (3) the amount of N fertilizer applied to the main crop and the ratoon crop of RR in the experiment was stated; (4) the amount of N absorbed and taken away by crops or the crop yield of the ratoon crop and the main crop was given; and (5) the detailed location of the experiment sites was given. A total of 782 studies fit the criteria and were included in this study, comprising over 72 experiments conducted in 16 provinces throughout China. If the same data appeared in multiple publications, they were entered into the study only once.

2.3. Data Calculation

2.3.1. Calculation of N-Use Efficiency and N Stored in Soil

The main external N inputs to the RR system in China included fertilizer N, atmospheric N deposition, biological N fixation, seed N, and N from irrigation water (irrigation N). The internal N cycle of the soil and a small amount of N input were not taken into account (straw returning to the field, soil organic matter humification, and mineralization). At the same time, NUE and N surplus were calculated. Since 2000, under the strict prohibition of the government and economic incentives, it has been assumed that all the straw returns to the soil [13,22,23]. Irrigation N has been considered for N surplus calculation in some studies, e.g., in greenhouse vegetables in the North China Plain (water-deficient area) [24]. However, RR is usually planted in an area with abundant rain (Figure 1 [21]), where both the amount of irrigation water and its N content are minor. Thus, irrigation N was not considered in this study.

The N partial factor productivity (PFPN) and NUE were calculated using the following equation [13,25]:

$$\text{PFPN} = \frac{\text{Yield}}{\text{N}_{\text{fer}}}$$

$$NUE = \frac{N_{har}}{N_{fer} + N_{dep} + N_{fix}}$$

where N_{har} is the grain N uptake, kg ha^{-1}; N uptake by grain was calculated by multiplying yield (kg ha^{-1}) by the N content of the grain (%). N_{fer}, N_{dep}, and N_{fix} represent the N input from fertilization, atmospheric deposition, and non-symbiotic N fixation, respectively (kg ha^{-1}). Details about the calculation of N_{har}, N_{dep}, and N_{fix} can be found in Tables S2–S4.

$N_{\Delta soil}$ was calculated as follows [26,27]:

$$N_{\Delta soil} = N_{fer} + N_{dep} + N_{fix} + N_{seedling} - N_{har} - N_{nit} - N_{vol} - N_{lea} - N_{run}$$

where $N_{\Delta soil}$ is the N stored in soil kg ha^{-1}. N_{fer}, N_{dep}, N_{fix}, and $N_{seeding}$ represent the N input from fertilization, atmospheric deposition, biological fixation, and seedlings, respectively (kg ha^{-1}). N_{har} is the N in harvested grain, kg ha^{-1}. N_{vol}, N_{lea}, and N_{run} represent the N output from NH_3 volatilization, N leaching, and N runoff, respectively (kg ha^{-1}). N_{nit} represents the N output from denitrification losses, which was estimated to be 21.6% of N fertilizer based on the mean value calculated from the published literature [28–31]. Details about the calculation of $N_{seeding}$, N_{vol}, N_{lea}, and N_{run} can be found in Tables S3 and S4.

2.3.2. Calculation of Nr Losses and N Surplus

Nr losses include NH_3 volatilization, N_2O, nitrate leaching, and runoff, but N_2 is not harmful to the environment, and, hence, it was not counted in the Nr losses [13] where crop seeds absorb only a small proportion of the total nutrient input [32]. Nr losses and N surplus were calculated as follows:

$$N_r \text{ losses} = N_{nit} + N_{vol} + N_{lea} + N_{run}$$

$$N_{sur} = N_{fer} + N_{dep} + N_{fix} - N_{har}$$

where N_{nit}, N_{vol}, N_{lea}, and N_{run} represent the N input from nitrification or denitrification loss, NH_3 volatilization, N leaching, and N runoff, respectively (kg ha^{-1}). N_{fer}, N_{dep}, N_{fix}, and N_{har} represent the N input from fertilization, atmospheric deposition, non-symbiotic N fixation, and grain N uptake, respectively (kg ha^{-1}), and grain N uptake was calculated by multiplying the dry matter content (kg ha^{-1}) by the N content of the grain (%). Details about the calculation of N uptake can be found in Table S2.

2.3.3. Optimized N Based on the Highest Yield, Highest NUE, and Grain N Uptake

Grain N uptake was calculated by multiplying the dry matter content (kg ha^{-1}) by the N content of the grain (%) [33]. Table S2 presents the details about the calculation of N uptake. The effect of N application on crop yield is divided into two stages: one is yield increase and the other is yield stabilization or even reduction with the increased N fertilization rate [34]. The diminishing marginal effect of N on yield can be observed empirically, which is mainly because of the cumulative effect of various physiological processes during plant growth [35]. Thus, a quadratic model was used to calculate the optimal N applications [36]. The optimal N application was calculated when the inflection point of the curve was met, following which the maximum yield was obtained. This method was used to calculate the optimal N application under the highest NUE (Table S6) and grain N uptake (Table S7) [33]. Therefore, the optimal N applications for the highest yield, the highest NUE, and the highest grain N uptake were defined as OPT-yield, OPT-NUE, and OPT-N uptake in this study, respectively. The un-optimized N management was defined as Un-OPT. To make the N application more in line with farmers' field management, outliers that exceeded three times of the average value were eliminated.

2.3.4. Data Analysis

Excel 2010 (Microsoft., Redmond, WA, USA) was used for data processing. SPSS 26.0 (IBM Corp., Chicago, IL, USA) was used for one-way ANOVA, and Arc Gis 10.0 (ESRI Inc., Redlands, CA, USA) and Excel 2010 was used for drawing.

3. Results

3.1. Comparison of N Application, Yields, and NUE in Different Regions

The results show that N application to the main crop was the highest in SEC (Figure 2). N application to the ratoon crop was the highest in NEC, indicating that a large amount of N fertilizer was used to obtain a high yield. The yield of the main crop in SEC was the highest (8.95 t ha^{-1}), and the yields of the ratoon crop in SEC and CC (4.66 and 4.60 t ha^{-1}) were higher than those in other regions ($p < 0.05$) (Figure 2). Total yield was defined as the sum of the yields of the main crop and the ratoon crop. The total yields in SEC and CC were significantly higher than those in SC, NEC, and SW, and the total yield in NEC was the lowest. The PFPN was the lowest in NEC, and the NUE in CC was 60%, which was 11−122% higher than that in the other four regions (Figure 2).

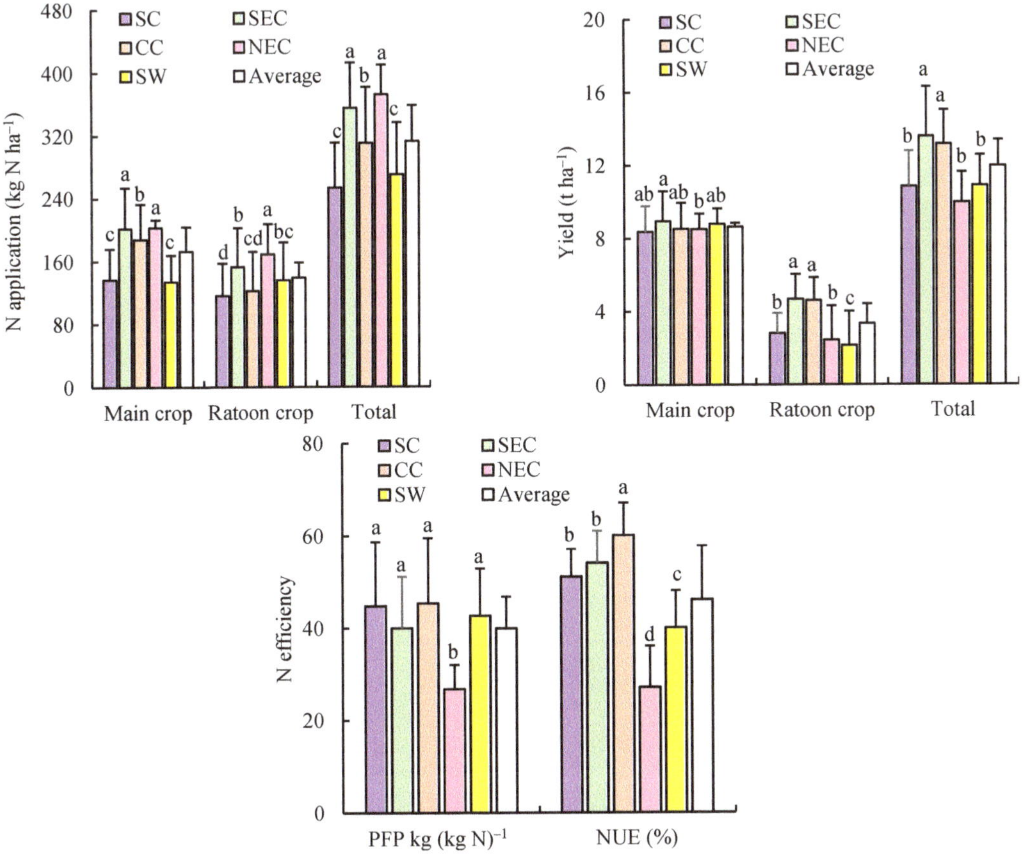

Figure 2. Estimation of yield, N application, and N efficiency of RR in different regions Values are means ± SD of three replicates. Different lowercase letters on bars indicate significant differences at $p < 0.05$.

3.2. N Stored in the Soil in Different Regions

N fertilizer is the main source of N input. The highest amount of N fertilizer was used in NEC, followed by SEC (Table 1). CC and NEC had the largest N deposition. Grain N in CC accounted for 60.81% of the total N output, while that in NEC accounted for only 43.80%. Besides grain N, NEC had the largest denitrification N loss and ammonia volatilization. SEC had the largest N output due to the largest crop uptake (57.21%). The NEC region showed the highest Nr losses, and the SC region showed the lowest Nr losses. Apparent N stored in the soil ($N_{\Delta soil}$) of SE, SEC, and CC was 20, 24, and 12 kg N ha^{-1}, respectively, while that of NEC and SW was 131 and 82 kg N ha^{-1}, respectively. The results indicate that the N input was close to the N output in planting areas, such as SE, SEC, and CC.

Table 1. Estimation of the seasonal N stored ($N_{\Delta soil}$) in soil in RR system in five areas in China.

Items	Sources (kg N ha^{-1})	SC (n = 44)	SEC (n = 147)	CC (n = 382)	NEC (n = 33)	SW (n = 157)
Input	N fertilizer	253 ± 57 ^{c}	355 ± 78 ^{a}	310 ± 72 ^{b}	371 ± 42 ^{a}	270 ± 66 ^{c}
	Deposition	41	43	47	47	38
	Biological fixation	25	25	25	25	25
	Seedling	3	3	3	3	3
	Sum	322	426	382	446	336
Output	Grain N uptake	189 ± 33 ^{c}	230 ± 45 ^{a}	225 ± 45 ^{a}	138 ± 17 ^{d}	132 ± 14 ^{b}
	Denitrification loss	55 ± 12 ^{c}	77 ± 17 ^{a}	67 ± 18 ^{b}	80 ± 8 ^{a}	58 ± 14 ^{c}
	NH_3 volatilization	48 ± 10 ^{c}	73 ± 21 ^{a}	61 ±17 ^{b}	74 ± 8 ^{a}	51 ± 11 ^{c}
	N leaching	5 ± 1 ^{c}	8 ± 4 ^{a}	6 ± 2 ^{b}	8 ± 1 ^{a}	6 ± 1 ^{c}
	N runoff	6 ± 1 ^{c}	15 ± 9 ^{a}	11 ± 4 ^{b}	15 ± 2 ^{a}	7 ± 2 ^{c}
	Sum	302	402	370	315	254
Nr losses		113	173	145	177	122
$N_{\Delta soil}$		20	24	12	131	82

Note: Deposition indicates atmospheric N deposition, and biological fixation indicates biological N fixation. Different letters indicate significant difference among treatments in the same site ($p < 0.05$); "±" followed by the standard deviation.

3.3. Correlation between N Application and Yield, NUE, and Grain N Uptake

As shown in Figure 3, the results indicate that the yield was significantly related to N application. When other conditions were constant, the yield first increased and then gradually decreased with the increased N application, with a turning point (the optimal N application). The equation $Y = -9 \times 10^{-5}x^2 + 0.0574x + 3.3637$ ($p < 0.01$) can express the relationship between yield and N application. Therefore, the RR yield attained the highest point (12.87 t ha^{-1}) when the N application rate was 319 kg ha^{-1}. Below a specific N application rate, NUE decreased when the N application rate exceeded 257 kg N ha^{-1} based on the relationship equation between NUE and N application ($Y = -0.0006x^2 + 0.3087x + 19.677$, $p < 0.01$). Therefore, NUE reached the highest point (59%) when the N application rate was 257 kg ha^{-1}. Grain N uptake generally increased with N application (Figure 2). Generally, grain N uptake showed a significant correlation with N application, which could be described using a quadratic equation ($Y = -0.0015x^2 + 0.9623x + 52.616$, $p < 0.01$). Moreover, the yield, NUE, and grain N uptake demonstrated a close relationship with N application in five typical Chinese RR regions (Tables S5–S7).

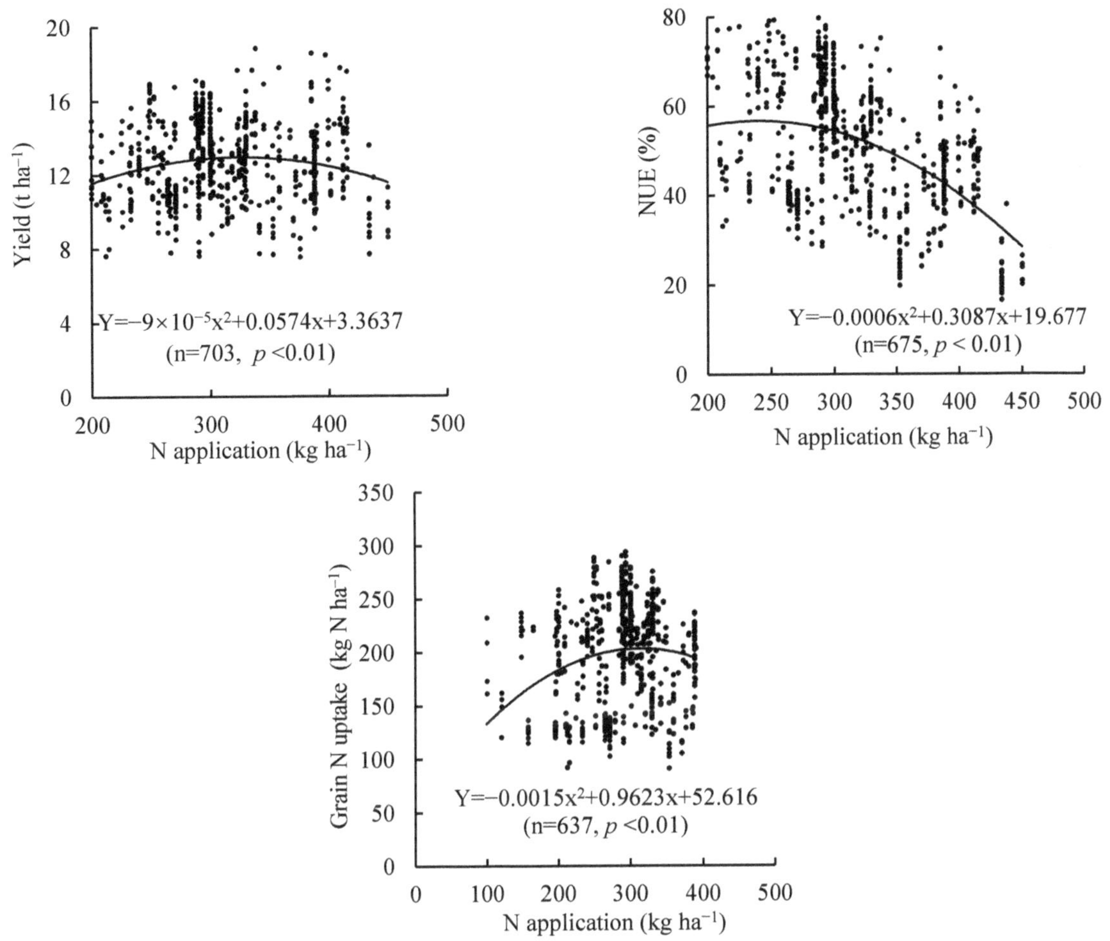

Figure 3. Relationship between N application and yield, NUE, and grain N uptake.

3.4. Performance under Optimized N Managements (OPTs) and Un-OPT Practice

The results show that the optimal N application was different in the five regions under the same indicator (Table 2). The yields of OPT and Un-OPT were 11.08–13.51 t ha^{-1} and 9.98–13.16 t ha^{-1}, respectively. The RR yield of OPT was 11% higher than that of Un-OPT. Compared with Un-OPT, the N surpluses of SEC, CC, NEC, and SW were reduced by 2–72 kg N ha^{-1} and 27–98 kg N ha^{-1} under OPT-yield and OPT-N uptake, respectively. After OPT-NUE, NUE was 22% higher than that of Un-OPT, and N surplus and Nr losses were also reduced in the five regions. Expressing Nr losses on a yield-scaled basis provides an indication of Nr losses per ton of grain yield. The average yield-scaled Nr losses for Un-OPT (12.35 kg N t^{-1}) were 6%, 24%, and 4% higher than those for OPT-yield, OPT-NUE, and OPT-N uptake, respectively.

Table 2. Yield, NUE, N surplus, and yield-scaled Nr loss responses to three optimal N applications in China.

	Zone	Optimal N Application (kg ha^{-1})	Yield (t ha^{-1})	NUE (%)	N Surplus (kg N ha^{-1})	Nr Losses (kg N ha^{-1})				Yield-Scaled Nr Losses (kg N t^{-1})
						NH_3	N_2O	L&R	Total	
OPT-yield	SC	289	11.23	50	181	54	62	13	129	11.50
	SEC	252	13.08	66	108	50	54	14	118	9.01
	CC	305	13.51	60	147	60	66	16	143	10.57
	NEC	395	13.23	31	322	78	85	23	187	14.12
	SW	279	11.08	39	211	52	60	13	126	11.37
	China	319	12.87	54	180	63	69	17	150	11.62
OPT-NUE	SC	158	8.88	62	89	32	34	8	74	8.35
	SEC	195	10.49	65	93	38	42	11	92	8.72
	CC	159	10.09	73	60	31	34	10	75	7.45
	NEC	390	11.04	39	283	77	84	23	184	16.70
	SW	167	6.75	47	129	33	36	8	78	11.54
	China	257	12.07	59	133	51	56	14	120	9.95
OPT-N uptake	SC	298	12.04	52	177	56	64	13	133	11.05
	SEC	284	13.82	64	128	56	61	15	133	9.61
	CC	285	14.01	65	122	56	62	15	133	9.52
	NEC	393	14.07	49	237	78	85	23	186	13.21
	SW	280	8.26	39	212	53	60	12	125	15.17
	China	321	12.69	53	183	64	69	18	150	11.86
Un-OPT	SC		10.85	50	156	48	55	11	114	10.49
	SEC		13.03	54	180	62	68	17	148	11.33
	CC		13.16	60	149	61	67	17	145	11.03
	NEC		9.98	27	335	74	80	21	175	17.57
	SW		10.98	40	212	53	60	12	125	11.41
	China		11.60	46	206	61	66	17	143	12.35

Note: Nr losses denote reactive N losses, NH_3 indicates NH_3 volatilization, N_2O indicates denitrification losses, L&R indicates the sum of N leaching and N runoff, and yield-scaled Nr losses indicate Nr losses/yield.

3.5. Assessment of N Management

The N input and harvested N of RR in China under OPT-yield, OPT-NUE, OPT-uptake, and Un-OPT are shown in Figure 4, and the desirable ranges for NUE (50–90%) that were suggested by the EU Nitrogen Expert Panel [35] are also shown in Figure 4. The N inputs of SEC and NEC under Un-OPT were exceeded by 400 kg N ha^{-1}yr^{-1}, but the N harvest in SEC was 40% higher than that in NEC. The N harvest of RR under OPT-yield, OPT-NUE, OPT-uptake, and Un-OPT were above the minimum productivity level (80 kg N ha^{-1} yr^{-1}) suggested by the EU Nitrogen Expert Panel [35], especially for the N harvest values in CC, which were much higher. The NUE values for OPT-yield, OPT-NUE, and OPT-N uptake were 17%, 28%, and 15% higher than those for Un-OPT, respectively, and the NUE values of RR under OPT-yield and OPT-uptake were within the desirable ranges (50–90%), showing that a high yield (high N harvest) was obtained together with a desirable NUE level.

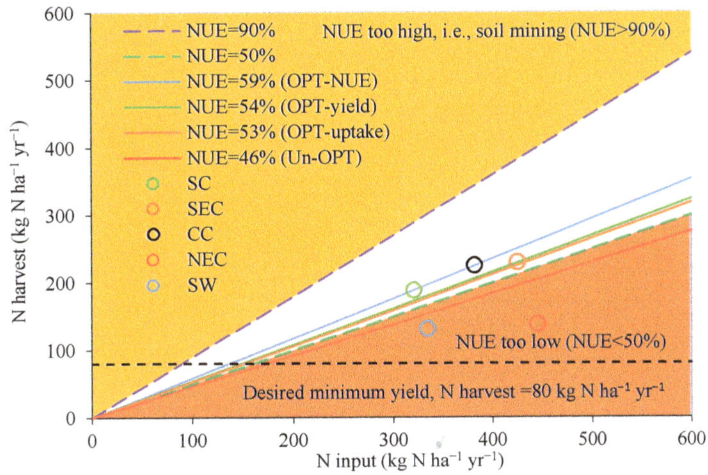

Figure 4. Comparison of N input and N harvest under different N managements of RR in China. N harvest indicates grain N uptake; N input includes fertilizer N, N deposition, biological N fixation, and seed N; the yellow and orange parts indicate high-NUE and low-NUE areas (data from Zhang C [13]); open circles denote data under Un-OPT in the five RR planting areas of China. The desirable ranges NUE = 90% and NUE = 50%, and the desired minimum yield level (N harvest = 80 kg $ha^{-1}yr^{-1}$) were suggested by the EU Nitrogen Expert Panel (EU Nitrogen Expert Panel 2015 [37]).

4. Discussion

4.1. Yield, NUE, Nr Losses, and N Surplus in Main RR Production Areas in China

Dense planting has been recommended as a promising practice to achieve higher grain yields [38,39]. Fujian is the main RR planting area in SEC. SEC had the highest yield (13.59 t ha^{-1}) at a higher planting density (27.15 $\times$ 10^4 hills ha^{-1}), followed by CC (13.16 t ha^{-1}; 25.33 $\times$ 10^4 hills ha^{-1}) (Table S8), indicating that SEC and CC were dominant in RR-growing areas.

There are several reasons for a low yield, and the specific reason in different planting areas was different, i.e., Sichuan, Guangxi, and Anhui provinces. The largest planting area of RR is Sichuan province in China [39], but it had a low RR yield (10.91 t ha $^{-1}$). The reasons for this are as follows: (i) the altitude in Sichuan rice planting areas is 200–800 m [40], and RR yield decreased when the altitude exceeded 350 m [41]; (ii) the average planting density was 21.65 $\times$ 10^4 hills ha^{-1} (Table S8), which resulted in low effective panicles and RR yield [39]; (iii) a high incidence of rice disease (e.g., sheath blight) decreased RR yield [39]. The low RR yield in Guangxi province was mainly caused by the frequently high temperature [42]. In Anhui province, rainstorms, floods, drought, hail, and typhoon disasters are frequent, causing serious losses to agricultural production [43].

Nr losses in the five regions ranged from 113 to 177 kg N ha^{-1} (average 146 kg N ha^{-1}). SC had the lowest Nr losses (113 kg N ha^{-1}), and NEC had the highest Nr losses (177 kg N ha^{-1}) (Table 1), which are higher than those of double-season rice under OPT in the Taihu region in China (102 kg N ha^{-1}) in the study conducted by Ju et al. [44]. The NUE of RR ranged from 27% to 60% (Figure 2), and the highest NUE was in CC (60%), which is close to that of the Chinese double-cropping system (68%) under OPT proposed by Zhang et al. [13]. Moreover, the average NUE was 47%, which is lower than the average predicted NUE (60%) of rice for 2050 [45], indicating that N application needs to be optimized for RR.

4.2. NUE and N Surplus under Three Optimal N Application Rates

Many methods (i.e., integrated soil–crop system management [46], response curves of N application and yield [47], and N balance management [48]) were used to determine the best N application amount, and the most common method was the recommended method based on the effective function of N application [12]. The relationship between yield and N application at specific locations or different scales has been examined in many studies [49,50], which include quadratic equations, quadratic-plus-plateau models, square roots, and exponential equations. The quadratic equation has been the method most commonly used to calculate the optimal N application in China [33,51]. The quadratic model between N application rate and yield (Table S5), NUE (Table S6), and crop N uptake (Table S7) can be established. Then, obvious inflection points and mutation points can be used to determine the optimal N application under different indicators. We can determine the minimum amount of N application needed to ensure a certain yield or gain [12,33,34]. The quadratic model recommended an optimal N application for RR of 319 kg ha^{-1} in order to obtain the highest yield (12.78 t ha^{-1}) under OPT-yield in this study (Table 2), which is lower than that in the study conducted by Cao et al. [8] (13.67 t ha^{-1}) under the optimal N application rate. This difference is mainly due to the fact that the quadratic model was selected in this study while the linear-plus-plateau model was used in the study of Cao et al. [8]. The theoretical optimal N surplus under the highest yield was 180 kg N ha^{-1}, which is higher than the N surplus benchmark (120 kg N ha^{-1}) determined by Zhang et al. [13]. This difference is mainly due to the fact that the crops researched were different. RR was studied in this study, while all the main Chinese rice-based systems (rice, double rice, rape-rice, and wheat-rice) were used in the study conducted by Zhang et al. [13], and different crops have different N surplus benchmarks. The highest NUE could be achieved with the lowest N application in this study (Table 2), which is consistent with the findings of Zhang et al. [12]. The highest NUE (59%) for Chinese RR estimated in our study was lower than the NUE (64%) for the rice–rice system proposed by Zhang et al. [13]. Crop N uptake was supposed to be an indicator for estimating the N utilization rate [52]. The N application rate under the highest grain N uptake was higher than that under the highest yield and NUE, and this result is similar to that of Zhang et al. [33].

The NUE of RR in China based on different indicators ranges from 53% to 59% (Table 2); this is close to the mean NUE target for 2050 suggested by Zhang et al. [45], in which higher NUE targets were set for rice (60%). N surplus (133–183 kg N ha^{-1}) (Table 3) based on different indicators in this study was higher than the average surplus target of all main grain crops in China (65 kg N ha^{-1}) and the worldwide average value for 2050 (53 kg N ha^{-1}) [45] (Table 3). The biggest differences between our study and that conducted by Zhang et al. [45] was based on data. First, the data of different crops were used in the study conducted by Zhang et al. [45], while the data of only RR in the five main RR regions were used in this study. Second, the data from the Food and Agriculture Organization (FAO) and International Fertilizer Industry Association (IFA) statistical databases were used by Zhang et al. [45], while data from on-farm experiments were obtained in this study (Table 3).

Table 3. Comparison of N surplus benchmarks between China and other countries/regions.

Regions	N Surplus (kg N ha^{-1})	Crops	References	Notes
The Netherlands	80	Arable land	[53]	N surplus benchmarks in 2003
Europe	80	All cropland	[37]	Overall mean N surplus benchmark
World	53	Rice	[45]	N surplus benchmarks for 2050
China	65	All cropland	[45]	N surplus benchmarks for China in 2050
China	120	Rice, rice–rice	[45]	N surplus benchmarks
China	180	Ratoon rice	This study	Average N surplus for the highest yield
China	138	Ratoon rice	This study	Average N surplus for the highest NUE
China	183	Ratoon rice	This study	Average N surplus for the highest grain N uptake

4.3. Policy Suggestions

Farms from some developed countries (e.g., the Netherlands and Europe) have achieved lower Nr losses than those under the fertilization plan [52], suggesting that our N surplus could be further reduced. The amount of N fertilizer needs to be reduced while the yield is maintained or improved in order to achieve the proposed N surplus for RR. The required improvements could be expressed as the full adoption of the "4R" of nutrient stewardship (right source, right rate, right time, and right place) [54]. Enhanced-efficiency fertilizers (e.g., controlled-release urea) can significantly increase rice yields by 26%, and reduce NH_3 volatilization (23–62%) and N surface runoff losses (8–58%) [55,56]. The rice nutrient expert system has been used to provide the correct N fertilizer amount based on the yield response of rice in the previous season, and it recommend a more accurate amount of N fertilization for rice [57]. For RR, the N fertilizer used in the first season had a significant effect on the yield of the main crop but little effect on the yield of the ratoon crop [8], while the N fertilizer used for bud promotion and seed promotion had significant effects on the yield of ratoon crops [58]. Therefore, the fertilization time should be precise. The deep placement of urea can better match the N demand of rice plants and effectively minimize NH_3 volatilization compared with broadcast [58,59]. New irrigation technology (e.g., dry–wet alternate irrigation) [60] and moldboard plowing with direct seeding [61] have also been found to realize higher yields with lower Nr losses, and they should also be used for RR. In addition, pest, weed, and disease control technologies also help farmers achieve high RR yields, for example, validamycin to eliminate pests (sheath blight) in RR, special herbicides to remove weeds (echinochloa crusgalli) in RR, and isoprothiolane to control disease (rice blast) in RR [21,39].

5. Conclusions

SEC and CC are the dominant regions for RR with higher yields and lower Nr losses. Hence, policy incentives should be implemented in these two regions for food security and environmental protection. Appropriate N surplus (180 kg N ha^{-1}) and NUE (54%) values under OPT-yield can not only increase yield but also reduce Nr losses. The "4R" of nutrient stewardship can be fully adopted to achieve N surplus in different regions under OPT-yield when the sustainable development of RR is encouraged in China.

Supplementary Materials: https://www.mdpi.com/article/10.3390/agriculture12071064/s1. References [13,21,42,43,62–67] are cited in the Supplementary Materials. Table S1: Information of zone, province and experiment sites; Table S2: Protein and nitrogen content of grain for ratoon rice in five areas of China; Table S3: Nutrient source (atmospheric deposition, biological fixation of nitrogen and rice seeding) into cropland; Table S4: Models for calculating reactive nitrogen (Nr) loss; Table S5: Relationship between N application and yield for RR in five areas of China; Table S6: Relationship between N application and NUE for RR in five areas of China; Table S7: Relationship between N application and grain N uptake for RR in five areas of China; Table S8: The accumulated temperature and planting density in different province.

Author Contributions: R.H.: conceptualization, methodology, writing—original draft, data curation, and writing—review and editing. Z.D.: validation and supervision. T.L.: conceptualization, methodology, and writing—review and editing. D.Z.: methodology. Y.T.: investigation and software. Y.C.: investigation. J.H.: supervision, funding acquisition, conceptualization, methodology, and validation. All authors have read and agreed to the published version of the manuscript.

Funding: This work is supported by the Engineering Research Center of Ecology and Agricultural Use of Wetland, the Ministry of Education (KF202109), and the Hubei special fund for agricultural science and technology innovation (2018skjcx01).

Institutional Review Board Statement: Not applicable.

Informed Consent Statement: Not applicable.

Data Availability Statement: The data that support the findings of this study are available from the corresponding author upon reasonable request.

Acknowledgments: We acknowledge the very helpful comments by C. Zhang (College of Tropical Crops, Hainan University) for the revision of this manuscript.

Conflicts of Interest: The authors declare that they have no known competing financial interests or personal relationships that could have appeared to influence the work reported in this paper.

References

1. Food Agriculture Organization of the United Nations. Food Outlook Biannual Report on Global Food Markets. 2022. Available online: https://www.fao.org/giews/reports/food-outlook/en/ (accessed on 7 July 2022).
2. Peng, S.; Tang, Q.; Zou, Y. Current status and challenges of rice production in China. *Plant Prod. Sci.* **2008**, *12*, 3–8. [CrossRef]
3. Liu, K.; Qin, J.; Zhang, B.; Zhao, Y. Physiological traits, yields and nitrogen translocation of ratoon rice in response to differ-ent cultivations and planting periods. *Afr. J. Agric. Res.* **2012**, *7*, 2539–2545. [CrossRef]
4. Yuan, S.; Cassman, K.G.; Huang, J.; Peng, S.; Grassini, P. Can ratoon cropping improve resource use efficiencies and profit-ability of rice in central China? *Field Crops Res.* **2019**, *234*, 66–72. [CrossRef]
5. Grafton, R.Q.; Williams, J.; Jiang, Q. Food and water gaps to 2050: Preliminary results from the global food and water system (GFWS) platform. *Food Secur.* **2015**, *7*, 209–220. [CrossRef]
6. National Bureau of Statics. 2021. Available online: https://data.stats.gov.cn/easyquery.htm?cn=C01 (accessed on 7 July 2022).
7. International Fertilizer Industry. 2021. Available online: https://fertilizer.org/public (accessed on 7 July 2022).
8. Cao, Y.X.; Zhu, J.Q.; Hou, J. Yield gap of ratoon rice and their influence factors in China. *Sci. Agric. Sin.* **2020**, *53*, 707–719. [CrossRef]
9. Zhang, L.; Xu, H.Q.; Li, L.L.; Chen, Y.W.; Zheng, H.B.; Tang, Q.Y.; Tang, J.W. Comparative study on CH4 emission from ratoon rice and double-cropping rice fields. *Sci. Agric. Sin.* **2019**, *52*, 2101–2113.
10. Jin, S.Q.; Fang, Z. Zero growth of chemical fertilizer and pesticide use: China's objectives, progress and challenges. *J. Resour. Ecol.* **2018**, *9*, 50–58. [CrossRef]
11. Yang, D.; Peng, S.; Zheng, C.; Xiang, H.; Wang, F. Effects of nitrogen fertilization for bud initiation and tiller growth on yield and quality of rice ratoon crop in central China. *Field Crops Res.* **2021**, *272*, 108286. [CrossRef]
12. Zhang, Y.T.; Wang, H.Y.; Lei, Q.L.; Zhang, J.Z.; Zhai, L.M.; Ren, T.Z.; Liu, H.B. Recommended methods for optimal nitro-gen application rate. *Sci. Agric. Sin.* **2018**, *51*, 117–127. [CrossRef]
13. Zhang, C.; Ju, X.; Powlson, D.; Oenema, O.; Smith, P. Nitrogen surplus benchmarks for controlling n pollution in the main cropping systems of China. *Environ. Sci. Technol.* **2019**, *53*, 6678–6687. [CrossRef]
14. Oenema, O.; Kros, H.; Vries, W.D. Approaches and uncertainties in nutrient budgets: Implications for nutrient management and environmental policies. *Eur. J. Agron.* **2003**, *20*, 3–16. [CrossRef]
15. Schr Der, J.J.; Scholefield, D.; Cabral, F.; Hofman, G. The effects of nutrient losses from agriculture on ground and surface water quality: The position of science in developing indicators for regulation. *Environ. Sci. Policy* **2004**, *7*, 15–23. [CrossRef]
16. Notaris, C.D.; Rasmussen, J.; Sørensen, P.; Olesen, J.E. Nitrogen leaching: A crop rotation perspective on the effect of N surplus, field management and use of catch crops. *Agric. Ecosyst. Environ.* **2018**, *255*, 1–11. [CrossRef]
17. Oenema, O.; Van Liere, L.; Schoumans, O. Effects of lowering nitrogen and phosphorus surpluses in agriculture on the quality of groundwater and surface water in the Netherlands. *J. Hydrol.* **2005**, *304*, 289–301. [CrossRef]
18. Hansen, B.; Thorling, L.; Dalgaard, T.; Erlandsen, M. Trend reversal of nitrate in Danish groundwater—A reflection of agricultural practices and nitrogen surpluses since 1950. *Environ. Sci. Technol.* **2011**, *45*, 228–234. [CrossRef]
19. Hartmann, T.E.; Yue, S.; Schulz, R.; Chen, X.; Zhang, F.; Müller, T. Nitrogen dynamics, apparent mineralization and balance calculations in a maize–wheat double cropping system of the North China Plain. *Field Crops Res.* **2014**, *160*, 22–30. [CrossRef]
20. Shi, Z.; Li, D.; Jing, Q.; Cai, J.; Jiang, D.; Cao, W.; Dai, T. Effects of nitrogen applications on soil nitrogen balance and nitro-gen utilization of winter wheat in a rice–wheat rotation. *Field Crops Res.* **2012**, *127*, 241–247. [CrossRef]
21. Shi, N.P.; Jiao, S.C. *Ratoon Rice Cultivation in China*; China Agriculture Press: Beijing, China, 1999.
22. Han, D.; Wiesmeier, M.; Conant, R.T.; Kühnel, A.; Sun, Z.; Knabner, I.K.; Hou, R.; Cong, P.; Liang, R.; Ouyang, Z. Large soil organic carbon increase due to improved agronomic management in the North China Plain from 1980s to 2010s. *Glob. Chang. Biol.* **2018**, *24*, 987–1000. [CrossRef]
23. Zhao, Y.; Wang, M.; Hu, S.; Zhang, X.; Shi, X. Economics- and policy-driven organic carbon input enhancement dominates soil organic carbon accumulation in Chinese croplands. *Proc. Natl. Acad. Sci. USA* **2018**, *115*, 4045. [CrossRef]
24. Ju, X.T.; Kou, C.L.; Zhang, F.S.; Christie, P. Nitrogen balance and groundwater nitrate contamination: Comparison among three intensive cropping systems on the North China Plain. *Environ. Pollut.* **2006**, *143*, 117–125. [CrossRef]
25. Liu, X.Y.; He, P.; Jin, J.Y.; Zhou, W.; Sulewski, G.; Phillips, S. Yield gaps, indigenous nutrient supply, and nutrient use efficien-cy of wheat in China. *Agron. J.* **2011**, *103*, 1452–1463. [CrossRef]
26. Ding, W.; Lei, H.; Xu, C.; Ke, H.; Li, H. Characteristics and spatial distribution of apparent nitrogen balance in the green-house vegetable cropping system in China. *J. Agric. Resour. Environ.* **2020**, *37*, 353–360. [CrossRef]
27. Wang, J.G.; Lin, S.; Li, B.G. Nitrogen cycling and management strategies in Chinese agriculture. *Sci. Agric. Sin.* **2016**, *49*, 503–517. [CrossRef]

28. Zhao, X.; Zhou, Y.; Wang, S.; Xing, G.; Shi, W.; Xu, R.; Zhu, Z. Nitrogen balance in a highly fertilized rice–wheat double-cropping system in southern China. *Soil Sci. Soc. Am. J.* **2012**, *76*, 1068. [CrossRef]
29. Li, X.B.; Xia, L.L.; Yan, X.Y. Application of membrane inlet mass spectrometry to directly quantify denitrification in flooded rice paddy soil. *Biol. Fertil. Soils* **2014**, *50*, 891–900. [CrossRef]
30. Xu, Y.B.; Cai, Z.C. Nitrogenous gas products of denitrification in subtropical soils. *Ecol. Environ. Sci.* **2014**, *23*, 932–937. [CrossRef]
31. Dash, C.J.; Sarangi, A.; Singh, D.K.; Singh, A.K.; Adhikary, P.P. Prediction of root zone water and nitrogen balance in an irrigated rice field using a simulation model. *Paddy Water Environ.* **2015**, *13*, 281–290. [CrossRef]
32. Yu, H.; Li, T.; Zhang, X. Nutrient budget and soil nutrient status in greenhouse system. *Agric. Sci. China* **2010**, *9*, 871–879. [CrossRef]
33. Zhang, Y.; Wang, H.; Lei, Q.; Luo, J.; Liu, H. Optimizing the nitrogen application rate for maize and wheat based on yield and environment on the Northern China Plain. *Total Environ.* **2018**, *618*, 1173–1183. [CrossRef]
34. Xia, Y.; Yan, X. Ecologically optimal nitrogen application rates for rice cropping in the Taihu Lake region of China. *Sustain. Sci.* **2012**, *7*, 33–44. [CrossRef]
35. Mortensen, J.R.; Beattie, B.R. Does choice of response function matter in setting maximum allowable N-application rates in Danish agriculture. *Cardon Res. Pap.* **2005**, 1. Available online: http://ag.arizona.edu/arec/pubs/researchpapers/2005-01mortensenbeattie.pdf (accessed on 7 July 2022).
36. Yan, L.I.; Yi, C.; Wu, C.; Xu, T.; Ji, X. Determination of optimum nitrogen application rates in Zhejiang Province, China, based on rice yields and ecological security. *J. Integr. Agr.* **2015**, *14*, 2426–2433.
37. Panel, E.N.E. *Nitrogen Use Efficiency (NUE) an Indicator for the Utilization of Nitrogen in Food Systems*; Wageningen University, Alterra: Wageningen, The Netherlands, 2015.
38. Yang, G.; Wang, X.; Nabi, F.; Wang, H.; Zhao, C.; Peng, Y.; Ma, J.; Hu, Y. Optimizing planting density and impact of panicle types on grain yield and microclimatic response index of hybrid rice (*Oryza sativa* L.). *Int. J. Plant Prod.* **2021**, *15*, 447–457. [CrossRef]
39. Xu, F.X.; Xiong, H.; Zhang, L.; Zhu, Y.C.; Jiang, P.; Guo, X.Y.; Liu, M. Progress in research of yield formation of ratooning rice and its high-yielding key regulation technologies. *Sci. Agric. Sin.* **2015**, *48*, 1702–1717. [CrossRef]
40. Deng, G.W.; Qing, Q.T.; Xu, J.X.; Sun, J. Integrated meteorological disaster risk regionalization of rice in Sichuan province. *Chin. J. Eco-Agric.* **2020**, *28*, 621–630. [CrossRef]
41. Fang, W.; Luo, W.Z.; Zhang, J.G.; Xiong, H.; Jiang, S.H.; Xu, S.H. Study on the ecological conditions and regional adaptability of ratooning rice in Sichuan province. Southwest China. *J. Agric. Sci.* **1994**, *17*, 15–18. [CrossRef]
42. Feng, L.Z. High Temperature Risk and Its Possible Effects on Rice Yield under Climate Change. Master's Thesis, Chinese Academy of Agricultural Sciences, Beijing, China, 2015. [CrossRef]
43. Xu, X.W.; Sun, M.Y.; Fang, Y.Y.; He, X.Q.; Xue, F.; Fu, W.; Mao, M. Impact of climatic change on rice pro-duction and response strategies in Anhui province. *J. Agro-Environ. Sci.* **2011**, *30*, 1755–1763. [CrossRef]
44. Ju, X.T.; Xing, G.X.; Chen, X.P.; Zhang, S.L.; Zhang, L.J.; Liu, X.J.; Cui, Z.L.; Yin, B.; Christie, P.; Zhu, Z.L. Reducing envi-ronmental risk by improving N management in intensive Chinese agricultural systems. *Proc. Natl. Acad. Sci. USA* **2009**, *106*, 3041–3046. [CrossRef]
45. Zhang, X.; Davidson, E.A.; Mauzerall, D.L.; Searchinger, T.D.; Dumas, P.; Shen, Y. Managing nitrogen for sustainable de-velopment. *Nature* **2015**, *528*, 51–59. [CrossRef]
46. Chen, X.; Cui, Z.; Vitousek, P.M.; Cassman, K.G.; Matson, P.A.; Bai, J.; Meng, Q.; Hou, P.; Yue, S.; Römheld, V. Integrated soil–crop system management for food security. *Proc. Natl. Acad. Sci. USA* **2011**, *108*, 6399–6404. [CrossRef]
47. Cui, Z.; Chen, X.; Zhang, F. Development of regional nitrogen rate guidelines for intensive cropping systems in China. *Agron. J.* **2013**, *105*, 1411–1416. [CrossRef]
48. Ju, X. Improvement and validation of theoretical N rate (TNR)-discussing the methods for N fertilizer recommendation. *Acta Pedol. Sin.* **2015**, *52*, 249–261. [CrossRef]
49. Valkama, E.; Salo, T.; Esala, M.; Turtola, E. Nitrogen balances and yields of spring cereals as affected by nitrogen fertiliza-tion in northern conditions: A meta-analysis. *Agric. Ecosyst. Environ.* **2013**, *164*, 1–13. [CrossRef]
50. Gaudin, A.C.; Janovicek, K.; Deen, B.; Hooker, D.C. Wheat improves nitrogen use efficiency of maize and soybean-based cropping systems. *Agric. Ecosyst. Environ.* **2015**, *210*, 1–10. [CrossRef]
51. Liu, H.; Wang, Z.; Yu, R.; Li, F.; Li, K.; Cao, H.; Yang, N.; Li, M.; Dai, J.; Zan, Y. Optimal nitrogen input for higher efficiency and lower environmental impacts of winter wheat production in China. *Agric. Ecosyst. Environ.* **2016**, *224*, 1–11. [CrossRef]
52. Pantoja, J.L.; Woli, K.P.; Sawyer, J.E.; Barker, D.W. Corn nitrogen fertilization requirement and corn–soybean productivity with a rye cover crop. *Soil Sci. Soc. Am. J.* **2015**, *79*, 1482–1495. [CrossRef]
53. Hanegraaf, M.C.; den Boer, D.J. Perspectives and limitations of the Dutch minerals accounting system (MINAS). *Eur. J. Agron.* **2003**, *20*, 25–31. [CrossRef]
54. Johnston, A.M.; Bruulsema, T.W. 4R nutrient stewardship for improved nutrient use efficiency. *Procedia Eng.* **2014**, *83*, 365–370. [CrossRef]
55. Zhang, S.; Shen, T.; Yang, Y.; Li, Y.C.; Wan, Y.; Zhang, M.; Tang, Y.; Allen, S.C. Controlled-release urea reduced nitrogen leaching and improved nitrogen use efficiency and yield of direct-seeded rice. *J. Environ. Manag.* **2018**, *220*, 191–197. [CrossRef]
56. Li, P.; Lu, J.; Wang, Y.; Wang, S.; Hussain, S.; Ren, T.; Cong, R.; Li, X. Nitrogen losses, use efficiency, and productivity of early rice under controlled-release urea. *Agric. Ecosyst. Environ.* **2018**, *251*, 78–87. [CrossRef]

57. Xu, X.; He, P.; Yang, F.; Ma, J.; Pampolino, M.F.; Johnston, A.M.; Zhou, W. Methodology of fertilizer recommendation based on yield response and agronomic efficiency for rice in China. *Field. Crop. Res.* **2017**, *206*, 33–42. [CrossRef]
58. Wang, Y.C. Effects of Nitrogen Management on Yield Formation of Ratoon Rice and the Related Mechanism. Ph.D. Thesis, Huazhong Agricultural University, Wuhan, China, 2019. [CrossRef]
59. Yao, Y.; Zhang, M.; Tian, Y.; Zhao, M.; Zhang, B.; Zhao, M.; Zeng, K.; Yin, B. Urea deep placement for minimizing NH_3 loss in an intensive rice cropping system. *Field Crops Res.* **2018**, *218*, 254–266. [CrossRef]
60. Zhang, L.; Jiang, P.; Gou, X.; Zhou, X.; Zhu, Y.; Liu, M.; Xiong, H.; Xu, F. Integrated water and nitrogen management practices to enhance yield and environmental goals in rice–ratoon rice systems. *Agron. J.* **2019**, *111*, 2821–2831. [CrossRef]
61. Asenso, E.; Zhang, L.; Tang, L.; Issaka, F.; Tian, K.; Li, J.; Hu, L. Moldboard plowing with direct seeding improves soil prop-erties and sustainable productivity in ratoon rice farmland in Southern China. *Sustainability* **2019**, *11*, 6499. [CrossRef]
62. Xu, W.; Luo, X.; Pan, Y.; Zhang, L.; Tang, A.; Shen, J.; Zhang, Y.; Li, K.; Wu, Q.; Yang, D. Quantifying atmospheric nitrogen deposition through a nationwide monitoring network across China. *Atmos. Chem. Phys.* **2015**, *15*, 12345–12360. [CrossRef]
63. Bouwman, L.; Goldewijk, K.K.; Van, D.H.K.W.; Beusen, A.H.W.; Van Vuuren, D.P.; Willems, J.; Rufino, M.C.; Stehfest, E. Exploring global changes in nitrogen and phosphorus cycles in agriculture induced by livestock production over the 1900–2050 period. *Proc. Natl. Acad. Sci. USA* **2013**, *110*, 20882–21195. [CrossRef]
64. Si, G.H.; Yuan, J.F.; Peng, C.L.; Xia, X.G.; Cheng, J.P.; Xu, X.Y.; Jia, P.A.; Xie, Y.Y.; Zhou, J.X. Nitrogen and phosphorus cycling characteristics and balance of the integrated rice-crayfish system. *Chin. J. Eco-Agric.* **2019**, *27*, 1309–1318.
65. Cui, Z.L.; Zhang, H.Y.; Chen, X.; Zhang, C.C.; Ma, W.Q.; Huang, C.D.; Zhang, W.F.; Mi, G.H.; Miao, Y.X.; Li, X.L.; et al. Pursuing sustainable productivity with millions of smallholder farmers. *Nature* **2018**, *555*, 363–366. [CrossRef]
66. Song, K.F.; Zhang, G.B.; Xu, H.; Ma, J. A review of research on influencing factors and sustainability of ratoon rice cultivation in China. *Acta Pedol. Sin.* **2020**, *57*, 1365–1377. [CrossRef]
67. Feng, D.Q.; Liu, X.C.; Liu, C.Z.; Lu, Z.W.; Wang, S.G.; Li, B.Y. Analysis and practice of climate suitability of ratoon rice in South area of Henan. *Shandong Agric. Sci.* **2012**, *44*, 41–44.

agriculture

MDPI

Article

Studying the Effect of Straw Returning on the Interspecific Symbiosis of Soil Microbes Based on Carbon Source Utilization

Yucui Ning [1], Xu Wang [1], Yanna Yang [1], Xu Cao [2], Yulong Wu [1], Detang Zou [3] and Dongxing Zhou [1,*]

1 College of Resources and Environmental Science, Northeast Agricultural University, Harbin 150030, China; yucuining@neau.edu.cn (Y.N.); s210201056@neau.edu.cn (X.W.); ynyang_neau@126.com (Y.Y.); a03190108@neau.edu.cn (Y.W.)

2 Institute of Microbiology, Heilongjiang Academy of Sciences, Harbin 150030, China; caoxu@hljas.cn

3 College of Agriculture, Northeast Agricultural University, Harbin 150030, China; zoudtneau@126.com

* Correspondence: zhouboshi@neau.edu.cn

Abstract: Heilongjiang province has made great contributions to ensuring the food security of China. Grain production has increased year by year, followed by a large amount of straw—especially the production of corn straw. Straw returning is the best treatment method from the perspective of ecology. This study simulated modern mechanized operation conditions from the field of soil biological characteristics to explore the impact of straw decomposition on the changes in the soil microbial community. In this study, in the black soil region of Northeast China (45°45′27″~45°46′33″ N, 126°35′44″~126°55′54″ E), the orthogonal experimental design was used to experiment for two years (2019–2020), using straw length, amount, and buried depth as returning factors. The carbon source utilization intensity algorithm that was developed by our team was used to extract a single carbon source. A compound mathematical model was constructed based on path analysis and grey relation analysis. This study analyzed the interspecific symbiotic relationship of soil microbes in the process of straw returning and explored the regulatory methods and schemes with which to promote straw decomposition. The results showed that in the first year after straw returning, the cumulative decomposition rate of straw could reach 55.000%; the supplement of the carbon source was glycyl-L-glutamic acid, which was helpful for the decomposition of straw. It was found that cyclodextrin should be added within 90–120 days after straw returning to promote decomposition. In the second year of straw returning, the cumulative decomposition rate of straw can reach 73.523% and the carbon sources α-D-lactose and D-galactonic acid γ-lactone should be supplemented appropriately to promote straw decomposition. This study provides an experimental basis for corn straw returning to the black soil of the cold regions, along with the scientific and technological support for the sustainable development of agriculture and a guarantee of national food security.

Keywords: straw returning; soil microbes; carbon source utilization; grey relational analysis; path analysis

Citation: Ning, Y.; Wang, X.; Yang, Y.; Cao, X.; Wu, Y.; Zou, D.; Zhou, D. Studying the Effect of Straw Returning on the Interspecific Symbiosis of Soil Microbes Based on Carbon Source Utilization. *Agriculture* **2022**, *12*, 1053. https://doi.org/10.3390/agriculture12071053

Academic Editors: Chengfang Li and Lijin Guo

Received: 6 July 2022
Accepted: 15 July 2022
Published: 19 July 2022

1. Introduction

As the main corn-producing area in China, the cold black soil region plays an important role in stabilizing the balance of grain supply and demand along with ensuring national food security [1,2]. However, the abandonment or random burning of corn straw has increased haze [3], the frequency of fires, and the waste of resources [4]. Therefore, determining how to efficiently deal with straw has become a critical concern.

Returning straw to the field can improve the soil environment [5], increase the content of soil organic matter [6], and enhance the ability of soil to retain water and fertilizer [7]. It can also supply necessary elements in plants [8], promote crop growth and development [9,10], and help with nitrogen fixation and emission reduction in the agricultural ecosystem [11,12]. Moreover, the rice yield can be effectively maintained by partially replacing mineral fertilizer with straw returning [13,14]. Recently, many scholars have conducted extensive research on straw returning to explore the best scheme of the process. These include studies on the degree of straw crushing, the amount of straw returning [15,16], the research and development of applied materials [17,18], the selection of farming methods [19,20], and the impact of soil types in the straw returning area on the straw decomposition effect [21,22]. Despite multiple studies, the research rarely involved studies on the regulation of interspecific symbiosis and the cooperation of soil microbes in the process of straw returning.

Based on this research gap, this study simulated the operating conditions of modern agricultural machinery, designed a three-factor orthogonal experiment using the amount, length, and buried depth of straw return as the factors, and carried out a two-year straw returning experiment in the cold black soil area. Using the carbon source utilization intensity algorithm that was developed by our team [23], the study extracted a single carbon source and analyzed the impact of straw returning on the carbon source utilization intensity of soil microbes. This study used the path analysis model (PA) and grey relational analysis model (GRA) to analyze the interspecific symbiotic relationship of soil microbes in the process of straw returning and find the regulatory methods and schemes with which to promote straw decomposition. This study provided scientific and technological support for the sustainable development of agriculture and to guarantee national food security.

2. Materials and Methods

2.1. Test Materials

The test was conducted at the teaching experimental base of Northeast Agricultural University (45°45′27″~45°46′33″ N, 126°35′44″~126°55′54″ E). The test area belongs to the temperate continental monsoon climate with an average annual temperature of 3.6 °C, annual precipitation of 500–600 mm, an average annual frost-free period of 135–140 days, and an effective accumulated temperature of 2700 °C [24]. The detailed meteorological data during the test are given in the Supporting Information.

The soil was typical black soil with a pH of 6.30 ± 0.06. It was composed of 39.06 ± 0.42 g/kg of organic matter; 2.20 ± 0.08 g/kg of total nitrogen; 2.71 ± 0.08 g/kg of total phosphorus; and 183.25 × 10^{-3} ± 0.16 × 10^{-3} g/kg of available potassium.

The tested straw was corn straw with total carbon of 479.0 ± 0.23 g/kg; total nitrogen of 13.16 ± 0.09 g/kg; total phosphorus of 4.56 ± 0.06 g/kg; and total potassium of 15.36 ± 0.07 g/kg. The C: N ratio was 36.13–36.78.

The mesh bag was cut from 100-mesh polyamide fiber, and the bag was 35 cm long and 25 cm wide.

2.2. Test Design

According to the three-factor five-level quadratic orthogonal rotation test design, the straw length, amount, and buried depth were taken as the test factors. Referring to the previous research results [24,25], the maximum and minimum values of each test factor are determined, that is, the actual value when the coding value is 1.682 and −1.682. Then, the actual value under other coding levels is determined according to the equivalent conversion between the coded value and the actual value. The test design result is shown in Table 1.

The straws with different weights and lengths were put into mesh bags (35 cm × 25 cm) and then soaked with water to enable the moisture content of the straw to reach 40%. The bags were buried in the soil horizontally. Each treatment was randomly arranged and repeated four times, with a total of 20 plots. Each plot was 15 m long and 1 m wide. During the straw returning period, no farming is carried out.

Table 1. Test design.

Treatments	Code Value			Actual Value		
	Test Factor A	Test Factor B	Test Factor C	Straw Length cm	Straw Amount kg/hm^2	Straw Buried Depth cm
1	1	1	1	20	6800	20
2	1	1	−1	20	6800	10
3	1	−1	1	20	2800	20
4	1	−1	−1	20	2800	10
5	−1	1	1	10	6800	20
6	−1	1	−1	10	6800	10
7	−1	−1	1	10	2800	20
8	−1	−1	−1	10	2800	10
9	1.682	0	0	25	4800	15
10	−1.682	0	0	5	4800	15
11	0	1.682	0	15	8000	15
12	0	−1.682	0	15	1600	15
13	0	0	1.682	15	4800	25
14	0	0	−1.682	15	4800	5
15	0	0	0	15	4800	15

According to the three-factors five-levels quadratic orthogonal rotation experimental design, fifteen groups of experiments were carried out. The straw amount in each experimental plot (1 m^2) is given in the Supporting Information.

Two kinds of decomposition tests were set up. The first was a one-year decomposition period while the second was a two-year decomposition period. In view of the climate impact of the cold black soil area, all the straws were buried in the spring on 6 May 2019. After 15 days of adaptation in the soil, 30 days cycles were taken for sampling in the one-year decomposition period until the end of autumn on 21 October. Thus, a total of five cycles were considered in the one-year decomposition period. The samples for the two-year decomposition period were taken on the same date of the next year (2020), as shown in Table 2.

Table 2. Sampling period.

Test Design	Straw Returning Time (Day)				
One-year decomposition period	30	60	90	120	150
Two-year decomposition period	A + 30	A + 60	A + 90	A + 120	A + 150

Note: A, year.

2.3. Sample Collection and Index Determination

2.3.1. Determination of Soil Microbial Community

According to the sampling method of rhizosphere soil, the soil around the mesh bag should be taken to the laboratory at 4 °C. The samples were activated at 25 °C for 24 h. After that, 10 g of sample was weighed and added to 90 mL of sterilized 0.85 mol/L NaCl solution. It then oscillated at 250 r/min for 30 min and was gradually diluted to 10^{-3} after standing for 10 min. The bench was clean in the vertical flow, followed by the inoculation of 150 μL of soil suspension into an ECO plate, and finally cultured in a constant temperature incubator at 28 °C. The absorbance value (OD value) at 590 nm was measured by taking 24 h as a culture cycle. The measurements of seven culture cycles (168 h) were continuously taken.

In the t culture cycle, the total change of 31 carbon sources in the ECO plate was given by the following:

$$y_t = \sum_{i=1}^{31} x_i^t, \text{ and } x_i^t = \left(OD_i^t - OD_i^{t-1}\right)^2 / \left|OD_i^{t-1}\right| \tag{1}$$

where y_t is the total change of 31 carbon sources; OD is the absorbance value of carbon source; and i indicates the type of carbon source, $i = 31$; $t = 1, 2, \cdots, 7$. The distribution of carbon sources is given in the Supporting Information.

Therefore, the utilization intensity of carbon source by microbes was given by

$$Z_i = \sum_{t=1}^{7} Q_i^t \tag{2}$$

where Q_i^t is the dimensionless data and $Q_i^t = x_i^t / y_t \times 100\%$.

2.3.2. Calculation of Straw Decomposition Rate

The straw, with the mesh bag, was placed into a sterile bag, stored at 4 °C and taken back to the laboratory. The mud and grassroots, which adhered to the mesh bag, were washed with deionized water and then dried in a constant temperature oven at 60 °C. The straw in each sampling period was accurately weighed and used to calculate the straw decomposition rate according to the Equation (3):

$$G_T = (M_0 - M_T) / M_0 \times 100\% \tag{3}$$

where G_T is the straw decomposition rate; M_0 is the initial dry weight of the straw; and M_T is the dry weight of the straw after T days of returning.

2.4. Data Analysis

2.4.1. Path Analysis Model (PA)

PA studies the direct effect, indirect effect, and total effect by decomposing the correlation between the independent variables and dependent variables [26]. In this study, the utilization intensity of the carbon source by soil microbes is the independent variable: $Z_1, Z_2, \cdots, Z_{31}$; the straw decomposition rate is the dependent variable: G. $R_{\alpha\beta}$ represents the simple correlation coefficients (spearman) of Z_α and Z_β; $R_{\alpha g}$ represents the correlation coefficient of Z_α and G; $P_{\alpha g}$ is the direct path coefficient, which indicates the direct effect Z_α on G when the other variables are fixed. $R_{\alpha g}$ can be decomposed into the following equations:

$$\begin{cases} P_{1g} + r_{12}P_{2g} + r_{13}P_{3g} + \cdots + r_{1k}P_{kg} = r_{1g} \\ r_{21}P_{1g} + P_{2g} + r_{23}P_{3g} + \cdots + r_{2k}P_{kg} = r_{2g} \\ r_{31}P_{1g} + r_{32}P_{2g} + P_{3g} + \cdots + r_{3k}P_{kg} = r_{3g} \\ \cdots \\ r_{k1}P_{1g} + r_{k2}P_{2g} + r_{k3}P_{3g} + \cdots + P_{kg} = r_{kg} \end{cases} \tag{4}$$

In this study, the absolute value of the path coefficient can be directly used to compare the importance of various microbial populations to straw decomposition. Among these, the direct path coefficient reflects the direct effect of this microbial population. Microbes can also affect the straw decomposition through the interaction with other microbial communities, which is expressed by the indirect path coefficient.

The indirect path of Z_α to dependent variable G through other variables Z_β is $R_{\alpha\beta}P_{\beta g}$, and the determination coefficient of Z_α to G is calculated as follows:

$$C^2_{(\alpha)} = P^2_{\alpha g} + 2\sum_{\alpha\neq\beta} P_{\alpha g}R_{\alpha\beta}P_{\beta g} = 2R_{\alpha g}P_{\alpha g} - P^2_{\alpha g} \quad (5)$$

where $C^2_{(\alpha)}$ is the determination coefficient.

Secondly, the path residual effect P_{Rg} is calculated. If the residual effect is minute (generally bounded by 0.20), it indicates that the PA included the main influencing factors; otherwise, variables need to be added to improve the model.

$$P_{Rg} = \sqrt{1 - \left(P_{1g}R_{1g} + P_{2g}R_{2g} + P_{3g}R_{3g} + \cdots + P_{kg}R_{kg}\right)} \quad (6)$$

In this study, the residual path coefficient was less than 20.00% as the judgment standard for extracting carbon sources $Z' = [Z_1, Z_2, Z_3, \cdots, Z_k]$.

2.4.2. Grey Relational Analysis Model (GRA)

GRA is a quantitative evaluation method based on grey system theory. It reflects the similarity of the development process between sequences through displacement difference. It can make up for the defect of the mathematical statistics method having a linear relationship with the sequence, which is irrelevant. It can overcome the deficiency of relying exclusively on the model for quantification and directly find the primary and secondary factors in the process of system development [27,28]. The specific process of GRA model construction was as follows:

In this study, the straw decomposition rate, G, is set as the parent sequence, and the carbon source extracted by path analysis, Z', is set as the sub-sequence.

Calculate the difference and take the absolute value, that is $\Delta_\eta(k) = \left|G_\mu(k) - Z'_\eta(k)\right|$.

Calculate the maximum and minimum values for all absolute values, that is: $\max_\eta \max_k \Delta_\eta(k)$ and $\min_\eta \min_k \Delta_\eta(k)$.

Calculate the relational coefficient according to the following formula.

$$\xi_{\mu\eta}(k) = \left\{\min_\eta \min_k \Delta_\eta(k) + \varepsilon\left[\max_\eta \max_k \Delta_\eta(k)\right]\right\} / \left\{\Delta_\eta(k) + \varepsilon\left[\max_\eta \max_k \Delta_\eta(k)\right]\right\} \quad (7)$$

where $\varepsilon \in \{0,1\}$ is the resolution coefficient. The smaller the ε value, the greater the difference between the relational coefficients and the stronger the discrimination ability. Referring to the previous research results [29], in this paper, $\varepsilon = 0.5$.

Calculation of grey comprehensive correlation degree (GCD): $\psi_{\alpha\beta} = \frac{1}{\rho}\sum_{k=1}^{\rho} \xi_{\mu\eta}(k)$.

3. Results and Analysis

3.1. Decomposition Rate of Straw with Different Returning Ways

As shown in Figure 1A, with the extension of straw returning time, the straw decomposition rate of each treatment group gradually increased. After 150 days of straw returning, in the T10 treatment, the cumulative decomposition rate of straw was the largest, 55.000%; in the two-year decomposition test ((A + 150) day), the cumulative decomposition rate of straw in the T10 treatment was still the largest, reaching 73.523%. In the process of straw decomposition, there was a trend of fast decomposition in the early stage and slow decomposition in the late stage. In the first two months of the one-year decomposition test, the monthly average decomposition rate was 9.310–11.000%; meanwhile, the monthly average decomposition rate of the last two months was 3.167–7.167%. The reason for this result is that, on the one hand, at the late stage of decomposition, the easily degradable organic matter in the straw gradually decreases, and the remaining part is mainly the difficult to decompose organic matter. On the other hand, it may be that the soil temperature decreases in the late stage of decomposition, resulting in the reduction in microbial activity, which is not conducive to the decomposition of straw [30]. As shown in Figure 1B,C, at the end of the test, in the two kinds of straw decomposition tests, the straw decomposition rate showed an inverted "U" shape with the increase in the coding value. High or low straw returning causes the imbalance of the soil carbon–nitrogen ratio, affects the number and activity of microbes, and leads to the reduction in straw decomposition rate [31]. Therefore, it is necessary to explore the evolution of soil microbial communities in the process of straw decomposition and find methods and schemes with which to promote straw decomposition.

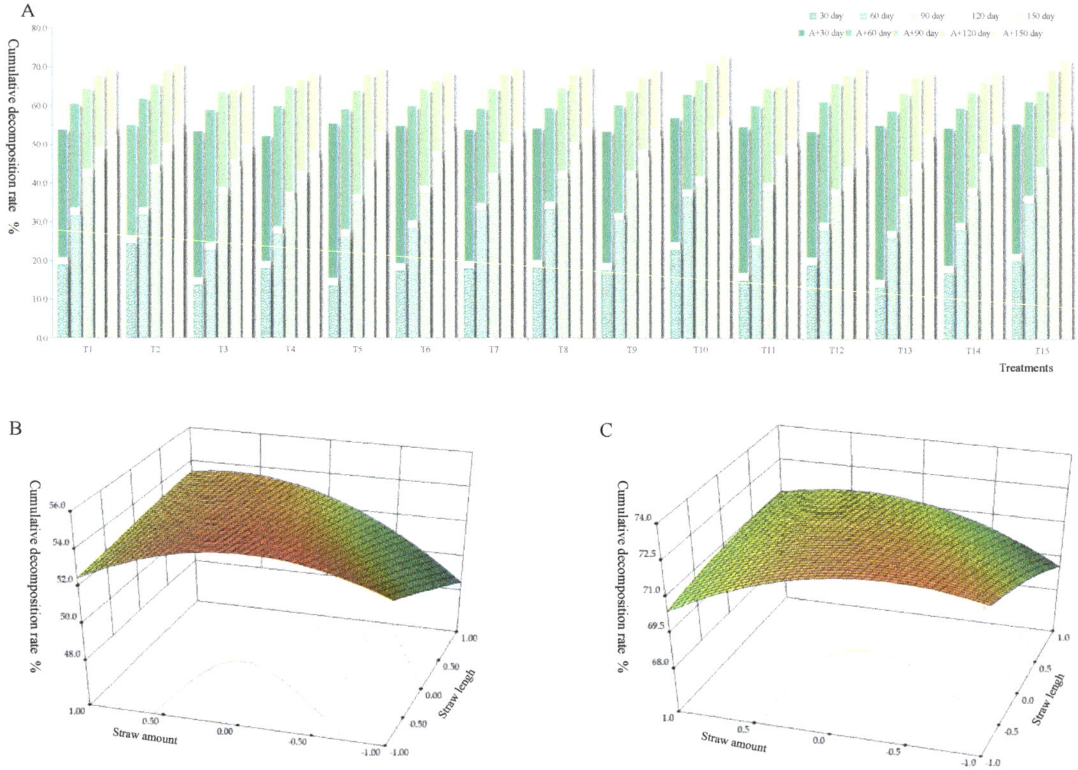

Figure 1. Cumulative decomposition rate of straw with different returning ways. (**A**) Including all straw returning periods; (**B**) after 150 days of straw returning, the change of straw decomposition rate caused by the interaction of straw amount and straw length; (**C**) after (A + 150) days of straw returning, the change of straw decomposition rate caused by the interaction of straw amount and straw length. Notes: The changes of straw decomposition rates caused by the interaction of straw amount and buried depth (Figure S1), and the interaction of straw length and buried depth (Figure S2) are given in Supporting Information.

It can be seen from Table S2 that the minimum value of the determination coefficient appeared in the 60 days of straw returning, which was 0.967, indicating that the relative contribution of the six variables which entered the PA to the straw decomposition rate had reached 96.7%, and the remaining path coefficient was 0.182, which met the judgment standard. The results showed that the PA was suitable for analyzing the relationship between straw decomposition and the soil microbial community in the cold black soil areas.

In the 30-day straw returning test group, the carbon sources L-arginine and *N*-acetyl-D-glucosamine had high direct path coefficients in the positive axis direction, which were 0.846 and 0.837, respectively. However, their total effect values, which were 0.315 and 0.101, respectively, were not high due to the counteraction of their negative indirect path effect. This was lower than that of the carbon source glycyl-L-glutamic acid, which had a total effect value of 0.371. In the 60-day group, the carbon source L-phenylalanine had the largest positive direct path effect of 0.437, with a total effect of 0.343. Although the carbon source D-malic acid had the largest positive indirect path effect of 0.741, due to the offset of the negative direct path effect of −0.581, its total effect was lower than that of the carbon source L-phenylalanine, which was 0.161. After 90 days of straw returning, the carbon source α-ketobutyric acid had the highest direct path coefficient of −1.239 and indirect path coefficient of 1.253, but the total effect value was only 0.014 due to the opposite effect between them. Simultaneously, the total effect of the carbon source *N*-acetyl-D-glucosamine was the largest in the negative direction with a total effect value of 0.512. The carbon source glycyl-L-glutamic acid had the largest total effect value of 0.379 in the positive direction. As shown in Figure 2, the carbon source α-ketobutyric acid played a major role through the indirect effect of the carbon source glycyl-L-glutamic acid with a value of 0.824.

After 120 days of straw returning, the carbon source L-asparagine ranked first with a positive direct effect of 0.549 and the carbon source α-cyclodextrin ranked second with a value of 0.536. However, the former counteracted the negative indirect effect of the carbon source D-glucosaminic acid, so its total effect value was lower than the latter. Simultaneously, as shown in Figure 2, the indirect effects between the carbon sources L-asparagine and α-cyclodextrin were negative and occupied large components, which were −0.107 and −0.110, respectively. In the last stage of the one-year straw returning test, the direct path effect of the carbon source glycyl-L-glutamic acid was the largest, which was 0.577. The indirect path effect of carbon source 4-hydroxy benzoic acid was the largest, which was 0.666, but the total effect value was negative at −0.469, due to the counteraction of the negative direct effect of −1.135. Although the indirect path effect of the carbon source D-xylose was also offset by the negative direct effect, its total effect value was the largest positive at 0.246.

As shown in Table 3, for the two-year straw returning test group, in the treatment of A + 30, the carbon source α-D-lactose had the maximum direct path effect of 0.506, the minimum indirect path effect of 0.052, and the maximum total path effect value of 0.558. Although the carbon source β-methyl-D-glucoside had a direct path effect of 0.500, its total effect value was only 0.062 due to the counteraction of its indirect path effect with a value of −0.439. In the treatments of A + 60 and A + 90, the total effect values of the carbon source D-galactonic acid γ-lactone were 0.714 and 0.648, respectively. These values ranked first in each group with a much higher total effect value than that of other carbon sources. This depended on them having the largest direct path effect. In the treatments of A + 150, however, the absolute values of the direct and the indirect path effect coefficients of each carbon source were large; due to the offset between positive and negative effects, only the carbon source tween 40 had a small positive total effect with a value of 0.059.

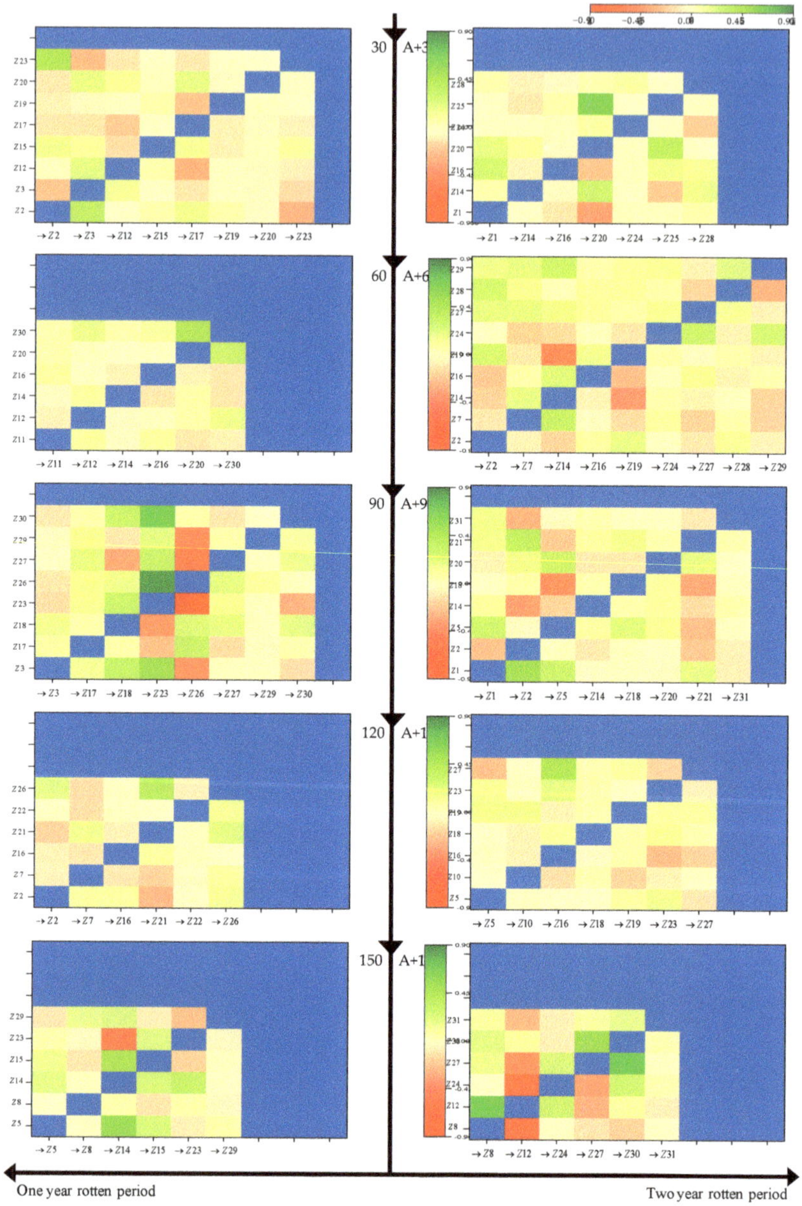

Figure 2. Indirect path effect between carbon sources. (A, year; Z0, water; Z1, β−methyl−D−glucoside; Z2, D−galactonic acid γ−lactone; Z3, L−arginine; Z4, pyruvic acid methyl ester; Z5, D−xylose; Z6, D−galacturonic acid; Z7, L−asparagine; Z8, tween 40; Z9, I−erythritol; Z10, 2−hydroxy benzoic acid; Z11, L−phenylalanine; Z12, tween 80; Z13, D−mannitol; Z14, 4−hydroxy benzoic acid; Z15, L−serine; Z16, α−cyclodextrin; Z17, N−acetyl−D−glucosamine; Z18, γ−hydroxybutyric acid; Z19, L−threonine; Z20, glycogen; Z21, D−glucosaminic acid; Z22, itaconic acid; Z23, glycyl−L−glutamic acid; Z24, D−cellobiose; Z25, glucose−1−phosphate; Z26, α−ketobutyric acid; Z27, phenylethyl−amine; Z28, α−D−lactose; Z29, D,L−α−glycerol phosphate; Z30, D−malic acid; Z31, putrescine.)

Table 3. The results of path analysis and grey correlation analysis.

Time (Day)	One-Year Decomposition Period					Two-Year Decomposition Period				
	Factors	Path Coefficient Directly	Path Coefficient Indirectly	Path Coefficient Total	*GCD*	Factors	Path Coefficient Directly	Path Coefficient Indirectly	Path Coefficient Total	*GCD*
30 A + 30	Z2	−0.646	0.182	−0.464	0.646	Z1	0.500	−0.439	0.062	0.659
	Z3	0.846	−0.531	0.315	0.758	Z14	−0.299	0.166	−0.133	0.685
	Z12	0.560	−0.255	0.305	0.702	Z16	−0.241	0.195	−0.047	0.678
	Z15	−0.192	0.206	0.014	0.662	Z20	−1.030	0.552	−0.478	0.678
	Z17	0.837	−0.736	0.101	0.681	Z24	−0.217	−0.215	−0.432	0.677
	Z19	0.354	−0.359	−0.005	0.666	Z25	−0.578	0.535	−0.043	0.667
	Z20	−0.282	0.346	0.064	0.709	Z28	0.506	0.052	0.558	0.760
	Z23	0.507	−0.135	0.371	0.710					
60 A + 60	Z11	0.437	−0.093	0.344	0.677	Z2	0.787	−0.073	0.714	0.805
	Z12	−0.511	0.093	−0.418	0.645	Z7	0.506	−0.205	0.301	0.757
	Z14	0.282	−0.138	0.144	0.674	Z14	0.765	−0.707	0.058	0.659
	Z16	0.359	−0.223	0.136	0.695	Z16	−0.339	−0.064	−0.403	0.664
	Z20	−0.962	0.331	−0.631	0.586	Z19	0.594	−0.211	0.383	0.712
	Z30	−0.581	0.741	0.161	0.679	Z24	−0.222	−0.034	−0.256	0.674
						Z27	−0.499	0.776	0.277	0.708
						Z28	0.316	0.124	0.440	0.715
						Z29	−0.538	0.919	0.380	0.740
90 A + 90	Z3	−0.374	0.374	0.000	0.690	Z1	−0.701	0.798	0.097	0.722
	Z17	−0.555	0.043	−0.512	0.651	Z2	1.296	−0.648	0.648	0.818
	Z18	−0.807	0.406	−0.401	0.705	Z5	−0.679	0.440	−0.239	0.672
	Z23	1.160	−0.780	0.379	0.751	Z14	0.367	−0.457	−0.091	0.724
	Z26	−1.239	1.253	0.014	0.722	Z18	0.348	−0.640	−0.292	0.699
	Z27	0.427	−0.237	0.190	0.727	Z20	−0.285	0.263	−0.022	0.735
	Z29	0.193	−0.203	−0.011	0.692	Z21	−0.818	0.602	−0.216	0.699
	Z30	−0.543	0.774	0.231	0.739	Z31	0.284	−0.270	0.014	0.674
120 A + 120	Z2	−0.437	0.012	−0.425	0.727	Z5	−0.663	0.102	−0.561	0.682
	Z7	0.549	−0.337	0.212	0.758	Z10	0.311	−0.373	−0.062	0.668
	Z16	0.536	−0.097	0.439	0.782	Z16	0.677	−0.475	0.202	0.696
	Z21	−0.677	0.087	−0.591	0.690	Z18	−0.280	0.262	−0.018	0.702
	Z22	0.198	−0.012	0.186	0.773	Z19	−0.362	0.429	0.067	0.681
	Z26	−0.323	0.338	0.015	0.732	Z23	−0.860	0.346	−0.514	0.624
						Z27	−0.294	0.101	−0.193	0.664
150 A + 150	Z5	−0.389	0.635	0.246	0.726	Z8	1.038	−0.979	0.059	0.716
	Z8	−0.522	−0.077	−0.599	0.701	Z12	−1.035	0.603	−0.432	0.690
	Z14	−1.135	0.666	−0.469	0.694	Z24	0.442	−0.729	−0.287	0.727
	Z15	−0.538	0.299	−0.239	0.726	Z27	−0.793	0.654	−0.139	0.735
	Z23	0.577	−0.418	0.159	0.724	Z30	−1.085	0.689	−0.397	0.738
	Z29	0.130	−0.104	0.026	0.708	Z31	−0.259	0.012	−0.246	0.701

Note: A, year. $Z_1, Z_2, \ldots, Z_{31}$ are the utilization intensity of the carbon source by soil microbes, the detailed information is given in Supporting Information. *GCD*: Grey comprehensive correlation degree.

3.2. Relational Analysis between Microbes and Straw Decomposition

In the 30-day straw returning group, the carbon source glycyl-L-glutamic acid with the largest total path effect value ranked second, and the carbon source glycogen had the same correlation degree of 0.710. Meanwhile, the carbon source L-arginine had the largest correlation degree of 0.758. In the 60-day straw returning group, the correlation degree of each carbon source entering the PA model was lower than 0.700 and the differences between the carbon sources were small. In the 90-day straw returning group, the correlation degree of the carbon source glycyl-L-glutamic acid was the largest, with a value of 0.751, indicating that it was closely related to the straw decomposition. Simultaneously, the correlation degree of the D-malic acid was the second largest, with a value of 0.739. This was similar to the total effect that was obtained by PA in the positive axis direction. A similar phenomenon occurred in the 120-day straw returning group where the correlation degree of carbon source α-cyclodextrin was the largest, with a value of 0.782. Meanwhile, the carbon sources N-acetyl-D-glucosamine and D-glucosaminic acid, with the maximum negative total effect in PA, had the minimum correlation degree with values of 0.651 and 0.690, respectively (Table 3). After 150 days of straw returning, the correlation degree of the carbon sources D-xylose and L-serine ranked first at 0.726, followed by the value of the correlation degree of the carbon source glycyl-L-glutamic acid at 0.724.

In the two-year straw returning test groups, in the treatments of A + 30, A + 60, and A + 90, the carbon sources with the largest positive direct effect had the largest correlation degrees, which were 0.760, 0.805, and 0.818, respectively. The correlation degree of the carbon source α-cyclodextrin with the largest positive direct effect of 0.696 was second only to the first carbon source γ-hydroxybutyric acid with a value of 0.702 in the treatment of A + 120. However, in the treatment of A + 150, D-malic acid, the carbon source with the largest positive indirect effect, had the highest correlation degree with a straw decomposition at a value of 0.738, while phenylethylamine, the carbon source with the second positive indirect effect, also had the second correlation degree of 0.735.

4. Discussion

Combined with the results of PA and GRA, the path map was drawn to analyze the interspecific symbiotic relationship of soil microbes during straw returning, and to find the methods and schemes for promoting straw decomposition.

In the one-year straw returning of the 30-day group, the carbon source L-arginine had the largest direct path effect value and correlation coefficient, but its total effect value was lower than that of the carbon source glycyl-L-glutamic acid. It can be seen from Figure 2 that the ranking of the correlation degree is affected through the indirect effect of the carbon source D-galactonic acid γ-lactone. Simultaneously, the carbon source D-galactonic acid γ-lactone had a negative maximum direct path effect and total effect. Therefore, when accelerating the decomposition of returning straw, it can be considered to reduce the input of the carbon source D-galactonic acid γ-lactone, and supplement glycyl-L-glutamic acid appropriately. As an amino acid carbon source, glycyl-L-glutamic acid plays an important role in the early stage of straw returning, which may be to balance the "carbon–nitrogen ratio" in the soil and provide suitable environmental conditions for the proliferation of microbial communities [32,33]. After 60 days of straw returning, the correlation degree difference between each carbon source entering the PA model and straw decomposition was small, indicating that the soil microbial community was in the stage of rapid reproduction and expanding population size at the time. This can be seen in Figure 3 showing that the direct demand for all kinds of carbon sources was large.

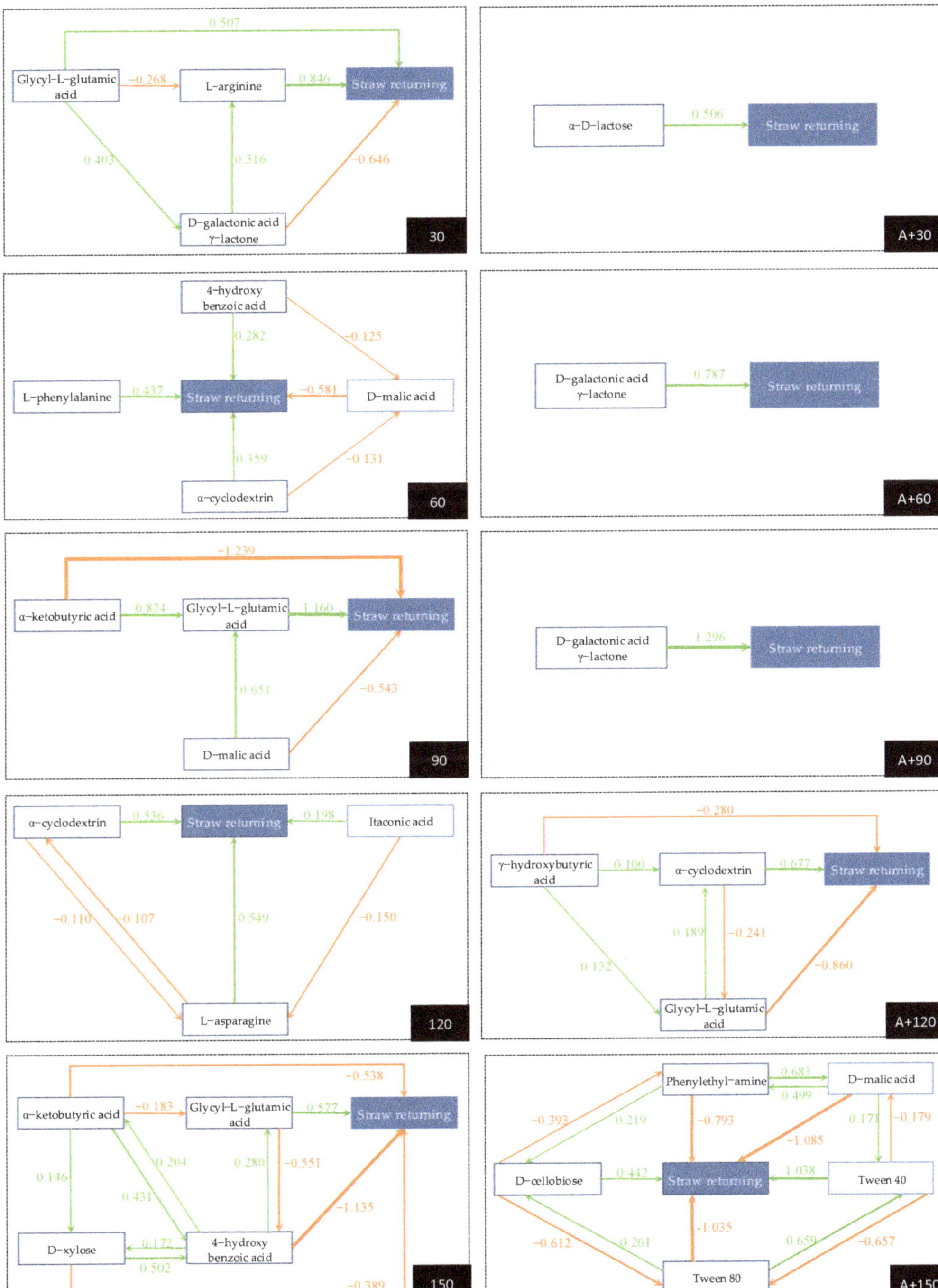

Figure 3. Path diagram.

In the 90-day straw returning group, although the results of GRA were consistent with the total effect that was obtained by PA in the positive axis direction, the first two were carbon source glycyl-L-glutamic acid and the carbon source D-malic acid. The former mainly promoted straw decomposition through the direct effect. While the latter promoted straw decomposition through an indirect effect on the former. Simultaneously, in the indirect impact of the carbon source α-ketobutyric acid on straw decomposition (Figure 2), the carbon source glycyl-L-glutamic acid played a major role. Therefore, glycyl-L-glutamic acid was the necessary carbon source for the soil microbial community within 60–90 days of straw returning. During the decomposition of straw, cellulose and hemicellulose, formed by hexose and pentose through a single bond, were decomposed first, followed by lignin, which was linked by benzene ring compounds through the δ bond and π bond [34]. It was found that the lignin-degrading microbes had a high demand for amino acid carbon sources [35,36]. Therefore, amino acid carbon sources should be supplemented in the later stage of straw returning to accelerate the straw decomposition.

As shown in Figure 3, after 120 days of straw returning, the direct path effect of itaconic acid on straw decomposition was offset by its indirect path effect through L-asparagine. Simultaneously, the antagonistic effect between the carbon source L-asparagine and the carbon source α-cyclodextrin, and its indirect path effect through the carbon source D-glucosaminic acid made the total effect value and the correlation degree of the carbon source L-asparagine lower than that of the carbon source α-cyclodextrin. On one hand, the existence of L-asparagine may inhibit the synthesis of some substances, thus slowing down the decomposition of straw by the microbial community. L-asparagine hydrolyzes the acylamino into aspartic acid and ammonia under the action of L-asparaginase [37]. Glucosamine can be used as the starting material for the asymmetric synthesis of various amino acids [38], and it is also a special component of the lipopolysaccharide of *Rhizobium leguminosarum*, which is crucial for the nitrogen cycle in organisms [39]. Alternatively, the unique external hydrophilic and internal hydrophobic structures of cyclodextrin can not only increase the biological activity of microbes to accelerate the straw decomposition [40], but also increase the permeability of the cell membrane to promote microbes to absorb nutrition more effectively [41]. Simultaneously, the cylindrical three-dimensional structure with one large side and another small side is conducive to the adsorption of ammonia [42]. Moreover, given the effect of cyclodextrin on the comprehensive improvement of the physical properties of soil [27,43], it can be supplemented to the soil within 90–120 days after the straw is returned to the field.

It can be seen from the path map that the carbon source glycyl-L-glutamic acid is also a necessary carbon source for the soil microbial community within 120–150 days of straw returning. However, it can be seen from Figure 2 that the indirect path effect of the carbon source 4-hydroxy benzoic acid through the other carbon sources was the largest at this stage, and the indirect path effects of the other carbon sources through 4-hydroxy benzoic acid were also at a high level. 4-hydroxy benzoic acid has strong allelopathy on rhizosphere microbes, can inhibit the function of root mitochondria [44], promote the growth of pathogens, and lead to the occurrence of soil-borne diseases [45,46]. Studies have found that straw returning well inhibits the soil-borne pathogens, including *Fusarium oxysporum* [47], *Rhizoctonia cerealis* [48], *Verticillium dahliae Kleb* [49], and *Plasmodiophora brassicae Woronin* [50]. Therefore, it is speculated that straw returning improves the living environment of microbes [51], promotes the proliferation of microbial populations that can use the carbon source 4-hydroxybenzoic acid in the soil to form dominant species, consumes 4-hydroxybenzoic acid in the soil, and alleviates the soil-borne diseases of crops. Yang et al. [52] confirmed that the insufficient ability of microbes to metabolize 4-hydroxybenzoic acid in the soil is an important factor that causes tobacco root rot. Zhang et al. [53] found that there are microbes which can metabolize 4-hydroxybenzoic acid in the specific disease inhibiting soil of tobacco bacterial wilt.

In the two-year straw returning test of A + 30 day treatment, seven carbon sources were mentioned to enter the PA model, but only the direct and the indirect path effect coefficients of the carbon source α-D-lactose were positive, and the correlation coefficient of α-D-lactose was the largest, indicating that α-D-lactose is the characteristic carbon source of straw decomposition at this stage. α-D-lactose, as the energy source of rod-shaped strain, can effectively improve the removal efficiency of lignin [54], and as a co-metabolic substrate, it can promote the degradation of polycyclic aromatic hydrocarbons by *Pseudomonas* [55]. Moreover, α-D-lactose can be used as an inducer to promote *Escherichia coli* to produce cyclodextrin glucosyltransferase, and then convert starch into cyclodextrin through a cyclization reaction [56,57]. Therefore, in the treatments of A + 30 and A + 60, the carbon sources α-D-lactose and α- cyclodextrin were extracted into the PA model (Table 3), while in the treatments of A + 60 and A + 90, the carbon source D-galactonic acid γ-lactone played an important role in straw decomposition, which may be related to lignin degradation. It was found that D-galactonic acid γ-lactone could be transformed into D-galactonic acid under the action of glucolactonase and enter the glycolysis process [57] to accelerate the degradation of lignocellulose [58,59].

Table 3 shows that for the treatments of A + 120 and A + 150, among the carbon sources that were extracted from the PA model, only three had a positive effect, but their total effect values were small, and the total effects between the other carbon sources and straw decomposition were negative. The results showed that the effect of the microbial community on straw decomposition was weakened at this stage. It may be possible that nutrients such as nitrogen, phosphorus, and potassium in the straw were released, and microbes no longer relied on the straw to provide the carbon source, nitrogen source, and energy. A long-term location returning test in Songnen Plain, conducted by Gong et al. [60], showed that after two years of straw returning, the degradation rates of cellulose and hemicellulose exceeded 80%, and lignin was low at 78.63%. The study by Chen et al. [61] confirmed that after half a year of straw returning, the release rate of nitrogen and phosphorus exceeded 70%, while the release rate of potassium was more than 90%.

Figure 2 shows that the indirect path effect between carbon sources increased, but the positive correlation ratio decreased, indicating that the symbiotic relationship between soil microbial species due to the straw decomposition decreased with the continuous decomposition of straw, which was consistent with the research results of Schmid et al. [62]. Simultaneously, Figure 3 shows that the complexity of the soil microbial network at this stage increased significantly. Tang et al. [63] found that the straw returning increased the network complexity to enhance the defense ability of crops against *Fusarium* wilt. Ma et al. [64] confirmed that straw returning can stimulate the growth of specific species clusters and inhibit the activity of pathogens by regulating the interaction between microbial populations.

5. Conclusions

In this study, straw length, amount, and buried depth were taken as the straw returning factors, and the two-year straw returning experiment was carried out by orthogonal design. The single carbon source was extracted by the carbon source utilization intensity algorithm, combined with a path analysis and grey correlation analysis to build a composite mathematical model to analyze the interspecific symbiotic relationship of soil microbes in the process of straw returning. The path map was drawn to explore the regulatory methods and schemes with which to promote straw decomposition.

It can be seen from the path map that in the black soil region of Northeast China (45°45′27″–45°46′33″ N, 126°35′44″~126°55′54″ E), in the first year of straw returning, the cumulative decomposition rate of straw can reach 55.000%. Further, supplementing the carbon source glycyl-L-glutamic acid to the soil was conducive to the decomposition of straw—especially within 90–120 days of straw returning, adding the carbon source cyclodextrin. In the second year of straw returning, the cumulative decomposition rate of straw could reach 73.523%, and carbon sources α-D-lactose and D-galactonic acid γ-lactone needed to be supplemented appropriately to promote straw decomposition. Additionally, 4-hydroxybenzoic acid degrading bacteria can be screened in the peripheral soil within 120–150 days of straw returning.

This results of this study provide an experimental basis for corn straw returning to the black soil of the cold regions, along with the scientific and technological support for the sustainable development of agriculture and a guarantee of national food security.

Supplementary Materials: The following are available online at https://www.mdpi.com/article/10.3390/agriculture12071053/s1, Figure S1: Straw decomposition rate caused by straw amount and buried delth: (A) 150 days of straw returning, (B) (A + 150) days of straw returning; Figure S2: Straw decomposition rate caused by straw length and buried delth: (A) 150 days of straw returning, (B): (A + 150) days of straw returning; Table S1: Test design (1 m^2); Table S2: The determination coefficient and residual path coefficient of PA; Table S3: The meteorological conditions of the experimental site during the test. Section S1: the distribution of carbon sources in ECO plates; Section S2: the straw amount in each experimental plot (1 m^2); Section S3: the changes of straw decomposition rates caused by the interaction of straw amount and buried depth; Section S4: the changes of straw decomposition rates caused by the interaction of straw length and buried depth; Section S5: the partial results of the path analysis (determination coefficient and residual path coefficient); Section S6: The meteorological conditions of the experimental site. Data S1: original data.

Author Contributions: Y.N.: writing—original draft preparation; data analysis; mathematical modeling. X.W. and Y.Y.: indices determination—soil microbial community. X.C. and Y.W.: indices determination—straw decomposition rate. D.Z. (Detang Zou): overall design; reviewing and editing. D.Z. (Dongxing Zhou): overall design; writing—reviewing and editing. All authors have read and agreed to the published version of the manuscript.

Funding: This study was supported by the National Natural Science Foundation of China (42107327), Heilongjiang Provincial Natural Science Foundation of China (YQ2021D002), and the Project funded by China Postdoctoral Science Foundation (2022M710650), Heilongjiang Postdoctoral Science Foundation (LBH-Z21120).

Institutional Review Board Statement: Not applicable.

Informed Consent Statement: This article does not contain any studies with human applicants or animals performed by any of the authors.

Data Availability Statement: All data generated or analysed during this study are included in the online version of this article as Supplementary Material.

Conflicts of Interest: The authors declare no conflict of interest.

References

1. Gu, Z.J.; Xie, Y.; Gao, Y.; Ren, X.Y.; Cheng, C.C.; Wang, S.C. Quantitative assessment of soil productivity and predicted impacts of water erosion in the black soil region of northeastern China. *Sci. Total Environ.* **2018**, *637*, 706–716. [CrossRef] [PubMed]
2. Zhang, S.; Liu, G.; Chen, S.L.; Rasmussen, C.; Liu, B.Y. Assessing soil thickness in a black soil watershed in Northeast China using random forest and field observations. *Int. Soil Water. Conse.* **2021**, *9*, 49–57. [CrossRef]
3. Shi, T.T.; Liu, Y.Q.; Zhang, L.B.; Hao, L.; Gao, Z.Q. Burning in agricultural landscapes: An emerging natural and human issue in China. *Landsc. Ecol.* **2014**, *29*, 1785–1798. [CrossRef]
4. Yin, H.J.; Zhao, W.Q.; Li, T.; Cheng, X.Y.; Liu, Q. Balancing straw returning and chemical fertilizers in China: Role of straw nutrient resources. *Renew. Sustain. Energy Rev.* **2018**, *81*, 2695–2702. [CrossRef]
5. Wang, M.Q.; Liu, Y.S.; Huang, Y.L.; Zhao, Y.Y.; Li, Z.X.; Han, Y.H. Research progress on effects of straw incorporation on soil micro-ecological environment. *Microbiol. China* **2022**, *49*, 807–816. [CrossRef]
6. Liu, C.; Lu, M.; Cui, J.; Li, B.; Fang, C.M. Effects of straw carbon input on carbon dynamics in agricultural soils: A meta-analysis. *Glob. Chang. Biol.* **2014**, *20*, 1366–1381. [CrossRef]
7. Nunes, C.; Joao, L.; Nogueirol, R.C.; Santos Menandro, L.M.; Bordonal, R.D.O.; Borges, C.D.; Cantarella, H.; Junqueira Franco, H.C. Agronomic and environmental implications of sugarcane straw removal: A major review. *GCB Bioenergy* **2017**, *9*, 1181–1195. [CrossRef]
8. Sarkar, S.; Skalicky, M.; Hossain, A.; Brestic, M.; Saha, S.; Garai, S.; Ray, K.; Brahmachari, K. Management of crop residues for improving input use efficiency and agricultural sustainability. *Sustainability* **2020**, *12*, 9808. [CrossRef]
9. Mabagala, F.S.; Geng, Y.H.; Cao, G.J.; Wang, L.C.; Wang, M.; Zhang, M.L. Silicon accumulation, partitioning and remobilization in spring maize (*Zea mays* L.) under silicon supply with straw return in Northeast China. *J. Plant Nutr.* **2020**, *44*, 1498–1514. [CrossRef]
10. Cai, L.J.; Zhang, J.T.; Liu, J.Q.; Gai, Z.J.; Guo, Z.H.; Zhao, G.F. Effects of long-term no-tillage straw returning on soil organic carbon and soybean yield in cold region. *Crops* **2021**, *6*, 189–192. [CrossRef]

11. Bordonal, R.D.O.; Nunes Carvalho, J.L.; Lal, R.; de Figueiredo, E.B.; de Oliveira, B.G.; Scala, N.L., Jr. Sustainability of sugarcane production in Brazil. A review. *Agron. Sustain. Dev.* **2018**, *38*, 13. [CrossRef]
12. Yan, C.; Yan, S.S.; Jia, T.Y.; Dong, S.K.; Ma, C.M.; Gong, Z.P. Decomposition characteristics of rice straw returned to the soil in Northeast China. *Nutr. Cycl. Agroecosys.* **2019**, *114*, 211–224. [CrossRef]
13. Chen, A.; Zhang, W.; Sheng, R.; Liu, Y.; Hou, H.; Liu, F.; Ma, G.; Wei, W.; Qin, H. Long-term partial replacement of mineral fertilizer with in situ crop residues ensures continued rice yields and soil fertility: A case study of a 27-year field experiment in subtropical China. *Sci. Total Environ.* **2021**, *787*, 147523. [CrossRef] [PubMed]
14. Huang, S.; Zeng, Y.; Wu, J.; Shi, Q.; Pan, X. Effect of crop residue retention on rice yield in China: A meta-analysis. *Field Crop. Res.* **2013**, *154*, 188–194. [CrossRef]
15. Zhao, J.L.; Wang, X.G.; Zhuang, J.; Cong, Y.J.; Lu, Y.; Guo, M.Z. Fine-crush straw returning enhances dry matter accumulation rate of maize seedlings in Northeast China. *Agronomy* **2021**, *11*, 1144. [CrossRef]
16. Xu, G.M.; Wang, X.C.; He, R.Y.; Ding, Q.S. Performance evaluation of rotary tillage straw returning based on composite indicators and measurement techniques. *Trans. Chin. Soc. Agric. Mach.* **2022**, *53*, 58–67. [CrossRef]
17. Ma, L.J.; Kong, F.X.; Wang, Z.; Luo, Y.; Lv, X.B.; Zhou, Z.G.; Meng, Y.L. Growth and yield of cotton as affected by different straw returning modes with an equivalent carbon input. *Field Crop. Res.* **2019**, *243*, 107616. [CrossRef]
18. Wang, D.; Sun, C.X.; Cui, M.; Shen, X.B.; Zhang, Y.L.; Xiao, J.H.; Liu, P.Y.; Zhang, Y.; Xie, H.T. An integrated analysis of transcriptome and metabolome provides insights into the responses of maize (*Zea mays* L.) roots to different straw and fertilizer conditions. *Environ. Exp. Bot.* **2022**, *194*, 104732. [CrossRef]
19. Ji, B.Y.; Hu, H.; Zhao, Y.L.; Mu, X.Y.; Liu, K.; Li, C.H. Effects of deep tillage and straw returning on soil microorganism and enzyme activities. *Sci. World J.* **2014**, *2014*, 451493. [CrossRef]
20. Li, Y.Z.; Song, D.P.; Dang, P.F.; Wei, L.N.; Qin, X.L.; Siddique, K.H.M. Combined ditch buried straw return technology in a ridge-furrow plastic film mulch system: Implications for crop yield and soil organic matter dynamics. *Soil Tillage Res.* **2020**, *199*, 104596. [CrossRef]
21. Zhang, H.; Liang, S.; Wang, Y.H.; Liu, S.W.; Sun, H.D. Greenhouse gas emissions of rice straw return varies with return depth and soil type in paddy systems of Northeast China. *Arch. Agron. Soil Sci.* **2021**, *67*, 1591–1602. [CrossRef]
22. Wang, S.; Zhu, H.Y.; Yang, Z.H.; Jiang, Z.Z.; Tao, Y.Z.; Liu, C.Z.; Faheem, J.M.; Li, M. Effects of straw returning modes on maize seedling growth under different soil conditions. *Chin. J. Ecol.* **2022**, *41*, 479–486. [CrossRef]
23. Ning, Y.C.; Zhou, H.R.; Zhou, D.X. Study on the microbial community in earthworm and soil under cadmium stress based on contour line analysis. *Environ. Sci. Pollut. Res.* **2019**, *26*, 20989–21000. [CrossRef] [PubMed]
24. Zhou, D.X.; Su, Y.; Ning, Y.C.; Rong, G.H.; Wang, G.D.; Liu, D.; Liu, L.Y. Estimation of the effects of maize straw return on soil carbon and nutrients using response surface methodology. *Pedosphere* **2018**, *28*, 411–421. [CrossRef]
25. Rong, G.H.; Ning, Y.C.; Cao, X.; Su, Y.; Li, J.; Li, L.; Liu, L.Y.; Zhou, D.X. Evaluation of optimal straw incorporation characteristics based on quadratic orthogonal rotation combination design. *J. Agric. Sci.* **2018**, *156*, 367–377. [CrossRef]
26. Coffman, D.L.; MacCallum, R.C. Using parcels to convert path analysis models into latent variable models. *Multivar. Behav. Res.* **2005**, *40*, 235–259. [CrossRef]
27. Hu, Y.F.; Zhang, M.K. Composition characteristics of aggregates in red sandstone soil and their responses to amendments. *Acta Agric. Jiangxi* **2019**, *31*, 88–93. [CrossRef]
28. Yang, Z.J.; Guo, X.T.; Sun, J.; Zhang, Y.L. Contextual and organizational factors in sustainable supply chain decision making: Grey relational analysis and interpretative structural modeling. *Environ. Dev. Sustain.* **2021**, *23*, 12056–12076. [CrossRef]
29. Liu, Y.L.; Huang, X.L.; Duan, J.; Zhang, H.M. The assessment of traffic accident risk based on grey relational analysis and fuzzy comprehensive evaluation method. *Nat. Hazards* **2017**, *88*, 1409–1422. [CrossRef]
30. Gao, F.; Jia, Z.K.; Lu, W.T.; Han, Q.F.; Yang, B.P.; Hou, X.Q. Effects of different straw returning treatments on soil water, maize growth and photosyntheticcharacteristics in the semi-arid area of Southern Ninaxia. *Acta Ecol. Sin.* **2011**, *31*, 777–783.
31. Zhang, P.; Wei, T.; Li, Y.L.; Wang, K.; Jia, Z.K.; Han, Q.F.; Ren, X.L. Effects of straw incorporation on the stratification of the soil organic C, total N and C:N ratio in a semiarid region of China. *Soil Tillage Res.* **2015**, *153*, 28–35. [CrossRef]
32. Ma, F.X.; Wang, Y.Y.; Yan, P.; Wei, F.; Sun, X.Z.; Liu, J.G. Effects of cotton straw incorporation on organic nitrogen fractions in long-term continuous cropping cotton field. *Ecol. Environ. Sci.* **2018**, *27*, 1459–1465. [CrossRef]
33. Long, Z.H.; Wang, J.; Hou, Z.A. Effects of cotton straw biochar returning and N application rate on soil organic nitrogen fractions in cotton field. *J. Shihezi Univ.* **2019**, *37*, 154–161. [CrossRef]
34. Zhang, Z.Q.; Zhang, A.Z.; Jiang, N. Effects of mixed fungal fermentation on degradation rate of cellulose and lignin of corn straw. *Chin. J. Anim. Nutr.* **2019**, *31*, 1385–1395. [CrossRef]
35. Zeng, Z.T.; Guo, X.Y.; Xu, P.A.; Xiao, R.; Huang, D.L.; Gong, X.M.; Cheng, M.; Yi, H.; Li, T.; Zeng, G.M. Responses of microbial carbon metabolism and function diversity induced by complex fungal enzymes in lignocellulosic waste composting. *Sci. Total Environ.* **2018**, *643*, 539–547. [CrossRef] [PubMed]
36. Jiang, L.L.; Han, G.M.; Lan, Y.; Liu, S.N.; Gao, J.P.; Yang, X.; Meng, J.; Chen, W.F. Corn cob biochar increases soil culturable bacterial abundance without enhancing their capacities in utilizing carbon sources in Biolog Eco-plates. *J. Integr. Agric.* **2017**, *16*, 713–724. [CrossRef]
37. Ghasemi, A.; Asad, S.; Kabiri, M.; Dabirmanesh, B. Cloning and characterization of Halomonas elongata L-asparaginase, a promising chemotherapeutic agent. *Appl. Microbiol. Biotechnol.* **2017**, *101*, 7227–7238. [CrossRef]

38. Pezzotti, F.; Therisod, H.; Therisod, M. Enzymatic synthesis of D-glucosaminic acid from D-glucosamine. *Carbohydr. Res.* **2005**, *340*, 139–141. [CrossRef]
39. Bhat, U.R.; Forberg, L.S.; Carlson, R.W. Structure of lipid a component of Rhizobium leguminosarum bv. phaseoli lipopolysaccharide. Unique nonphosphorylated lipid A containing 2-amino-2-deoxygluconate, galacturonate, and glucosamine. *J. Biol. Chem.* **1994**, *269*, 14402–14410. [CrossRef]
40. Wu, N.; Li, X.F.; Huang, G.S.; Pan, P.; Wang, C.; Liu, X.Y.; Zeng, M. Adsorption and biodegradation functions of novel microbial embedding polyvinyl alcohol gel beads modified with cyclodextrin: A case study of benzene. *Environ. Technol.* **2018**, *40*, 1948–1958. [CrossRef]
41. Tribak, M.; Ocampo, J.A.; Garcia-Romera, I. Production of xyloglucanolytic enzymes by Trichoderma viride, Paecilomyces farinosus, Wardomyces inflatus, and Pleurotus ostreatus. *Mycologia* **2002**, *94*, 404–410. [CrossRef]
42. Du, L.L.; Li, G.X.; Yuan, J.; Yuan, J.B. Effect of additives on NH_3 and H_2S emissions during kitchen waste composting. *Trans. Chin. Soc. Agric. Eng.* **2015**, *31*, 195–200. [CrossRef]
43. Yan, J.L.; Zhang, M.K.; Wang, D.Z. Different amendments: Effect on soil physical properties of newly reclaimed land in low hilly region. *Chin. Agric. Sci. Bull.* **2021**, *37*, 67–73. [CrossRef]
44. Zhang, G.W.; Yang, C.Q.; Liu, R.X.; Ni, W.C. Effects of p-hydroxybenzoic acid and phloroglucinol on mitochondria function and root growth in cotton (*Gossypium hirsutum L.*) seedling roots. *Chin. J. Appl. Ecol.* **2018**, *29*, 231–237. [CrossRef]
45. Sun, S.S.; Gong, B.; Wen, D.; Wang, X.F.; Wei, M.; Yang, F.J.; Li, Y.; Shi, Q.H. Effect of exogenous melatonin on physiological and biochemical characteristics of cucumber radicles under p-hydroxybenzoic acid. *Chin. J. Appl. Ecol.* **2016**, *27*, 897–903. [CrossRef]
46. Zhao, X.S.; Qi, Y.Z.; Yan, C.M.; Zhen, W.C. Allelopathy of six organic acids on wheat sheath blight in the soil of winter wheat-summer maize double cropping straw returning system. *Sci. Agric. Sin.* **2020**, *53*, 3095–3107. [CrossRef]
47. Tang, L.L.; Nie, S.R.; Li, W.H.; Fan, C.; Wang, S.Q.; Wu, F.Z.; Pan, K. Wheat straw increases the defense response and resistance of watermelon monoculture to Fusarium wilt. *BMC Plant Biol.* **2019**, *19*, 551. [CrossRef]
48. Qi, Y.Z.; Zhen, W.C.; Li, H.Y. Allelopathy of decomposed maize straw products on three soil-born diseases of wheat and the analysis by GC-MS. *J. Integr. Agric.* **2015**, *14*, 88–97. [CrossRef]
49. Zhang, Q.; Li, Y.B.; Teng, L.P.; Yue, J. Allelopathy of different decomposed liquids of cotton stalk on fusarium oxysporum and verticillium dahliae. *J. Agro-Environ. Sci.* **2012**, *31*, 1696–1701.
50. Han, Z.; Di, C.Q.; Rahman, M.; Gao, D.M.; Wu, F.Z.; Pan, K. Repeated application of rice straw stabilizes soil bacterial community composition and inhibits clubroot disease. *Agriculture* **2021**, *11*, 108. [CrossRef]
51. Chen, L.J.; Zhou, J.H.; Chen, G.; Liu, L.; Tan, J. Research progress of influence and mechanism of field straw residue incorporation on soil-borne diseases in crops. *Crop Res.* **2018**, *32*, 535–540. [CrossRef]
52. Yang, L.P.; Yao, X.Y.; Li, Q.; Jiang, Q.P. An analysis of function diversity of soil microbes in relation to the outbreak of tobacco root rot. *Plant Dr.* **2020**, *33*, 36–41. [CrossRef]
53. Zhang, S.T.; Liu, X.J.; Zhou, L.H.; Deng, L.Y.; Zhao, W.A.; Liu, Y.; Ding, W. Alleviating soil acidification could increase disease suppression of bacterial wilt by recruiting potentially beneficial rhizobacteria. *Microbiol. Spectr.* **2022**, *10*, e0233321. [CrossRef]
54. Raj, A.; Reddy, M.M.K.; Chandra, R.; Purohit, H.J.; Kapley, A. Biodegradation of kraft-lignin by Bacillus sp isolated from sludge of pulp and paper mill. *Biodegradatioin* **2007**, *18*, 783–792. [CrossRef]
55. Liu, X.C.; Wang, J.; Guo, S.H.; Chen, C.M. Effects of co-substrates and inorganic salts on degradation of crude oil by oil degrading bacteria. *Environ. Prot. Chem. Ind.* **2008**, *3*, 218–221. [CrossRef]
56. Cheng, J.; Wu, D.; Chen, S.; Wu, J.; Chen, J. Effect of complex and synthetic medium on extracellular production of α-cyclodextrin glycosyltransferase in *E. coli*. *China Biotechnol.* **2010**, *30*, 36–42. [CrossRef]
57. Hayward, M.R.; AbuOun, M.; Woodward, M.J.; Jansen, V.A.A. Temperature and oxygen dependent metabolite utilization by salmonella enterica serovars derby and mbandaka. *PLoS ONE* **2015**, *10*, e0120450. [CrossRef]
58. Makonde, H.M.; Mwirichia, R.; Osiemo, Z.; Boga, H.I.; Klenk, H.P. 454 Pyrosequencing-based assessment of bacterial diversity and community structure in termite guts, mounds and surrounding soils. *SpringerPlus* **2015**, *4*, 471. [CrossRef]
59. Sun, X.X.; Wei, J.H.; Li, J.J.; Ni, J.F. Whole-genome analysis of the dominant bacterium Dysgonomonas macrotermitis in the hindgut of Macrotermes barneyi. *Acta Microbiol. Sin.* **2018**, *58*, 995–1003. [CrossRef]
60. Gong, Z.P.; Deng, N.Z.; Song, Q.L.; Li, Z.T. Decomposing characteristics of maize straw returning in Songnen Plain in long-time located experiment. *Trans. Chin. Soc. Agric. Eng.* **2018**, *34*, 139–145. [CrossRef]
61. Chen, L.J.; Chen, R.; Zhou, J.H.; Yan, C.B.; Liu, L.; Li, Q.; Zhang, Y. Decomposition characteristics of three crop straws and the effects of their decomposed liquids on phytophthora nicotianae. *Chin. Tob. Sci.* **2021**, *42*, 33–39. [CrossRef]
62. Schmid, C.A.O.; Schroder, P.; Armbruster, M.; Schloter, M. Organic amendments in a long-term field trial-consequences for the bulk soil bacterial community as revealed by network analysis. *Microb. Ecol.* **2018**, *76*, 226–239. [CrossRef]
63. Tang, L.L.; Xia, Y.; Kou, J.M.; Wu, F.Z.; Li, W.H.; Pan, K. Control of fusarium wilt by wheat straw is associated with microbial network changes in watermelon rhizosphere. *Sci. Rep.* **2020**, *10*, 12736. [CrossRef]
64. Ma, L.; Li, Y.; Wei, J.L.; Li, Z.S.; Zhou, X.L.; Zheng, F.L.; Wu, X.B.; Wang, L.; Liu, Z.H.; Tan, D.S. Effects of long-term straw returning on fungal community, enzyme activity and wheat yield in a fluvo-aquic soil. *Environ. Sci.* 2022. [CrossRef]

agriculture

Article

Biochar-Based Fertilizer Improved Crop Yields and N Utilization Efficiency in a Maize–Chinese Cabbage Rotation System

Huan Zhao [1,2], Tingting Xie [3], Houjun Xiao [2] and Ming Gao [1,*]

[1] College of Resources and Environment, Southwest University, Chongqing 400716, China; zhaohuancnm@163.com
[2] Guizhou Institute of Soil and Fertilizer, Guizhou Academy of Agricultural Sciences, Guiyang 550006, China; xiao-hjnky@163.com
[3] Guizhou Testing Technology Research and Application Center, Guiyang 550014, China; x18198613293@163.com
* Correspondence: gaoming@swu.edu.cn; Tel.: +86-023-6825-1249

Abstract: Optimizing fertilization strategies is crucial for obtaining high crop yields and efficient N utilization. This study aimed to understand the potential increase in crop yield and the N utilization efficiency under biochar-based fertilizer (BF) in a maize–Chinese cabbage rotation system. Biochar-based slow-release fertilizer (BF) is an important nutrient-efficient management strategy. The yields and growth-related traits of the crops, N utilization efficiency, quality, and dynamic changes in soil inorganic N in a maize-cabbage rotation system were investigated in a pot experiment under three N fertilizer application strategies in 2019–2020; the maize stage included (1) zero-N fertilizer, i.e., control (N 0 g pot^{-1}); (2) NPK (N 5.25 g pot^{-1}); and (3) BF (N 5.25 g pot^{-1}). The Chinese cabbage stage included (1) zero-N fertilizer, i.e., control (N 0 g pot^{-1}); (2) NPK (N 6.25 g pot^{-1}); and (3) BF (N 6.25 g pot^{-1}). Compared with the CK and NPK treatments, the BF treatment had the highest average maize and Chinese cabbage yields at 86.99 g plant^{-1} and 498.88 g plant^{-1}, respectively. BF improved the plant height, stem diameter, and ear height of maize and the leaf length, leaf width, and leaf number of Chinese cabbage, as well as increased the N utilization efficiency of maize and cabbage. BF increased the starch content of maize grain and the amino acid, sugar, and vitamin C contents of cabbage. In the critical growth stages of maize and Chinese cabbage, BF application increased the content of soil inorganic N, which coincided with the nutrient requirements in the critical growth stages of the crops. Overall, BF is an effective method to improve crop yield and N utilization in the maize–Chinese cabbage rotation systems and is a fertilization strategy with broad applicability prospects.

Keywords: biochar; maize–cabbage system; yield; crop quality; N utilization efficiency; soil inorganic N

Citation: Zhao, H.; Xie, T.; Xiao, H.; Gao, M. Biochar-Based Fertilizer Improved Crop Yields and N Utilization Efficiency in a Maize–Chinese Cabbage Rotation System. *Agriculture* **2022**, *12*, 1030. https://doi.org/10.3390/agriculture12071030

Academic Editors: Chengfang Li and Lijin Guo

Received: 14 June 2022
Accepted: 13 July 2022
Published: 14 July 2022

1. Introduction

Along with the growth of the world's population, global food demands are on the rise [1], with the expectation that per capita food requirements will nearly double by 2050 [2]; however, there is increasing concern over the mounting burden of food production [3]. In response to these challenges, it is evident that an increase in crop yield per acre is vital. The most consumed crop nutrient is nitrogen (N) [4], and N fertilization is an important agricultural technology to increase crop production per unit of land, which is important for economic and social progress [5]. Unfortunately, the amount of conventional N fertilizer used in agricultural production is increasing while plant N fertilization efficiency is underperforming, and crop yield potential is not realized [6,7]. The type and quantity of fertilizer affects not only crop yields but also soil physicochemical properties, which in turn have a significant effect on soil fertility and productivity [8,9]. Because of leaching, volatilization, denitrification, fixation, erosion, and runoff, there will be inefficiencies in crop nutrient use and environmental pollution, in addition to increases in fertilizer prices, with negative economic and environmental impacts [3,10–12]. Therefore, a balanced strategy is required to maintain crop yield and N utilization efficiency while also minimizing nutrient losses during crop production.

In modern agricultural production, the application of slow-release fertilizers (SRFs) is vital technology to ensure that crops are produced in a sustainable and high-quality manner [13,14]. Nevertheless, SRFs are extremely expensive, so they are not used widely. Biochar-based slow-release fertilizers can solve the nutrient deficiencies of biochar, the high nutrient loss rate of traditional chemical fertilizers, and low crop nutrient utilization [15,16], primarily due to the benefits of biochar nutrient retention, carbon sequestration, emission reduction, and soil improvement, while realizing the functions of nutrient adsorption and slow-release, reducing nutrient loss [17,18]. As a result of their wide availability and extreme cost-effectiveness across the globe, biochar-based resources can be used as cost-effective and climate-smart nutrient carriers for the formulation of slow-release fertilizers [19]. Biochar-based slow-release fertilizers (BFs) are a new type of fertilizer derived from combining biochar and chemical fertilizers through a specific process [20]. According to research, BFs can reduce the number of chemical fertilizers and application times, improve crop nutrient efficiency, increase crop yield and quality, and improve soil physical and chemical properties, which allow for improvements in the utilization rate of agricultural waste [21,22]. It is becoming increasingly popular to use biochar-based slow-release fertilizers in agriculture due to their high fertilizer efficiency and low environmental impact [3].

Crop rotation is one of the most effective ways to maximize agricultural economic benefits and productivity per unit of arable area, but different crop rotation systems, soil types, and fertilizer levels affect crop yield differently [23–25]. Waxy maize, also called sticky maize, can be processed or consumed directly, and its value is high, economically, nutritionally, and processing wise [26]. Vegetables are of great economic importance, and their cultivation continues to increase [27]. The rotation of maize and vegetables is a widespread practice in Southwest China [12]. Crop rotation systems positively impact land-use efficiency and crop yield. Nevertheless, inadequate nutrient uptake by plants and environmental effects caused by unreasonable N fertilization are common problems. There have been few reports of biochar-based slow-release fertilizers in the yellow soil maize-cabbage rotation system in Guizhou, China.

Here, we performed a two-year study (2019–2020) using pot experiments focusing on the effect of biochar-based fertilizers on a maize–Chinese cabbage rotation system. However, there is little information available on BF application for crop yield and changes in soil inorganic N dynamics, particularly in yellow soil under maize–vegetable (Chinese cabbage) rotation systems. Therefore, we aimed to (1) investigate the effects of BF application on maize-cabbage yields and biological traits; (2) determine maize-cabbage N uptake and utilization efficiency under BF application; and (3) clarify the effects of BF application on the maize–Chinese cabbage rotation system and the dynamic changes in soil NH_4^+-N and NO_3^--N.

2. Materials and Methods

2.1. Site Description

Pot experiments were conducted at the Institute of Soil Fertilization, Guizhou Academy of Agricultural Sciences (1060 m above sea level, 106°07′ E, 26°11′ N), Guizhou, China. The pot experiment took place from 2019 to 2020. The area has a subtropical monsoon climate, with an average annual temperature of 15.3 °C and yearly rainfall of 1100–1200 mm. Before the experiment, the soil chemistry was determined; the pH was 7.29, and the SOM content was 25.24 g kg^{-1}. The available nitrogen (N, alkaline hydrolysis N), phosphorus (P, Olsen-

P), and potassium (K, ammonium acetate-extractable K) contents were 78.40, 9.85, and 85.76 mg kg^{-1}, respectively.

2.2. Experimental Design

The experiment consisted of three treatments, and each treatment was repeated three times, according to a completely random design. The diameter and height of each plastic pot were 38 cm and 40.5 cm, respectively. Each pot was filled with 25 kg of soil. The soil samples were collected locally from 0 to 20 cm in a field experiment at the Institute for Soil Fertilization, Guizhou Academy of Agricultural Sciences. The soil was typic yellow soil (entisol); one maize or one cabbage plant was planted per pot. The experimental treatments in the annual maize stage included (1) zero-N fertilizer, i.e., control (N 0 g pot^{-1}, P_2O_5 2.75 g pot^{-1}, and K_2O 5.25 g pot^{-1}); (2) NPK (N 5.25 g pot^{-1}, P_2O_5 2.75 g pot^{-1}, and K_2O 5.25 g pot^{-1}); and (3) BF (N 5.25 g pot^{-1}, P_2O_5 2.75 g pot^{-1}, and K_2O 5.25 g pot^{-1}). The experimental treatments in the annual Chinese cabbage stage included (1) zero-N fertilizer, i.e., control (N 0 g pot^{-1}, P_2O_5 3.75 g pot^{-1}, and K_2O 6.25 g pot^{-1}); (2) NPK (N 6.25 g pot^{-1}, P_2O_5 3.75 g pot^{-1}, and K_2O 6.25 g pot^{-1}); and (3) BF (N 6.25 g pot^{-1}, P_2O_5 3.75 g pot^{-1}, and K_2O 6.25 g pot^{-1}). The specific maize and Chinese cabbage nutrient requirements were obtained from "Experimental Research and Statistical Analysis" [28].

According to the NPK fertilization strategy, N fertilizer was divided into basal fertilizer (60%) and topdressing fertilizer (40%). Urea was top-dressed in the NPK fertilization treatment during the maize jointing and cabbage rosette growth stages. P and K were applied as basal fertilizers in the zero-N fertilizer and NPK fertilization treatments. According to the BF treatment strategy, BF was used as a basal fertilizer in a one-time application. Mineral N, P, K, and BF refer to urea, superphosphate, potassium sulfate, and biochar-based slow-release fertilizer, respectively. BF is a slow-release biochar-based fertilizer (N 15%, P_2O_5 10%, and K_2O 15%), which is a new type of fertilizer made by mixing biochar and other chemical fertilizers through a particular process with a total nutrient content $\geq$ 40% and a carbon content (calculated as C) $\geq$ 6%, produced by Qinfeng Zhongcheng New Biomass Materials (Nanjing) Co., Ltd in Nanjing, China. Biochar was produced by pyrolyzing maize straw at a high temperature of 450 °C for 2 h under anaerobic conditions.

In Southwest China, the rotation of maize and vegetables is widely practiced, which was used in this experiment. The pot experiments used a maize (*Zea mays*)–Chinese cabbage (*Brassica campestris* L. spp. *pekinensis*) rotation system. The first season of maize was transplanted on April 19 and harvested on 10 September 2019; the second season of Chinese cabbage was transplanted on 20 October 2019 and harvested on 10 January 2020; the third season of maize was transplanted on 20 April, and harvested on 15 September 2020; the fourth season of Chinese cabbage was transplanted on 25 October 2020 and harvested on 15 January 2021. The experiment was performed under field conditions. As the maize and Chinese cabbage were transplanted, basal fertilizer was applied to the topsoil at a depth of 20 cm. The processes of transplanting and fertilizing are almost simultaneous. A standard chemical program was used to control weeds, insects, and diseases in the plots during the various crop growth stages of the maize and Chinese cabbage. No irrigation was applied during the growing season. The management of maize and Chinese cabbage in this pot experiment was the same as that in the fields.

2.3. Sampling and Measurements

Samples of fresh soil between 0 and 20 cm deep were collected at the seedling stage, jointing stage, heading stage, and harvest stage of maize and at the seedling stage, rosette stage, and harvest stage of cabbage. Surface debris was removed, and the samples were bagged, transported in a careful manner, and maintained at room temperature (25 °C). NH_4^+-N and NO_3^--N were measured using indigo blue colorimetry and ultraviolet spectrophotometry, respectively, according to Bao [29].

The starch, sugar, and crude protein contents of maize and the contents of amino acids, sugar, and vitamin C in the edible parts of cabbage were analyzed after the 2020 harvests.

The nutritional qualities of the maize grain and edible cabbage parts were determined according to Bao [29]. 3,5-Dinitrosalicylic acid was used to detect reducing sugars in maize grain and cabbage; the spectrophotometric method was used to determine the amino acids in cabbage; the Kjeldahl method was used to determine the crude protein content; anthrone colorimetry was used to determine the maize grain starch, and 2,6-dichloro-indophenol titration was used to determine the vitamin C content in cabbage. After being air-dried, weighed, and sieved through a 0.15 mm sieve, the total N of maize and cabbage samples was determined using the micro-Kjeldahl method [29].

2.4. Calculations

The plant height, stem diameter, and ear height of maize, and the leaf length, leaf width, and leaf number of Chinese cabbage were used as the representative maize and cabbage growth-related traits. The plant height and ear height of maize and the leaf length and leaf width of Chinese cabbage were measured using a ruler. The stem diameter of maize was measured using a Vernier caliper. Based on the methods reported by Hartmann et al. (2015) [30], the N fertilizer use efficiency (NUE) was calculated as follows: NUE = (UN − UN0)/FN × 100, where UN is the total N uptake of plants under fertilization (g plant^{-1}); UN0 is the total N uptake of plants without N fertilization (g plant^{-1}); FN is the amount of N fertilizer applied (g plant^{-1}). The sum of NH_4^+-N and NO_3^--N is inorganic N. Inorganic N includes NH_4^+-N and NO_3^--N.

2.5. Statistical Analyses

The significance of differences between soil and plant indicators was measured by one-way ANOVA. Duncan's multiple ranges (SSR) test was used to check the significance of treatment effects at $p < 0.05$. The statistical analysis was conducted using SPSS Version 16.0 (SPSS Inc., Chicago, IL, USA). The figures and tables were compiled using Excel 2016 and Origin 22.0.

3. Results

3.1. Crop Yields and Growth-Related Traits

As shown in Figure 1, the N treatments affected the crop yield in the maize–Chinese cabbage rotation (2019–2020). The yield of maize and Chinese cabbage in 2019–2020 under the BF treatment was the highest at 70.43 g plant^{-1}–103.55 g plant^{-1} and 484.00 g plant^{-1}–513.76 g plant^{-1}, respectively. Compared to zero-N, the yields of maize grain and Chinese cabbage significantly increased under the NPK and BF treatments. Compared with the NPK treatment, the BF treatment significantly improved the maize grain yield by 35.26% and 17.78% (Figure 1a,b) and the Chinese cabbage yield by 32.78% and 59.55%, respectively, in 2019 and 2020.

Figures 2 and 3 show the growth-related traits of maize and Chinese cabbage in 2019–2020 under different fertilization treatments. Compared to the zero-N treatment, the results of the variance analysis showed that the NPK treatment had significant effects on the ear height of maize in 2019 (Figure 2c), the leaf length and leaf number of Chinese cabbage in 2019–2020, and the leaf width of Chinese cabbage in 2020 (Figure 3a,b,d–f), and the BF treatment had significant effects on the plant height, ear height, and stem diameter of maize in 2019–2020 (Figure 2a–f) and the leaf length, leaf width, and leaf number of Chinese cabbage in 2019–2020 (Figure 3a–f). Figures 2 and 3 show that the growth-related traits of maize and Chinese cabbage in 2019–2020 under the BF treatment were significantly higher than those under the zero-N treatment. Compared to the NPK treatment, the BF treatment significantly improved the plant height, ear height, and stem diameter of maize by 7.65–8.77%, 4.20–9.35%, and 17.92–40.73%, respectively, in 2019–2020 (Figure 2a–f). The leaf width and leaf number of Chinese cabbage under the BF treatment increased by 15.96–19.84% and 9.09–20.00%, respectively, compared to the NPK treatment (Figure 3c–f).

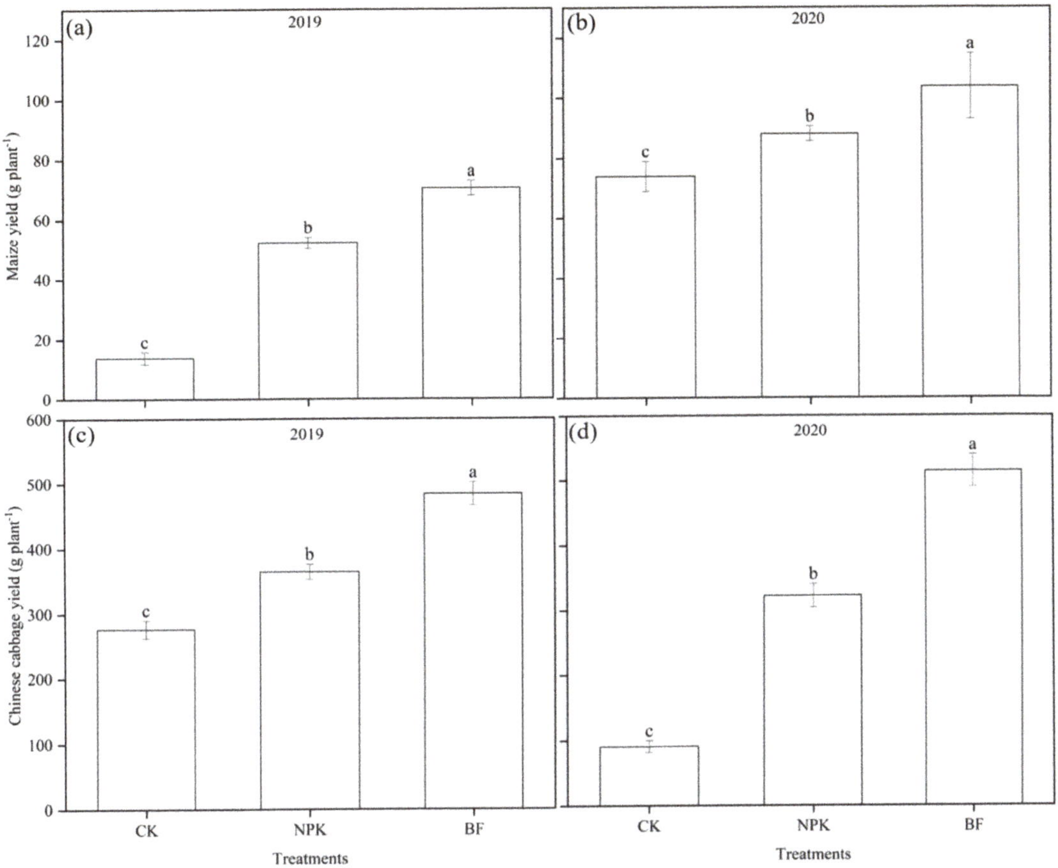

Figure 1. Yields of maize and cabbage in 2019–2020 under different N treatments. (**a**), Maize yield in 2019. (**b**), Maize yield in 2020. (**c**), Chinese cabbage yield in 2019. (**d**), Chinese cabbage yield in 2020. The error bars show the standard deviations of the means (n = 3). Different lowercase letters indicate significant differences among treatments ($p < 0.05$).

3.2. Crop N Uptake and Utilization Efficiency

Figure 4 shows the N uptake of maize and Chinese cabbage in 2019–2020 under different fertilization treatments. Compared to the zero-N treatment, the N uptake of maize grain and Chinese cabbage increased with different types of fertilizer (Figure 3). Compared to the NPK treatment, the BF treatment significantly improved the N uptake of maize in 2019 (Figure 4a) but not of Chinese cabbage in 2020 (Figure 4d). There was no significant difference between the NPK and BF treatments in terms of the N uptake of maize in 2020 (Figure 4b) or the N uptake of Chinese cabbage in 2019 (Figure 4c).

Figure 5 shows the N utilization efficiency of maize and Chinese cabbage in 2019–2020 under the NPK and BF treatments. Compared with the NPK treatment, the BF treatment had the highest average N utilization efficiency in maize and Chinese cabbage, at 44.31 kg kg^{-1} and 40.73 kg kg^{-1}, respectively (Figure 5a–d).

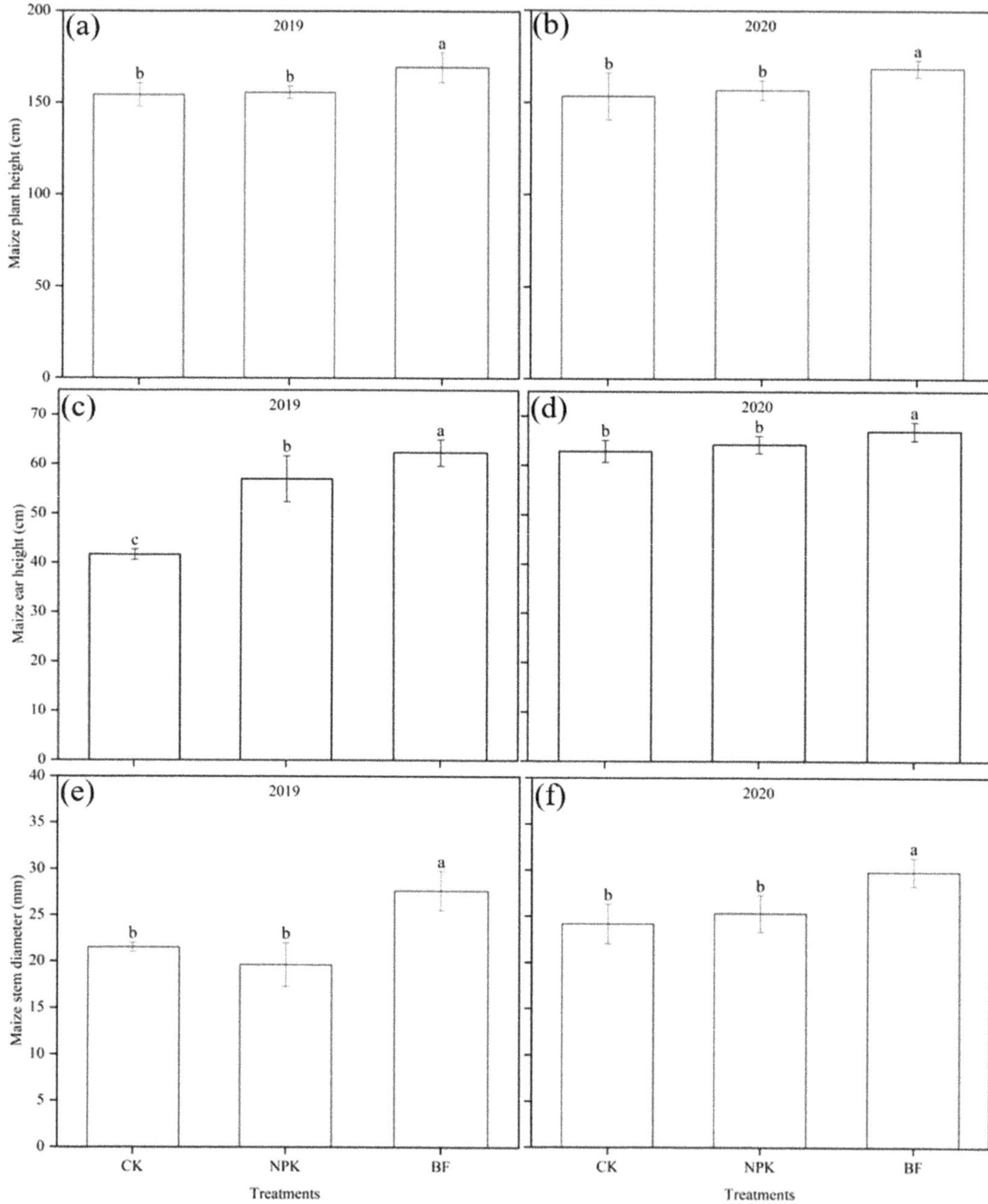

Figure 2. Growth-related traits of maize in 2019–2020 under different N treatments. (**a**), Plant height of maize in 2019. (**b**), Plant height of maize in 2020. (**c**), Ear height of maize in 2019. (**d**), Ear height of maize in 2020. (**e**), Stem diameter of maize in 2019. (**f**), Stem diameter of maize in 2020. The error bars show the standard deviations of the means ($n = 3$). Different lowercase letters indicate significant differences among treatments ($p < 0.05$).

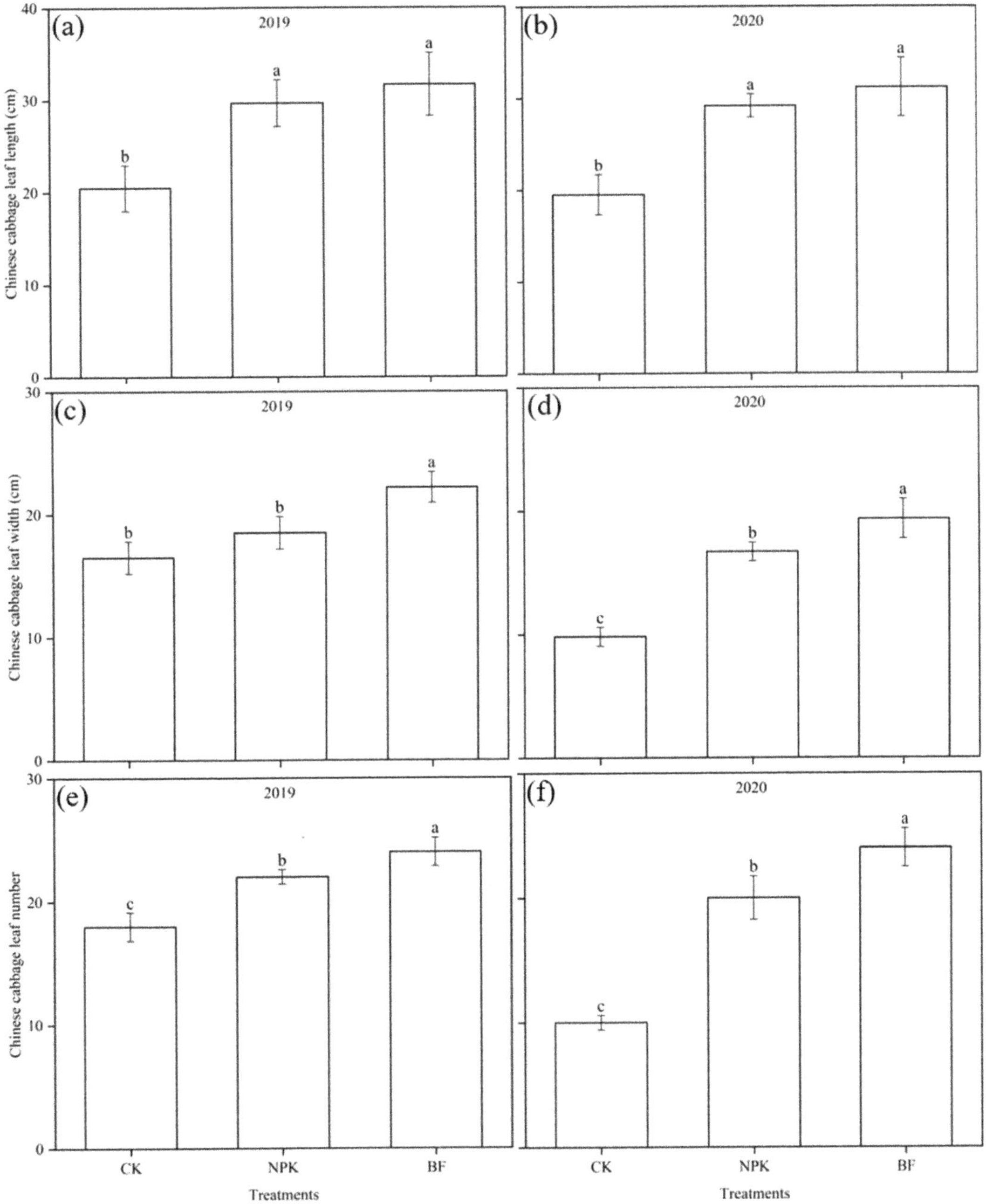

Figure 3. Growth-related traits of Chinese cabbage in 2019–2020 under different N treatments. (**a**), Leaf length of Chinese cabbage in 2019. (**b**), Leaf length of Chinese cabbage in 2020. (**c**), Leaf width of Chinese cabbage in 2019. (**d**), Leaf width of Chinese cabbage in 2020. (**e**), Leaf number of Chinese cabbage in 2019. (**f**), Leaf number of Chinese cabbage in 2020. The error bars show standard deviations of the means (n = 3). Different lowercase letters indicate significant differences among treatments ($p < 0.05$).

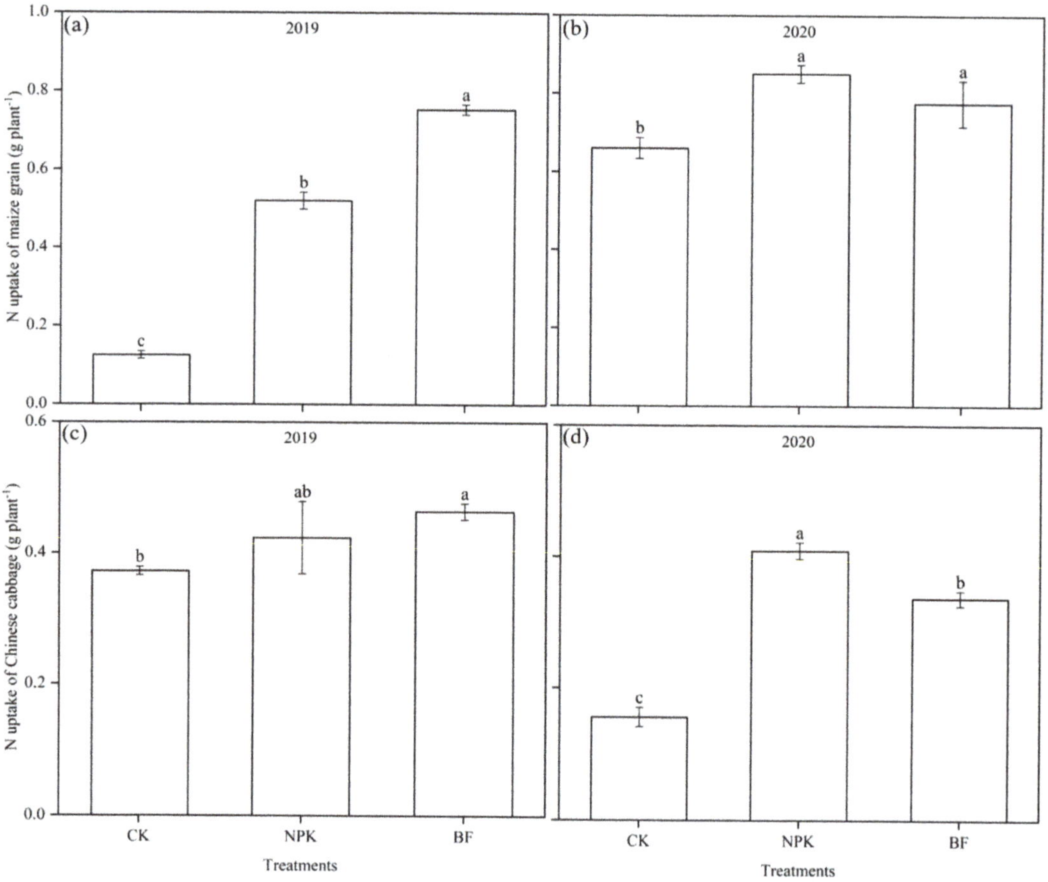

Figure 4. N uptake of maize and Chinese cabbage in 2019–2020 under different N treatments. (**a**), N uptake of maize in 2019. (**b**), N uptake of maize in 2020. (**c**), N uptake of Chinese cabbage in 2019. (**d**), N uptake of Chinese cabbage in 2020. The error bars show the standard deviations of the means ($n = 3$). Different lowercase letters indicate significant differences among treatments ($p < 0.05$).

3.3. Crop Quality

Figure 4 shows the nutritional quality of maize and cabbage under the different types of fertilizers. Compared with the zero-N treatment, the contents of maize grain sugar and crude protein under the NPK treatment significantly improved by 16.21% and 10.95%, respectively (Figure 6b,c), and the sugar content of Chinese cabbage under the NPK treatment significantly improved by 65.89% (Figure 6e). Compared to the zero-N treatment, the contents of maize grain starch, sugar, and crude protein under the BF treatment significantly improved by 6.56%, 15.36%, and 12.58%, respectively (Figure 6a–c). The contents of Chinese cabbage amino acids, sugar, and vitamin C under the BF treatment improved significantly in comparison to the zero-N treatment by 17.98%, 91.23%, and 15.43%, respectively (Figure 6d–f). In comparison with the NPK treatment, the content of maize grain starch under the BF treatment significantly improved by 4.91% (Figure 6a), and the contents of Chinese cabbage amino acids, sugar, and vitamin C under the BF treatment significantly increased by 13.84%, 15.28%, and 19.94%, respectively (Figure 6d–f).

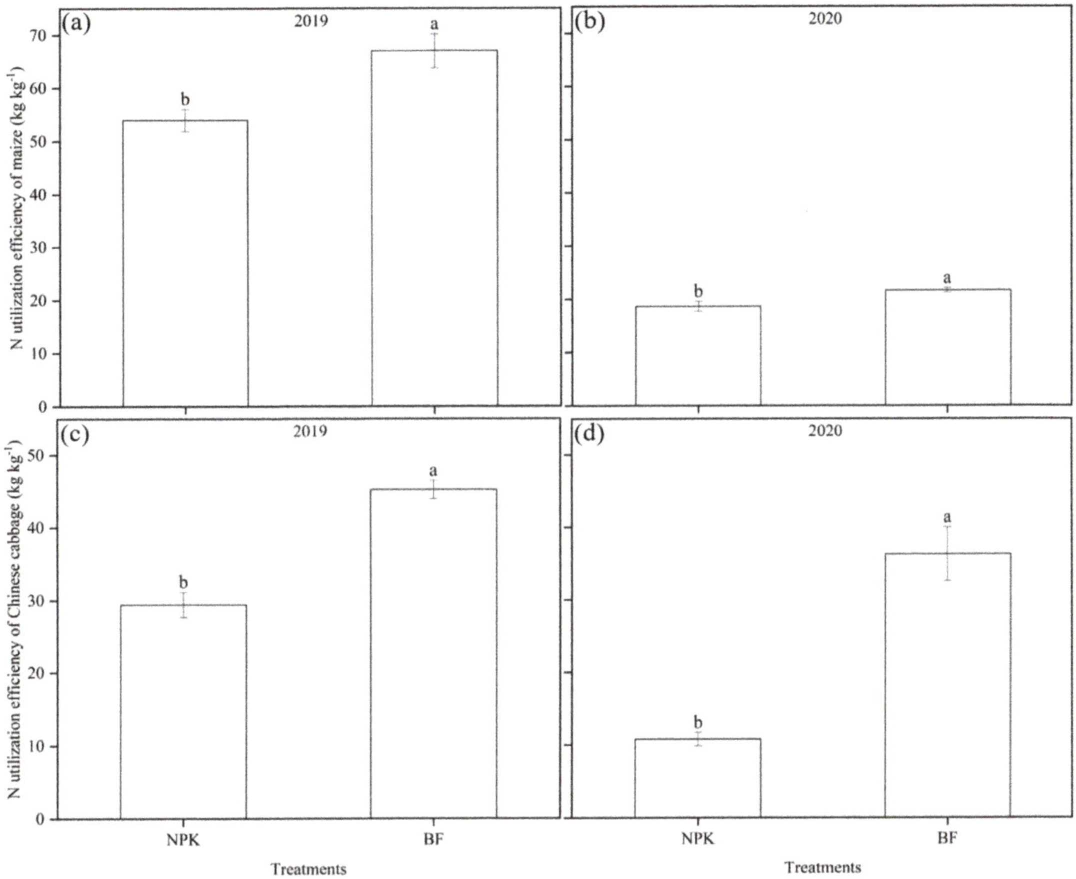

Figure 5. N utilization efficiency of maize and Chinese cabbage in 2019–2020 under different N treatments. (**a**), N utilization efficiency of maize in 2019. (**b**), N utilization efficiency of maize in 2020. (**c**), N utilization efficiency of Chinese cabbage in 2019. (**d**), N utilization efficiency of Chinese cabbage in 2020. The error bars show the standard deviations of the means (n = 3). Different lowercase letters indicate significant differences among treatments ($p < 0.05$).

3.4. Dynamic Changes in Soil NH_4^+-N, NO_3^--N, and Inorganic N

Figure 7a shows the dynamic changes in soil NH_4^+-N under different fertilization treatments of maize and cabbage at different growth stages. With the development of the maize and cabbage growth stages, soil NH_4^+-N showed a mild fluctuation and a declining trend under zero-N fertilizer. The 2020 cabbage harvest stage had the lowest content of soil NH_4^+-N (Figure 7a). The content of soil NH_4^+-N under the NPK and BF treatments was higher than that under the zero-N treatment in each growth stage of maize and cabbage. In both BF and NPK treatment soils, the trend of NH_4^+-N increased during the rotation cycle of maize and cabbage and then decreased. At all growth stages of the continuous maize and cabbage rotation, the content of soil NH_4^+-N under the BF treatment was higher than that under the NPK treatment.

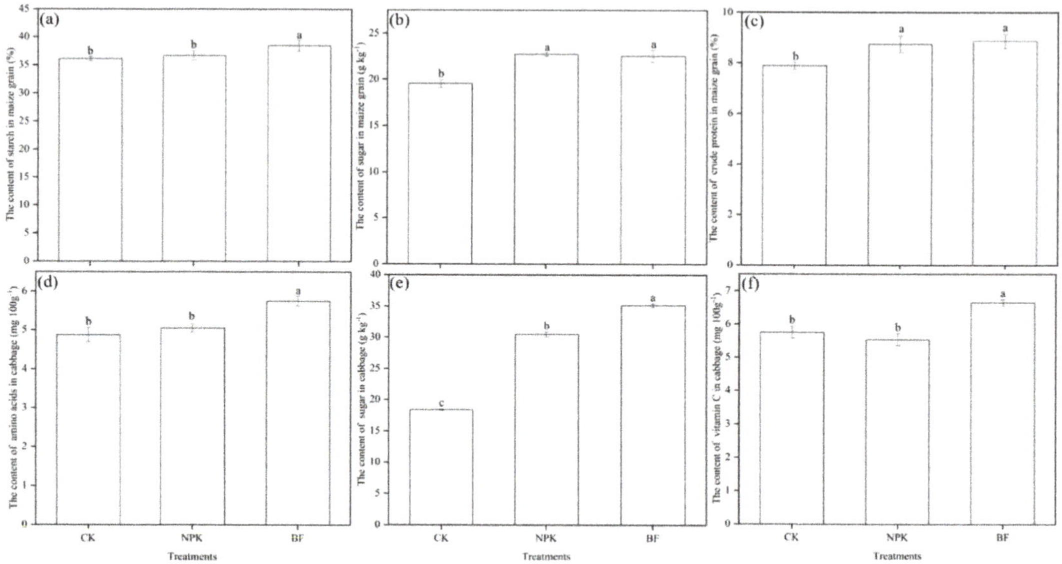

Figure 6. Quality of the maize and cabbage under different N treatments in 2020. (**a**), The content of starch in maize grain. (**b**), The sugar content in maize grain. (**c**), The content of crude protein in maize grain. (**d**), The amino acid content in cabbage. (**e**), The sugar content in cabbage. (**f**), The content of vitamin C in cabbage. The error bars show the standard deviations of the means (n = 3). Different lowercase letters indicate significant differences among treatments ($p < 0.05$).

Figure 7b shows dynamic changes in soil NO_3^--N under different fertilization treatments of maize and cabbage at different growth stages. With the development of the maize and cabbage growth stages, the soil NO_3^--N showed a mild fluctuation and an increasing trend in the zero-N fertilizer treatment, whereas the opposite was true for the NH_4^+-N change trend. Compared to the zero-N treatment, the soil NO_3^--N under the NPK fertilizer treatment was significantly increased only at the 2019 maize jointing and heading stages, the 2019 cabbage harvesting stage, the 2020 maize heading and harvesting stages, and the 2020 cabbage seedling and rosette stages. At all growth stages of the continuous maize and cabbage rotation, the soil NO_3^--N under the BF treatment was higher than that under the zero N treatment. The soil NO_3^--N under the BF treatment (except for the 2019 maize jointing stage) was higher than that under the NPK treatment in each stage of the maize and cabbage rotation.

Figure 7c shows dynamic changes in soil inorganic N under different fertilization treatments of maize and cabbage at different growth stages. Soil inorganic N under the zero-N treatment at all stages was low and tended to fluctuate slightly. At all growth stages of the continuous maize and cabbage rotation, the soil inorganic N under the NPK and BF treatments was higher than that under the zero N treatment. The soil inorganic N under the BF treatment (except for the 2019 maize jointing and 2020 cabbage rosette stages) was higher than that under the NPK treatment at each stage of the maize and cabbage rotation.

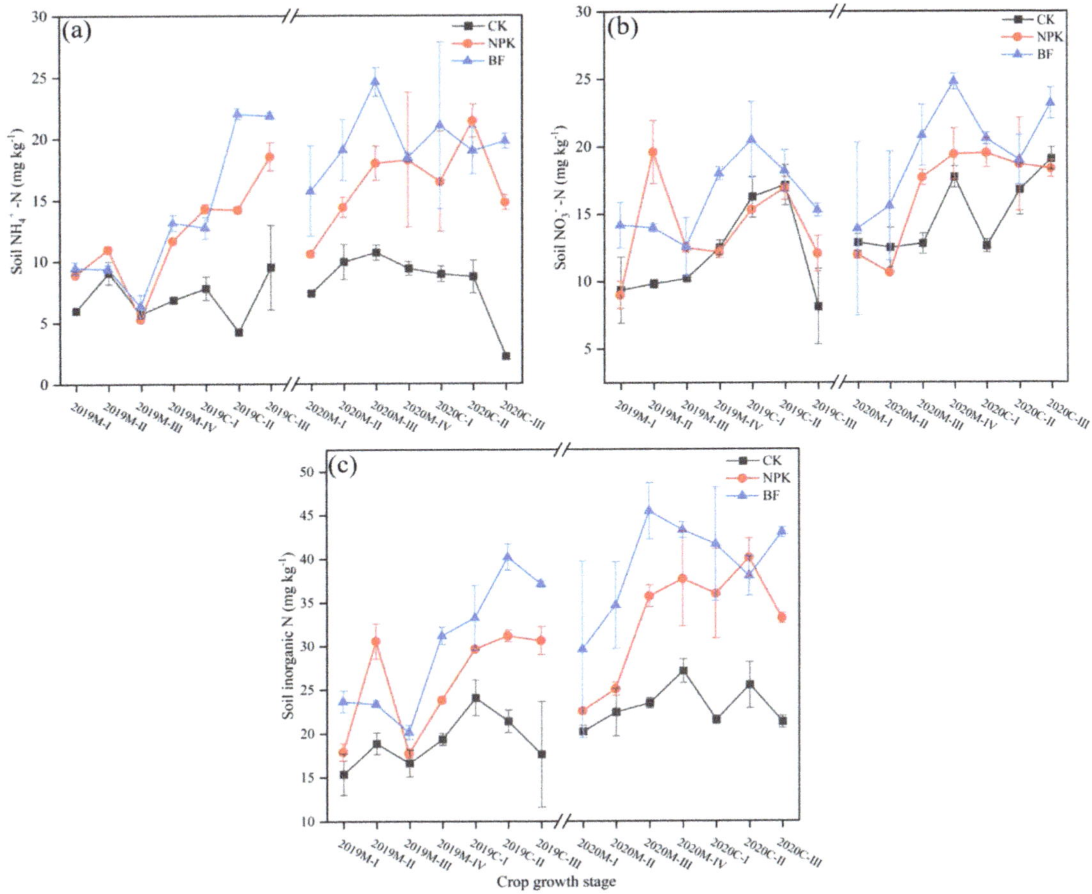

Figure 7. Dynamic changes in the soil NH_4^+-N, NO_3^--N, and inorganic N levels under different fertilization treatments of maize and cabbage at different growth stages between 2019 and 2020. (**a**), The dynamic changes in soil NH_4^+-N (**b**), the dynamic changes in soil NO_3^--N (**c**), The dynamic changes in soil inorganic N. 2019M-I, 2019M-II, 2019M-III, and 2019M-IV represent the seedling stage, jointing stage, heading stage, and harvest stage of maize, respectively, in 2019. 2019C-I, 2019C-II, and 2019C-III represent the seedling stage, rosette stage, and harvest stage of cabbage, respectively, in 2019. 2020M-I, 2020M-II, 2020M-III, and 2020M-IV represent the seedling stage, jointing stage, heading stage, and harvest stage of maize, respectively, in 2020. 2020C-I, 2020C-II, and 2020C-III represent the seedling stage, rosette stage, and harvest stage of cabbage, respectively, in 2020. The error bars show the standard deviations of the means (n = 3).

4. Discussion

To maximize crop yield, fertilization is a critical agricultural practice, and crop yield is strongly influenced by the type and number of fertilizers used [31,32]. Changes in the maize grain and cabbage yield under the different N fertilization treatments (Figure 1) showed a significant but strong effect of N fertilizer on maize grain and cabbage production, which were higher under the NPK and BF treatments than under zero N. Crop yield is affected by the number of fertilizers as well as the type [33]. On the whole, reasonable N fertilizer use appropriately increases crop yields [12,34]. The application of BF enhanced crop productivity compared to conventional fertilization and no fertilizer application, according to Melanie et al. 2022 [22]. The results of this study show that in comparison with the NPK treatment, the BF application rates improved the yields of maize grain and Chinese cabbage over the two rotation cycles (2019–2020). Similar results have also been found in other studies examining the effects of carbon-based fertilizers on crop yields [3,15]. BF maintains the high availability of soil nutrients during crop growth while improving crop nutrient utilization efficiency, thus promoting crop yield.

The most critical agronomic traits of maize are plant height, ear height, and stem diameter, all of which are directly related to crop lodging resistance, biomass production, and yield [35–38]. In general, different effects were observed for each of the agronomic traits of maize (plant height, ear height, and stem diameter) when N fertilization was applied, as shown in Figure 2. In comparison with the zero-N treatment, the NPK fertilizer application only improved maize ear height, which resulted in poor maize lodging resistance. In contrast, BF increased plant height, ear height, and stem diameter of maize simultaneously and increased maize lodging resistance more effectively. Compared with the zero-N treatment, the NPK and BF applications significantly increased the Chinese cabbage leaf length, leaf width, and leaf number (Figure 3). Chinese cabbage leaf length, leaf width, and leaf number improved more with BF application than with NPK fertilization (Figure 3). Hence, biochar-based slow-release fertilizer is more effective in improving Chinese cabbage biomass production. Biochar has improved the growth characteristics of different crops [39].

In agricultural production, efficient N management is crucial to minimize N losses and improve N uptake and N use efficiency [40,41]. It is important to note that inefficient fertilizers can lose large amounts of nutrients through leaching, volatilization, denitrification, immobilization, erosion, and runoff, reduce crop nutrient efficiency, and cause environmental pollution. Crops with high N uptake efficiency have high yields, and crop yield is correlated with N uptake [42,43]. Agricultural development requires slow-release fertilizer, especially when nutrient losses are high. Slow-release fertilizer has the advantage of improving the nutrient absorption efficiency of crops, resulting in more uniform fertilizer release during the growing season through one-time fertilization and less excessive absorption of nutrients by crops [44]. Compared to the zero-N treatment, the NPK and BF applications improved the N uptake of the maize grain and Chinese cabbage (Figure 4). Compared to NPK application, the BF treatment significantly increased the N utilization efficiency of maize and Chinese cabbage (Figure 5). Thus, BF was able to meet the N nutrient demands of maize and cabbage better than NPK. This occurred because carbon-based slow-release fertilizer contains biochar that adsorbs N through chemical and physical adsorption. The efficiency of different N sources depends on the type of N fertilizer used [45]. Carbon-based fertilizer, as a combined product of nutrients and biochar, can enhance the positive interactions between N fertilizer and biochar in soils [46]. N from pure urea is an amide N, and all N in BF is an efficient N form [4]. Biochar is considered a key property for retaining soil nutrients in a form that is suitable for crop use [47]. The pore structure is one of the main reasons for the improvement of the slow-release performance of biochar-based slow-release fertilizers [48]. A study revealed that by using an infiltration method of urea and biochar, urea penetrated adequately into the pores of the biochar and was evenly distributed on its surface; this method met the N requirements of crops better than the direct mixing of biochar and urea [4]. Biochar's high organic carbon content, large surface area, high microporosity, and range of functional groups also help plants retain nutrients [16].

Crop quality is important for human health and economic value and optimizing the management of N fertilizer can significantly enhance crop quality [49,50]. Maize grain consists mainly of starch (70%) and protein (10%), while the quality of grains is largely determined by the amount and composition of the protein [51]. The NPK treatment significantly increased the maize grain sugar and crude protein contents compared to the zero-N treatment (Figure 6b,c). However, the BF treatment not only significantly increased the maize grain sugar and crude protein contents but also significantly increased the maize grain starch content (Figure 6a–c). There was a significant increase in maize grain starch in the BF treatment compared to the NPK treatment (Figure 6a). Studies have demonstrated that the appropriate application of N fertilizer can maximize the nutritional value of maize [52,53]. Moreover, the quality of the nutritional components of vegetables is very important for human health [54]. Compared to the zero-N treatment, the NPK treatment significantly increased the cabbage sugar content (Figure 6e). The BF treatment significantly improved not only the cabbage sugar but also the amino acid and vitamin C contents (Figure 6d–f). Additionally, there were significant improvements in the contents of amino acids, sugar, and vitamin C in cabbage under the BF treatment compared to the NPK treatment (Figure 6d–f).

The different N treatments had varying effects on the dynamics and mechanisms of soil inorganic N and N forms in the maize and cabbage rotation system (Figure 7). Compared with the zero-N treatment, the addition of N fertilizer enhanced NH_4^+-N, NO_3^--N, and inorganic N in the soil during each growth stage of maize and cabbage, improving the utilization of crop N as well as the yield and quality. The N forms under the zero-N treatment showed small fluctuations. In stark contrast, the N forms in the NPK and BF treatments fluctuated upward with a wide fluctuation range. We found an interesting phenomenon in which, as the maize and cabbage rotation continued, conventional N fertilizer mainly had a strong and positive impact on soil NH_4^+-N (Figure 7a), whereas the effect on NO_3^--N was relatively weak (Figure 7b). In contrast, NH_4^+-N and NO_3^--N continued to increase in the soil treated with BF. Biochar-based slow-release fertilizers occur in the plant-absorbable form of N, so they have different effects on soil N forms. Studies have found that biochar use significantly increases plant uptake of added NO_3^--N while reducing the uptake of added NH_4^+-N [55]. An optimal N rate regime balances crop demand with soil availability [56]. Soil NO_3^--N accumulation is significantly related to crop yield and aboveground biomass [57]. Therefore, biochar-based slow-release fertilizer has a higher crop N use efficiency and higher yield potential than conventional N fertilizer.

In general, the N release rate of the common fertilizer at the earlier stage of maize and cabbage growth was higher. Nevertheless, it was relatively lower at the later stage of crop growth. In contrast, the N nutrient release rate of the biochar-based slow-release fertilizer at the earlier stage of crop growth was lower, but it was relatively higher at the later stage of development. These nutrient release dynamics are more in line with the nutrient demand regulation of crop growth, and inorganic N was high, as shown in Figure 7c. Therefore, the application of BF can significantly improve the efficient uptake and utilization of nutrients by crops, thus increasing crop yield levels and improving quality. Most plants grow best with a mixture of NH_4^+-N and NO_3^--N, while NH_4^+-N as a sole N source may inhibit their growth [41,58]. Urea, a conventional N fertilizer, is rapidly released into the soil upon application. The N in urea is easily hydrolyzed to ammonium N, the majority of which is then converted to nitrate by rapid nitrification, which can negatively affect the soil environment, groundwater, and atmosphere through leaching, runoff, and volatilization [45,59–61]. Figure 7c shows that in 2019, traditional N fertilizer sharply increased the soil inorganic N levels after the jointing stage of maize and then sharply decreased. In contrast, in the 2019–2020 growing season, the soil inorganic N level continued to increase gradually under the BF treatment, which also indicated the slow release and long duration of N nutrients from the slow-release fertilizer. Biochar can reduce the leaching of dinitrate and ammonium in the soil, thus significantly increasing the soil N content [62]. Biochar-based fertilizer is an effective slow-release fertilizer that reduces soil inorganic N loss in a maize and cabbage rotation system.

5. Conclusions

The biochar-based slow-release fertilizer significantly improved the crop yield, growth-related traits, quality, and N utilization efficiency in this maize–Chinese cabbage rotation system. Biochar-based slow-release application increased soil NH_4^+-N, NO_3^--N, and inorganic N. The biochar-based slow-release nutrient pattern more closely matched crop nutrient needs as each stage of plant growth advanced. The application of biochar-based slow-release fertilizer is a nutrient-efficient management strategy for maize–Chinese cabbage rotation systems, increasing crop productivity while reducing negative environmental impacts and promoting sustainable agriculture. We expect that the selection of efficient slow-release fertilizers in maize and vegetable rotations will sustainably increase N efficiency and crop production potential in the hilly area of Southwest China.

Author Contributions: Conceptualization, H.Z., T.X. and H.X.; methodology, H.Z.; software, H.Z.; validation, H.Z., H.X. and T.X.; formal analysis, H.Z.; investigation, H.Z. and T.X.; writing—original draft preparation, H.Z.; writing—review and editing, H.Z. and M.G.; visualization, H.Z.; supervision, M.G. All authors have read and agreed to the published version of the manuscript.

Funding: This study was financially supported by the National Key Research and Development Plan of China (Grant No. 2018YFD0200705).

Institutional Review Board Statement: Not applicable.

Informed Consent Statement: Not applicable.

Data Availability Statement: The data will be provided upon request by the corresponding author.

Conflicts of Interest: The authors declare no conflict of interest.

References

1. Azeem, B.; KuShaari, K.; Man, Z.B.; Basit, A.; Thanh, T.H. Review on Materials & Methods to Produce Controlled Release Coated Urea Fertilizer. *J. Control. Release* **2014**, *181*, 11–21. [CrossRef] [PubMed]
2. Brown, M.E.; Hintermann, B.; Higgins, N. Markets, Climate Change, and Food Security in West Africa. *Environ. Sci. Technol.* **2009**, *43*, 8016–8020. [CrossRef] [PubMed]
3. Rashid, M.; Hussain, Q.; Khan, K.S.; Alwabel, M.I.; Hayat, R.; Akmal, M.; Ijaz, S.S.; Alvi, S.; Obaid-ur-Rehman. Carbon-Based Slow-Release Fertilizers for Efficient Nutrient Management: Synthesis, Applications, and Future Research Needs. *J. Soil Sci. Plant Nutr.* **2021**, *21*, 1144–1169. [CrossRef]
4. Xiang, A.H.; Qi, R.Y.; Wang, M.F.; Zhang, K.; Jiang, E.C.; Ren, Y.Z.; Hu, Z.W. Study on the Infiltration Mechanism of Molten Urea and Biochar for a Novel Fertilizer Preparation. *Ind. Crops Prod.* **2020**, *153*, 112558. [CrossRef]
5. Zhang, X.; Davidson, E.A.; Mauzerall, D.L.; Searchinger, T.D.; Dumas, P.; Shen, Y. Managing Nitrogen for Sustainable Development. *Nature* **2015**, *528*, 51–59. [CrossRef]
6. Gao, K.; Yu, Y.F.; Xia, Z.T.; Yang, G.; Xing, Z.L.; Qi, L.T.; Ling, L.Z. Response of Height, Dry Matter Accumulation and Partitioning of Oat (*Avena Sativa*, L.) to Planting Density and Nitrogen in Horqin Sandy Land. *Sci. Rep.* **2019**, *9*, 7961. [CrossRef]
7. Wu, X.; Hu, H.; Li, S.J.; Zhao, J.N.; Li, J.; Zhang, G.L.; Li, G.; Xiu, W.M. Chemical Fertilizer Reduction with Organic Material Amendments Alters Co-Occurrence Network Patterns of Bacterium-Fungus-Nematode Communities under the Wheat–Maize Rotation Regime. *Plant Soil* **2022**, *473*, 605–623. [CrossRef]
8. Saha, S.; Gopinath, K.A.; Mina, B.L.; Gupta, H.S. Influence of Continuous Application of Inorganic Nutrients to a Maize–Wheat Rotation on Soil Enzyme Activity and Grain Quality in a Rainfed Indian Soil. *Eur. J. Soil Biol.* **2008**, *44*, 521–531. [CrossRef]
9. Francioli, D.; Schulz, E.; Lentendu, G.; Wubet, T.; Buscot, F.; Reitz, T. Mineral vs. Organic Amendments: Microbial Community Structure, Activity and Abundance of Agriculturally Relevant Microbes Are Driven by Long-Term Fertilization Strategies. *Front. Microbiol.* **2016**, *7*, 1446. [CrossRef]
10. González, M.E.; Cea, M.; Medina, J.; González, A.; Diez, M.C.; Cartes, P.; Monreal, C.; Navia, R. Evaluation of Biodegradable Polymers as Encapsulating Agents for the Development of a Urea Controlled-Release Fertilizer Using Biochar as Support Material. *Sci. Total Environ.* **2015**, *505*, 446–453. [CrossRef]
11. Trinh, M.V.; Tesfai, M.; Borrell, A.; Nagothu, U.S.; Bui, T.P.L.; Quynh, V.D.; Thanh, L.Q. Effect of Organic, Inorganic and Slow-Release Urea Fertilisers on CH_4 and N_2O Emissions from Rice Paddy Fields. *Paddy Water Environ.* **2017**, *15*, 317–330. [CrossRef]
12. Xie, J.; Wang, J.; Hu, Q.J.; Zhang, Y.; Wan, Y.; Zhang, C.M.; Zhang, Y.Q.; Shi, X.J. Optimal N Management Improves Crop Yields and Soil Carbon, Nitrogen Sequestration in Chinese Cabbage-Maize Rotation. *Arch. Agron. Soil Sci.* **2022**, 1–14. [CrossRef]
13. Wang, C.; Lv, J.; Coulter, J.A.; Xie, J.M.; Yu, J.H.; Li, J.; Zhang, J.; Tang, C.N.; Niu, T.H.; Gan, Y.T. Slow-Release Fertilizer Improves the Growth, Quality, and Nutrient Utilization of Wintering Chinese Chives (*Allium tuberosum* Rottler Ex Spreng.). *Agronomy* **2020**, *10*, 381. [CrossRef]
14. Zhao, Y.X.; Fan, Z.; Chen, Y.R.; Huang, X.X.; Zhai, S.; Sun, S.C.; Tian, X.F. A Bio-Based Hydrogel Derived from Moldy Steamed Bread as Urea-Formaldehyde Loading for Slow-Release and Water-Retention Fertilizers. *ACS Omega* **2021**, *6*, 33462–33469. [CrossRef]
15. Ndoung, O.C.N.; de Figueiredo, C.C.; Ramos, M.L.G. A Scoping Review on Biochar-Based Fertilizers: Enrichment Techniques and Agro-Environmental Application. *Heliyon* **2021**, *7*, e08473. [CrossRef]
16. Mousavi, S.M.; Srivastava, A.K.; Cheraghi, M. Soil Health and Crop Response of Biochar: An Updated Analysis. *Arch. Agron. Soil Sci.* **2022**, 1–26. [CrossRef]

17. Khan, M.A.; Kim, K.W.; Mingzhi, W.; Lim, B.K.; Lee, W.H.; Lee, J.Y. Nutrient-Impregnated Charcoal: An Environmentally Friendly Slow-Release Fertilizer. *Environmentalist* **2008**, *28*, 231–235. [CrossRef]
18. El-Naggar, A.; Lee, S.S.; Rinklebe, J.; Farooq, M.; Song, H.; Sarmah, A.K.; Zimmerman, A.R.; Ahmad, M.; Shaheen, S.M.; Ok, Y.S. Biochar Application to Low Fertility Soils: A Review of Current Status, and Future Prospects. *Geoderma* **2019**, *337*, 536–554. [CrossRef]
19. Gao, Y.; Shao, G.C.; Yang, Z.; Zhang, K.; Lu, J.; Wang, Z.Y.; Wu, S.Q.; Xu, D. Influences of Soil and Biochar Properties and Amount of Biochar and Fertilizer on the Performance of Biochar in Improving Plant Photosynthetic Rate: A Meta-Analysis. *Eur. J. Agron.* **2021**, *130*, 126345. [CrossRef]
20. Marcińczyk, M.; Oleszczuk, P. Biochar and Engineered Biochar as Slow-And Controlled-Release Fertilizers. *J. Clean. Prod.* **2022**, *339*, 130685. [CrossRef]
21. Wen, P.; Wu, Z.S.; Han, Y.J.; Cravotto, G.; Wang, J.; Ye, B.C. Microwave-Assisted Synthesis of a Novel Biochar-Based Slow-Release Nitrogen Fertilizer with Enhanced Water-Retention Capacity. *ACS Sustain. Chem. Eng.* **2017**, *5*, 7374–7382. [CrossRef]
22. Melo, L.C.A.; Lehmann, J.; da Carneiro, J.S.S.; Camps-Arbestain, M. Biochar-Based Fertilizer Effects on Crop Productivity: A Meta-Analysis. *Plant Soil* **2022**, *472*, 45–58. [CrossRef]
23. Liu, J.; Liu, H.; Huang, S.M.; Yang, X.Y.; Wang, B.R.; Li, X.Y.; Ma, Y.B. Nitrogen Efficiency in Long-Term Wheat–Maize Cropping Systems under Diverse Field Sites in China. *Field Crops Res.* **2010**, *118*, 145–151. [CrossRef]
24. Zhang, M.; Tian, Y.H.; Zhao, M.; Yin, B.; Zhu, Z.L. The Assessment of Nitrate Leaching in a Rice–Wheat Rotation System Using an Improved Agronomic Practice Aimed to Increase Rice Crop Yields. *Agric. Ecosyst. Environ.* **2017**, *241*, 100–109. [CrossRef]
25. Zhang, T.; Chen, A.Q.; Liu, J.; Liu, H.B.; Lei, B.K.; Zhai, L.M.; Zhang, D.; Wang, H.Y. Cropping Systems Affect Paddy Soil Organic Carbon and Total Nitrogen Stocks (in Rice-Garlic and Rice-Fava Systems) in Temperate Region of Southern China. *Sci. Total Environ.* **2017**, *609*, 1640–1649. [CrossRef] [PubMed]
26. Luo, M.J.; Shi, Y.X.; Yang, Y.; Zhao, Y.X.; Zhang, Y.X.; Shi, Y.M.; Kong, M.S.; Li, C.H.; Feng, Z.; Fan, Y.L.; et al. Sequence Polymorphism of the Waxy Gene in Waxy Maize Accessions and Characterization of a New Waxy Allele. *Sci. Rep.* **2020**, *10*, 15851. [CrossRef]
27. Xie, Y.X.; Li, J.P.; Guo, X.; Zhao, J.; Yang, B.; Xiao, W.W.; Yang, H.F. Health Status among Greenhouse Workers Exposed to Different Levels of Pesticides: A Genetic Matching Analysis. *Sci. Rep.* **2020**, *10*, 8714. [CrossRef]
28. Bai, H.Y.; Xiao, J.Z. *Experimental Research and Statistical Analysis*; World Book Press: Xian, China, 1998. (In Chinese)
29. Bao, S. *Methods for Soil Agricultural and Chemical Analysis*; China Agriculture Press: Beijing, China, 2000. (In Chinese)
30. Hartmann, T.E.; Yue, S.C.; Schulz, R.; He, X.K.; Chen, X.P.; Zhang, F.S.; Müller, T. Yield and N Use Efficiency of a Maize–Wheat Cropping System as Affected by Different Fertilizer Management Strategies in a Farmer's Field of the North China Plain. *Field Crops Res.* **2015**, *174*, 30–39. [CrossRef]
31. Manna, M.C.; Swarup, A.; Wanjari, R.H.; Mishra, B.; Shahi, D.K. Long-Term Fertilization, Manure and Liming Effects on Soil Organic Matter and Crop Yields. *Soil Tillage Res.* **2007**, *94*, 397–409. [CrossRef]
32. Körschens, M.; Albert, E.; Armbruster, M.; Barkusky, D.; Baumecker, M.; Behle-Schalk, L.; Bischoff, R.; Čergan, Z.; Ellmer, F.; Herbst, F.; et al. Effect of Mineral and Organic Fertilization on Crop Yield, Nitrogen Uptake, Carbon and Nitrogen Balances, as Well as Soil Organic Carbon Content and Dynamics: Results from 20 European Long-Term Field Experiments of the Twenty-First Century. *Arch. Agron. Soil Sci.* **2013**, *59*, 1017–1040. [CrossRef]
33. Shi, W.; Bian, R.J.; Li, L.Q.; Lian, W.L.; Liu, X.Y.; Zheng, J.F.; Cheng, K.; Zhang, X.H.; Drosos, M.; Joseph, S.; et al. Assessing the Impacts of Biochar-blended Urea on Nitrogen Use Efficiency and Soil Retention in Wheat Production. *GCB Bioenergy* **2022**, *14*, 65–83. [CrossRef]
34. Qian, L.; Chen, L.; Joseph, S.; Pan, G.X.; Li, L.Q.; Zheng, J.W.; Zhang, X.H.; Zheng, J.F.; Yu, X.Y.; Wang, J.F. Biochar Compound Fertilizer as an Option to Reach High Productivity but Low Carbon Intensity in Rice Agriculture of China. *Carbon Manag.* **2014**, *5*, 145–154. [CrossRef]
35. Ma, Q.H. The Expression of Caffeic Acid 3-O-Methyltransferase in Two Wheat Genotypes Differing in Lodging Resistance. *J. Exp. Bot.* **2009**, *60*, 2763–2771. [CrossRef]
36. Miclaus, M.; Balacescu, O.; Has, I.; Balacescu, L.; Has, V.; Suteu, D.; Neuenschwander, S.; Keller, I.; Bruggmann, R. Maize Cytolines Unmask Key Nuclear Genes That Are under the Control of Retrograde Signaling Pathways in Plants. *Genome Biol. Evol.* **2016**, *8*, 3256–3270. [CrossRef] [PubMed]
37. Wei, L.; Zhang, X.; Zhang, Z.H.; Liu, H.H.; Lin, Z.W. A New Allele of the Brachytic2 Gene in Maize Can Efficiently Modify Plant Architecture. *Heredity* **2018**, *121*, 75–86. [CrossRef] [PubMed]
38. Chen, Y.Q.; Zhou, Q.Q.; Tian, R.M.; Ma, Z.H.; Zhao, X.F.; Tang, J.H.; Fu, Z.Y. Proteomic Analysis Reveals That Auxin Homeostasis Influences the Eighth Internode Length Heterosis in Maize (*Zea mays*). *Sci. Rep.* **2018**, *8*, 7159. [CrossRef]
39. Ali, L.; Manzoor, N.; Li, X.Q.; Naveed, M.; Nadeem, S.M.; Waqas, M.R.; Khalid, M.; Abbas, A.; Ahmed, T.; Li, B.; et al. Impact of Corn Cob-Derived Biochar in Altering Soil Quality, Biochemical Status and Improving Maize Growth under Drought Stress. *Agronomy* **2021**, *11*, 2300. [CrossRef]
40. Abbasi, M.K.; Tahir, M.M.; Sadiq, A.; Iqbal, M.; Zafar, M. Yield and Nitrogen Use Efficiency of Rainfed Maize Response to Splitting and Nitrogen Rates in Kashmir, Pakistan. *Agron. J.* **2012**, *104*, 448–457. [CrossRef]
41. Abbasi, M.K.; Tahir, M.M.; Rahim, N. Effect of N Fertilizer Source and Timing on Yield and N Use Efficiency of Rainfed Maize (*Zea Mays* L.) in Kashmir–Pakistan. *Geoderma* **2013**, *195*, 87–93. [CrossRef]

42. Azeez, J.O.; Adetunji, M.T.; Lagoke, S.T.O. Response of Low-Nitrogen Tolerant Maize Genotypes to Nitrogen Application in a Tropical Alfisol in Northern Nigeria. *Soil Tillage Res.* **2006**, *91*, 181–185. [CrossRef]
43. Iqbal, A.; He, L.; Khan, A.; Wei, S.Q.; Akhtar, K.; Ali, I.; Ullah, S.; Munsif, F.; Zhao, Q.; Jiang, L.G. Organic Manure Coupled with Inorganic Fertilizer: An Approach for the Sustainable Production of Rice by Improving Soil Properties and Nitrogen Use Efficiency. *Agronomy* **2019**, *9*, 651. [CrossRef]
44. Abou-Zied, S.T.; Sibak, H.A.; Abd El-Latif, A.L.; Essa, R.E. Diffusion Rate and Activity Index of Some Slow Release Nitrogen Fertilizer. *Egypt. J. Soil Sci.* **2018**, *58*, 255–266. [CrossRef]
45. Van Drecht, G.; Bouwman, A.F.; Knoop, J.M.; Beusen, A.H.W.; Meinardi, C.R. Global Modeling of the Fate of Nitrogen from Point and Nonpoint Sources in Soils, Groundwater, and Surface Water. *Glob. Biogeochem. Cycles* **2003**, *17*, 1115. [CrossRef]
46. Zheng, J.F.; Han, J.M.; Liu, Z.W.; Xia, W.B.; Zhang, X.H.; Li, L.Q.; Liu, X.Y.; Bian, R.J.; Cheng, K.; Zheng, J.W.; et al. Biochar Compound Fertilizer Increases Nitrogen Productivity and Economic Benefits but Decreases Carbon Emission of Maize Production. *Agric. Ecosyst. Environ.* **2017**, *241*, 70–78. [CrossRef]
47. Rasse, D.P.; Weldon, S.; Joner, E.J.; Joseph, S.; Kammann, C.I.; Liu, X.Y.; O'Toole, A.; Pan, G.X.; Kocatürk-Schumacher, N.P. Enhancing Plant N Uptake with Biochar-Based Fertilizers: Limitation of Sorption and Prospects. *Plant Soil* **2022**, *475*, 213–236. [CrossRef]
48. An, X.F.; Wu, Z.S.; Yu, J.Z.; Cravotto, G.; Liu, X.C.; Li, Q.; Yu, B. Copyrolysis of Biomass, Bentonite, and Nutrients as a New Strategy for the Synthesis of Improved Biochar-Based Slow-Release Fertilizers. *ACS Sustain. Chem. Eng.* **2020**, *8*, 3181–3190. [CrossRef]
49. Zhang, J.Z.; Li, B.S.; Zhang, J.L.; Christie, P.; Li, X.L. Organic Fertilizer Application and Mg Fertilizer Promote Banana Yield and Quality in an Udic Ferralsol. *PLoS ONE* **2020**, *15*, e0230593. [CrossRef]
50. Maignan, V.; Bernay, B.; Géliot, P.; Avice, J.C. Biostimulant Effects of Glutacetine®and Its Derived Formulations Mixed with N Fertilizer on Post-Heading N Uptake and Remobilization, Seed Yield, and Grain Quality in Winter Wheat. *Front. Plant Sci.* **2020**, *11*, 607615. [CrossRef]
51. Sánchez-Rodríguez, A.R.; Rey, M.D.; Nechate-Drif, H.; Castillejo, M.Á.; Jorrín-Novo, J.V.; Torrent, J.; del Campillo, M.C.; Sacristán, D. Combining P and Zn Fertilization to Enhance Yield and Grain Quality in Maize Grown on Mediterranean Soils. *Sci. Rep.* **2021**, *11*, 7427. [CrossRef]
52. Rehman, A.; Farrukh Saleem, M.F.; Ehsan Safdar, M.E.; Hussain, S.; Akhtar, N. Grain Quality, Nutrient Use Efficiency, and Bioeconomics of Maize under Different Sowing Methods and NPK Levels. *Chil. J. Agric. Res.* **2011**, *71*, 586–593. [CrossRef]
53. Simić, M.; Dragičević, V.; Drinic, S.M.; Vukadinović, J.; Kresović, B.; Tabaković, M.; Brankov, M. The Contribution of Soil Tillage and Nitrogen Rate to the Quality of Maize Grain. *Agronomy* **2020**, *10*, 976. [CrossRef]
54. Fan, S.K.; Zhu, J.; Tian, W.H.; Guan, M.Y.; Fang, X.Z.; Jin, C.W. Effects of Split Applications of Nitrogen Fertilizers on the Cd Level and Nutritional Quality of Chinese Cabbage. *J. Zhejiang Univ. Sci. B* **2017**, *18*, 897–905. [CrossRef] [PubMed]
55. Kalu, S.; Oyekoya, G.N.; Ambus, P.; Tammeorg, P.; Simojoki, A.; Pihlatie, M.; Karhu, K. Effects of Two Wood-Based Biochars on the Fate of Added Fertilizer Nitrogen—A ^{15}N Tracing Study. *Biol. Fertil. Soils* **2021**, *57*, 457–470. [CrossRef]
56. Thiyagarajan, T.M.; Stalin, P.; Dobermann, A.; Cassman, K.G.; ten Berge, H.F.M. Soil N Supply and Plant N Uptake by Irrigated Rice in Tamil Nadu. *Field Crops Res.* **1997**, *51*, 55–64. [CrossRef]
57. Guo, B.B.; Liu, B.C.; He, L.; Wang, Y.Y.; Feng, W.; Zhu, Y.J.; Jiao, N.Y.; Wang, C.Y.; Guo, T.C. Root and Nitrate-N Distribution and Optimization of N Input in Winter Wheat. *Sci. Rep.* **2019**, *9*, 18018. [CrossRef]
58. Mahmood, T.; Kaiser, W.M. Growth and Solute Composition of the Salt-Tolerant Kallar Grass [*Leptochloa Fusca* (L.) Kunth] as Affected by Nitrogen Source. *Plant Soil* **2003**, *252*, 359–366. [CrossRef]
59. Galloway, J.N.; Townsend, A.R.; Erisman, J.W.; Bekunda, M.; Cai, Z.; Freney, J.R.; Martinelli, L.A.; Seitzinger, S.P.; Sutton, M.A. Transformation of the Nitrogen Cycle: Recent Trends, Questions, and Potential Solutions. *Science* **2008**, *320*, 889–892. [CrossRef]
60. Chen, J.; Lü, S.Y.; Zhang, Z.; Zhao, X.X.; Li, X.M.; Ning, P.; Liu, M.Z. Environmentally Friendly Fertilizers: A Review of Materials Used and Their Effects on the Environment. *Sci. Total Environ.* **2018**, *613*, 829–839. [CrossRef] [PubMed]
61. Gao, Y.X.; Song, X.; Liu, K.X.; Li, T.G.; Zheng, W.K.; Wang, Y.; Liu, Z.G.; Zhang, M.; Chen, Q.; Li, Z.L.; et al. Mixture of Controlled-Release and Conventional Urea Fertilizer Application Changed Soil Aggregate Stability, Humic Acid Molecular Composition, and Maize Nitrogen Uptake. *Sci. Total Environ.* **2021**, *789*, 147778. [CrossRef]
62. Agegnehu, G.; Srivastava, A.K.; Bird, M.I. The Role of Biochar and Biochar-Compost in Improving Soil Quality and Crop Performance: A Review. *Appl. Soil Ecol.* **2017**, *119*, 156–170. [CrossRef]

agriculture

MDPI

Article

Effect of Maize (*Zeal mays*) and Soybean (*Glycine max*) Intercropping on Yield and Root Development in Xinjiang, China

Wenwen Wei, Tingting Liu, Lei Shen, Xiuyuan Wang, Shuai Zhang and Wei Zhang *

College of Agriculture, Shihezi University, Shihezi 832003, China; 20212012015@stu.shzu.edu.cn (W.W.); 20202012015@stu.shzu.edu.cn (T.L.); 20192012011@stu.shzu.edu.cn (L.S.); 20192012027@stu.shzu.edu.cn (X.W.); 20212112022@stu.shzu.edu.cn (S.Z.)
* Correspondence: bluesky2002040@shzu.edu.cn

Abstract: Intercropping is a breakthrough in land-use optimization. This work aimed to study the effects of intercropping patterns on the growth, yield, root morphological characteristics, and interspecific competition of maize and soybean, as well as provide a reference for the development of intercropping patterns of maize and soybean in Northwest China. Three different cropping patterns were designed: monocropping maize, monocropping soybean, and maize-soybean intercropping. Agronomic traits, intercropping indicators such as land equivalent ratio (LER), aggressivity (A), competition ratio (CR), and actual yield loss (AYL), as well as root morphological characteristics were assessed. The results showed that, compared with monocropping, the intercropping maize plant height increased by 6.07–8.40%, and the intercropping soybean plant height increased by 35.27–38.94%; the root length density (RLD) of intercropping maize was higher than that of monocropping maize, the RLD of intercropping soybean was lower than that of monocropping soybean, in the 0–40 cm soil layer the intercropping increased maize RLD by 1.79–7.44% while the soybean RLD was reduced by 3.06–9.46%; the aggressivity of maize was greater than 0 and the competition ratio was greater than 1, which was the dominant species; the maize/soybean land equivalent ratio was 1.18–1.26, which improved the land utilization rate. Therefore, the effect of increasing yield can be achieved by changing the maize and soybean planting method, which is beneficial to the ecological strategy of sustainable development in the northwest region.

Keywords: interspecific competition; land equivalent ratio; planting pattern; root length density; root morphological characteristics

Citation: Wei, W.; Liu, T.; Shen, L.; Wang, X.; Zhang, S.; Zhang, W. Effect of Maize (*Zeal mays*) and Soybean (*Glycine max*) Intercropping on Yield and Root Development in Xinjiang, China. *Agriculture* **2022**, *12*, 996. https://doi.org/10.3390/agriculture12070996

Academic Editor: Jochen Mayer

Received: 22 June 2022
Accepted: 8 July 2022
Published: 10 July 2022

1. Introduction

Ensuring food security is the foundation of economic development and social stability [1]. In the face of a growing global population, food security and food sovereignty are seriously threatened [2]. China has the largest population and is also the largest agro-based country in the world [3]. Under the enormous pressure of the increasing population, how to ensure food security is an urgent problem needing to be solved. The global spread of COVID-19 has complicated the international equilibrium of grain production and trade, is disrupting China's food security in the short term, while critical quantitative variables such as grain production and grain consumption per capita have declined. Land-saving technological progress will contribute the most to the arable land area per capita of wheat and other grains in the long run [4]. The volume of China's grain imports has increased, and the number of exports has fallen. Therefore, the yield of staple grain, oil, and protein crops must be enhanced to satisfy food demands for daily dietary energy requirements [5,6].

Intercropping is a widely used agricultural system of cultivating two or more crops simultaneously in one field during the same or part of their growing season [7]. About one third of China's arable land adopts the multi-species model and contributes half of the total output of all crops, which is an important part of China's agricultural heritage [8]. Historically, intercropping has contributed greatly to crop production in Chinese agriculture [9]. Compared with monocropping cropping systems, intercropping can increase the total yield per unit of land area and can greatly promote crop production due to the more efficient use of one or more resources in time and space [10,11]. Intercropping has received increased attention in recent years due to its clear agro-ecological advantages.

Maize (*Zea mays* L.) and soybean (*Glycine max* L. Merill) are important grain and oil crops in China. Intercropping can improve the stability of farmland ecosystems while land production efficiency and has become an increasingly popular planting method [12]. Due to the low economic benefits of soybeans, China's soybean planting area has been declining year by year, and the domestic soybean market mainly relies on imports. In the past 10 years, China's soybean consumption has remained more than 80% dependent on the international market, becoming the world's largest soybean importer country [13]. Due to the limited arable land resources in China, it is impossible to significantly increase the area of arable land for soybean cultivation. Therefore, while the maize planting area continues to increase, the development of the maize/soybean compound planting model can achieve a win-win situation for maize and soybean yields. The No.1 Central Document in 2020 clearly pointed out that it is necessary to stabilize grain production and increase support for the promotion of new agronomics for maize and soybean intercropping [14]. The No. 1 Central Document in 2022 also clearly proposes to promote corn and soybean strip compound planting in Huanghuaihai, northwest and southwest regions [15]. According to the report of the Ministry of Agriculture and Rural Affairs of China, the national soybean and maize belt compound planting area reached 467,000 hm^2 in 2021 and it will be attempted to increase the soybean and maize belt compound planting area by 1,000,000 hm^2 in 2022 [16].

Maize/soybean intercropping has long been widely practiced in China and has played an important role in enhancing crop production and increasing the income of farmers [17,18]. At present, the research on maize/soybean intercropping mode mostly focuses on the research on population yield [19], photosynthetic physiology [20], water use efficiency [21], and nutrient use efficiency [22]. However, there are relatively few studies on the root morphology of the intercropping system under the combination of maize and soybean, especially the response relationship of root morphology under the intercropping condition is still unclear.

The goal of the present study was to simultaneously evaluate root morphology growth characteristics and yield in a maize/soybean intercropping system compared with sole-cropping. At present, there have been few studies on the dynamic changes and interspecific competition of maize and soybean root systems in different growth stages of intercropping, and most of the previous studies on the underground part of crops used destructive sampling. In the experiment, the root canal method was used to sample the crop roots without damage, so that we could more comprehensively understand the law of dynamic changes of the crop root system and further explore the change law of interspecific competition. The purpose of this study was to understand the following.

(i) Which planting mode is advantageous for root growth of crop in this area?
(ii) Which planting mode of intercropping or sole-cropping provide better yield advantages?

We hypothesized that (i) this type of intercropping system would negatively affect the root growth of crops through underground competition. However, we also hypothesized that (ii) this intercropping system would have an overall positive effect on productivity by improving the efficiency of land resource use.

2. Materials and Methods

2.1. Experimental Site

The field experiments were conducted from 2019 to 2020 at the Agricultural Research Station of Shihezi University in Xinjiang Uygur Autonomous Region, China (44°19′ N, 86°03′ E). This area has a temperate continental climate. The mean annual temperature is 8.1 °C, the annual sunshine duration ranges from 2418 to 2732 h, the annual rainfall ranges from 180 to 270 mm, and the annual evaporation ranges from 1000 to 1500 mm. The soil has a sandy loam texture with a pH of 7.6, 13.260 g kg^{-1} organic matter, 0.890 g kg^{-1} total N, 0.023 g kg^{-1} quick-acting phosphorus, 0.259 g kg^{-1} quick-acting potassium, and 0.058 g kg^{-1} alkali nitrogen.

2.2. Plant Materials and Experimental Design

Field experiments were conducted during the 2019 and 2020 cropping seasons. Maize (*Zea mays*) and soybean (*Glycine max*) were sown on the same days: 28 April 2019, and 30 April 2020. The maize sowing depth was 4 cm, and the soybean sowing depth was 3 cm; a sub-membrane drip irrigation planting method was used. Three treatments (monocropping maize, monocropping soybean, and maize intercropping with soybean) were established. There were a total of 3 plots, 3 repetitions, the test plot was 20 m long, 16 m wide, making a total of 320 m^2. The maize variety was kws3654, and the soybean variety was new soybean No. 1.

In monocropping maize, the row spacing and plant spacing were 60 and 30 cm, respectively, and the planting density was 5.5×10^4 plants hm^{-2}, whereas in monocropping soybean, the row and plant spacings were 40 and 30 cm, and the planting density was 8.3×10^4 plants hm^{-2}. In maize intercropping with soybean, the row spacing and plant spacing of corn were 60 cm and 30 cm, and the planting density was 2.8×10^4 plants hm^{-2}. the row and plant spacing of soybean was 30 cm, and the planting density was 5.5×10^4 plants hm^{-2}. The experiment was carried out under field conditions, while irrigation, fertilization, and crop management were carried out according to local methods, based on fully ensuring the needs of crop growth and development.

2.3. Weather Conditions

The weather conditions during the study are shown in Figure 1. The annual precipitation was about 211 mm. The highest monthly value was recorded in May (28 mm on average), whereas the lowest values were recorded in January (9 mm) and February (9 mm). The highest average monthly temperature was recorded in July (32 °C), whereas the lowest was recorded in January (−21 °C).

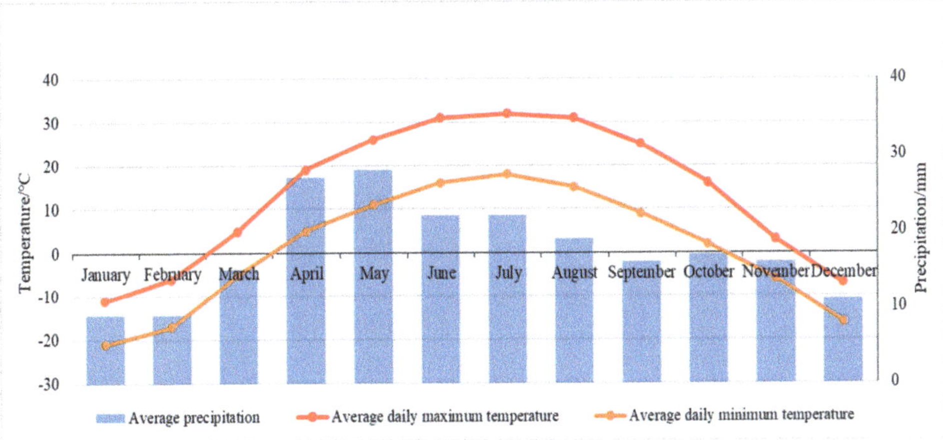

Figure 1. Average monthly precipitation and average temperatures in Shihezi.

2.4. Data Sampling

The data were collected from 19 May 2019 and from 20 May 2020. The sampling time and the corresponding growth period of sampling are shown in Table 1.

Table 1. Sampling time and corresponding crop growth stage.

	2019	2020	Maize	Soybean
I	19 May	20 May	Seedling stage	Seedling stage
II	4 Jun	6 Jun	Jointing stage	Branching stage
III	30 Jun	2 Jul	Large bell mouth stage	Flowering stage
IV	14 Jul	16 Jul	Silking stage	Pod setting stage
V	28 Jul	30 Jul	Grain filling stage	Drumming stage
VI	12 Aug	14 Aug	Maturation stage	Maturation stage

2.5. Plant Height

In each plot, five adjacent plants with similar growth and vigor were selected; the height from the base to the top of the plant was measured, and the average value was calculated.

2.6. Chlorophyll Content (SPAD)

In each plot, five plants with similar growth and vigor were selected, and the SPAD value of the leaves was measured by a hand-held chlorophyll analyzer SPAD-502 (Beijing, China). The 4th leaf was measured at the seedling stage of maize, the 9th leaf was measured at the jointing stage, the three ear leaves were measured after the large bell mouth stage, and the top expanded leaf was measured for soybean. The middle of the leaves was measured at each stage, avoiding the veins, and three points on each leaf were measured and the average value calculated.

2.7. Root Morphological Characteristics

A CI–600 image acquisition instrument (Shanghai, China) was used to capture root images. The embedded angle of the micro-root canal was 45° from the ground, as shown in Figure 2. We collected a soil sample from a depth of 0–20 cm and from a depth of 20–40 cm.

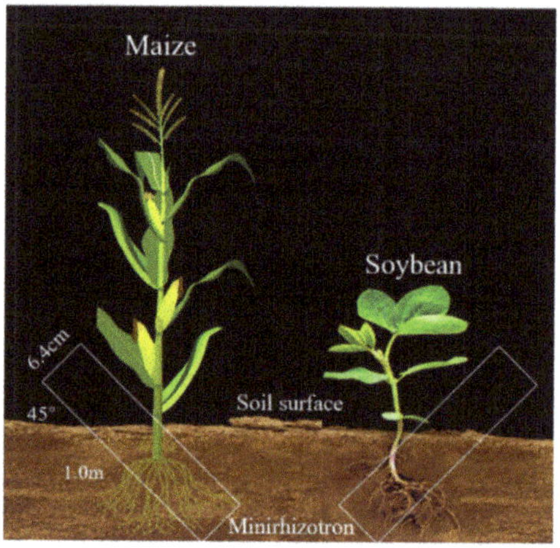

Figure 2. Schematic diagram of micro-root canal layout.

2.8. Root Length Density

The distribution of roots in different soil layers can be indirectly reflected by the root length density, which is given by

$$\text{RLDv} = \frac{\text{RL}}{W \times H \times D} \times \sin 45 \quad (1)$$

where RL is the length of the thin root at the observation interface (mm), W is the width of the image taken by the instrument (cm), H is the length of the image (cm), and D (cm) represents the thickness of the soil layer of the observation interface (D = 0.2, 0.4 m). The RLD of maize and soybean were obtained according to Equation (1).

2.9. Yield and Competition Index

During the harvesting period of soybean and maize, three replicated sampling plots (1 m × 1 m area) were randomly selected from each treatment. The number of maize plants, the number of ears per plant, the number of grains per ear were counted and 1000-seed weighed. The number of soybean plants were counted, all pods per soybean harvested, and 1000-seed per pod and the harvested soybeans after drying at 70 °C were weighed. The theoretical yield of soybean was calculated by the actual yield of the sample plot and plot area.

The land equivalent ratio (LER) is used as an indicator of land productivity for the intensification of the evaluated alternatives [23]. If the value of LER is greater than one, the intercropping system favors the crop growth and yield of the intercropped species; if the LER value is less than one, the intercropping system reduces the growth and yield of the intercropped species. The LER was obtained as follows:

$$\text{LER} = \frac{Y_{mi}}{Y_m} + \frac{Y_{si}}{Y_s} \quad (2)$$

where Y_m and Y_{mi} are the monocropping and intercropping maize yields, respectively. Y_s and Y_{si} are the monocropping and intercropping soybean yields, respectively. LER > 1 signifies that the intensification alternative is more productive than the sum of the sole crops of the component species.

Actual yield loss is the proportionate yield loss or gain of intercrops in comparison to the respective sole crop [24], where:

$$\text{AYL}_m = \frac{Y_{mi}/Z_{mi}}{Y_m/Z_m} - 1,\ \text{AYL}_s = \frac{Y_{si}/Z_{si}}{Y_s/Z_s} - 1,\ \text{AYL} = \text{AYL}_m + \text{AYL}_s \quad (3)$$

Here, Z_m and Z_s represent the proportion of maize and soybean planting in monocropping, respectively, Z_{mi} and Z_{si} represent the planting proportion of maize and soybean in intercropping, respectively, AYL_m and AYL_s represent the actual yield loss of maize and soybean in the intercropping system, respectively, and AYL represents the actual yield losses in intercropping systems. AYL > 0, indicates that the intercropping system has the advantage of intercropping, and AYL < 0, indicates that the intercropping system has no yield advantage.

Aggressivity refers to the degree to which the relative yield increase of a crop in an intercropping system is greater than the yield increase of another crop [25]. The specific calculation of the aggressivity of a crop is as follows:

$$A_m = \frac{Y_{mi}}{Y_m Z_{mi}} - \frac{Y_{si}}{Y_s Z_{si}},\ A_s = \frac{Y_{si}}{Y_s Z_{si}} - \frac{Y_{mi}}{Y_m Z_{mi}}. \quad (4)$$

Here, A_m and A_s represent the encroachment power of maize and soybean in the intercropping system, respectively. A_m = 0, indicates that the two crops have the same competitiveness; A_m > 0, indicates that the competitiveness of maize is higher than that of soybean; A_s > 0, indicates that the competitiveness of soybean is higher than that of maize.

Competitive ratio is the ability of a crop in an intercropping system to compete relative to another crop [26]. where:

$$CR_m = \frac{E_m}{E_s} \times \frac{Z_{si}}{Z_{mi}},\quad CR_s = \frac{E_s}{E_m} \times \frac{Z_{mi}}{Z_{si}}. \quad (5)$$

Here, CR_m and CR_s represent the competition ratios of maize and soybean in the intercropping system, respectively. CR_m > 1, indicates that maize is more competitive than soybean; CR_s > 1, indicates that soybean is more competitive than maize.

2.10. Data Analysis

An analysis of variance was used to perform data analysis using SPSS 19.0 (SPSS Inc., Chicago, IL, USA). The average values were compared using least significant differences (LSD) at the 0.05 level. Origin 2018 (Northampton, MA, USA) was used to draw the figures.

3. Results

3.1. Plant Height

The plant heights of monocropping and intercropping maize and soybean increased with the advancement of the growth period. In the later stage of crop growth, the crop height tended to be stable, showing an overall "S"-shaped growth curve, with a "slow, fast and slow" growth trend (Figure 3).

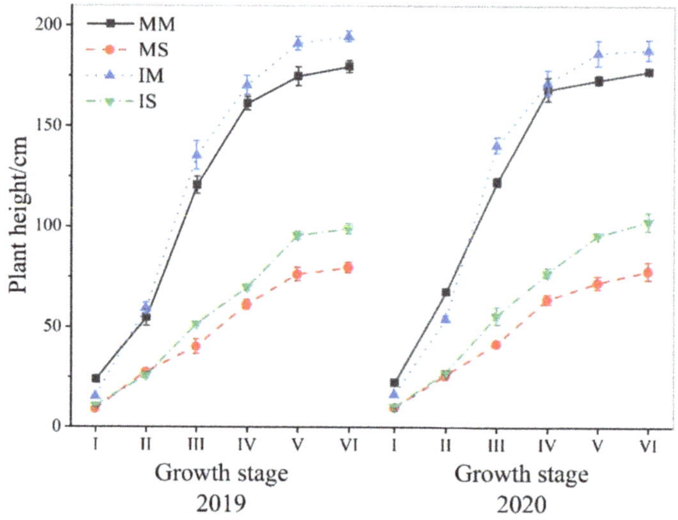

Figure 3. Dynamic changes of plant height of maize and soybean in monoculture and intercropping during 2019 and 2020. Abbreviations: MM—monocropping maize, MS—monocropping soybean, IM—intercropping maize, IS—intercropping soybean. In maize, I–VI mean seedling stage, jointing stage, large bell mouth stage, silking stage, grain filling stage, maturation stage, respectively. In soybean, I–VI mean seedling stage, branching stage, flowering stage, pod setting stage, drumming stage, maturation stage, respectively.

The cropping pattern significantly affected the plant height. The height of monocropping and intercropping maize increased rapidly from the seedling stage to the silking stage and increased significantly from the jointing stage to the large bell mouth stage. Compared with monocropping, intercropping significantly increased the height of maize at the large bell mouth stage, grain filling stage, and mature stage. In 2019, the plant height of intercropping maize increased by 12.44%, 9.40%, and 8.40% at the large flare stage, grain filling stage, and mature stage, respectively; in 2020, it increased by 15.32%, 7.82%, and 6.07%, respectively. The 2-year results showed that the intercropping of maize and soybean increased the height of maize by 6.07–8.40%.

The height of monocropping and intercropping soybean increased rapidly from the emergence stage to the pod setting stage and increased significantly from the seedling stage to the branching stage. Compared with monocropping, intercropping significantly increased soybean height at the flowering, drumming, and maturity stages. In 2019, intercropping increased soybean plant height by 28.09%, 25.15%, and 35.27% at flowering, drumming, and mature stages, respectively; in 2020, it increased by 33.60%, 32.15%, and

38.94%, respectively. The 2-year results showed that the intercropping of maize and soybean increased soybean height by 35.27–38.94%.

3.2. SPAD Values

The SPAD values of monocropping and intercropping maize and soybean showed a trend of first increasing and then decreasing gradually with the advancement of the growth period (Figure 4). In 2019, the monocropping and intercropping maize increased rapidly from the seedling stage to the large bell mouth stage and reached a peak in the large bell mouth stage, after which the SPAD value gradually decreased. Intercropping significantly increased the SPAD value of the 2019 maize grain filling stage by 15.35% ($p = 0.01$). In 2020, the monocropping and intercropping maize peaked at the jointing stage and then gradually decreased. Intercropping significantly increased the SPAD value of maize at the jointing stage by 8.50% ($p = 0.045$). In 2019, monocropping and intercropping soybeans reached the peak at the branching stage; intercropping significantly reduced the SPAD value of soybeans at the branching stage by 3.01% ($p = 0.027$). In 2020, the SPAD value of mono-crop soybeans peaked at the branching stage; the SPAD value of intercropped soybeans peaked at the flowering stage. Intercropping significantly decreased the SPAD value of soybean by 15.27% at the branching stage ($p = 0.011$).

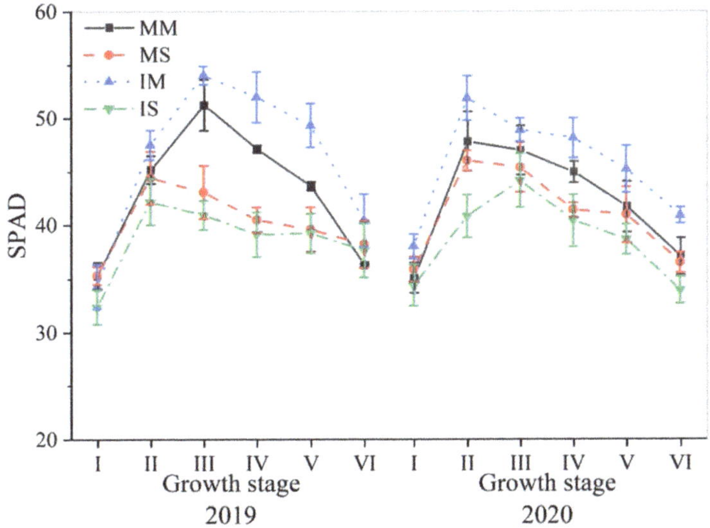

Figure 4. Dynamic changes in the SPAD values of maize and soybean in monoculture and under intercropping during 2019 and 2020. Abbreviations: MM—monocropping maize, MS—monocropping soybean, IM—intercropping maize, IS—intercropping soybean. In maize, I–VI mean seedling stage, jointing stage, large bell mouth stage, silking stage, grain filling stage, maturation stage, respectively. In soybean, I–VI mean seedling stage, branching stage, flowering stage, pod setting stage, drumming stage, maturation stage, respectively.

3.3. Root Morphological Characteristics

The root system is the main organ for crops to absorb nutrients and water, and the interaction between crops is closely related to the spatial distribution of the root parameters. At the same soil depth, the root length, root surface area, and root volume of intercropped maize were higher than those of monocropping maize, and the root length, root surface area, and root volume of monocropping soybean were higher than those of intercropped soybean (Figure 5). Under the mono intercropping mode, the length, volume, and surface area of crop roots in the 0–20 cm soil layer were greater than those in the 20–40 cm soil layer, that is, the root parameters showed a downward trend with the increase in soil depth. The root parameters of maize and soybean were more concentrated in the 0–20 cm soil layer.

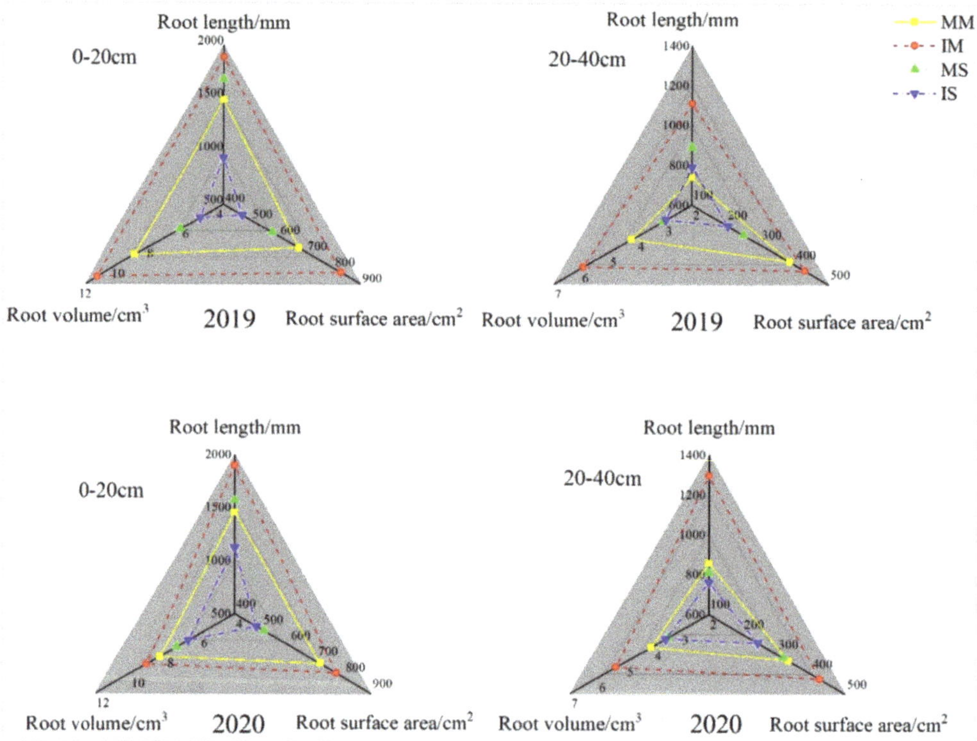

Figure 5. Dynamic changes of root morphological characteristics of maize and soybean in monoculture and intercropping during 2019 and 2020. Abbreviations: MM—monocropping maize, MS—monocropping soybean, IM—intercropping maize, IS—intercropping soybean.

In 2-year, in the 0–20 cm soil layer, compared with monocropping, intercropping increased the root length, root surface area and root volume of maize by 28.79%, 15.48% and 16.67%, respectively; and decreased the root length, root surface area and root volume of soybean by 59.52%, 14.51%, and 15.71%, respectively. In the 20–40 cm soil layer, compared with monocropping, intercropping increased the root length, root surface area and root volume of maize by 50.69%, 19.34%, and 36.66%, respectively; and decreased the root length, root surface area and root volume of soybean by 59.39%, 2.69%, and 4.36%, respectively.

3.4. RLD

With the advancement of the crop growth period, the RLD value showed an increasing trend. At the same soil depth, the RLD value of intercropped maize was higher than that of monocropping maize, and the RLD value of monocropping soybean was higher than that of intercropped soybean. At different soil depths, the RLD value of crops in the 0–20 cm soil layer was higher than that of the 20–40 cm soil layer. That is, with the increase in the soil depth, the RLD value gradually decreased (Figure 6).

In 2019, in the 0–20 cm soil layer, the RLD of intercropping maize was 1.79% higher than that of monocropping, but the RLD of intercropping soybean was 7.61% lower than that of monocropping. The RLD of soybean decreased by 9.46% compared with that under monocropping. In 2020, in the 0–20 cm soil layer, the RLD of intercropping maize was 7.44%

higher than that under monocropping, but the RLD of intercropping soybean was 3.06% lower than that under monocropping. The RLD of soybean decreased by 8.81% compared with that under monocropping.

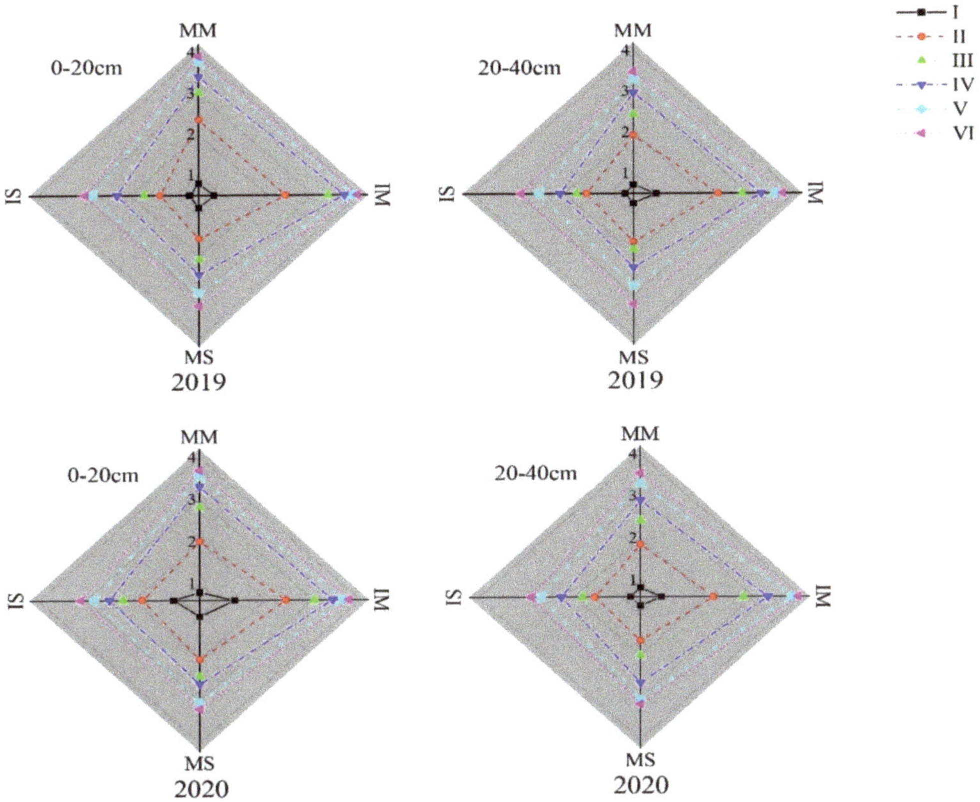

Figure 6. Dynamic changes of RLD of maize and soybean in monoculture and intercropping during 2019 and 2020. Abbreviations: MM—monocropping maize, MS—monocropping soybean, IM—intercropping maize, IS—intercropping soybean. In maize, I–VI mean seedling stage, jointing stage, large bell mouth stage, silking stage, grain filling stage, maturation stage, respectively. In soybean, I–VI mean seedling stage, branching stage, flowering stage, pod setting stage, drumming stage, maturation stage, respectively.

3.5. Correlation

Correlation analysis showed that root morphology, root length density, and soil depth were negatively correlated (Table 2). In monocropping maize treatments, root length, root surface area, and root volume were significantly negatively correlated with soil depth. In intercropping maize, root surface area was significantly negatively correlated with soil depth, and root length was significantly negatively correlated with soil depth. In monocropping soybean, root length was significantly negatively correlated with soil depth, and root surface area, root volume, and root length density were significantly negatively correlated. In intercropping soybean treatment, root surface area was significantly negatively correlated with soil depth, and root length density was significantly negatively correlated with soil depth.

Table 2. Correlation analysis between soil depth and root morphological characteristics in monocropping maize and soybean and maize/soybean.

	Index	RL	RSA	RV	RLD	SD
MM	RL	1				
	RSA	0.966 *	1			
	RV	0.986 *	0.976 *	1		
	RLD	0.899	0.813	0.921	1	
	SD	−0.992 **	−0.991 **	−0.993 **	−0.875	1
IM	RL	1				
	RSA	0.973 *	1			
	RV	0.904	0.972 *	1		
	RLD	0.979 *	0.916	0.842	1	
	SD	−0.982 *	−0.994 **	−0.942	−0.924	1
MS	RL	1				
	RSA	0.964 *	1			
	RV	0.954 *	0.935 *	1		
	RLD	0.979 *	0.889	0.928	1	
	SD	−0.992 **	−0.966 *	−0.984 *	−0.966 *	1
IS	RL	1				
	RSA	0.887	1			
	RV	0.964 *	0.956 *	1		
	RLD	0.981 *	0.939	0.961 *	1	
	SD	−0.887	−0.994 **	−937	−0.951 *	1

Note: * means $p < 0.05$ significant level, ** means $p < 0.01$ extremely significant level. Abbreviations: MM—monocropping maize, MS—monocropping soybean, IM—intercropping maize, IS—intercropping soybean, RL—root length, RSA—root surface area, RV—root volume, RLD—root length density, SD—soil depth.

3.6. Yield Composition

Compared with monocropping, the yield of intercropping maize was higher (Table 3). The number of grains per panicle and the 1000-grain weight were significantly higher than those in the monocrop planting mode, but the difference in the number of panicles was not significant. In 2019, compared with monocropping maize, the number of ears, kernels per ear and the 1000-grain weight of intercropping maize increased by 30.66%, 4.30%, and 7.67%, respectively. In 2020, compared with monocropping maize, the corresponding values of intercropped maize increased by 34.35%, 8.06%, and 6.96%, respectively. In 2-year, intercropping increased maize yield by 49.39–58.10%.

Table 3. Monocropping and intercropping maize yield and yield components in 2019 and 2020.

	Treatment	Number of Spikes per Plant (Piece)	Ear Grain Numbers (Grain)	1000-Seed Weight (g)	Yield (kg·hm^{-2})
2019	Monocropping Maize	1.37 ± 0.03 a	318.00 ± 7.79 a	405.48 ± 5.07 a	9715.83 a
	Intercropping Maize	1.79 ± 0.17 a	331.67 ± 10.34 b	436.58 ± 8.35 b	7257.40 b
2020	Monocropping Maize	1.31 ± 0.09 a	294.33 ± 6.93 a	407.32 ± 4.77 a	8637.82 a
	Intercropping Maize	1.76 ± 0.28 a	318.04 ± 7.55 b	435.66 ± 7.63 b	6828.10 b

Note: Means followed by different letters are significantly different at 0.05 levels.

The yield of intercropped soybeans was lower than that of monocropping soybeans (Table 4). The number of pods per plant and the 1000-grain weight of intercropped soybeans were significantly lower than those of monocropping soybeans, and the difference in the number of seeds per pod between the intercropping and monocropping treatments was not significant. The number of pods per plant, the number of grains per pod, and the 1000-grain weight of intercropping soybeans in 2019 were all lower than those under monocropping, by 27.80%, 4.56%, and 4.91%, respectively. In 2020, the number of pods per intercropped soybean plant, the number of grains per pod, and the 1000-grain weight decreased by 22.32%, 5.69%, and 3.41%, respectively, compared with monocropping. In 2-year, intercropping reduced soybean yield by 29.24–34.48%.

Table 4. Monocropping and intercropping soybean yield and yield components in 2019 and 2020.

	Treatment	Pods per Plant (Piece)	Seeds per Plant (Grain)	1000-Seed Weight (g)	Yield ($kg \cdot hm^{-2}$)
2019	Monocropping Soybean	32.66 ± 9.32 a	2.85 ± 0.08 a	232.59 ± 3.70 a	1796.93 b
	Intercropping Soybean	23.58 ± 5.79 b	2.72 ± 0.03 a	221.16 ± 8.71 b	780.16 a
2020	Monocropping Soybean	31.58 ± 8.80 a	2.81 ± 0.06 a	238.35 ± 3.44 a	1755.543 b
	Intercropping Soybean	24.03 ± 6.03 b	2.65 ± 0.05 a	230.23 ± 5.99 b	823.13 a

Note: Means followed by different letters are significantly different at 0.05 levels.

3.7. Land Equivalent Ratio and Actual Yield Loss

The LER is used as an indicator to measure the yield advantage, and the LER is calculated from the monocropping and intercropping yields [14]. The land equivalent ratio of the intercropping system was 1.18–1.26, i.e., the monocropping needs to increase the land area by 18–26% to achieve the same yield as the intercropping, showing the obvious intercropping advantage. The actual yield loss of maize in the intercropping system was greater than 0, and the actual yield loss of soybean was less than 0, Y > 0, indicating that the maize/soybean intercropping system has intercropping advantages (Table 4).

3.8. Aggressivity and Competitive Ratio

Aggressivity measures the intercrop competition using the simple difference between the extents to which crops a and b vary from their respective expected yields. This study showed that $A_m > 0$, indicates that the competitiveness of maize is higher than that of soybean, and maize as the dominant species. $CR_m > 1$, compared with soybean, maize had a higher competition ratio in the intercropping system, suggesting that maize was more competitive than soybean in the intercropping system (Table 5).

Table 5. Yield and competition index.

	LER	AYL	A_m	CR_m	CR_s
2019	1.181	0.149	+0.839	3.441	0.291
2020	1.259	0.289	+0.873	3.372	0.297

Abbreviations: LER—land L equivalent ratio, AYL—actual yield loss, A_m—aggressivity of maize, CR_m—competitive ratio of maize, CR_s—competitive ratio of soybean.

4. Discussion

4.1. Agronomic Traits of Crops

Plant height is one of the basic indicators used in morphological observations and reflects the growth and development of crops and the rate and robustness of plant growth [27]. In the maize–soybean intercropping system, the shading by the taller maize crop modifies the light environment experienced by the lower soybean crop in terms of both light quantity (PAR—photosynthetically active radiation) and quality (R:FR ratio). These changes are affected by the intercropping configuration and crop architecture and cause changes in both plant height and growth of the soybean crop [28]. This study showed that intercropping increased the plant height of maize and soybean. Intercropped maize is a high-level crop that was less affected by soybean in the later growth stage, and the competition for light, water, and nutrients was greater than that of soybean. With the advancement of the growth period, the degree of shading of maize increased, and the plants underwent a series of shading reactions to adapt to shading stress, resulting in the preferential supply of soybean photosynthates to stem elongation, thereby increasing plant height. Liu, et al. [28] also found that the internode length, plant height, and specific leaf area of intercropped soybean increased due to the reduction of the R: FR ratio of photosynthetically active radiation at the top of the intercropping soybean canopy.

The absorption and utilization of light energy by plants directly affect the growth and development of crops, and the most direct effect of light on crops is photosynthesis. There is a significant positive correlation between the crop SPAD value and the photosynthetic capacity [29]. The increase or decrease of SPAD value affects the content of chlorophyll, and the color of leaves will also change accordingly. The change of leaf color can basically reflect the nitrogen nutritional status of the plant and the nutritional status of nitrogen is also reflected in the change of SPAD value [30]. In intercropped maize and soybean, soybean can supply part of the required nitrogen for maize through its own nitrogen fixation function, improve the efficiency of maize's absorption and utilization of nitrogen, and further increase the chlorophyll content of plant leaves [31]. This study found that the SPAD value of intercropping maize was significantly different from that of monocropping maize, and the SPAD value of intercropping maize was stronger than that of monocropping maize. The SPAD value of intercropped soybean was higher than that of monoculture soybean. Compared with monocropping, intercropping of maize and soybean can maintain a higher level of SPAD, can effectively promote photosynthesis, and is conducive to increasing yield in the later period.

4.2. Root Morphological Characteristics

The yield advantages of intercropping systems are due to both above- and below-ground interactions between the intercropped species [32]. The root system is the main organ for the absorption, transmission, storage, and utilization of underground resources such as water and nutrients in an intercropping compound system. Root systems are key areas of crop resource competition and compensation in intercropping systems and are important contributors to yield formation [33]. Li, et al. [34] and Shinano, et al. [35] showed that root morphologies affect intercrop competition in intercropping systems. The intercropping of broad bean and maize changed the root morphology of crops and increased the effective space for crop water and nutrient absorption [36]. Ren, et al. [37] showed that the intercropping of soybeans expanded the ecological niche of the maize root system in the horizontal and vertical directions, and the root length density and root surface area were positively correlated with nitrogen absorption while promoting the vitality of the maize root system. In legume and Gramineae intercropping, legumes promote Gramineae nitrogen absorption through rhizobia nitrogen fixation and nitrogen transfer, thereby increasing the root growth of Gramineae crops [38]. The results showed that compared with the single cropping mode, the root parameters were improved in the intercropping mode, indicating that the intercropping mode effectively improved the root morphology. Among them, the root length and root surface area increased most obviously, which may be due to the relatively low planting density of the intercropping mode, which gave the root system more growth space and promoted the extension of the root system. Among them, the root length and root volume increased significantly in shallow soil. After the intercropping of maize and soybean, the root morphology of maize (shallow root system) and soybean root system (deep root system) were induced to change, giving full play to the complementarity of the root space niche.

The crop growth and final yield of an intercropping system are closely related to the distribution of roots, which determines the uptake and utilization of water and nutrients. The distribution of roots in different soil layers is reflected by the RLD [39]. Root distribution plays an important role in intercropping dominance. Studies of Gao, et al. [40] have shown that the RLD and root surface area density (RSAD) of peanuts in an intercropping system were lower than those of monocropping peanuts, and the RLD and RSAD of intercropped maize were still higher than those of monocropping maize. This study showed that maize and soybean intercropping had a significant effect on RLD compared with monocropping. The RLD of intercropping maize was higher than that of monocropping maize, intercropping promotes root proliferation of maize crops, and maize has intercropping advantages in yield due to the increase of root length density. The RLD of intercropping soybean was lower than that of monocropping soybean, and the distribution of soybean roots was inhibited. This result is inconsistent with the first hypothesis that the corn intercropping soybean system has a positive effect on the root distribution of corn and a negative effect on the root distribution of soybean through underground competition.

4.3. Yield and Land Productivity

Intercropping can optimize the population structure through reasonable crop collocation and appropriate cultivation techniques as well as to give full play to the advantages of the mutual benefits between species to achieve high crop yield and high efficiency [41]. Maize/soybean intercropping has a greater impact on crop yield, but maize and soybean have different performances. The proportion of maize and soybean in intercropping was 1:2, the yield of intercropping maize was equivalent to 49.39–58.10% of the yield of monocropping, and the yield of intercropping soybean was only equivalent to 29.24–34.48% of the yield of monocropping, indicating that intercropping significantly increased the yield of maize, the soybean yield decreased. The nitrogen element supply, fixed by bean plants in the intercropping system, encourages the root system of maize plants to expand their reach therefore it had an impact on raising maize yields [42]. According to Gard and Mckibben [43], an intercropping system has a certain yield reduction effect compared with monocropping. The reduction in seed yield due to intercropping could be the result of interspecific competition and the depressive effect of maize, a C_4 species, on soybean, a C_3 crop. Crops with the C_4 photosynthetic pathway such as maize are known to be dominant when intercropped with C_3 crops such as soybean [44]. Further, the reduction in intercropped soybean could be due to shading by the taller maize plants [45]. In intercropping systems, shorter crops experience shading from taller crops, thus increasing plant height, and decreasing yield [46]. Research has shown that the yield of soybean is inhibited by maize, and the appropriate nitrogen application rate cannot alleviate the inhibitory effect on soybean [47]. It may be that soybean is not sensitive to nitrogen fertilizer, and maize responds quickly to nitrogen fertilizer, resulting in soybean being inhibited by maize, while maize shows yield advantage [48].

Yield advantage is usually assessed and quantified by calculating the land equivalent ratio (LER) in intercropping [49]. In this study, the LER values of corn–soybean intercropping were 1.18–1.26. This means that an additional 18–26% of land area was needed for a monoculture cropping system to produce the equal production as an intercropping system. Hafid et al. [50] stated that the increase in land productivity was caused by choosing the right combination of plants and cropping systems and the existence of a relationship or mutualism symbiosis between plants which were planted in an intercropping way. This symbiosis is closely related to the need for nitrogen for the main plant which was fulfilled from the attached plants through its ability to fix nitrogen from the air. On the other hand, plants that are tolerant to shade can live under stands. The combination of cereal crops and legumes was the best combination. This result supports the second hypothesis that such an intercropping system would have an overall positive impact on productivity by increasing the efficiency of land resource use.

4.4. Interspecific Competition

The competition ratio of maize in the intercropping system was greater than 1, and the aggressivity of maize was greater than 0, indicating that in the symbiotic period of maize and soybean, soybean was at a competitive disadvantage in the intercropping system, and maize was a competitive crop. Previous studies have shown that there is strong interspecific competition among different crops in the intercropping system, and the resource competitiveness of Gramineae crops is higher than that of legumes [51]. According to Banik, et al. [24], the AYL index can give more precise information than the other indices on the inter- and intra-specific competition of the component crops and the behavior of each species involved in the intercropping systems. Quantification of yield loss or gain due to association with other species or the variation of the plant population could not be obtained through partial LERs, whereas partial AYL shows the yield loss or gain by its sign as well as its value. The actual yield loss of maize was greater than 0, indicating that maize has a yield advantage in the intercropping system. The actual reduction of soybean yield is less than 0, indicating that soybean has no yield advantage in the intercropping system, which is consistent with the fact that maize is a competitive crop in the intercropping system, and soybean is a competitive disadvantage crop. The result of interspecific competition was that the actual yield loss of the intercropping system was greater than 0, indicating that the intercropping of maize and soybean had yield advantage because the yield of maize was increased, and the yield of soybean was unchanged or decreased. This is consistent with previous studies on the intercropping of wheat and peas [52], oats and wild peas [26], millet and soybeans [53], which showed that in the legume and Gramineae intercropping system, the yield of grasses increased and the yield of legumes decreased, since the crops in the intercropping system have differences in competitiveness [52]. The intercropping of tall crops (maize) and dwarf crops (soybeans) is caused by the increase of above-ground light interception of maize and the improvement of underground nutrient and water use efficiency [54]. The biological characteristics of soybean are different from those of maize, and it is in a disadvantageous position in the competition for soil water and nutrient absorption and the competition for light interception [55].

5. Conclusions

Maize/soybean intercropping has effects on crop growth, yield, and root morphology. The growth parameters (plant height, relative chlorophyll content) of maize and soybean in intercropping system were better than with monocropping. The yield components of intercropping maize in terms of the number of spikes per plant, ear grain numbers, and 1000-seed weight were higher than those of monocropping, however, in contrast to soybean, monocropping soybean had higher yield parameters than intercropping. The RLD of intercropping maize increased compared to monocropping, indicating greater root growth. The intercropping of maize and soybean has yield advantages; the land equivalent ratio was between 1.18 and 1.26, the aggressivity of maize was between 0.84 and 0.87, and the competition ratio was between 3.37 and 3.44. The reason for improving the yield of the intercropping population is the increase of maize yield and the higher competitiveness of maize for resources than soybean, and maize is the dominant species. Maize and soybean intercropping can improve land use efficiency and crop yield and should be properly promoted to increase maize/soybean productivity.

Author Contributions: Conceptualization, W.W.; Data curation, W.W. and X.W.; Formal analysis, W.W.; Funding acquisition, W.Z.; Methodology, W.W.; Project administration, W.Z.; Resources, W.Z.; Supervision, W.Z.; Visualization, W.W. and T.L.; Writing—original draft, W.W.; Writing—review and editing, T.L., L.S. and S.Z. All authors have read and agreed to the published version of the manuscript.

Funding: This work was financially supported by the Innovation and Development Project of Shihezi University (CXFZ202008), by the National Natural Science Foundation of China (Project Nos. 31460335 and 31560376).

Institutional Review Board Statement: Not applicable.

Informed Consent Statement: Informed consent was obtained from all subjects involved in the study.

Data Availability Statement: The datasets used and/or analyzed during the current study are available from the first author upon reasonable request.

Conflicts of Interest: The authors declare that they have no known competing financial interests or personal relationships that could have appeared to influence the work reported in this paper.

References

1. Li, Y.; Zhang, W.; Ma, L.; Wu, L.; Shen, J.; Davies, W.J.; Dou, Z. An analysis of China's grain production: Looking back and looking forward. *Food Energy Secur.* **2015**, *3*, 19–32. [CrossRef]
2. Campbell, B.M.; Beare, D.J.; Bennett, E.M.; Hall-Spencer, J.M.; Ingram, J.S.; Jaramillo, F.; Shindell, D. Agriculture production as a major driver of the Earth system exceeding planetary boundaries. *Ecol. Soc.* **2017**, *22*, 8. [CrossRef]
3. Zhang, Z.; Ping, X.-U.; Duan, Z. Food security should be the ultimate goal of agricultural modernization in China. *Chin. J. Eco-Agric.* **2015**, *23*, 1215–1219. [CrossRef]
4. Lin, X.; Qi, L.; Pan, H.; Sharp, B. COVID-19 Pandemic, Technological Progress and Food Security Based on a Dynamic CGE Model. *Sustainability* **2022**, *14*, 1842. [CrossRef]
5. Grassini, P.; Cassman, K.G. High-yield maize with large net energy yield and small global warming intensity. *Proc. Natl. Acad. Sci. USA* **2012**, *109*, 1074–1079. [CrossRef]
6. Ingram, J. Perspective: Look beyond production. *Nature* **2017**, *544*, S17. [CrossRef]
7. Zhang, D.; Sun, Z.; Feng, L.; Bai, W.; Zhang, L. Maize plant density affects yield, growth and source-sink relationship of crops in maize/peanut intercropping. *Field Crop. Res.* **2020**, *257*, 107926. [CrossRef]
8. Ai, P.-R.; Ma, Y.-J.; Ma, L. Study on evaporation variation of jujube trees under drip irrigation of jujube and cotton intercropping in an arid area. *Acta Ecol. Sin.* **2018**, *38*, 4761–4769. [CrossRef]
9. Tong, P.-Y. Achievements and perspectives of tillage and cropping systems in China. *Crop. System. Cult. Technol.* **1994**, *77*, 1–5.
10. Inal, A.; Gunes, A.; Zhang, F.; Cakmak, I. Peanut/maize intercropping induced changes in rhizosphere and nutrient concentrations in shoots. *Plant Physiol. Biochem.* **2007**, *45*, 350–356. [CrossRef]
11. Zhang, L.-Z.; Van der Werf, W.; Zhang, S.-P.; Li, B.; Spiertz, J.H.J. Growth, yield and quality of wheat and cotton in relay strip intercropping systems. *Field. Crop. Res.* **2007**, *103*, 178–188. [CrossRef]
12. Zhao, D.; Yuan, J.; Hou, Y.; Li, T.; Liao, Y. Tempo-spatial dynamics of AMF under maize soybean intercropping. *Chin. J. Eco-Agric.* **2020**, *28*, 631–642. [CrossRef]
13. Zhai, T.; Wu, L. Study on development situation and revitalization strategy of soybean industry in China from an open perspective. *Soybean Sci.* **2020**, *39*, 472–478.
14. Central Committee of the Communist Party of China, State Council. 2020. Available online: http://www.gov.cn/zhengce/2020-02/05/content_5474884.htm (accessed on 22 June 2022).
15. Central Committee of the Communist Party of China, State Council. 2022. Available online: http://www.81.cn/yw/2022-02/22/content_10134209.htm (accessed on 20 June 2022).
16. Ministry of Agriculture and Rural Affairs of the People's Republic of China. 2022. Available online: http://news.china.com.cn/2022-01/20/content_78001838.html (accessed on 20 June 2022).
17. Li, L.; Zhang, L.; Zhang, F. Crop mixtures and the mechanisms of overyielding. In *Encyclopedia of Biodiversity*, 2nd ed.; Levin, S.A., Ed.; Academic Press: Waltham, MA, USA, 2013; pp. 382–395.
18. Li, C.; Hoffland, E.; Kuyper, T.W.; Yu, Y.; Zhang, C.; Li, H.; Zhang, F.; Werf, W.V.D. Syndromes of production in intercropping impact yield gains. *Nat. Plants* **2020**, *6*, 653–660. [CrossRef]
19. Zhu, Y.; Gao, F.; Cao, P.; Wang, L. Effect of plant density on population yield and economic output value in maize-soybean intercropping. *J. Appl. Ecol.* **2015**, *26*, 1751–1758.
20. Liu, X.; Rahman, T.; Song, C.; Yang, F.; Su, B.; Cui, L.; Yang, W. Relationships among light distribution, radiation use efficiency and land equivalent ratio in maize-soybean strip intercropping. *Field Crop. Res.* **2018**, *224*, 91–101. [CrossRef]
21. Yin, W.; Chai, Q.; Zhao, C.; Yu, A.; Fan, Z.; Hu, F.; Coulter, J.A. Water utilization in intercropping: A review. *Agric. Water Manag.* **2020**, *241*, 106335. [CrossRef]
22. Rodriguez, C.; Carlsson, G.; Englund, J.E.; Flöhr, A.; Pelzer, E.; Jeuffroy, M.H.; Jensen, E.S. Grain legume-cereal intercropping enhances the use of soil-derived and biologically fixed nitrogen in temperate agroecosystems. A meta-analysis. *Eur. J. Agron.* **2020**, *118*, 126077. [CrossRef]
23. Mead, R.; Willey, R.W. The Concept of a 'Land Equivalent Ratio' and Advantages in Yields from Intercropping. *Exp. Agric.* **1980**, *16*, 217–228. [CrossRef]
24. Banik, P.; Sasmal, T.; Ghosal, P.K. Evaluation of mustard (*Brassica compestris* Var. Toria) and legume intercropping under 1:1 and 2:1 row-replacement series systems. *J. Agron. Crop. Sci.* **2000**, *185*, 9–14. [CrossRef]
25. Agegnehu, G.; Ghizaw, A.; Sinebo, W. Yield performance and land-use efficiency of barley and faba bean mixed cropping in Ethiopian highlands. *Eur. J. Agron.* **2006**, *25*, 202–207. [CrossRef]
26. Dhima, K.V.; Lithourgidis, A.S.; Vasilakoglou, I.B.; Dordas, C.A. Competition indices of common vetch and cereal intercrops in two seeding ratio. *Field Crop. Res.* **2007**, *100*, 249–256. [CrossRef]
27. Verreynne, J.S.; Rabe, E.; Theron, K.I. The effect of combined deficit irrigation and summer trunk girdling on the internal fruit quality of 'Marisol' Clementines. *Sci. Hortic.* **2001**, *91*, 25–37. [CrossRef]
28. Liu, X.; Rahman, T.; Song, C.; Su, B.; Yang, F.; Yong, T.; Yang, W. Changes in light environment, morphology, growth and yield of soybean in maize-soybean intercropping systems. *Field Crop. Res.* **2017**, *200*, 38–46. [CrossRef]
29. Li, Y.-J.; Ma, L.-S.; Wu, P.-T.; Zhao, X.-N.; Chen, X.-L.; Gao, X.-D. Yield, yield attributes and photosynthetic physiological characteristics of dryland wheat (*Triticum aestivum* L.)/maize (*Zea mays* L.) strip intercropping. *Field Crop. Res.* **2020**, *248*, 107656. [CrossRef]

30. Li, G.-H.; Xue, L.-H.; You, J. Spatial distribution of leaf N content and SPAD value and determination of the suitable leaf for N diagnosis in rice. *Sci. Agric. Sin.* **2007**, *40*, 1127–1134.
31. Bethlenfalvay, G.J.; Reyes-Solis, M.G.; Camel, S.B.; Ferrera-Cerrato, R. Nutrient transfer between the root zones of soybean and maize plants connected by a common mycorrhizal mycelium. *Physiol. Plant.* **1991**, *82*, 423–432. [CrossRef]
32. Gao, Y.; Duan, A.-W.; Qiu, X.-Q.; Liu, Z.-G.; Sun, J.-S.; Zhang, J.-P.; Wang, H.-Z. Distribution of roots and root length density in a maize/soybean strip intercropping system. *Agric. Water. Manag.* **2010**, *98*, 199–212. [CrossRef]
33. Xia, H.-Y.; Zhao, J.-H.; Sun, J.-H.; Bao, X.-G.; Christie, P.; Zhang, F.-S.; Li, L. Dynamics of root length and distribution and shoot biomass of maize as affected by intercropping with different companion crops and phosphorus application rates. *Field Crop. Res.* **2013**, *150*, 52–62. [CrossRef]
34. Li, L.; Tilman, D.; Lambers, H.; Zhang, F.-S. Plant diversity and overyielding: Insights from belowground facilitation of intercropping in agriculture. *New Phytol.* **2014**, *203*, 63–69. [CrossRef]
35. Shinano, T.; Osaki, M.; Yamada, S.; Tadano, T. Comparison of root growth and nitrogen absorbing ability between Gramineae and Leguminosae during the vegetative stage. *Soil Sci. Plant. Nutr.* **1994**, *40*, 485–495. [CrossRef]
36. Li, Y.-Y.; Pang, F.-H.; Sun, J.-H.; Li, L.; Cheng, X. Effects of root barrier between intercropped maize and faba bean and nitrogen (N) application on the spatial distributions and morphology of crops' roots. *J. China Agric. Univ.* **2010**, *15*, 13–19.
37. Ren, Y.; Zhang, L.; Yan, M.; Zhang, Y.; Chen, Y.; Palta, J.A.; Zhang, S. Effect of sowing proportion on above-and below-ground competition in maize–soybean intercrops. *Sci. Rep.* **2021**, *11*, 15760. [CrossRef]
38. Wahla, I.H.; Ahmad, R.I.A.Z.; Ehsanullah, A.A.; Jabbar, A.B.D.U.L. Competitive functions of components crops in some barley based intercropping systems. *Int. J. Agric. Biol.* **2009**, *11*, 69–72.
39. Adiku, S.G.K.; Ozier-Lafontaine, H.; Bajazet, T. Patterns of root growth and water uptake of a maize-cowpea mixture grown under greenhouse conditions. *Plant Soil* **2001**, *235*, 85–94. [CrossRef]
40. Gao, Y.-L.; Sun, Z.-X.; Bai, W.; Feng, L.-S.; Cai, Q.; Feng, C.; Zhang, Z. Spatial distribution characteristics of root system and the yield in maize-peanut intercropping system. *J. Maize Sci.* **2016**, *24*, 79–87. [CrossRef]
41. Yang, F.; Huang, S.; Gao, R.; Liu, W.; Yong, T.; Wang, X.; Yang, W. Growth of soybean seedlings in relay strip intercropping systems in relation to light quantity and red:far-red ratio. *Field Crop. Res.* **2014**, *155*, 245–253. [CrossRef]
42. Ceunfin, S.; Prajitno, D.; Suryanto, P.; Putra, E.T.S. Penilaian kompetisi dan keuntungan hasil tumpangsari jagung kedelai di bawah tegakan kayu putih. *Savana Cendana* **2017**, *2*, 1–3. [CrossRef]
43. Gard, L.E.; McKibben, G.E. "No-till" crop production proving a most promising conservation measure. *Outlook Agric.* **1973**, *7*, 149–154. [CrossRef]
44. Hiebsch, C.K.; Tetio-Kagho, F.; Chirembo, A.M.; Gardner, F.P. Plant Density and Soybean Maturity in a Soybean-Maize Intercrop. *Agrono. J.* **1995**, *87*, 965–969. [CrossRef]
45. Muoneke, C.O.; Ogwuche, M.; Kalu, B.A. Effect of maize planting density on the performance of maize/soybean intercropping system in a guinea savannah agroecosystem. *Afr. J. Agric. Res.* **2007**, *2*, 667–677. [CrossRef]
46. Wu, Y.; Gong, W.; Yang, F.; Wang, X.; Yong, T.; Yang, W. Responses to shade and subsequent recovery of soya bean in maize-soya bean relay strip intercropping. *Plant Prod. Sci.* **2016**, *19*, 206–214. [CrossRef]
47. Zhang, W.; Li, S.; Shen, Y.; Yue, S. Film mulching affects root growth and function in dryland maize-soybean intercropping. *Field Crop. Res.* **2021**, *271*, 108240. [CrossRef]
48. Zhang, R.; Meng, L.; Li, Y.; Wang, X.; Ogundeji, A.O.; Li, X.; Li, S. Yield and nutrient uptake dissected through complementarity and selection effects in the maize/soybean intercropping. *Food Energy Secur.* **2021**, *10*, 379–393. [CrossRef]
49. Vandermeer, J.H. *The Ecology of Intercropping*; Cambridge University Press: Cambridge, UK, 1992.
50. Hafid, H.; Syaiful, S.A.; Fattah, A.; Djufry, F. The effect of the number of rows and varieties of soybean on growth and yield in intercropping with corn. *IOP Conf. Ser. Earth Environ. Sci.* **2021**, *648*, 012204. [CrossRef]
51. Ndakidemi, P.A. Manipulating legume/cereal mixtures to optimize the above and below ground interactions in the traditional African cropping systems. *Afr. J. Biotechnol.* **2006**, *525*, 2526–2533. [CrossRef]
52. Lithourgidis, A.S.; Vlachostergios, D.N.; Dordas, C.A.; Damalas, C.A. Dry matter yield, nitrogen content, and competition in pea–cereal intercropping systems. *Eur. J. Agron.* **2011**, *34*, 287–294. [CrossRef]
53. Li, Z.; Wang, H.-F.; Wang, Y.-Y.; Yang, J.; Yu, B.-X.; Huang, S. Impact of millet and soybean intercropping on their photosynthetic characteristics and yield. *J. Agric. Sci. Technol.* **2020**, *22*, 168. [CrossRef]
54. Wang, Y.; Zhao, Z.; Li, J.; Zhang, M.; Zhou, S.; Wang, Z.; Zhang, Y. Does maize hybrid intercropping increase yield due to border effects? *Field Crop. Res.* **2017**, *214*, 283–290. [CrossRef]
55. Zhang, X.-C.; Wang, H.-L.; Yu, X.-F.; Hou, H.-Z.; Fang, Y.-J.; Ma, Y.-F. The study on the effect of potato and beans intercropping with whole field plastics mulching and ridge-furrow planting on soil thermal-moisture status and crop yield on semi-arid area. *Sci. Agric. Sin.* **2016**, *49*, 468–481.

agriculture

Article

Effects of Water-Saving Irrigation on Direct-Seeding Rice Yield and Greenhouse Gas Emissions in North China

Xiaoning Hang [1,2], Frederick Danso [3], Jia Luo [1], Dunxiu Liao [1,2], Jian Zhang [1,2,*] and Jun Zhang [3,*]

1 Chongqing Academy of Agricultural Sciences, Chongqing 401329, China; hangxiaoning@163.com (X.H.); luojia925@163.com (J.L.); xiuchai2006@163.com (D.L.)
2 Agriculture and Rural Sub-Center of Chongqing Ecological Environment Monitoring Center, Chongqing 401329, China
3 Key Laboratory of Crop Physiology & Ecology, Institute of Crop Sciences, Chinese Academy of Agricultural Sciences, Beijing 100081, China; dansotodanso@gmail.com
* Correspondence: zhangjian99815@sina.com (J.Z.); zhangjun@caas.cn (J.Z.)

Abstract: Rice cultivation consumes more than half of the planet's 70% freshwater supply used in agricultural production. Competing water uses and climate change globally are putting more pressure on the limited water resources. Therefore, water-saving irrigation (WSI) is recommended for rice production in water scares areas. The impact of WSI techniques on direct-seeding rice production and greenhouse gas emissions in North China is becoming increasingly important in the era of climate change. Therefore, we conducted a two-year field experiment on directly seeded rice to assess the impact of traditional flooding irrigation (CK) and three water saving irrigation (WSI) methods, including drip irrigation with an irrigation amount of 50 mm (DI_1) and 35 mm (DI_2) at each watering time and furrow wetting irrigation (FWI), on rice yield and greenhouse emissions. Generally, the WSI techniques decreased the number of rice panicles per m^{-2}, spikelet per panicle, 1000-grain weight and rice yield compared to CK. Rice yield and yield components of (DI_1) were significantly higher than (DI_2). The adoption of either (DI_1) or (FWI) showed insignificant variation in terms of rice yield and its yield components measured except for 1000-grain weight. The water productivity was 88.9, 16.4 and 11.4% higher in the FWI plot than the CK, DI_1 and DI_2 plots, respectively. The WSI decreased cumulative CH_4 emission significantly by 73.0, 84.7 and 64.4% in DI_1, DI_2 and FWI, respectively, in comparison with CK. The usage of DI_2 triggered 1.4 and 2.0-fold more cumulative N_2O emission compared to DI_1 and FWI, respectively. Area-scaled emission among the water-saving irrigation methods showed no significance. The yield-scaled emission in DI_1 and DI_2 and FWI were 101, 67.5 and 102%, respectively, significantly lower than CK. The adoption of FWI produced an acceptable rice yield with the lowest yield-scaled emission and highest water productivity among the irrigation practices. Our experiment demonstrates that dry direct-seeding with furrow irrigation can impact triple-wins of sustainable rice yield, high water-use efficiency and low GHG emissions in North China.

Keywords: rice production; CH_4; N_2O; water productivity; global warming

Citation: Hang, X.; Danso, F.; Luo, J.; Liao, D.; Zhang, J.; Zhang, J. Effects of Water-Saving Irrigation on Direct-Seeding Rice Yield and Greenhouse Gas Emissions in North China. *Agriculture* **2022**, *12*, 937. https://doi.org/10.3390/agriculture12070937

Academic Editor: Jose L. Gabriel

Received: 24 April 2022
Accepted: 24 June 2022
Published: 28 June 2022

1. Introduction

As the most important staple food of the world, rice represents 19% of human calorific intake [1]. Global population is projected by 2050 to reach 9 billion, and a 50% increase in rice production may be needed for the impending demands [2]. Globally, approximately 70% of the planet's freshwater supply is consumed through agricultural production [3]. In recent times, the sustainability of irrigated rice systems are under threat, owing to agricultural intensification, depleting water reserves and limited water availability across the globe [4]. Rice cultivation, accounting for 40% of the agricultural freshwater usage, worsening climatic conditions, rising population and competing water uses constraints farmers access to adequate and timely supply of water [5]. Therefore, for sustainable rice cultivation, it is essential that water is managed appropriately. China is among the largest rice producers and the second major user of water for irrigation globally [6]. In recent times, large-scale rice production has moved northward [7]. The cultivated area and total production in North China have increased by 101.1% and 143.2%, respectively since 1990, and account for 18.8% and 20.4% of Chinese total rice sown area and production in 2012, respectively. The expected socioeconomic growth, associated water resource demand and consumption through rice production can be reasonably projected to increase exponentially in North China. From the findings of Jiang et al. [8], the continuous adoption of traditional irrigation practices that use huge volumes of water and accounts for over 60% of water use for producing rice across China may not be sustainable in North China, where water shortage is severe. Additionally regional and seasonal water shortages caused by drought and future climate change scenarios will make water shortage more severe in the region and threaten rice production [9,10]. Although globally, the production of rice contributes only 1.5% of the overall anthropogenic greenhouse gas (GHG), this portion is considerably greater in rice-producing nations [11]. A substantial quantity of greenhouse gas (GHG) is released into the environment with current practices of rice production that consume vast amounts of water [5]. Therefore, target to limit global warming to 1.5 °C will be compromised due to insufficient agricultural emission reductions [12]. Accordingly, several water-saving irrigation (WSI) know-hows have been developed and disseminated in China, such as alternate wetting and drying, soil saturated cultivation, drip irrigation, bed-furrow base irrigation and non-flooded mulching cultivation to replace the traditional flood irrigation [13,14]. The choice of these WSI may impact rice growth and greenhouse gas (GHG) emissions. The adoption of WSI can cause a reduction in rice yield [15], maintain or even increase rice yield [16]. Compared with continuous flooding, WSI, which involves one or several drainage methods that minimize CH_4 production, demonstrates an important prospect to reduce CH_4 emissions [14,17], though it may trigger substantial N_2O emissions caused by wet-dry cycles of the soil [18]. In recent times, water-saving irrigation of drip irrigation in combination with plastic film mulch, furrow wetting irrigation and intermittent irrigation has been integrated with dry direct-seeding of rice in North China. Study of the integrated effects of rice planting techniques with water-saving irrigation on the yield of rice and GHG emissions is limited. Therefore, measurement of rice yield and GHG emission could provide additional confirmation to elucidate the integrated impact of dry direct-seeding of rice and WSI measures in North China. Therefore, using a two-year field experiment, three water-saving irrigation methods under the dry direct-seeding system in North China were appraised. Our objectives were to evaluate the effects of the improved planting technique and water management practice on rice yield and yield components—CH_4 and N_2O emissions.

2. Materials and Methods

2.1. Experimental Location

The field experimentation was set up in the Yellow River Irrigation Area at the Ling Wu experimental Farm in 2014 and 2015, Yinchuan City (38°12′ N latitude, 106°27′ E longitude), Ningxia Province, China (Figure 1a). The soil type was an irrigating warped soil with the basic chemical properties: organic matter 12.2 g kg^{-1}, total salt 1.2 g kg^{-1}, total N 0.8 g kg^{-1}, available N 57.8 mg kg^{-1}, available P 26.5 mg kg^{-1} and available K 141.1 mg kg^{-1}. The experimental site is characterized by a temperate arid climate with mean annual temperature and precipitation of 8.5 °C and 200 mm, respectively. The precipitation and air temperatures data obtained from Ling Wu meteorological department during the rice growing seasons in 2015 are shown in Figure 1. Rainfall occurred between June–August and was almost lacking in the course of rice-seed emergence in May. Total rainfall from the seeding stage to maturity stage was 256 mm and 213 mm in 2014 and 2015, respectively. The lowest and the highest daily mean air temperatures were 13.1 °C on 5 May and 27.6 °C on 12 August in 2014, respectively, and 13.3 °C on 14 May and 27.9 °C on 10 August in 2015, respectively. From June until August, air temperature was relatively lower than the optimal temperature required for rice growth.

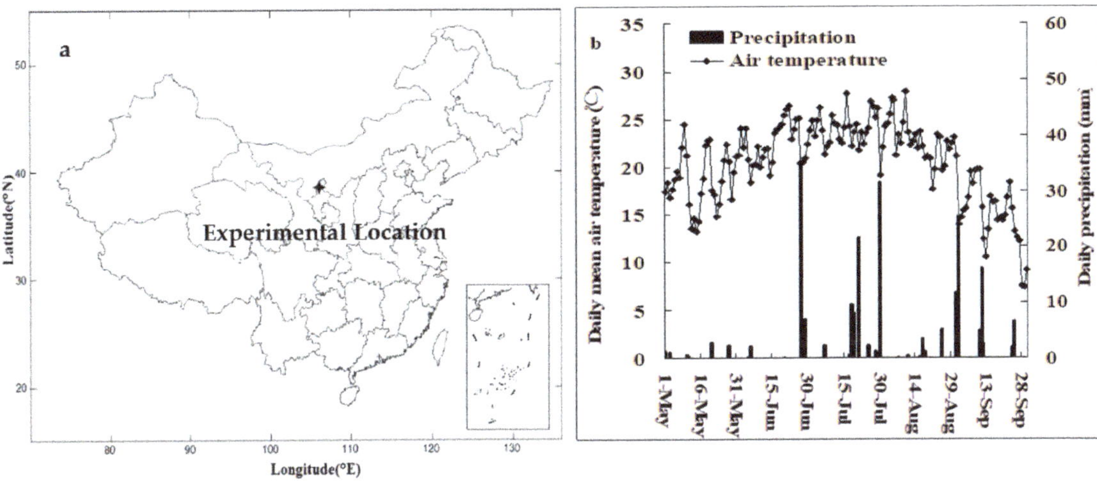

Figure 1. Experimental location (**a**) and daily mean air temperature and daily precipitation (**b**) of rice cropping seasons in 2015.

2.2. Experimental Design

The field experiment was a randomized block design in three replications and consisted of four irrigation treatments, namely: (1) Traditional flood irrigation (CK); (2) Drip irrigation under plastic film mulching with 50 mm irrigation amount at each watering time when the relative soil water content (RSWC) was less than 100% (DI_1); (3) Drip irrigation under plastic film mulching with 35 mm irrigation amount at each watering time at the same time of DI_1 (DI_2) and (4) Furrow wetting irrigation (FWI). The replicate plot sizes of 15 m × 20 m were separated by 30 cm-wide soil ridges covered with plastic film to inhibit water and nutrient exchange between plots.

2.3. Water and Crop Management

Land preparation in all the treatments was carried out by ploughing and leveling the soil under dry conditions. The rice variety, Ningjing 31, was directly seeded on 1st May, and harvested between 24–28 September for all the treatments in 2014 and 2015 (Table 1). Based on the local agronomic practices for higher rice yield, similar fertilization rates were adopted for the treatments. The N fertilizer was applied as urea at a rate of 240 kg N ha^{-1}, 40% as basal application before seeding, 30% at the tillering stage and 30% at the panicle initiation stage. Basal phosphorus fertilizer of calcium superphosphate was applied at 112.5 kg ha^{-1} P_2O_5, whiles no K fertilizer was added during rice growth (Table 1).

All treatments were flooded with 100 mm of water on 1st May after direct seeding (Figure 2). Subsequently, only the CK followed the traditional continuous flooding. The drip system for DI_1 and DI_2 consisted of a small pump, a water meter, a control head unit, PVC mainline, polyethylene mains and laterals (Xinjiang Tianye Company, Shihezi, China). DI_1 drip irrigated received 50 mm water amount at each irrigating time when the relative soil water content (RSWC) was 0.1 m and below 100%. A similar irrigation schedule was implemented in DI_2 except that it received 35 mm of water at each irrigating time. In the furrow wetting irrigation (FWI) treatment, the plots were maintained at moist condition the whole period of rice growth. Each replicate plot of FWI, prior to direct-seeding, was divided into five strips (three meters in width) and separated by furrows (25 cm width and 30 cm in depth). After direct-seeding on the strips, the furrows were filled with water to maintain a constant wet condition on the strips. No obvious water level was retained on the seedling strips during the entire growth period. The water flow of CK and FWI were measured by separated flume flow meter. TDR100 was used to test the RSWC.

Table 1. Mode and timing of experimental field management practices in the four irrigation regimes.

Practice	CK	DI_1 and DI_2		FWI
Land preparation and seed sowing method	Ploughing, Dry direct seeding	Ploughing, Dry direct seeding		Ploughing and furrowing, Dry direct seeding
Fertilization amount and timing	N fertilizer: 240 kg N ha^{-1} as urea, 40% applied before seeding, 30% at tillering stage, 30% at panicle stage; P fertilizer: 112.5 kg P_2O_5 ha^{-1} as $Ca(H_2PO_4)_2$, applied before seeding. All fertilizers were applied by hand onto the soil surface.	N fertilizer: 240 kg N ha^{-1} as urea, 40% applied before seeding, 30% at tillering stage, 30% at panicle stage; P fertilizer: 112.5 kg P_2O_5 ha^{-1} as $Ca(H_2PO_4)_2$, applied before seeding. All fertilizers were dissolved in the irrigation water and applied through drip water flow during watering.		N fertilizer: 240 kg N ha^{-1} as urea, 40% applied before seeding, 30% at tillering stage, 30% at panicle stage; P fertilizer: 112.5 kg P_2O_5 ha^{-1} as $Ca(H_2PO_4)_2$, applied before seeding. All fertilizers were hand applied directly to the soil surface.
Plastic film mulching	None	Plastic film mulching before seeding		None
Irrigation methods	Continuous flooding	Drip irrigation with 50 mm at each watering time when RSWC was below 100%	Drip irrigation with 35 mm at the watering time when RSWC was below 100%	The furrows constantly supplied with water to maintain moist condition in the strips during the entire rice growing period
Seeding and harvesting dates	Direct seeding on 1 May; Harvested on 28 September	Direct seeding on 1 May; Harvested on 24–26 September		Direct seeding on 1 May; Harvested on 27 September
Total irrigation amount	1270 mm	700 mm	520 mm	625 mm

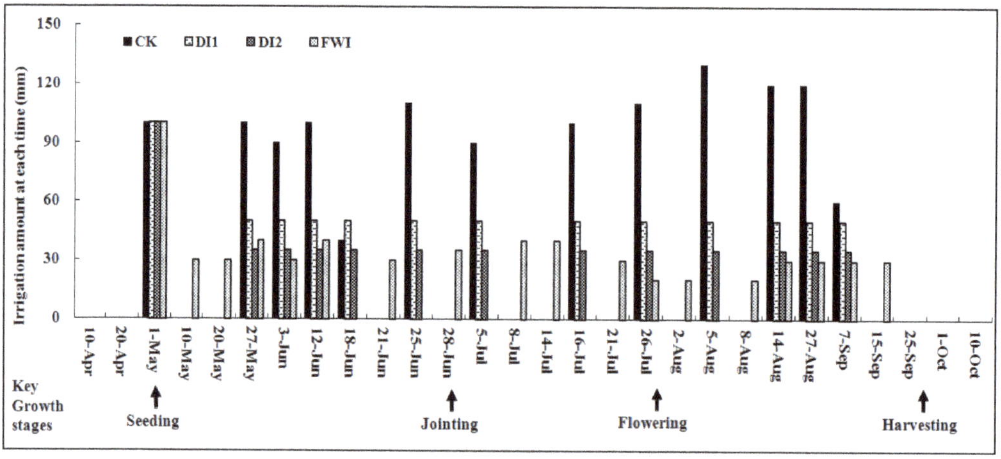

Figure 2. Irrigation at each watering period during the rice cropping seasons.

Irrigation times for CK, DI_1, DI_2 and FMI were 13, 13, 13 and 18 days, respectively (Figure 2). The total irrigation amounts were 1270, 700, 520 and 625 mm in the CK, DI_1, DI_2 and FMI plots, respectively (Table 1). All treatments were subjected to same pesticide and herbicide applications rates according to the local standards for high yields and pest control.

2.4. Greenhouse Gas Sampling

The static closed chamber and gas chromatography methods were adopted to sample and measure CH_4 and N_2O every 10 days in 2015 [19]. Polyvinyl chloride (PVC) chambers in accordance with the rice height and fitted with a battery-operated fan for thorough gas mixture in the head space were used. Collected gases were analyzed to obtain the concentrations of CH_4 and N_2O using a gas chromatograph (Agilent 7890A, Santa Clara, CA, USA) mounted with a flame ionization detector (FID) and an electron capture detector (ECD) to detect CH_4 and N_2O, respectively. The CH_4 and N_2O fluxes were calculated as:

$$G = (\Delta C/\Delta t) \times (V/A) \times \alpha$$

where G is the gas flux rate (g N_2O-N or CH_4-C ha^{-1} d^{-1}), $\Delta C/\Delta t$ designates the increase of gas concentration in the chamber (g L^{-1} d^{-1}), V is the chamber volume (L), A is area enclosed by the chamber (ha), and α is a conversion coefficient for elemental C (α = 0.749) or N (α = 0.636). The slope of the mixing ratio of four sequential samples was used in the determination of both CH_4 and N_2O fluxes. Cumulative CH_4 and N_2O emissions were computed using the formula described by Cai et al. [20].

The area-scaled GHG emission was converted to CO_2 equivalent (CO_2-eq) as follows:

$$\text{Area-scaled GHG emission (kg } CO_2\text{-eq ha}^{-1} \text{ yr}^{-1}) = 25 \times CH_4 + 298 \times N_2O$$

where, CH_4 and N_2O represent the seasonal cumulative emissions. Yield-scaled GHG emission was computed by dividing area-scaled emission by yield of rice [21].

2.5. Yield and Yield Components Measurement

A one m^2 rice plant at physiological maturity was harvested for yield determination. Grain yield was adjusted to 14% moisture content using the formula:

$$\text{Yield} = GW \times (100 - GMC)\%/(100 - 14)\%$$

where: GW = Grain weight. GMC = Grain moisture content.

Number of panicles was evaluated by counting the total panicle number per 1 m^2 per plot. Spikelet per panicle was evaluated by counting both the filled and unfilled spikelets per 1 m^2 randomly taken from each plot. Dry weight of 1000 grains from three replicates samples of filled grains per plot were obtained by drying at 70 °C in the oven for 72 h to constant dry weight.

2.6. Statistical Analyses

The data was analyzed using analysis of one-way variance (SPSS 23.0 for windows) to test the differences among the treatments. The least significant difference (LSD) test was used to compared treatment means ($p < 0.05$). Microsoft Excel 2003 was used to compute the standard deviation of the means.

3. Results

3.1. Rice Plant Growth and Grain Yield

Differences that were significant at the rice growth stages and biomass production were recorded between irrigation treatments (Table 2). Water-saving irrigation advanced rice heading and maturity stage, resulting in a reduction in the length of the rice growth period. Compared to CK, the primary heading stage was advanced by 2, 1 and 1 day in 2014, and 3, 3 and 2 days in 2015 in the DI_1, DI_2 and FWI plots, respectively. Consequently, the length of rice growth was shortened by 2, 2 and 1 day(s) in 2014, and 2, 4 and 1 day(s) in 2015 in DI_1, DI_2 and FWI plots, respectively.

Table 2. Impact of irrigation on rice growth stages and aboveground biomass at pre- and post-anthesis phases.

Treatment	Heading Stage		Maturity Stage		Biomass Production	
	Date (MM-DD)	Advanced Day(s)	Date (MM-DD)	Advanced Day(s)	Pre-Anthesis Period (t ha^{-1})	Post-Anthesis Period (t ha^{-1})
2014						
CK	07-29	-	09-25	-	9.4 ± 0.2	5.2 ± 0.2
DI$_1$	07-27	2	09-23	2	8.5 ± 0.3	4.1 ± 0.2
DI$_2$	07-28	1	09-23	2	7.6 ± 0.1	3.2 ± 0.3
FWI	07-28	1	09-24	1	8.6 ± 0.3	4.5 ± 0.3
2015						
CK	07-31	-	09-26	-	9.6 ± 0.1	5.3 ± 0.1
DI$_1$	07-28	3	09-24	2	8.4 ± 0.4	3.9 ± 0.6
DI$_2$	07-28	3	09-22	4	7.0 ± 0.2	2.6 ± 0.4
FWI	07-29	2	09-25	1	8.2 ± 0.2	4.3 ± 0.5

CK (Traditional flood irrigation); DI$_1$ (Drip irrigation under plastic film mulching with 50 mm irrigation); DI$_2$ (Drip irrigation under plastic film mulching with 35 mm irrigation); FWI (Furrow wetting irrigation).

Water-saving irrigation (WSI) practices significantly decreased rice biomass production (Tables 2 and 3). The lowest aboveground biomass production was found in the DI$_2$ plots. As compared to the CK, the pre-anthesis aboveground biomass production over two study years was 11.1%, 23.2% and 11.6% lower in the DI$_1$, DI$_2$ and FWI plots, respectively while the post-anthesis aboveground biomass production was 23.8%, 44.8% and 16.2% lower in the DI$_1$, DI$_2$ and FWI plots, respectively (Table 2). Consequently, the adoption of water-saving irrigation resulted in a reduction of 15.6%, 30.2 and 13.2% relative to the CK in the DI$_1$, DI$_2$ and FWI plots, respectively (Table 3). Rice yields ranging from 5.9 to 8.7 t ha^{-1} produced significant differences in the different irrigation treatments (Table 3). The highest yield was found in the CK plot and the lowest existed in the DI$_2$ plot in both years. The choice of DI$_1$, DI$_2$ and FWI produced 10.3%, 32.1% and 8.1% lower rice yield in comparison with CK in 2014, and 11.8%, 34.7% and 10.2% lower in 2015. Non-significant yield differences were noted amid the adoption of CK and FWI in 2014 but were significant in 2015 (Table 3). Water-saving significantly decreased rice panicles per area, with DI$_2$ recording the lowest. The choice of DI$_1$ significantly lowered number of panicles compared to CK. Spikelets per panicle and the 1000-grain weight showed significant variation among the irrigation treatments. Noticeable was the significantly lower spikelets and 1000-grain weight in DI$_2$ plots.

Table 3. Rice yield and yield components as impacted by water-saving irrigation.

Treatment	Rice Yield (t ha^{-1})	Number of Panicles (m^{-2})	Spikelets Panicle^{-1}	1000-Grain Weight (g)
2014				
CK	8.7 ± 0.2 a	430.2 ± 8.6 a	98.9 ± 1.6 a	24.8 ± 0.2 a
DI$_1$	7.8 ± 0.3 b	412.5 ± 7.9 b	97.4 ± 1.2 a	24.2 ± 0.1 b
DI$_2$	6.2 ± 0.3 c	356.8 ± 12.6 c	95.3 ± 1.4 b	22.1 ± 0.2 c
FWI	8.2 ± 0.3 ab	416.2 ± 11.3 ab	98.1 ± 1.5 a	24.4 ± 0.1 b
2015				
CK	8.5 ± 0.1 a	442.0 ± 6.2 a	100.0 ± 2.1 a	24.7 ± 0.1 a
DI$_1$	7.5 ± 0.2 b	403.7 ± 8.4 b	96.6 ± 1.7 ab	23.7 ± 0.1 c
DI$_2$	5.9 ± 0.3 c	301.0 ± 10.1 c	94.3 ± 1.7 b	21.8 ± 0.1 d
FWI	7.9 ± 0.2 b	411.0 ± 11.6 b	99.2 ± 2.0 a	24.1 ± 0.1 b

CK (Traditional flood irrigation); DI$_1$ (Drip irrigation under plastic film mulching with 50 mm irrigation); DI$_2$ (Drip irrigation under plastic film mulching with 35 mm irrigation); FWI (Furrow wetting irrigation). Different letters in the same column shows significant differences at $p < 0.05$.

3.2. CH_4 and N_2O Emission Fluxes and Seasonal Emission Ratios

Similar patterns of CH_4 fluxes existed in the irrigation methods (Figure 3a). The maximum emission fluxes occurred during rice heading and flowering stages, and the lowest occurred during the seedling and maturity stages amongst the treatments. The variations of CH_4 emission fluxes were similar with the seasonal changes of air temperature (Figure 1b). However, differences of significance in the mean peak CH_4 emission fluxes between CK and the other irrigation methods were noted (Figure 3a). No variation of significance in the flux peak existed in the three water-saving methods. The peak mean CH_4 emission was noted in the CK plots. The mean flux value was 267, 537 and 191% more in the CK plot compared to those of DI_1, DI_2 and FWI plots, respectively ($p < 0.05$). Seasonal variation patterns of N_2O fluxes were variable (Figure 3b). The highest flux peaks were noted in the DI_1, DI_2 treatments while the lowest occurred in the CK plots. The flux in CK was 55.5, 305.1 and 82.5% lower than those in the DI_1, DI_2 and FWI treatments, respectively. The flux of the total emission at CO_2-eq scale was 147, 140 and 126% lower in the DI_1, DI_2 and FWI treatments compared to the CK plot (Figure 3c). The adoption of DI_1 and DI_2 recorded higher emission ratios at the pre-anthesis stage compared to FWI and CK (Figure 3d). At the post-anthesis stage a lower emission ratio was noted in DI_1 and DI_2 in comparison with FWI and CK.

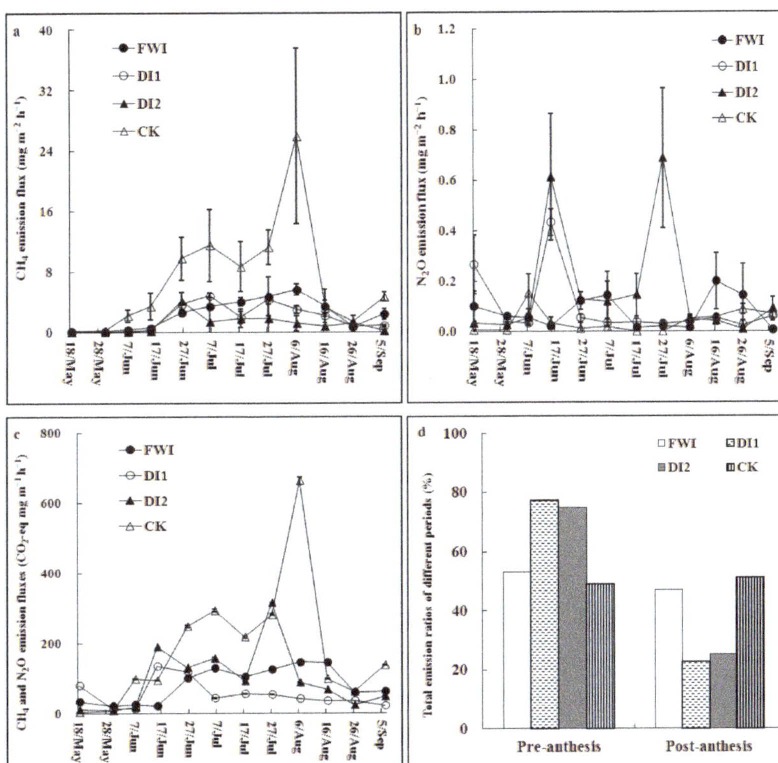

Figure 3. Differences in CH_4 (**a**), N_2O (**b**), CO_2 equivalent of CH_4 and N_2O (**c**) emission fluxes and emission ratios of pre- and post-anthesis periods (**d**) in irrigation plots.

3.3. Water Productivity and Area and Yield-Scaled Emissions

The irrigation methods exhibited significantly different water productivity levels ($p < 0.05$) (Table 4). The adoption of DI_1, DI_2 and FWI showed increased water productivity compared to CK. The highest value of water productivity was noted in the FWI plot, whereas the lowest was detected in the CK plot. The water productivity was 88.9, 16.4 and 11.4% higher in the FWI plot than those in the CK, DI_1 and DI_2 plots, respectively. Using the WSI significantly decreased cumulative emission of CH_4 by 73.0, 84.7 and 64.4% in DI_1, DI_2 and FWI, respectively, compared to CK (Table 4). Also, among the water-saving irrigation, significant differences were noted, with DI_2 recording 43.6 and 57.2%, lower cumulative CH_4 than DI_1 and FWI, respectively. Significantly, cumulative N_2O emission was 2.8, 4.1 and 2.0-fold more in DI_1, DI_2 and FWI than CK. The usage of DI_2, triggered a 1.4 and 2.0-fold more cumulative N_2O emission compared to DI_1 and FWI, respectively. The area-scaled emission in the CK was 129, 141 and 116% higher ($p < 0.05$) than those in the DI_1 and DI_2 and FWI plots, respectively. Area-scaled emission amidst the WSI methods recorded no significant variation, though area-scaled emission between WSI and the CK were significantly different. The yield-scaled emission in DI_1 and DI_2 and FWI were 101.0, 67.5 and 102.0%, respectively, significantly less than CK ($p < 0.05$). Among the WSI, significant differences in yield-scaled emission were observed, with the lowest yield-scaled emission found in the FWI plot.

Table 4. Impact of water-saving irrigation on cumulative CH_4, N_2O emissions, area and yield-scaled emissions and water productivity.

Treatment	CH_4 (kg CO_2-eq ha^{-1})	N_2O (kg CO_2-eq ha^{-1})	Area-Scaled Emission (kg CO_2-eq ha^{-1})			Yield-Scaled Emission (kg CO_2-eq t^{-1})	Water Productivity (kg m^{-3})
			Pre-Anthesis	Post-Anthesis	Total		
CK	5212.7 ± 1288.5 a	364.0 ± 41.4 c	2924.2 ± 998.7 a	2652.5 ± 411.7 a	5576.7 ± 1309.1 a	656.1 ± 130.2 a	0.66 ± 0.02 c
DI_1	1406.7 ± 148.3 c	1033.1 ± 221.0 ab	1415.7 ± 175.2 b	410.2 ± 31.7 d	2439.8 ± 121.8 b	325.5 ± 24.4 c	1.08 ± 0.04 b
DI_2	792.8 ± 101.7 d	1517.4 ± 271.3 a	2195.8 ± 235.4 a	728.3 ± 142.7 c	2310.2 ± 182.2 b	391.6 ± 27.1 b	1.13 ± 0.03 ab
FWI	1853.1 ± 187.7 b	731.4 ± 129.4 b	1369.5 ± 135.7 b	1215.0 ± 134.7 b	2584.5 ± 221.2 b	325.3 ± 19.4 c	1.26 ± 0.05 a

CK (Traditional flood irrigation); DI_1 (Drip irrigation under plastic film mulching with 50 mm irrigation); DI_2 (Drip irrigation under plastic film mulching with 35 mm irrigation); FWI (Furrow wetting irrigation). Different letters in the same column shows significant differences at $p < 0.005$.

4. Discussion

Compared to the traditional continuous flooding, water-saving irrigation (WSI) could increase water productivity [22,23] and maintain or increase rice grain yield [24], although some studies have reported contrary findings [14,25]. The results of this study indicated that the adoption of WSI amplified water-use efficiency but caused a reduction in rice yield (Tables 2 and 3). A substantial decline in water application may adversely impact rice yield due to sensitivity to non-saturated soil environments [26]. This was very prominent in the drip irrigation with 35 mm irrigation (DI_2) arising primarily from limited water for rice biomass and panicle per area development and consequently affecting rice yield (Table 3). This also supports the assertion that irrigation volumes impact WSI [14,24]. Although water-saving irrigation caused rice yield reduction, the drop was significant in DI_2 water-saving irrigation methods. He et al. [27] established that yield reduction occurs in extreme water-saving irrigation, owing to inadequate tillers and spikes. The lowest reduction in yield was in FWI, which produced the highest water productivity value (Table 4). This arises due to hastened canopy closure and decreased partial stomatal closure for the period of soil drying cycles, helping to minimize evapotranspiration [28,29], and less percolation of water into the soil [27]. Therefore, the choice of FWI may offer an alternative for maintaining yields while minimizing water consumption. Previous studies show that high ground water mitigates the influence of water-saving irrigation on the growth of rice at the post-anthesis stage [30,31]. In our study, lower groundwater table and precipitation during rice growing season in the two study years could have exacerbated water limitation for rice growth, negatively impacted panicles, spikelet numbers and grain filling, and subsequently significantly reduced the 1000-grain weight (Table 3).

No obvious increases in CH_4 emission were recorded at the rice tillering stage (Figure 3a). The non-optimal and relatively lower air temperature of 19.2 °C during rice tillering may have hindered methanogenic activities that stimulate CH_4 production during rice growth [32]. The peak flux of CH_4 was noticeable at the heading stage for all the treatments, similar to previous works of Chen et al. [13]. Rising daily mean air temperature of more than 25 °C at the rice booting and heading stages, well-developed aerenchyma for CH_4 emitting, and increased rice growth that stimulated root-derived exudation for methanogenic activities [15] may explain the peak flux occurring at the heading stage (Figure 3a). Studies show that soil water status affects CH_4 formation and emission [17]. In our study, though significant differences were noted in the CH_4 emission from the WSI, it did not trigger an exponential increase in CH_4 emission in comparison to the CK. Compared to continuous flooding irrigation, WSI irrigation had a superior prospect to decrease CH_4 emissions in line with alterations in soil water dynamics [33]. Evidently, a reduction in water use corresponded with a decline in the emission of CH_4, especially in DI_2. This supports the assertion that WSI shows a significant potential to mitigate CH_4 emissions [14]. In comparison with the traditional flooding irrigation, the adoption of WSI substantially triggered N_2O emission arising from one or more drainage events and the wet-dry cycles to suppress CH_4 production during rice growth [15]. Similar to previous studies [34,35], the adoption of continuous flooding demonstrated higher CH_4 emission compared to WSI. The cycle of continuous dry-wet cycles and the smaller amounts of water available in the WSI might have negatively affected CH_4 production [36] by inhibiting the formation of soil reductive conditions. A reduction in soil water content via WSI is presumed as a favorable preference for CH_4 mitigation. Among the water-saving irrigation practices, the reductions in CH_4 emissions in the drip irrigation plots (DI_1 and DI_2) were significantly higher than that in the furrow wetting irrigation plot (Table 4). This was expounded by the fact that lower soil moisture content, in both DI_1 and DI_2, stifled the emission of CH_4 to a very low-level during rice growth. Our observations support a previous study by Katayanagi et al. [37], who reported a 73% mitigation of CH_4 emission via WSI during rice cultivation. Thus, however, these higher reductions in CH_4 emissions could not compensate for the higher increases in N_2O emissions in the drip irrigation plots. Consequently, the CO_2-eq emissions of CH_4 and N_2O were similar among the three water-saving irrigation methods. Since the reductions in rice yield were higher in the drip irrigation fields compared to that of FWI field, the lowest yield-scaled CO_2-eq emission was found in the FWI field.

5. Conclusions

Sustainable water management in direct-seeded rice highlights the importance of adopting water-saving irrigation to reduce GHG emission, increase water productivity and sustain rice yield. In contrast to continuous flooding, WSI caused a decline in CH_4 emissions while essentially triggering N_2O emission increases. The highest water productivity and rice yield, lower area and yield-scaled emission among the WSI were observed in the adoption of furrow wetting irrigation (FWI). For sustainable direct-seeded rice production under water-saving irrigation in North China, furrow wetting irrigation (FWI) is recommended to sustain rice yield and minimize greenhouse gas emissions.

Author Contributions: Literature search, X.H. and F.D.; figures, X.H. and J.L.; study design, X.H. and J.Z. (Jian Zhang); investigation, X.H. and J.L.; data analysis, D.L. and J.Z. (Jian Zhang); writing—review and editing, X.H., J.Z. (Jian Zhang) and J.Z. (Jun Zhang) All authors have read and agreed to the published version of the manuscript.

Funding: This work was supported by the National Key R&D Program of China (2016YFD0300907).

Institutional Review Board Statement: Not applicable.

Informed Consent Statement: Not applicable.

Data Availability Statement: The data presented in this study are available on request from the authors.

Conflicts of Interest: The authors declare no conflict of interest.

References

1. Livsey, J.; Katterer, T.; Vico, G.; Lyon, S.W.; Lindborg, R.; Scaini, A.; Da, C.T.; Manzoni, S. Do alternative irrigation strategies for rice cultivation decrease water footprints at the cost of long-term soil health? *Environ. Res. Lett.* **2019**, *14*, 074011. [CrossRef]
2. Maraseni, T.N.; Deo, R.C.; Qu, J.; Gentle, P.; Neupane, P.R. An international comparison of rice consumption behaviours and greenhouse gas emissions from rice production. *J. Clean. Prod.* **2018**, *172*, 2288–2300. [CrossRef]
3. Campbell, B.M.; Beare, D.J.; Bennett, E.M.; Hall-Spencer, J.M.; Ingram, J.S.I.; Jaramillo, F.; Ortiz, R.; Ramankutty, N.; Sayer, J.A.; Shindell, D. Agriculture production as a major driver of the Earth system exceeding planetary boundaries. *Ecol. Soc.* **2017**, *22*, 8. [CrossRef]
4. Liu, J.; Yang, H.; Gosling, S.N.; Kummu, M.; Flörke, M.; Hanasaki, N.; Wada, Y.; Zhang, X.; Zheng, C. Water scarcity assess-ments in the past, present, and future. *Earth's Future* **2017**, *5*, 545–559. [CrossRef] [PubMed]
5. Enriquez, Y.; Yadav, S.; Evangelista, G.K.; Villanueva, D.; Burac, M.A.; Pede, V. Disentangling Challenges to Scaling Alter-nate Wetting and Drying Technology for Rice Cultivation: Distilling Lessons from 20 Years of Experience in the Philip-pines. *Front. Sustain. Food Syst.* **2021**, *5*, 675818. [CrossRef]
6. Dalin, C.; Qiu, H.; Hanasaki, N.; Mauzerall, D.L.; Rodriguez-Iturbe, I. Balancing water resource conservation and food security in China. *Proc. Natl. Acad. Sci. USA* **2015**, *112*, 4588–4593. [CrossRef]
7. Xu, C.C.; Zhou, X.Y.; Li, F.B.; Fang, F.P. Northward research of Chinese rice production. *Issues Agric. Econ.* **2013**, *7*, 5. (In Chinese)
8. Jiang, S.; Wang, J.; Zhao, Y.; Shang, Y.; Gao, X.; Li, H.; Wang, Q.; Zhu, Q. Sustainability of water resources for agriculture considering grain production, trade and consumption in China from 2004 to 2013. *J. Clean. Prod.* **2017**, *149*, 1210–1218. [CrossRef]
9. Wang, H.; Bouman, B.A.M.; Zhao, D.; Wang, C.; Moya, P.F. Aerobic rice in northern China: Opportunities and challenges. In *Water-Wise Rice Production*; Bouman, B.A.M., Hengsdijk, H., Hardy, B., Bindraban, P.S., Tuong, T.P., Ladha, J.K., Eds.; International Rice Research Institute: Los Banos, Philippines, 2022; pp. 143–154.
10. Schewe, J.; Heinke, J.; Gerten, D.; Haddeland, I.; Arnell, N.W.; Clark, D.B.; Dankers, R.; Eisner, S.; Fekete, B.M.; Colón-González, F.J.; et al. Multimodel assessment of water scarcity under climate change. *Proc. Natl. Acad. Sci. USA* **2014**, *111*, 3245–3250. [CrossRef]
11. Wassmann, R.; Pasco, R.; Zerrudo, J.; Ngo, D.M.; Vo, T.B.T.; Sander, B.O. Introducing a new tool for greenhouse gas calcula-tion tailored for cropland: Rationale, operational framework and potential application. *Carbon Manag.* **2019**, *10*, 79–92. [CrossRef]
12. Leahy, S.; Clark, H.; Reisinger, A. Challenges and Prospects for Agricultural Greenhouse Gas Mitigation Pathways Consistent with the Paris Agreement. *Front. Sustain. Food Syst.* **2020**, *4*, 69. [CrossRef]
13. Chen, W.; Wang, Y.; Zhao, Z.; Cui, F.; Gu, J.; Zheng, X. The effect of planting density on carbon dioxide, methane and ni-trous oxide emissions from a cold paddy field in the Sanjiang Plain, northeast China. *Agric. Ecosyst. Environ.* **2013**, *178*, 64–67. [CrossRef]
14. Xu, Y.; Ge, J.; Tian, S.; Li, S.; Nguy-Robertson, A.L.; Zhan, M.; Cao, C. Effects of water-saving irrigation practices and drought resistant rice variety on greenhouse gas emissions from a no-till paddy in the central lowlands of China. *Sci. Total Environ.* **2015**, *505*, 1043–1052. [CrossRef] [PubMed]
15. Li, X.; Yuan, W.; Xu, H.; Cai, Z.; Yagi, K. Effect of timing and duration of midseason aeration on CH4 and N2O emissions from irrigated lowland rice paddies in China. *Nutr. Cycl. Agroecosystems* **2011**, *91*, 293–305. [CrossRef]
16. Li, S.X.; Wang, Z.H.; Stewart, B.A. Responses of Crop Plants to Ammonium and Nitrate N. *Adv. Agron.* **2013**, *118*, 205–397. [CrossRef]
17. Li, J.L.; Li, Y.E.; Wan, Y.F.; Wang, B.; Waqas, M.A.; Cai, W.W.; Guo, C.; Zhou, S.H.; Su, R.S.; Qin, X.B.; et al. Combination of modified nitrogen fertilizers and water saving irrigation can reduce greenhouse gas emissions and in-crease rice yield. *Geoderma* **2017**, *315*, 1–10. [CrossRef]
18. Wang, J.Y.; Jia, J.X.; Xiong, Z.Q.; Khalil, M.A.K.; Xing, G. Water regime–nitrogen fertilizer–straw incorporation interaction: Field study on nitrous oxide emissions from a rice agroecosystem in Nanjing, China. *Agric. Ecosyst. Environ.* **2011**, *141*, 437–446. [CrossRef]
19. Zou, J.W.; Huang, Y.; Jiang, J.Y.; Zheng, X.H.; Sass, R.L. A 3-year field measurement of methane and nitrous oxide emissions from rice paddies in China: Effects of water regime, crop residue, and fertilizer application. *Glob. Biogeochem. Cycles* **2005**, *19*, GB2021. [CrossRef]
20. Cai, Y.; Ding, W.; Luo, J. Nitrous oxide emissions from Chinese maize—Wheat rotation systems: A 3-year field measurement. *Atmos. Environ.* **2013**, *65*, 112–122. [CrossRef]
21. Feng, J.; Chen, C.; Zhang, Y.; Song, Z.; Deng, A.; Zheng, C.; Zhang, W. Impacts of cropping practices on yield-scaled green-house gas emissions from rice fields in China: A meta-analysis. *Agric. Ecosyst. Environ.* **2013**, *164*, 220–228. [CrossRef]
22. Fan, M.S.; Liu, X.J.; Jiang, R.F.; Zhang, F.S.; Lu, S.H.; Zeng, X.; Christie, Z.P. Crop yields, internal nutrient efficiency, and changes in soil properties in rice—Wheat rotations under non-flooded mulching cultivation. *Plant Soil.* **2005**, *277*, 265–276. [CrossRef]
23. Zhang, Z.; Zhang, S.; Yang, J.; Zhang, J. Yield, grain quality and water use efficiency of rice under non-flooded mulching cultivation. *Field Crop. Res.* **2008**, *108*, 71–81. [CrossRef]
24. Liu, L.; Chen, T.; Wang, Z.; Zhang, H.; Yang, J.; Zhang, J. Combination of site-specific nitrogen management and alternate wetting and drying irrigation increases grain yield and nitrogen and water use efficiency in super rice. *Field Crop. Res.* **2013**, *154*, 226–235. [CrossRef]

25. Linquist, B.A.; Anders, M.M.; Adviento-Borbe, M.A.A.; Chaney, R.L.; Nalley, L.L.; da Rosa, E.F.F.; Van Kessel, C. Reducing greenhouse gas emissions, water use, and grain arsenic levels in rice systems. *Glob. Chang. Biol.* **2015**, *21*, 407–417. [CrossRef] [PubMed]
26. Kumar, A.; Nayak, A.K.; Pani, D.R.; Das, B.S. Physiological and morphological responses of four different rice cultivars to soil water potential based deficit irrigation management strategies. *Field Crop. Res.* **2017**, *205*, 78–94. [CrossRef]
27. He, G.; Wang, Z.; Cui, Z. Managing irrigation water for sustainable rice production in China. *J. Clean. Prod.* **2019**, *245*, 118928. [CrossRef]
28. Kadiyala, M.D.M.; Jones, J.W.; Mylavarapu, R.S.; Li, Y.C.; Reddy, M.D. Identifying irrigation and nitrogen best management practices for aerobic rice–maize cropping system for semi-arid tropics using CERES-rice and maize models. *Agric. Water Manag.* **2015**, *149*, 23–32. [CrossRef]
29. Du, T.S.; Kang, S.Z.; Zhang, J.H.; Davies, W.J. Deficit irrigation and sustainable water-resource strategies in agriculture for China's food security. *J. Exp. Bot.* **2015**, *66*, 2253–2269. [CrossRef]
30. Kreye, C.; Dittert, K.; Zheng, X.; Zhang, X.; Lin, S.; Tao, H.; Sattelmacher, B. Fluxes of methane and nitrous oxide in wa-ter-saving rice production in north China. *Nutr. Cycl. Agroecosyst.* **2007**, *77*, 293–304. [CrossRef]
31. Yang, S.; Peng, S.; Hou, H.; Xu, J. Controlled irrigation and drainage of a rice paddy field reduced global warming potential of its gas emissions. *Arch. Agron. Soil Sci.* **2013**, *60*, 151–161. [CrossRef]
32. Chan, A.S.K.; Parkin, T.B. Methane Oxidation and Production Activity in Soils from Natural and Agricultural Ecosystems. *J. Environ. Qual.* **2001**, *30*, 1896–1903. [CrossRef]
33. Scheer, C.; Grace, P.R.; Rowlings, D.W.; Payero, J. Nitrous oxide emissions from irrigated wheat in 561 Australia: Impact of irrigation management. *Plant Soil.* **2012**, *359*, 351–362. [CrossRef]
34. Ahmad, S.; Li, C.; Dai, G.; Zhan, M.; Wang, J.; Pan, S.; Cao, C. Greenhouse gas emission from direct seeding paddy field under different rice tillage systems in central China. *Soil Tillage Res.* **2009**, *106*, 54–61. [CrossRef]
35. Tyagi, L.; Kumari, B.; Singh, S. Water management—A tool for methane mitigation from irrigated paddy fields. *Sci. Total Environ.* **2010**, *408*, 1085–1090. [CrossRef] [PubMed]
36. Estop-Aragonés, C.; Knorr, K.H.; Blodau, C. Belowground in situ redox dynamics and methanogenesis recovery in a de-graded fen during dry-wet cycles and flooding. *Biogeosciences* **2013**, *10*, 421–436. [CrossRef]
37. Katayanagi, N.; Furukawa, Y.; Fumoto, T.; Hosen, Y. Validation of the DNDC-Rice model by using $CH_4$and N_2O flux data from rice cultivated in pots under alternate wetting and drying irrigation management. *Soil Sci. Plant Nutr.* **2012**, *58*, 360–372. [CrossRef]

agriculture

Article

Effect of No-Tillage Management on Soil Organic Matter and Net Greenhouse Gas Fluxes in a Rice-Oilseed Rape Cropping System

Huabin Zheng [1], Xianliang Tang [1], Jiabin Wei [1], Huaqin Xu [2,*], Yingbin Zou [1] and Qiyuan Tang [1]

[1] College of Agronomy, Hunan Agricultural University, Changsha 410128, China; hbzheng@hunau.edu.cn (H.Z.); yuebao2022@126.com (X.T.); yeshangtianxia@163.com (J.W.); ybzou123@126.com (Y.Z.); qytang@hunau.edu.cn (Q.T.)

[2] College of Resources and Environment, Hunan Agricultural University, Changsha 410128, China

* Correspondence: xu7541@hunau.edu.cn; Tel.: +86-731-8461-8076

Abstract: No-tillage (NT) management is considered a leading approach for sustaining crop production and improving soil and environmental quality. Based on a long-term no-tillage experiment in a rice–oilseed rape cropping system, we examined differences in soil organic matter (SOM), soil microbial carbon (C) and nitrogen (N) content, and methane (CH_4) and carbon dioxide (CO_2) fluxes between NT and conventional tillage (CT) management. SOM under NT was 21.0 g kg^{-1}, and a significant difference was detected between 2004 and 2016. SOM increased under NT and CT by averages of 0.60 and 0.32 g kg^{-1} $year^{-1}$, respectively. Soil microbial C and N content were higher under CT than under NT. However, soil C:N ratios under NT were 17.4 and 9.7% higher than the CT, respectively, whereas soil microbial C:N ratios under NT were on average 9.47 and 9.70% higher. In addition, about 70% of CO_2 net uptake and over 99% of net CH_4 emissions occurred during the rice season in May–September in the rice–oilseed rape cropping system. Annual cumulative CH_4 and daytime net ecosystem CO_2 exchange (NEE) under NT was 1813.9 g CO_2 equiv. m^{-2}, 10.8% higher than that under CT. Our results suggest that a higher soil microbial C:N ratio and NEE (CH_4 and daytime CO_2) could contribute to increasing SOM/C in the surface soil under NT management.

Keywords: C:N ratio; net ecosystem exchange; soil microbial carbon; soil microbial nitrogen

Citation: Zheng, H.; Tang, X.; Wei, J.; Xu, H.; Zou, Y.; Tang, Q. Effect of No-Tillage Management on Soil Organic Matter and Net Greenhouse Gas Fluxes in a Rice-Oilseed Rape Cropping System. *Agriculture* **2022**, *12*, 918. https://doi.org/10.3390/agriculture12070918

Academic Editor: Ryusuke Hatano

Received: 7 May 2022
Accepted: 27 May 2022
Published: 24 June 2022

1. Introduction

In recent years, a farming method with low labor requirements and simpler operation, no-tillage (NT), had been more and more widely promoted in modern agricultural production. In China, about 7 million hm^2 of cultivated land are under NT management, accounting for 5.5% of the total area under conservation agriculture worldwide [1]; cropland NT management started in 1970, with cereal crops including rice (*Oryza sativa* L.), corn (*Zea mays* L.), wheat (*Triticum aestivum* L.), and economical crops including soybeans (*Glycine max* L.) and oilseed rape (*Brassica campestris* L.). NT management is considered a leading approach to sustaining crop production and addressing soil and environmental quality concerns [1–3].

Studies have found that NT can improve crop yield [4–6]. The reason may be that NT can improve soil water content [6,7]; increase soil nitrogen, phosphorus, and other nutrients [8,9]; reduce soil nutrient leaching [10]; and improve soil enzyme activity [11] and increase microbial biomass [12,13]. The conclusion of NT on improving environmental quality is still controversial. A previous study comparing multiple tillage methods found that NT can reduce CH_4 emissions and global warming potential (GWP) of rice fields, but increase N_2O emissions [14]; compared with conventional tillage (CT), NT increases CH_4 emissions in winter and decrease N_2O and CO_2 emissions in summer [15]; and compared with reduce tillage (RT), it was found that NT can increase SOC and decrease greenhouse gas intensity (GHGI) [16]. However, some studies have shown that no-tillage significantly increases the CH_4 emission from rice fields and enhances the net comprehensive warming potential (GWP) and greenhouse gas intensity (GHGI) of rice fields [17], and significant increase in N_2O emissions [18]. In addition, many studies discussed the impact of NT on carbon (C) sequestration [2] and economy [19].

Previous studies have mainly focused on dryland and dryland crops, and there is also great controversy over whether NT can reduce greenhouse gas emissions, so it is necessary to study the impact of NT management on soil organic carbon (SOC), especially under dryland rice rotation systems. The dynamics of soil microbial carbon, nitrogen and net greenhouse gases can more accurately determine the impact of no-tillage on greenhouse gas emissions. Our hypotheses are that no-tillage increases soil nutrient content and reduces greenhouse gas emissions than conventional tillage under dry-wet rotation. To test our hypothesis, based on a fixed experiment of no-tillage management in a rice–oilseed rape cropping system, we measured CH_4 and CO_2 fluxes and net ecosystem exchange (NEE) using an ultraportable greenhouse gas analyzer and a static transparent chamber, and measured the changes of SOC and dynamic changes in soil microbial C and N and to evaluate the effect of NT management on CH_4 and daytime CO_2 NEE.

2. Materials and Methods

2.1. Experimental Design and Field Management

Starting in 2004, a long-term experiment was conducted based on a rice–oilseed rape cropping system at Changsha (28°11′ N, 113°04′ E), Hunan Province, China. The study site has a moist subtropical monsoon climate with a mean annual temperature of about 17.0 °C, mean annual rainfall of about 1355 mm, and about 1677 h mean annual sunshine [20]. The soil of the experimental field was clayey soil with 1.5% organic matter and 0.14% total N [20]. Liangyoupeijiu, the first super-hybrid rice variety in China, and Xiangzayou 6, a hybrid oilseed rape cultivar, were used in the experiment [21].

In this fixed experiment, a randomized block design was established with four different tillage and cultivation treatments, including conventional tillage (CT) and transplanting (CTTP), NT and transplanting (NTTP), CT and direct seeding (CTDS), and NT and direct seeding (NTDS). Each field plot was 30 × 30 m, with four replicates. More detailed information about the treatments and other management practices are reported in Huang et al. (2011). Based on this experiment, we measured soil greenhouse gas emissions (CO_2 and CH_4) in the CTTP and NTTP treatments in situ and in real time during 2015–2016. Soil microbe C and N, as well as soil chemical properties in the different layers, were also measured.

2.2. Soil Chemical Properties

In 2004, soil samples from the 0–20-cm soil layer were used to determine the soil fundamental fertility; in 2016, soil samples in the 0–60-cm soil profiles were used to determine soil chemical properties. Soil profiles were divided into six layers (0–10, 10–20, 20–30, 30–40, 40–50, and 50–60 cm). Soil samples by the five-point method were dried naturally and then used to determine soil organic matter (SOM) by $K_2Cr_2O_7$-concentrated H_2SO_4 and heating. Soil total N was determined by the Kjeldahl method, which involved two steps: (1) digestion of the sample to convert organic N into NH_4^+-N and (2) determination of NH_4^+-N in the digest. The soil C:N ratio was calculated by dividing the SOC concentration by the total N concentration. Soil total P was conducted using the H_2SO_4-H_2O_2-HF method and determined using colorimetric method. Soil total K was conducted using micro-diffusion and determined using flame spectrophotometry. Soil total P was determined by molybdenum antimony-D-iso-ascorbic acid colorimetry (MADAC). NH_4OAc-extractable K of soil samples was determined by flame spectrophotometry.

2.3. Real-Time CH_4 and CO_2 Flux Measurements

Real-time CH_4 and CO_2 fluxes were determined using the static chamber method with an ultraportable greenhouse gas analyzer (CH_4, CO_2, H_2O; Los Gatos Research (LGR,

San José, CA, USA). The static chamber was a square box with a side length of 50 cm and height of 120 cm. A fluted base matching that of the static chamber was planted in the soil in advance. Sampling was conducted at 9:00–11:00 a.m. and 15:00–17:00 p.m. on sunny days, and testing within the plot took 5 min. The sampling dates were 20, 37, 57, 77, and 102 days after rice transplanting in the rice season and 127, 159, 187, 242, and 239 days after rice transplanting in the oilseed rape season.

Temperature was recorded accurately in the static chamber and in the 0–3-cm soil layer. Plants were sampled from a 0.24-m^2 area within each plot on the sampling date. Plant samples were separated into leaf, straw, and grains by hand, the volume was determined using the drainage method. The drainage method was that plant samples was immersed in the fixed volume vessel (1000, 2000, and 5000 mL), and collected and measured the water volume by other volume vessel. Lastly, the effective volume of the chamber was used to determine the volume of plants in the chamber.

2.4. Daytime and Seasonal CO_2 and CH_4 Net Ecosystem Exchange

Daytime CO_2 (F, g m^{-2} d^{-1}) and CH_4 net ecosystem exchange (F, g m^{-2} d^{-1}) were calculated as follows:

$$F = \frac{P \times V}{R \times A \times (T + 273.15)} \times \frac{dc}{dt} \times \frac{M \times S}{10^6} \quad (1)$$

where P is atmospheric pressure under standard conditions (101.2237 × 10^3 Pa); V is the effective volume in the chamber (m^3), i.e., the difference between the volume of static chamber and the volume of the plant, fan, and thermometer; R is the gas constant (8.3144 J mol^{-1} K^{-1}); A is the covering area of the chamber (m^2); T is the average temperature at the testing time inside the chamber (°C); dc/dt is the rate of change in CO_2 and CH_4 concentrations (mg dm^{-3} s^{-1}); M is the CO_2 or CH_4 relative molecular mass (g mol^{-1}); and S is the duration of the sampling day (s).

Seasonal emissions in daytime CO_2 and CH_4 were calculated as follows:

$$T = \sum [(F_i + F_{i+1})/2 \times d] \quad (2)$$

where T (g m^{-2}) is the total seasonal emissions, F_i and F_{i+1} are the measured fluxes on two consecutive sampling days, and d is the number of days between the two sampling dates.

2.5. Soil Microbe C and N

In 2016, the five-point method was used to sample soil in the 0–20-cm soil layer. The sampling dates were prior to rice planting, at 20, 37, 57, 77, 102, 119, and 159 days after rice transplanting during the rice season, and 187, 239, 281, and 314 days after rice transplanting in the oilseed rape season. Fresh soil samples were used directly to determine soil microbial C and N using the chloroform fumigation–incubation and K_2SO_4 extraction methods. Here, soil microbial carbon (mg kg^{-1}) = EC/0.33; soil microbial nitrogen (mg kg^{-1}) = EN × 5.0, where the soil microbial C and N coefficients are 0.33 and 5.0, respectively, and EC and EN are the differences in organic C and N in the K_2SO_4 extraction solution between fumigation and non-fumigation.

2.6. Data Analyses

Means of the Indexes were compared using Microsoft Excel 2007 software (Microsoft Corporation, Redmond, WA, USA) and Fisher's least significant difference (LSD) method. We performed a factorial analysis of variance and a least squares difference to test for statistically significant differences between the NT and CT using Statistix 8.0. (Analytical software, Tallahassee, FL, USA).

3. Results

3.1. Soil Chemical Properties

Under no-tillage (NT) and conventional tillage (CT) management, soil organic carbon (SOC) increased by averages of 0.60 and 0.32 g kg^{-1} year^{-1}, respectively (Table 1). In 2016, SOC under NT was 21.0 g kg^{-1}, and a significant difference was observed between 2004 and 2016. Soil total N decreased by averages of 19 mg kg^{-1} year^{-1} under NT and 26 mg kg^{-1} year^{-1} under CT. Soil total N and P under NT increased by 5.7 and 5.1%, respectively, compared with those under CT. There was no significant difference between 2004 and 2016. Soil C:N ratios under NT and CT were 17.4 and 15.8, respectively, with that under NT being 9.7% higher than that under NT (Figure 1). Soil total K decrease by an average of 0.51 g kg^{-1} year^{-1} under NT and 0.43 g kg^{-1} year^{-1} under CT; there was a significant difference in these parameters between 2004 and 2016. Soil NaOH hydrolysable N and Olsen P increased across years, and soil NH_4OAc extractable K decreased dramatically across years, by 8.28 mg kg^{-1} year^{-1} under NT and 7.51 mg kg^{-1} year^{-1} under CT. Soil chemical properties decreased as soil layer depth increased (Table 2).

Table 1. Variation of 0–20 cm soil layer soil chemical properties between the NT and CT treatment from 2004 to 2016.

Year	SOC (g·kg^{-1})		STN(g·kg^{-1})		STP(g·kg^{-1})		STK(g·kg^{-1})		SNN(mg·kg^{-1})		SOP(mg·kg^{-1})		SNK(mg·kg^{-1})	
	CT	NT	CT	NT	CT	NT	CT	NT	CT	NT	CT	NT	CT	NT
2004	15.0		1.40		1.18		18.10		137.0		38.4		113.0	
2016	18.2	21.0	1.15	1.21	1.17	1.23	13.79	12.99	154.8	154.7	52.3	57.6	37.9	30.2
Year	ns	*	ns	ns	ns	ns	*	*	ns	ns	ns	ns	*	*
Tillage #	*		ns		ns		ns		ns		ns		ns	

Soil sample with three replications using the five-point method ($n = 3$). # Significant difference ($p < 0.05$) between 2004 and 2016; SOC, soil organic matter; STN, soil total N; STP, soil total P; STK, soil total K; SNN, soil NaOH hydrolysable N; SOP, soil olsen P; SNK, soil NH_4OAc extractable K. * are significantly different according to LSD at $p < 0.05$. ns are not significantly different according to LSD at $p = 0.05$.

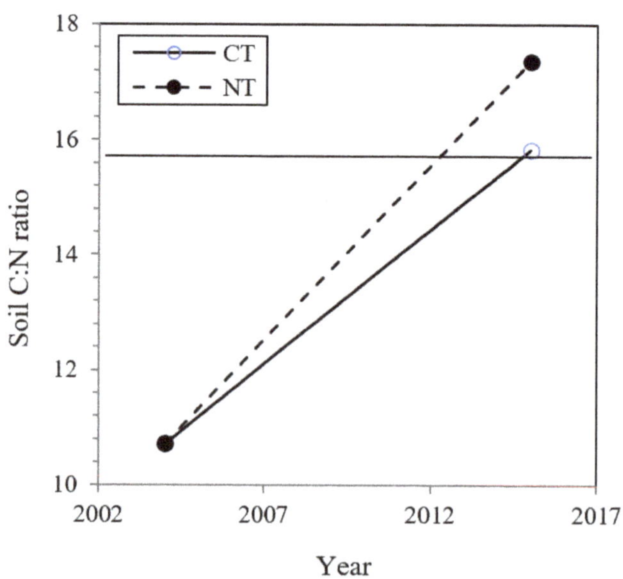

Figure 1. Variation in the soil carbon: nitrogen (C:N) ratio under CT and NT management in a rice–oilseed rape cropping system from 2004 to 2016.

Table 2. Variation of 0–60 cm soil layer soil chemical properties in 2016.

Soil Layer (cm)	SOC (g·kg^{-1})		STN (g·kg^{-1})		STP (g·kg^{-1})		STK (g·kg^{-1})		SNN (mg·kg^{-1})		SOP (mg·kg^{-1})		SNK (mg·kg^{-1})	
	CT	NT	CT	NT	CT	NT	CT	NT	CT	NT	CT	NT	CT	NT
0~10	18.45	21.97	1.46	1.45	1.25	1.25	12.85	12.90	186.21	189.69	57.50	60.47	43.94	36.87
10~20	18.00	20.10	0.83	0.97	1.09	1.21	14.72	13.08	123.31	119.64	47.14	54.77	31.80	23.55
20~30	16.87	17.67	0.68	0.69	0.80	0.95	15.56	14.85	97.05	105.75	28.03	40.57	19.18	19.90
30~40	16.17	15.96	0.86	0.74	0.62	0.69	16.08	16.38	104.47	95.33	17.79	25.67	20.80	20.17
40~50	13.94	14.02	0.81	0.87	0.55	0.52	14.63	15.81	84.14	75.54	13.48	12.54	16.70	19.13
50~60	9.38	9.47	0.65	0.62	0.39	0.44 *	17.89	20.67 *	60.50	70.70 **	4.42	6.07 *	17.21	19.22

SOC, soil organic matter; STN, soil total N; STP, soil total P; STK, soil total K; SNN, soil NaOH hydrolysable N; SOP, soil olsen P; SNK, soil NH_4OAc extractable K. *, ** are significantly different according to LSD at $p = 0.05$ and 0.01 between the NT and CT, otherwise, there was no significant difference between the NT and CT.

3.2. Daytime and Annual CH_4 and CO_2 Net Ecosystem Exchange

Daytime CO_2 net ecosystem exchange (NEE) was 14.5% higher in the rice season and 5.9% higher in the oilseed season under NT than under CT (Figure 2a). During the rice season, the maximum daytime CO_2 NEE values under NT and CT were 37.28 g m^{-2} d^{-1} and 34.82 g m^{-2} d^{-1} at 57 days after rice transplanting. During the oilseed season, daytime CO_2 NEE was highest from 222 to 316 days after rice transplanting. There was a dramatic difference in daytime CO_2 NEE between the rice and oilseed seasons (Figure 3). Daytime CO_2 NEE during the rice season was 70.0% under NT and 68.4% under CT, whereas that during the oilseed season was 30.0% under NT and 21.6% under CT.

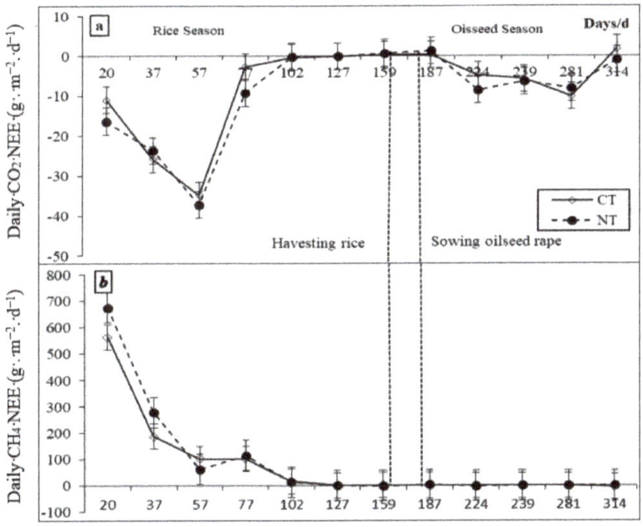

Figure 2. Changes in daily CO_2 (**a**) and CH_4 (**b**) net ecosystem exchange (NEE) under conventional tillage (CT) and no-tillage (NT) management in a rice–oilseed rape cropping system. The bars mean the standard error.

The ranges of CH_4 NEE were −0.38 to 674.70 mg m^{-2} d^{-1} during the rice season and −4.28 to 3.29 mg m^{-2} d^{-1} during the oilseed season (Figure 2b). CH_4 NEE during the rice season was 15.0 g m^{-2} season^{-1} under NT, accounting for 99.6% of total CH_4 NEE. CH_4 NEE during the rice season was 12.9 g m^{-2} season^{-1} under CT, 16.6% lower than that under NT in the rice–oilseed rape cropping system.

Total CH_4 emissions were 376.9 g CO_2 equiv. m^{-2} under NT and 323.4 g CO_2 equiv. m^{-2} under CT. In the rice–oilseed rape cropping system, annual CH_4 and daytime CO_2 NEE under NT was 1813.9 g CO_2 equiv. m^{-2}, 10.8% higher than that under CT.

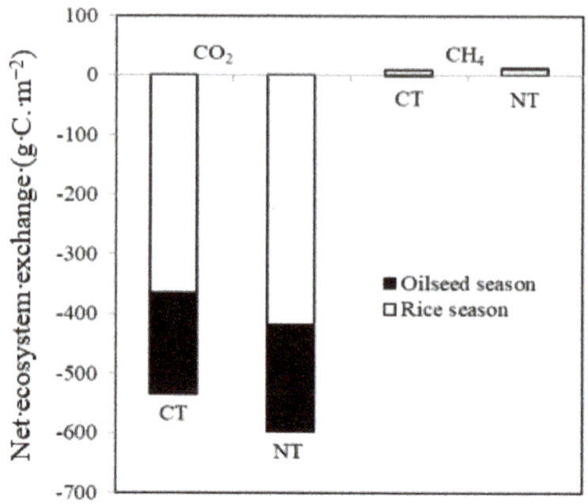

Figure 3. Seasonal CH_4 and CO_2 net ecosystem exchange (NEE) under CT and NT management in a rice-oilseed rape cropping system.

3.3. Soil Microbial C and N

The annual variation in soil microbial carbon (C) and nitrogen (N) was consistent between NT and CT management (Figure 4a,b). Soil microbial C and N under CT were higher by averages of 16.0 and 32.9%, respectively, than those under NT, whereas the soil microbial C:N ratio under CT was lower by an average of 9.7% than that under NT (Figure 4c).

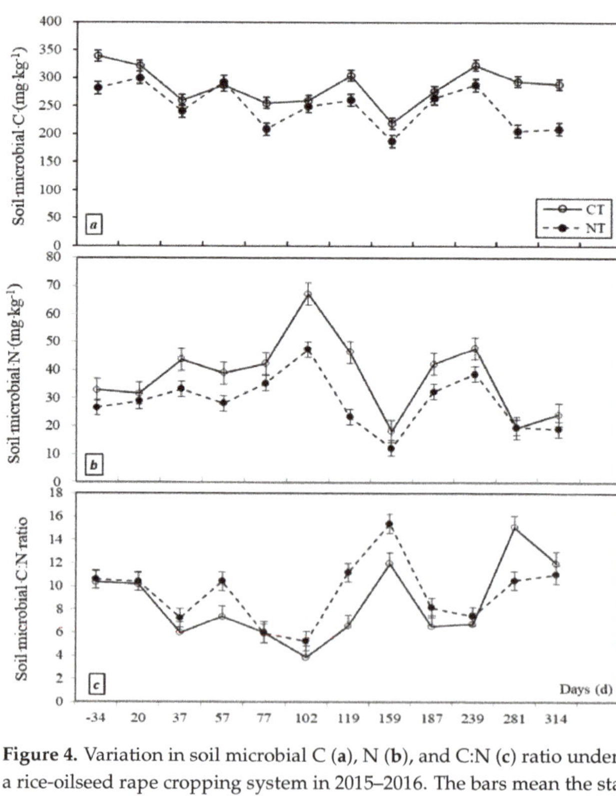

Figure 4. Variation in soil microbial C (**a**), N (**b**), and C:N (**c**) ratio under CT and NT management in a rice-oilseed rape cropping system in 2015–2016. The bars mean the standard error.

4. Discussion

Rice fields are a complex agro-ecological area, and greenhouse gases and soil fertility will be constantly changing and changing by various activities, in which farming methods can directly change the soil. Yonemura et al. (2014) reported that no-tillage (NT) management was generally effective in mitigating total global warming potential through reduced soil respiration and N_2O emission in temperate regions [22]; Li et al. (2014) found that NT can reduce soil disturbance, increase the stability of aggregates, and facilitate the formation of refractory carbon, thereby reducing soil CO_2 emissions [23]. Compared with conventional tillage (CT), NT can also enhance CH_4 oxidative and methanotrophic activity, ultimately reducing CH_4 emissions. In this study, we found that NT did not decrease the emission of CO_2 and CH_4. During the growth of crops, the absorption of CO_2 and the release of CH_4 by NT were higher than those of CT. This may be mainly due to the slow entry of nitrogen fertilizer into the soil under NT and the higher C:N ratio, which weakened the microbial activity and the ability to compete for nutrition, resulting in the weakening of the decomposition of CH_4 and the weakening of the competition for microbial nutrition for crops. On the contrary, the growth was better than that of CT. Finally, it may increase the CO_2 absorption in the daytime and the CO_2 emission at night.

The soil carbon:nitrogen (C:N) ratio is an important soil fertility indicator due to the close interactive relationship between soil organic carbon (SOC) and total N [24]. Wan et al. (2015) reported that the soil C:N ratio was an important factor influencing soil microbial community structure in subtropical coniferous and broadleaf forest plantations [25]. Under NT management, positive changes in soil physical properties appear to be closely related to positive effects of NT on SOC accumulation [7]. Our study found that although the soil microbial carbon and soil microbial nitrogen of CT were slightly higher than those of NT, which may be caused by CT directly changing the soil structure and making it easier for nutrients to enter the soil, but the soil microbial C:N ratio of CT was lower than that of NT, indicating that the ability of microorganisms to compete for fat may be reduced under NT. Therefore, further studies are needed to explain fully the phenomenon.

Previous studies have reported that the NT management can increase SOC with soil physical structure [26], fertilizer N input [27], and crop rotation [28], and that cropping frequency and fertilizer N input in association with NT resulted in increased SOC [29], and these just explain why NT under crop rotation and fertilization in this study could increase the SOC content in soil more than CT. Xiao et al. (2020) found that long-term NT was more beneficial to SOC increase in soil surface [30]. Many other studies have reported that NT increases SOC only in the upper 10 cm of the soil [2,31]. Even in long-term (>30 years) tillage studies, NT appears to increase wet aggregate stability only in the upper 10 cm of the soil [13,31], this is consistent with our research results. In the rice–rapeseed planting system, reasonable fertilizer nitrogen input and NT treatment can significantly increase the SOC content in the upper soil layer, especially in the 0-10 cm soil layer.

In addition to studying soil physical properties (reviewed by Blanco-Canqui and Ruis, 2018), chemical properties [32], and green-house gas (GHG) emissions, there are many reports that have looked at the effects on GHG emissions from other perspectives. For example, Kulmány et al. (2022) reported that moisture content, air temperature and pressure all play important roles in CO_2 emissions in NT systems [33]; and Yonemura et al. (2014) report that CH_4 emissions increased significantly under NT in a wet, temperate climate [22]. Therefore, in addition to soil physical properties, chemical properties and microbial activities, greenhouse gas emissions or absorption under NT management are also affected by climate, temperature, and other external factors. Most of the current studies have studied the impact of NT on greenhouse gases by measuring several related indicators, which is too one-sided and subjective, resulting in great differences in research conclusions. How to more comprehensively explore the impact of tillage methods on greenhouse gas emissions is worth pondering. Such as through the actual field combined with computer model comprehensive study of NT is how to affect greenhouse gas production [34].

5. Conclusions

In a fixed experiment in a moist subtropical monsoon climate, we found that soil organic matter (SOM) was 21.0 g kg^{-1} under no-tillage (NT), with a significant difference observed between 2004 and 2016. Under NT and conventional tillage (CT), SOM increased by averages of 0.60 and 0.32 g kg^{-1} $year^{-1}$, respectively. Soil microbial carbon (C) and nitrogen (N) under CT were higher than those under NT; however, the soil C:N ratio was 17.4 under NT, 9.7% higher than that under CT, and the soil microbial C:N ratio under NT was an average of 9.47, 9.7% higher than under CT on average. Using an ultraportable greenhouse gas analyzer and static transparent chamber method of circulation gas recovery, we found that about 70% of daytime CO_2 net uptake and over 99% of CH_4 net emission occurred during the rice season in a rice–oilseed rape cropping system, and annual CH_4 and daytime CO_2 net ecosystem exchange under NT was 1813.9 g CO_2 equiv. m^{-2}, 10.8% higher than that under CT. Consequently, long-term no-tillage could increase soil organic matter/carbon in the surface soil, and increase annual net ecosystem exchange in a rice–oilseed rape cropping system, especially CH_4 emission; more attention is needed on to how to reduce CH_4 emission under the background of climate change.

Author Contributions: Conceptualization, H.Z., H.X. and Y.Z.; methodology, H.Z.; validation, H.Z. and X.T.; investigation, H.Z. and J.W.; writing—original draft preparation, H.Z.; writing—review and editing, H.Z. and H.X.; supervision, Y.Z.; project administration, H.X. and Q.T.; funding acquisition, Q.T. All authors have read and agreed to the published version of the manuscript.

Funding: This research was fund by the Earmarked Fund for China Agriculture Research System (Grand No. CARS-01-27).

Institutional Review Board Statement: Not applicable.

Informed Consent Statement: Not applicable.

Data Availability Statement: The datasets generated during and/or analyzed during the current study are available from the corresponding author on reasonable request.

Acknowledgments: We are thankful to anonymous reviewers and editors for their helpful comments and suggestions.

Conflicts of Interest: The authors declare that they have no competing interests.

References

1. Kassam, A.; Friedrich, T.; Derpsch, R.; Kienzle, J. Overview of the worldwide spread of conservation agriculture. *Field Actions Sci. Rep.* **2015**, *8*, 1–11.
2. Palm, C.; Blanco-Canqui, H.; DeClerck, F.; Gatere, L.; Grace, P. Conservation agriculture and ecosystem services: An overview. *Agric. Ecosyst. Environ.* **2014**, *187*, 87–105. [CrossRef]
3. Mitchell, J.; Harben, R.; Sposito, G.; Shrestha, A.; Munk, D.; Miyao, G.; Sothard, R.; Ferris, H.; Horwath, W.R.; Kueneman, E.; et al. Conservation agriculture: Systems thinking for sustainable farming. *Calif. Agric.* **2016**, *70*, 53–56. [CrossRef]
4. Yin, M.H.; Li, Y.N.; Chen, P.P.; Xu, L.Q.; Shen, S.L.; Wang, X.Y. Effect of no-tillage on maize yield in northern region of China—A meta-analysis. *Sci. Agric. Sin.* **2018**, *51*, 843–854. (In Chinese)
5. Canisares, L.P.; Grove, J.; Miguez, F.; Poffenbarger, H. Long-term no-till increases soil nitrogen mineralization but does not affect optimal corn nitrogen fertilization practices relative to inversion tillage. *Soil Tillage Res.* **2021**, *213*, 105080. [CrossRef]
6. Sanz-Cobena, A.; Lassaletta, L.; Aguilera, E.; del Prado, A.; Garnier, J.; Billen, G.; Iglesias, A.; Sanchez, B.; Guardia, G.; Abalos, D.; et al. Strategies for greenhouse gas emissions mitigation in Mediterranean agriculture: A review. *Agric. Ecosyst. Environ.* **2017**, *238*, 5–24. [CrossRef]
7. Blanco-Canqui, H.; Ruis, S.J. No-tillage and soil physical environment. *Geoderma* **2018**, *326*, 164–200. [CrossRef]
8. Cook, R.L.; Trlica, A. Tillage and fertilizer effects on crop yield and soil properties over 45 years in southern Illinois. *Agron. J.* **2016**, *108*, 415–426. [CrossRef]
9. Yuan, J.; Sadiq, M.; Rahim, N.; Li, G.; Yan, L.; Wu, J.; Xu, G. Tillage Strategy and Nitrogen Fertilization Methods Influences on Selected Soil Quality Indicators and Spring Wheat Yield under Semi-Arid Environmental Conditions of the Loess Plateau, China. *Appl. Sci.* **2022**, *12*, 1101. [CrossRef]
10. Amin, M.M.; Al Minhaj, A.; Islam, D.; Bhowmik, B.; Hasan, M.M.; Islam, M.N. Mulch and no-till impacts on nitrogen and phosphorus leaching in a maize field under sub-tropic monsoon climate. *Environ. Chall.* **2021**, *5*, 100346. [CrossRef]

11. Pandey, D.; Agrawal, M.; Bohra, J.S. Effects of conventional tillage and no tillage permutations on extracellular soil enzyme activities and microbial biomass under rice cultivation. *Soil Tillage Res.* **2014**, *136*, 51–60. [CrossRef]
12. Briedis, C.; de Moraes Sa, J.C.; Lal, R.; Tivet, F.; de Oliveira Ferreira, A.; Franchini, J.C.; Schimiguel, R.; da Cruz Hartman, D.; dos Santos, J.Z. Can highly weathered soils under conservation agriculture be C saturated? *Catena* **2016**, *147*, 638–649. [CrossRef]
13. Rodrigues, L.A.T.; Dieckow, J.; Giacomini, S.; Ottonelli, A.S.; Zorzo, G.P.P.; Bayer, C. Carbon sequestration capacity in no-till soil decreases in the long-term due to saturation of fine silt plus clay-size fraction. *Geoderma* **2022**, *412*, 115711. [CrossRef]
14. Tang, H.; Li, C.; Shi, L.; Cheng, K.; Wen, L.; Li, W.; Xiao, X. Effects of Short-Term Tillage Managements on CH_4 and N_2O Emissions from a Double-Cropping Rice Field in Southern of China. *Agronomy* **2022**, *12*, 517. [CrossRef]
15. Vilakazi, B.S.; Zengeni, R.; Mafongoya, P.; Ntsasa, N.; Tshilongo, J. Seasonal Effluxes of Greenhouse Gases Under Different Tillage and N Fertilizer Management in a Dryland Maize Mono-crop. *J. Soil Sci. Plant Nutr.* **2021**, *21*, 2873–2883. [CrossRef]
16. Lenka, S.; Lenka, N.K.; Rao, A.S.; Raghuwanshi, J.; Singh, B.; Saha, J.K.; Patra, A.K. Tillage and nutrient management influence net global warming potential and greenhouse gas intensity in soybean-wheat cropping system. *Indian J. Exp. Biol. IJEB* **2022**, *60*, 207–214.
17. Cheng, C.; Zeng, Y.J.; Yang, X.X.; Huang, S.; Luo, K.; Shi, Q.H.; Pan, X.H.; Shang, Q.Y. Effects of different farming methods on net warming potential and greenhouse gas intensity of rice fields. *J. Environ. Sci.* **2015**, *35*, 1887–1895. (In Chinese)
18. Guo, L.; Zhang, L.; Liu, L.; Sheng, F.; Cao, C.; Li, C. Effects of long-term no tillage and straw return on greenhouse gas emissions and crop yields from a rice-wheat system in central China. *Agric. Ecosyst. Environ.* **2021**, *322*, 107650. [CrossRef]
19. González-Sanchez, E.J.; Kassam, A.; Basch, G.; Streit, B.; Holgado-Cabrera, A.; Trivino-Tarradas, P. Conservation agriculture and its contribution to the achievement of agri-environmental and economic challenges in Europe. *AIMS Agric. Food* **2016**, *1*, 387–408. [CrossRef]
20. Huang, M.; Zou, Y.B.; Feng, Y.H.; Cheng, Z.W.; Mo, Y.L.; Ibrahim, M.D.; Xia, B.; Jiang, P. No-tillage and direct seeding for super hybrid rice production in rice–oilseed rape cropping system. *Eur. J. Agron.* **2011**, *34*, 278–286. [CrossRef]
21. Liang, S.M.; Xie, R.Z.; Li, C.S.; Yang, J.Z.; Tang, Y.L.; Wu, C.; Wang, L.B.; Li, S.K. Effects of Tillage Systems on Fields in Chengdu Plain II: The Evaluation of Soil Quality. *Sci. Agric. Sin.* **2011**, *44*, 738–744. (In Chinese)
22. Yonemura, S.; Nouchi, I.; Nishimura, S.; Sakurai, G.; Togami, K.; Yagi, K. Soil respiration, N_2O, and CH_4 emissions from an andisol under conventional-tillage and no-tillage cultivation for 4 years. *Biol. Fertil. Soils* **2014**, *50*, 63–74. [CrossRef]
23. Li, Y.C.; Hou, C.C.; Li, Y.; Guo, Z.J. Effects of no-tillage and straw mulching on greenhouse gas emissions from farmland soil. *J. Ecol. Environ.* **2014**, *23*, 1076–1083. (In Chinese)
24. Lou, Y.L.; Xu, M.G.; Chen, X.N.; He, X.H.; Zhao, K. Stratification of soil organic C, N and C:N ratio as affected by conservation tillage in two maize fields of China. *Catena* **2012**, *95*, 124–130. [CrossRef]
25. Wan, X.H.; Huang, Z.Q.; He, Z.M.; Yu, Z.P.; Wang, M.H.; Davis, M.R.; Yang, Y.S. Soil C:N ratio is the major determinant of soil microbial community structure in subtropical coniferous and broadleaf forest plantations. *Plant Soil* **2015**, *387*, 103–116. [CrossRef]
26. Devine, S.; Markewitz, D.; Hendrix, P.; Coleman, D. Soil aggregates and associated organic matter under conventional tillage, no-tillage, and forest succession after three decades. *PLoS ONE* **2014**, *9*, e84988. [CrossRef]
27. Congreves, K.A.; Hooker, D.C.; Hayes, A.; Verhallen, E.A.; van Eerd, L.L. Interaction of long-term nitrogen fertilizer application, crop rotation, and tillage system on soil carbon and nitrogen dynamics. *Plant Soil* **2017**, *410*, 113–127. [CrossRef]
28. Halvorson, J.J.; Liebig, M.A.; Archer, D.W.; West, M.S.; Tanaka, D.L. Impacts of crop sequence and tillage management on soil carbon stocks in south-central North Dakota. *Soil Sci. Soc. Am. J.* **2016**, *80*, 1003–1010. [CrossRef]
29. Triberti, L.; Nastri, A.; Baldoni, G. Long-term effects of crop rotation, manure and mineral fertilisation on carbon sequestration and soil fertility. *Eur. J. Agron.* **2016**, *74*, 47–55. [CrossRef]
30. Xiao, L.; Zhou, S.; Zhao, R.; Greenwood, P.; Kuhn, N.J. Evaluating soil organic carbon stock changes induced by no-tillage based on fixed depth and equivalent soil mass approaches. *Agric. Ecosyst. Environ.* **2020**, *300*, 106982. [CrossRef]
31. Kibet, L.C.; Blanco-Canqui, H.; Jasa, P. Long-term impacts on soil organic matter components and related properties on a Typic Argiudoll. *Soil Tillage Res.* **2016**, *155*, 78–84. [CrossRef]
32. Singh, K.; Mishra, A.K.; Singh, B.; Singh, R.P.; Dhar, P.D. Tillage effects on crop yield and physicochemical properties of sodic soils. *Soil Use Manag.* **2014**, *30*, 414–422. [CrossRef]
33. Kulmány, I.M.; Giczi, Z.; Beslin, A.; Bede, L.; Kalocsai, R.; Vona, V. Impact of environmental and soil factors in the prediction of soil carbon dioxide emissions under different tillage systems. *Ecocycles* **2022**, *8*, 27–39. [CrossRef]
34. Zhang, X.; Sun, Z.; Liu, J.; Ouyang, Z.; Wu, L. Simulating greenhouse gas emissions and stocks of carbon and nitrogen in soil from a long-term no-till system in the North China Plain. *Soil Tillage Res.* **2018**, *178*, 32–40. [CrossRef]

agriculture

Article

Integrated Organic-Inorganic Nitrogen Fertilization Mitigates Nitrous Oxide Emissions by Regulating Ammonia-Oxidizing Bacteria in Purple Caitai Fields

Daijia Fan [1,2], Cougui Cao [1] and Chengfang Li [1,*]

[1] MOA Key Laboratory of Crop Ecophysiology and Farming System in the Middle Reaches of the Yangtze River, College of Plant Science & Technology, Huazhong Agricultural University, Wuhan 430070, China; daijia.fan@outlook.com (D.F.); ccg@mail.hzau.edu.cn (C.C.)

[2] Institute of Plant Nutrition, Resources and Environment, Beijing Academy of Agriculture and Forestry Sciences, Beijing 100097, China

* Correspondence: lichengfang@mail.hzau.edu.cn; Fax: +86-027-8728-2131

Abstract: Purpose Nitrogen (N) fertilizer application in agricultural soil is a primary anthropogenic nitrous oxide (N_2O) source. Currently, the effect of the N fertilizer type on N_2O emissions from upland soil has been rarely reported. To this end, impacts of various types of N fertilizer on N_2O emissions in purple caitai (*Brassica campestris* L. ssp. *chinensis* var. *purpurea*) fields are investigated in this work. The field experiment was carried out with four treatments, including inorganic N fertilization (I), organic N fertilization (O), integrated organic-inorganic N fertilization (I+O) and no fertilization (CK). The nitrifier/denitrifier abundance was determined using absolute real-time quantitative PCR. Compared with I and O, I+O significantly increased dissolved organic C content, microbial biomass C and microbial biomass N by 24–63%, 12–38% and 13–36% on average, respectively. Moreover, the seasonal cumulative N_2O-N emissions and fertilizer-induced N_2O emission factor under I+O were significantly lower than those under I and O by 17–29% and 23–39%, respectively. The results indicate that N fertilizer type significantly affects the N_2O emissions, and the integrated organic-inorganic N fertilization can mitigate the N_2O emissions primarily by inhibiting the nitrification mediated by ammonia-oxidizing bacteria in purple caitai fields. Integrated organic-inorganic N fertilization is an ideal N fertilization regime to enhance soil fertility and yield and reduce N_2O emissions in the upland fields.

Keywords: ammonia-oxidizing bacteria; integrated fertilization regime; N_2O emission factor; N_2O flux; purple caitai fields

Citation: Fan, D.; Cao, C.; Li, C. Integrated Organic-Inorganic Nitrogen Fertilization Mitigates Nitrous Oxide Emissions by Regulating Ammonia-Oxidizing Bacteria in Purple Caitai Fields. *Agriculture* **2022**, *12*, 723. https://doi.org/10.3390/agriculture12050723

Academic Editor: Manuel Ângelo Rosa Rodrigues

Received: 20 April 2022
Accepted: 18 May 2022
Published: 20 May 2022

1. Introduction

Nitrous oxide (N_2O) is a bigger contributor to global warming compared with CO_2 [1]. The N_2O concentration in the atmosphere has risen by 20% since 1860 [1]. In addition, N_2O is a dominant ozone-depleting substance and could remain the most threatening throughout the twenty-first century if its emissions are not controlled [2]. Considering that about 50–60% of N_2O emissions are derived from agricultural soils [3], it is imperative to adopt proper agricultural practices to reduce N_2O emissions.

Chemical nitrogen (N) fertilizers have been excessively applied worldwide to improve soil fertility and crop yields, consequently causing environmental losses such as soil degradation, water eutrophication and GHG emissions [4]. In recent years, it has been established that integrated inorganic-organic N fertilizer application could improve soil aggregation, soil structure and carbon (C) sequestration [5]. It has been found that N_2O emissions is affected by the type and composition of the fertilizer [6]. Specifically, numerous studies have revealed that organic N fertilizers can lead to less N_2O emissions than inorganic N fertilizers [7]. However, some other researchers observed greater N_2O emissions under organic N fertilization than under mineral N fertilization [8]. In addition, Ding et al. [9] proposed use of compost and urea to reduce N_2O emissions, whereas Agegnehu et al. [10] discovered that organic amendments incorporated with conventional N fertilizer could significantly increase N_2O emissions. These diverse and inconsistent conclusions indicate that the dynamic responses of N_2O emissions to organic and inorganic N fertilization are still elusive and require further research.

The N_2O in the soil is mainly biologically produced by nitrification and denitrification [11]. Nitrification is the oxidation of ammonia first to nitrite and then to nitrate, which is predominantly performed by aerobic nitrifying microorganisms including ammonia-oxidizing bacteria (AOB) and archaea (AOA) [12]. The initial step of nitrification is catalyzed by ammonia monooxygenase (AMO) [13]. Owing to its importance in the energy-generating metabolism, *amoA* is primarily applied as a marker gene in nitrification studies [14]. Reversely, denitrification is a facultative anaerobic pathway during which nitrate is reduced to nitrite and free nitrogen [15]. Nitrite reductase (NIR) is an essential enzyme that converts nitrite to N_2O, and the most widely used marker genes for NIR are *nirK* and *nirS* [16]. All these microbial processes are highly susceptible to environmental parameters and agricultural practices including fertilization. Therefore, quantifying the microbial functional genes in the process of nitrification and denitrification can provide important information for the mitigation of N_2O emissions [17].

Purple caitai (*Brassica campestris* L. ssp. *chinensis* var. *purpurea*), also known as zicaitai, is a traditional vegetable crop widely planted in the south of China and has become increasingly popular due to its high nutrient content [18]. Usually, high rates of N fertilizer are applied to promote purple caitai growth and development, inevitably leading to a considerable amount of N_2O emissions from the soil. This environmental impact varies with different types of N fertilizers as mentioned above, which, however, has been scarcely studied in purple caitai fields. Hence, this paper studied the impact of various types of N fertilizers on N_2O emissions in purple caitai fields. We hypothesized that the combined application of inorganic/organic N fertilizers could reduce N_2O emissions through decreasing nitrification and denitrification.

2. Material and Methods

2.1. Experimental Site

The experimental site is located in Huazhong Agricultural University, Wuhan City, Hubei Province, China (30°28′21″ N latitude, 114°20′48″ E longitude), with an average annual temperature of 16.3 °C and an average annual precipitation of 1163 mm (rainfall mostly occurs between May and August) from 1961 to 2010 (Figure 1). The mean daily temperature and precipitation during the purple caitai growing seasons in 2016 and 2017 are shown in Figure 1. The total precipitation during the experimental period was 287.9 mm in 2016 and 66.1 mm in 2017, respectively. The soil is classified as Alisols with a clay loam texture (FAO soil clarification). The main soil properties before the experiment (measured in September 2016) were pH 7.03, organic C 9.13 g kg^{-1}, total N 1.15 g kg^{-1}, total phosphorus (P) 0.39 g kg^{-1}, and total potassium (K) 8.67 g kg^{-1}.

2.2. Experimental Design

An upland field experiment on purple caitai was carried out during the 2016 and 2017 growing seasons. The seedlings of purple caitai (HSCT, *Brassica campestris* L. ssp. *chinensis* var. *purpurea*) were transplanted in October and harvested in March of the next year, followed by a fallow season. Four treatments, including no fertilization (CK), inorganic N fertilization (I), organic N fertilization (O) and integrated organic-inorganic N fertilization (I+O) were implemented, and each treatment had three replications. Each plot was 12 m^2 in area. Each treatment of the experimental field has been planted with purple caitai under the same N fertilization since 2011.

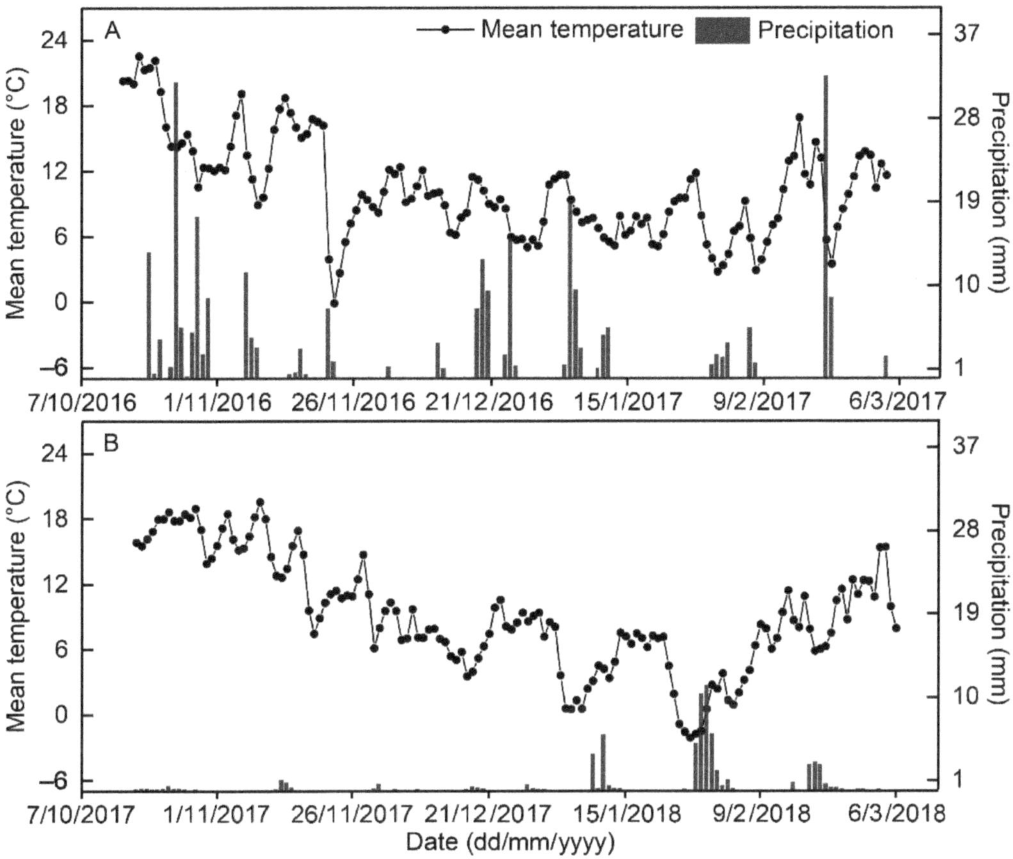

Figure 1. Changes in average daily temperature and rainfall during purple caitai growing seasons in 2016 (**A**) and 2017 (**B**).

The soil was moldboard plowed to a depth of 15 cm beforehand. Compound fertilizer ($N:P_2O_5:K_2O$ = 15%:15%:15%), urea (46% N), calcium superphosphate (17% P_2O_5), potassium chloride (60% K_2O) and pelleted organic fertilizer ($N:P_2O_5:K_2O$ = 10%:3%:2%, living bacteria count $\geq 2 \times 10^7$ CFUs g^{-1}; Compound Bio-NH_4^+-fertilizer, Wuhan Heyuan Green Organism Co., Ltd., Wuhan, China) were selectively incorporated into the topsoil (0–20 cm depth) to provide 225 kg N ha^{-1}, 112.5 kg P_2O_5 ha^{-1} and 112.5 kg K_2O ha^{-1} for the plots under fertilization treatments throughout the growing seasons. The P and K fertilizers were used as basal fertilizers. As for the N fertilizer, 50%, 100% and 75%, respectively, were applied to the plots under I, O and I+O treatments as basal fertilizer, while the rest was applied at the bolting stage as topdressing fertilizer. No fertilizer was applied to the CK plots. The specific fertilization regimes are shown in Table 1. After basal fertilization, 20-day-old seedlings of purple caitai were manually transplanted on the same day. The field was only irrigated once to 3 mm immediately after the transplanting of seedlings. No irrigation was conducted after this time.

Table 1. Fertilization regimes of different treatments.

Treatment	Basal Fertilizer (15 October 2016 and 17 October 2017)	Topdressing Fertilizer (14 December 2016 and 16 December 2017)
I	750 kg compound fertilizer ha^{-1}	245 kg urea ha^{-1}
O	2250 kg organic fertilizer ha^{-1} + 112.5 kg KCl ha^{-1} + 264.5 kg calcium superphosphate ha^{-1}	–
I+O	1125 kg organic fertilizer ha^{-1} + 375 kg compound fertilizer ha^{-1} + 56 kg KCl ha^{-1} + 133 kg calcium superphosphate ha^{-1}	122.3 kg urea ha^{-1}
CK	–	–

I, inorganic N fertilization; O, organic N fertilization; I+O, integrated organic-inorganic N fertilization; CK, no fertilization.

2.3. Gas Sampling

Soil N_2O fluxes throughout the growing seasons of purple caitai were measured by the static chamber-gas chromatography method [19] (Li et al., 2013). In brief, the sampling chamber was a 1.1 m high steel barrel with a diameter of 0.3 m. Chamber bases with a groove of the ring were installed in each plot. The air-tightness of the chambers during the gas sampling was ensured by filling water in the groove of the ring. After shutting down the chamber, the air was mixed using four fans on the top of the chamber. A 30 mL syringe was used to extract the gas from the barrel through the three-way valve and inject it into a 30 mL glass vial that had been vacuumed beforehand. The sampling interval was 10 min, and the sampling time was 0 min, 10 min, 20 min and 30 min. At the same time, the barrel height and air temperature inside the chamber were recorded. The sampling was carried out after every N fertilization event or every ten days otherwise (from 15 October 2016 to 4 March 2017 and from 17 October 2017 to 6 March 2018) between 8:30 am and 11:00 am, with four samples being successively collected per plot at the interval of 10 min (at 0 min, 10 min, 20 min and 30 min).

The N_2O concentrations were measured by gas chromatography (Shimadzu, GC-14B, Tokyo, Japan) as described by Li et al. [19]. The N_2O flux was calculated according to the method of Li et al. [19], and seasonal cumulative N_2O emissions were measured according to the following equations:

$$CE = \sum_i^n ((F_i + F_{i+1})/2 \times 24 \times D_i)/10^5 \times 2 \times \frac{14}{44} \quad (1)$$

where CE is the cumulative N_2O emissions over the whole growing season of purple caitai (kg ha^{-1}), F_i and F_{i+1} represent the N_2O fluxes measured on two adjacent sampling dates (μg m^{-2} h^{-1}), D_i represents the length of the ith sampling interval (d), 14 represents the relative atomic mass of N, 44 is the relative molecular mass of N_2O, and n represents the total number of sampling intervals.

The fertilizer-induced N_2O emission factor (EF_{N2O}) was calculated as the difference in seasonal cumulative N_2O-N emissions between N fertilizer treatments and CK divided by the total amount of fertilized N [1].

2.4. Soil Sampling and Measurement

Five soil cores were collected from each plot on the same date as the gas sampling. After the removal of plant debris and stones, the soil cores from the same plot were mixed and homogenized into a composite sample. These composite samples were then divided into subsamples for chemical and biological analysis.

Chemical analysis was carried out for the soil samples obtained throughout the growing seasons of purple caitai (at an interval of one month). The contents of total C (TC) and N (TN) were measured using a FlashEA 1112 element analyzer (Thermo Fisher Scientific, Waltham, MA USA). The dissolved organic C (DOC) was extracted from a soil–water solution using the suction filtration method [20], and the microbial biomass C (MBC) and microbial biomass N (MBN) were extracted using the chloroform fumigation-extraction method [21]. Soil DOC, MBC and MBN contents were measured using the Walkley–Black method [22]. The soil NH_4^+–N and NO_3^-–N were extracted by dissolving 20 g of fresh soil with 100 mL of 1 mol L-1 KCl solution and filtered through Whatman #1 filter paper after being shaken for one hour [23] (Zaman et al., 2009). The NH_4^+–N and NO_3^-–N contents in the soil extracts were then analyzed using a flow injection analyzer. The gross nitrification and denitrification rates were measured using the method proposed by Yao et al. [20]. In brief, fresh soils were amended with ammonium sulfate or KNO_3. The treated soil was thoroughly homogenized in the bottle and added deionized water. The bottles were covered by a polyethylene film with tiny holes and incubated at 30 °C. After 15 days of incubation, the treated soil was extracted with 2 mol L^{-1} KCl, and mineral N (NH_4^+ and NO_3^-) was determined. The treated soils from the other bottles were incubated under anaerobic conditions at 30 °C for 5 days with NO_3^-, extracted with KCl and determined using the continuous-flowing analyzer. Soils without ammonium sulfate or KNO_3 were taken as the controls. The rates of nitrification and denitrification were calculated as the differences in mineral N concentrations or NO_3^- contents between the 0 and 7 or 5 day samples divided by the amounts of added ammonium sulfate or KNO_3.

2.5. Measurement of Yield and Calculation of Yield-Scaled N_2O Emissions

When the length of the red tender stems of the vegetable was more than 40 cm, the red tender stems from the plots were harvested and weighed.

The yield-scaled N_2O emissions (t CO_2-eq. t^{-1} yield) were calculated as the ratio of cumulative N_2O emissions (converted into CO_2 equivalents) to purple caitai yields.

2.6. Absolute Real-Time Quantitative Polymerase Chain Reaction (PCR) Analysis

Genomic DNA was extracted from the composite soil samples and then analyzed using a NanoDrop spectrophotometer (NanoDrop Technologies Inc., Wilmington, DE, USA). Absolute real-time quantitative PCR was carried out in 96-well PCR optical plates in triplicate per sample. The PCR protocol is shown in Table S1.

The proper dilution factor of the template DNA was determined by running quantitative PCR with different dilutions of template DNA in order to avoid PCR inhibition. Using a threshold cycle of 31 as the detection limit, a melting curve analysis was carried out to examine the specificity of the amplified products. Standards for all assays were prepared and then serially diluted for the construction of standard curves.

2.7. Statistical Analysis

One-way fixed effects ANOVA or two-way repeated measures ANOVA using the general linear model of SPSS software was conducted. Before ANOVA, the normality and homoscedasticity of data were tested, and the datasets that did not pass the tests were subjected to log transformation. One-way or two-way repeated measures ANOVA was conducted. If the ANOVA result was significant, the least significant difference test or Tukey's HSD test was carried out. The significance of the difference was defined as $p \leq 0.05$. A similarity percentage (SIMPER) analysis was performed to analyze the relative contribution of each functional gene to the variations in nitrifier/denitrifier abundance among different treatments with Past (Øyvind Hammer, University of Oslo, Oslo, Norway; Version 3.26). The partial correlation coefficient (r) was calculated using SPSS [24].

3. Results

3.1. Soil Chemical Properties

Compared with CK, N fertilization treatments significantly enhanced the TC and TN contents at the harvest stages by 23–45% and 31–57%, respectively (Table S2). Furthermore, the N fertilizer type imposed a significant impact on TC, with significantly higher TC under I+O than under I and O by 7–18%. However, I+O only significantly increased TN in the

2017 growing season by 16–17% compared with I and O, while no significant differences were found between I, O and I+O in the 2016 growing season (Table S2).

The NH_4^+–N, NO_3^-–N, DOC, MBC and MBN contents under N fertilization treatments varied similarly in the 2016 and 2017 growing seasons of purple caitai, and all peaked immediately after the application of basal and topdressing N fertilizers (15 October and 14 December 2016; 17 October and 16 December 2017) (Figure 2).

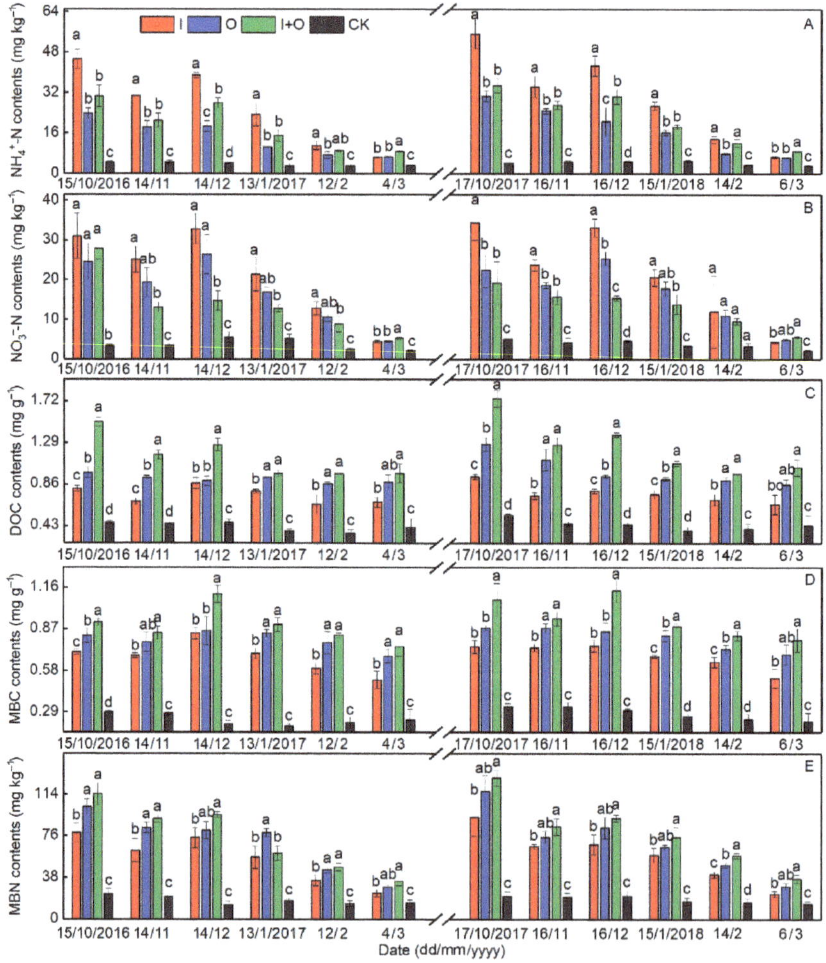

Figure 2. Soil chemical properties including NH_4^+–N (**A**), NO_3^-–N (**B**), DOC (**C**), MBC (**D**) and MBN (**E**) contents under different treatments throughout two growing seasons of purple caitai. Different letters indicate significant differences at the level of 0.05. DOC, dissolved organic C; MBC, microbial biomass C; MBN, microbial biomass N; I, inorganic N fertilization; O, organic N fertilization; I+O, integrated organic/inorganic N fertilization; CK, no fertilization.

According to two-way repeated measures ANOVA, N fertilization treatments increased the contents of NH_4^+–N and NO_3^-–N as compared with CK, (Figure 2A,B). Moreover, there were significant differences in the contents of soil NH_4^+–N and NO_3^-–N among N fertilization treatments (Figure 2A,B). The NH_4^+–N contents from four sampling stages (15 October 2016 to 13 January 2017 and 17 October 2017 to 15 January 2018) under I+O were significantly lower than those under I by 21–37%. The NH_4^+–N contents on 14 December 2016, 4 March and 16 December 2017, and 14 February and 6 March 2018 under I+O 35–54% ($p < 0.05$) higher than under O. Meanwhile, the NO_3^-–N contents from four sampling stages (15 October 2016 to 13 January 2017 and 17 October 2017 to 15 January 2018) under I+O were also significantly lower than those under I by 31–55%. The NO_3^-–N contents on 12 December 2016 and 16 December 2017 under I+O were significantly lower than those under O by 40–44%.

As for the DOC, MBC and MBN contents, they were all significantly higher under N fertilization treatments than under CK (Figure 2C–E). Specifically, among N fertilization treatments, I+O caused higher DOC and MBC contents than I and O. In addition, in the 2016 growing season, the MBN contents were significantly higher under I+O than under I by 34% on average, while no significant differences were found between I+O and O. However, in the 2017 growing season, the MBN contents under I+O were significantly higher than under I and O by 13–36% on average.

3.2. Soil Nitrification and Denitrification Rates, and Abundances of Nitrifier and Denitrifier Genes

The nitrification and denitrification rates throughout two growing seasons of purple caitai showed similar variations to the N_2O fluxes (Figure 3).

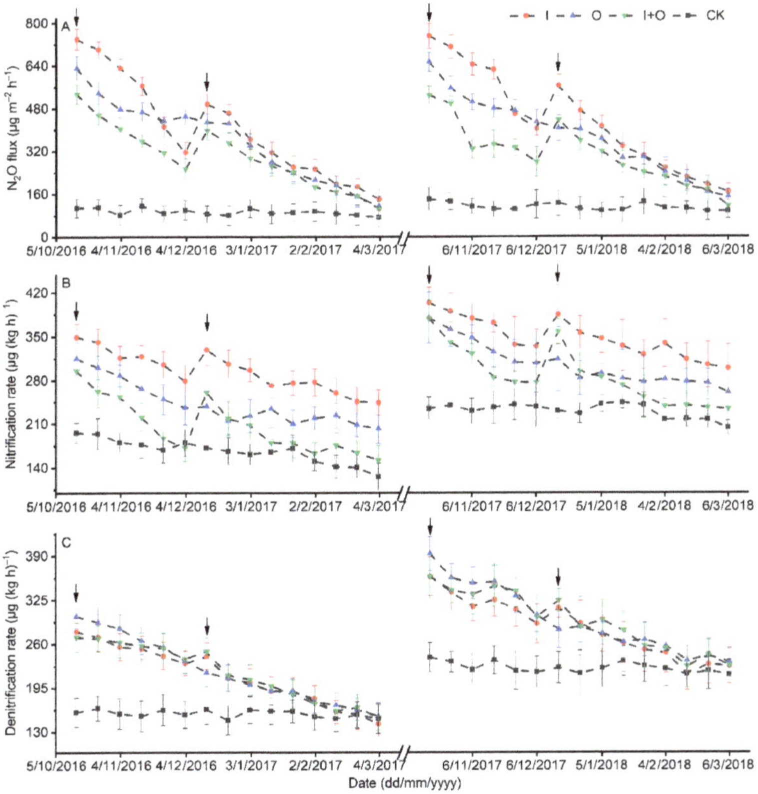

Figure 3. The N_2O fluxes (**A**), gross rates of nitrification (**B**) and denitrification (**C**) in soil under different treatments throughout two growing seasons of purple caitai. Arrows indicate N fertilization. I, inorganic N fertilization; O, organic N fertilization; I+O, integrated organic/inorganic N fertilization; CK, no fertilization.

Both the nitrification and denitrification rates under N fertilization treatments peaked immediately after the application of basal and topdressing fertilizers. The nitrification rates were 125.6–403.4 μg kg^{-1} h^{-1}, and the denitrification rates were 141.9–392.7 μg kg^{-1} h^{-1}. N fertilization treatments increased the nitrification and denitrification rates by 24–77% and 26–40% on average compared with CK, respectively. Moreover, N fertilizer types were found to have a significant impact on the nitrification rate. The nitrification rate under I+O was significantly lower than that under I and O by 6–30% on average. However, no significant differences in denitrification rates were found among I, O and I+O.

As illustrated in Figure 4, the abundances of nitrifier and denitrifier genes peaked immediately after the application of basal and topdressing fertilizers. N fertilization treatments increased the abundance of nitrifiers and denitrifiers compared with CK (Figure 4). Besides, the abundances of nitrifiers and denitrifiers were also significantly affected by N fertilizer type. The abundance of the *AOB-amoA* gene under I+O was lower than I and O, while there was no significant difference in *AOA-amoA* gene abundance among I, O and I+O. In addition, the abundances of both *nirK* and *nirS* genes under I+O were significantly lower than those under I and O by 16–35% and 24–44%, respectively.

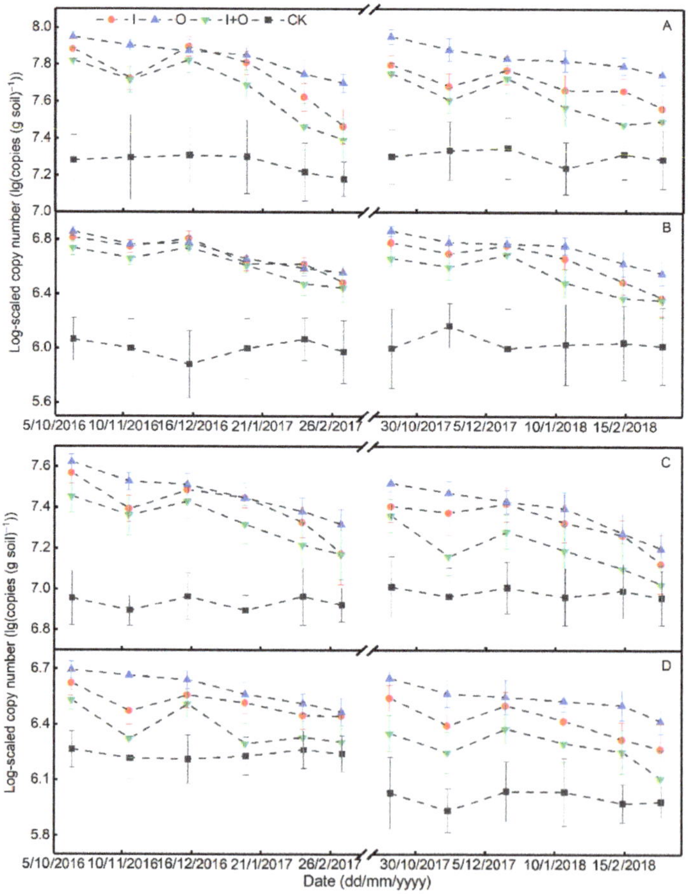

Figure 4. The abundance of nitrifier genes, including *AOB-amoA* (**A**) and *AOA-amoA* (**B**), and denitrifier genes including *nirK* (**C**) and *nirS* (**D**) at log-scale in soil under different treatments throughout two growing seasons of purple caitai. AOB, ammonia-oxidizing bacteria; AOA ammonia-oxidizing archaea; I, inorganic N fertilization; O, organic N fertilization; I+O, integrated organic/inorganic N fertilization; CK, no fertilization.

3.3. N_2O Emissions

The N_2O fluxes under N fertilization treatments showed an identical trend in the 2016 and 2017 growing seasons of purple caitai and peaked immediately after the application of basal and topdressing N fertilizers (Figure 3A). The N_2O fluxes were 139.2–750.0 μg m^{-2} h^{-1} under I, 113.2–652.8 μg m^{-2} h^{-1} under O, 103.7–533.7 μg m^{-2} h^{-1} under I+O, and 72.9–139.9 μg m^{-2} h^{-1} under CK.

The N fertilizer type showed a significant impact on the seasonal cumulative N_2O emissions in purple caitai fields (Table 2). N fertilization treatments increased the cumulative N_2O-N emissions by 176–332% compared with CK. Among N fertilization treatments, I+O caused the lowest seasonal cumulative N_2O-N emissions, which were significantly lower than those under I and O. In addition, the type of N fertilizers showed significant effects on the EF_{N2O} in purple caitai fields as well (Table 2). The EF_{N2O} under I+O was significantly lower than that under I and O in two growing seasons.

Table 2. Seasonal cumulative N_2O–N emissions, EF_{N2O}, yield and yield-scaled N_2O emission under different treatments in two growing seasons of purple caitai.

Year	Treatment	CE (kg ha^{-1})	EF_{N2O} (%)	Yield (t ha^{-1})	Yield-Scaled N_2O Emission (t CO_2-eq. t^{-1} Yield)
2016	I	8.54 ± 0.05 a	2.92 ± 0.07 a	17.09 ± 0.65 b	0.13 ± 0.00 b
	O	7.63 ± 0.09 b	2.51 ± 0.10 b	13.08 ± 0.84 c	0.16 ± 0.01 a
	I+O	6.33 ± 0.13 c	1.94 ± 0.12 c	18.65 ± 1.05 a	0.09 ± 0.00 c
	CK	1.98 ± 0.13 d	–	6.76 ± 0.59 d	0.08 ± 0.00 d
2017	I	9.24 ± 0.16 a	3.05 ± 0.08 a	15.76 ± 1.29 b	0.16 ± 0.01 b
	O	7.96 ± 0.16 b	2.48 ± 0.08 b	11.69 ± 0.73 c	0.18 ± 0.01 a
	I+O	6.58 ± 0.21 c	1.86 ± 0.10 c	17.74 ± 0.76 a	0.10 ± 0.00 c
	CK	2.38 ± 0.02 d	–	6.42 ± 0.14 d	0.10 ± 0.00 c

Different small letters between treatments in a line mean significant differences at $p < 0.05$. CE stands for seasonal cumulative N_2O–N emissions. EF_{N2O} stands for fertilizer-induced N_2O emission factor. Different letters in the same column indicate significant differences at the level of 0.05. I, inorganic N fertilization; O, organic N fertilization; I+O, integrated organic-inorganic N fertilization; CK, no fertilization.

3.4. Correlation among Soil Chemical Properties, Nitrifier/Denitrifier Abundance, Nitrification/Denitrification Rate and N_2O Flux

The results showed that the *AOB-amoA* gene was the major contributor to the variations in nitrifier abundance among fertilization treatments, while *nirK* and *nirS* genes had roughly the same contribution to the variations in denitrifier abundance among fertilization treatments (Table 3). N_2O flux was positively related to nitrification rate, while it showed no significant correlation with denitrification rate (Table 4). A SIMPER analysis was carried out to estimate the respective contributions of each functional gene to the variations in nitrifier/denitrifier abundance among different fertilization treatments.

Table 3. Respective contribution of each nitrifier (*AOB-amoA* and *AOA-amoA*) and denitrifier (*nirK* and *nirS*) gene to community abundance variations (%) among fertilization treatments during two growing seasons of purple caitai.

Growing Season	Treatments	*AOB-amoA*	*AOA-amoA*	*nirK*	*nirS*
2016	I vs. O	38.9	14.0	25.8	21.3
	I vs. I+O	39.9	12.9	24.0	23.2
	O vs. I+O	38.2	10.1	23.3	28.4
2017	I vs. O	36.7	15.3	23.5	24.5
	I vs. I+O	37.8	10.6	26.6	25.0
	O vs. I+O	35.0	13.7	23.5	27.8

AOB, ammonia-oxidizing bacteria; AOA ammonia-oxidizing archaea; I, inorganic N fertilization; O, organic N fertilization; I+O, integrated organic-inorganic N fertilization.

Table 4. Partial correlation coefficients among soil chemical properties, nitrifier/denitrifier abundance and nitrification/denitrification rates in two growing seasons of purple caitai.

Growing Season	Variable	*AOB-amoA*	*AOA-amoA*	*nirK*	*nirS*	NR	DNR	N_2O Flux
	NH_4^+–N	ns	ns	ns	ns	0.42 **	ns	0.39 **
	NO_3^-–N	0.29 *	0.47 **	0.30 *	0.32 *	0.25 **	ns	0.47 **
	DOC	ns	−0.21 *	ns	ns	−0.52 **	ns	−0.34 *
	MBC	−0.52 *	ns	ns	ns	−0.46 **	ns	−0.58 **
	MBN	−0.65 *	ns	ns	ns	−0.40 **	ns	−0.45 **
2016	*AOB-amoA*	1	0.61 **	0.77 **	0.60 **	0.72 **	ns	0.46 **
	AOA-amoA	-	1	0.52 **	0.62 **	0.22 *	ns	ns
	nirK	-	-	1	0.62 **	0.33 *	ns	0.40 **
	nirS	-	-	-	1	0.36 **	0.28 *	0.35 **
	NR	-	-	-	-	1	ns	0.56 **
	DNR	-	-	-	-	-	1	ns
	NH_4^+–N	ns	ns	ns	ns	0.46 **	ns	0.34 **
	NO_3^-–N	ns	0.38 **	0.42 **	0.35 **	0.31 *	ns	0.35 **
	DOC	ns	−0.31 *	ns	ns	−0.47 *	ns	−0.39 **
	MBC	−0.48 *	ns	ns	ns	−0.49 *	ns	−0.47 **
	MBN	−0.72 **	ns	ns	ns	−0.54 *	ns	−0.34 *
2017	*AOB-amoA*	1	0.68 **	0.68 **	0.72 **	0.74 **	ns	0.58 *
	AOA-amoA	-	1	0.49 **	0.70 **	ns	ns	ns
	nirK	-	-	1	0.69 **	0.35 *	ns	0.36 **
	nirS	-	-	-	1	0.31 **	ns	0.27 *
	NR	-	-	-	-	1	ns	0.49 **
	DNR	-	-	-	-	-	1	ns

DOC, dissolved organic C; MBC, microbial biomass C; MBN, microbial biomass N; AOB, ammonia-oxidizing bacteria; AOA ammonia-oxidizing archaea; NR, nitrification rate; DNR, denitrification rate. ns, not significant; * $p \leq 0.05$; ** $p \leq 0.01$.

Partial correlation coefficients were used to measure the correlations among soil chemical properties, nitrifier/denitrifier abundance, nitrification/denitrification rate and N_2O flux in the 2016 and 2017 growing seasons of purple caitai (Table 4). There was a weak positive correlation between *AOB-amoA* and NO_3^-–N in the 2016 growing season, while there were negative correlations between *AOB-amoA* and MBC/MBN in both growing seasons. *AOA-amoA* was in positive relation with NO_3^-–N but in negative relation with DOC, though the correlations were not strong. *nirK* and *nirS* were both weakly positively correlated with NO_3^-–N. Moreover, nitrification rate was in a moderate and a weak positive relation with NH_4^+–N and NO_3^-–N, respectively, but in a moderate negative relation with DOC, MBC and MBN. There was a strong positive relation between nitrification rate and *AOB-amoA*, while the correlations between nitrification rate and the other three functional genes were all weakly positive. However, no significant correlations of denitrification rate with other variables, except for the weak positive one with *nirS* in the 2016 growing season, were observed.

3.5. Yield and Yield-Scaled N_2O Emission

The N fertilization significantly increased the yield by 82–172% compared with CK (Table 2). Among N fertilization treatments, I+O caused the highest yield in both 2016 and 2017, followed by I and O. Moreover, N fertilizer types significantly affected the yield-scaled N_2O emissions. Among N fertilization treatments, I+O treatment resulted in the lowest yield-scaled N_2O emission.

4. Discussion

4.1. Effect of N Fertilizer Type on Soil Properties

The N fertilizer type had a great influence on the soil properties in the purple caitai fields. The contents of available N in topsoil (0–20 cm) under integrated organic-inorganic N fertilization were significantly lower than that under inorganic N fertilization (Figure 2A,B), which was directly induced by the higher and faster mineral N input from inorganic fertilizers. Besides, the application of integrated organic and inorganic N fertilizers significantly increased the storage of labile soil organic C (SOC) and N (Figure 2) as compared with the single application of organic or inorganic N fertilizer. Research has been reported on the combined application of organic and inorganic N fertilizers, which could stimulate soil microbial activity compared with the single application of organic or inorganic N fertilizer [25,26], which thereby boosts organic matter mineralization and facilitate SOC turnover and nutrient availability in soil. Moreover, compared with organic N fertilization, integrated organic-inorganic N fertilization could lead to an increase in labile SOC and N pool due to a stronger priming effect of root exudates on soil organic matter owing to the promoted plant growth under faster mineral N input [27]. So, higher DOC, MBC and MBN were observed under I+O than I and O (Figure 1), which also suggests that the application of integrated organic-inorganic fertilizers could help maintain soil fertility by enhancing the labile organic C and N. Similar results were reported by Ali et al. [28], who indicated that the combined application of inorganic-organic N fertilizers could improve soil quality with more labile SOC fractions. Interestingly, we observed higher DOC immediately after the N application. This may be because this study was preceded by continuous cultivation of purple caitai under different N fertilization for 5 years, thus leading to higher initial DOC contents under I+O at the beginning of this study.

4.2. Effect of N Fertilizer Type on the Abundance and Activity of Nitrifiers/Denitrifiers

Higher abundance and greater activity of nitrifiers/denitrifiers were observed immediately after the application of basal and topdressing fertilizers during the two growing seasons as well as under N fertilization treatment compared with CK (Figures 3B,C and 4). Moreover, the activity and abundance of nitrifiers/denitrifiers are positively correlated with the N fertilization intensity [29], and thus N addition significantly increased AOB abundance [30].

The abundances of *AOB-amoA*, *nirK* and *nirS* genes under I+O were all significantly lower than those under I and O, whereas there was no significant difference in abundance of *AOA-amoA* genes among various N fertilization treatments (Figure 4), indicating that AOA are less sensitive to the variations in N fertilizer type than other N-cycling microbial communities, which is consistent with the results of SIMPER analysis (Table 3). Similarly, many studies have shown that N fertilization can induce obvious changes in the AOB community but not in the AOA community [31]. Moreover, different N fertilizer types only altered the activity of nitrifiers but not that of denitrifiers (Figure 3B,C), contradicting the variations in denitrifier abundance (Figure 4). This inconsistency was probably because of the coupled effects of the denitrification-controlling abiotic factors such as oxygen (O_2) content and organic C/N substrates [32].

The MBC and MBN contents under I+O were higher than those under I and O (Figure 2D,E), implying that the integrated N fertilization could enhance the microbial abundance in soil. It has been noted that heterotrophic bacteria usually have higher abundance and activity than nitrifying bacteria [33]. Thus, with more O_2 being consumed by other microorganisms, the growth of AOB was inhibited, resulting in the lower abundance of AOB, which agrees well with the negative correlation between AOB and MBC/MBN (Table 4). Besides, the lower NH_4^+–N availability under I+O compared with that under I may further inhibit the nitrification (Figure 2A). Moreover, the nitrification rate was only strongly correlated with AOB (Table 4), so the decline in AOB abundance was most responsible for the significantly lower nitrification rate under I+O compared with that under I and O. The abundances of *nirK* and *nirS* genes under I+O were both significantly lower than those under I and O (Figure 4C,D) probably owing to the significantly lower NO_3^-–N contents (Figure 2B and Table 4), given that the availability of N oxides is key to denitrification [34]. However, the inhibiting effects of the decreased denitrifier abundance on denitrification rate under I+O might have been offset by the stimulating effects of O_2 depletion under higher microbial activity and labile organic C amendment due to higher DOC contents (Figure 2C–E), considering that most denitrifiers are heterotrophic anaerobes [35].

4.3. Effect of N Fertilizer Type on N_2O Emissions

The N_2O fluxes in the soil under N fertilization treatments peaked immediately following N fertilization in both growing seasons (Figure 3A), which agrees well with previously reported results [36]. This is possibly due to a drastic boost in the nitrification rates (Figure 3B and Table 4). Moreover, the N_2O emissions from CK ranged from 1.98 kg ha^{-1} to 2.38 kg ha^{-1} during the purple caitai growing seasons. The emissions are higher than 0.68 kg ha^{-1}–1.28 kg ha^{-1} of background N_2O emissions from Chinese vegetable soils [37], which could be attributed to the relatively higher air temperature and soil N availability in this study. Moreover, the total seasonal precipitation ranged from 66.1 to 287.9 mm during the experimental periods. Moreover, high rainfall and air temperature and neutral soil pH (7.03) can be beneficial to soil nitrification and denitrification [37,38], thus inducing relatively large N_2O from CK in this study.

The integrated organic-inorganic fertilization significantly decreased the EF_{N2O} (Table 2). The EF_{N2O} in this study was within the range of 1.86–3.05% (Table 2), which is larger than the default IPCC value [1] as well as other estimates in Chinese croplands [39]. Aside from the possible systemic error caused by the choice of a linear regression model for flux calculation as well as the inaccuracies in measurements of sampling time, temperature and chamber volume [19], a high EF_{N2O} value might result from the regional discrepancies such as temperature and soil type [40]. Besides, N_2O emission factors actually increase with N additions [41]. Hence, the relatively high N inputs in our study might have led to higher EF_{N2O}.

In line with previously reported results [42], the combined application of organic and inorganic N fertilizer significantly reduced the N_2O emissions compared with the single application (Figure 3A and Table 2). As soil N_2O emissions from nitrification and denitrification depend on soil available N, N fertilizer application is an important driver of N_2O emissions in the soil [42]. Application of inorganic N fertilizer alone can increase soil mineral-N contents from the applied N fertilizers in excess of crop requirements [42]. Higher contents of NH_4^+–N and NO_3^-–N (Figure 2) under I than under I+O support this view in this study. Therefore, higher N_2O emissions from I than from I+O were found (Table 2). Although O caused lower soil NH_4^+–N contents than I+O (Figure 2), higher nitrification under O than under I+O (Figure 4) due to improved soil porosity under O [26] (Ali et al. 2014) ultimately results in more N_2O emissions under O. Studies proposed that nitrification imposes a greater effect on N_2O emissions [38]. In this study, there was a significant positive relation between N_2O flux and nitrification rate, while no significant correlation existed between N_2O flux and denitrification rate (Table 4). Besides, the nitrification rates significantly declined under I+O, whereas N fertilizer type had no significant impact on the denitrification rates (Figure 3B,C). The above results indicate that the declined nitrification plays a more important role than denitrification in lowering N_2O emissions under I+O and the integrated organic-inorganic N fertilization diminishes N_2O emissions primarily through the regulation of nitrification. In conclusion, integrated organic-inorganic N fertilization can be recommended as the optimum fertilization regime for enhancing soil fertility and mitigating N_2O emissions in purple caitai fields.

4.4. Effect of N Fertilizer Type on Yield and Yield-Scaled N_2O Emission

The effects of combined inorganic/organic N fertilizers on crop yield have been frequently studied [25,26]. Studies reported that combined inorganic/organic N fertilizers could significantly increase crop yield compared with the single N application. Similar results were observed in our study (Table 2). Enhanced yield in the present study may be ascribed to improved N use efficiency through increased soil N supply and improved soil microbial activity (Figure 2) [25]. Moreover, compared with I, O resulted in a lower purple caitai yield. This may be because organic N fertilizer alone does not provide enough available N for vegetable growth (Figure 2) due to low mineralization relative to chemical N fertilizer [26].

The yield-scaled N_2O emission, an index for evaluating the source or sink of soil N_2O per ton of yield [3], not only considers crop yield but also incorporates N_2O emissions. Therefore, the index can be used to investigate the relationship between agronomic productivity and environmental sustainability (e.g., greenhouse gas emissions) in agricultural production [37]. In this study, among N fertilization treatments, I+O showed the lowest yield-scaled N_2O emissions due to the highest yield and the lowest N_2O emissions (Table 2), suggesting that combined inorganic/organic N fertilization could reduce soil N_2O emissions and contribute to high purple caitai productivity. Moreover, the combination could improve the soil fertility (Figure 2), and thus it can be concluded that the combined inorganic/organic N fertilization is a sustainable agricultural technology for reducing soil N_2O emissions while improving soil fertility and yield of purple caitai.

5. Conclusions

The response of nitrous oxide (N_2O) emissions to the combined application of inorganic and organic nitrogen (N) fertilizers from upland soils remains still unclear. In this view, this study hypothesized that the combined application could mitigate N_2O emissions from purple caitai fields compared with the single application of inorganic or organic N fertilizers by decreasing nitrification and denitrification. The results showed that compared with the single application, the combined application significantly improved soil N availability, promoted microbial biomass, and diminished the N_2O emissions. Moreover, partial correlation and similarity percentage analyses revealed that the ammonia-oxidizing bacteria (AOB) community was the major contributor to N_2O production, and the integrated fertilization mitigated the N_2O emissions primarily through inhibiting the nitrification by AOB in purple caitai fields. Therefore, we recommend the integrated organic-inorganic N fertilization as an optimum N fertilization practice to enhance the soil fertility and yield and mitigate the N_2O emissions in the upland fields.

Supplementary Materials: The following supporting information can be downloaded at: https://www.mdpi.com/article/10.3390/agriculture12050723/s1, Table S1: Primers and protocols for RT-qPCR. Table S2: TC and TN contents in soil under different treatments at the harvest stages in the 2016 and 2017 growing seasons of purple caitai.

Author Contributions: C.L. conceived and designed the research; D.F. provided experimental data; D.F. conducted the measurement; D.F. and C.L. performed statistical analysis; D.F. wrote the manuscript; C.L. and C.C. commented on and revised the manuscript. All authors have read and agreed to the published version of the manuscript.

Funding: This work was funded by the Key Research and Development Project of Hubei Province (2021BBA224).

Institutional Review Board Statement: Not applicable for studies not involving humans or animals.

Informed Consent Statement: This paper does not contain any studies with human participants and/or animals. Informed consent was obtained from all individual participants included in this study.

Data Availability Statement: The data presented in this study are available on request from the corresponding author.

Conflicts of Interest: The authors declare that they have no conflict of interest.

References

1. IPCC. *2013 Climate Change—The Physical Science Basis, Contribution of Working Group I to the Fifth Assessment Report of the Intergovernmental Panel on Climate Change*; Stocker, T.F., Qin, D., Plattner, G.K., Tignor, M., Allen, S.K., Boschung, J., Nauels, A., Eds.; Cambridge University Press: Cambridge, UK; New York, NY, USA, 2013.
2. Ravishankara, A.R.; Daniel, J.S.; Portmann, R.W. Nitrous oxide N_2O. The dominant ozone-depleting substance emitted in the 21st century. *Science* **2009**, *326*, 123–125. [CrossRef]
3. Mosier, A.; Kroeze, C.; Nevison, C.; Oenema, O.; Seitzinger, S.; van Cleemput, O. Closing the global N_2O budget, nitrous oxide emissions through the agricultural nitrogen cycle. *Nutr. Cycl. Agroecosyst.* **1998**, *52*, 225–248. [CrossRef]
4. Schindler, D.W.; Hecky, R.E. Eutrophication, More Nitrogen Data Needed. *Science* **2009**, *324*, 721–722. [CrossRef] [PubMed]
5. Wen, Y.; Xiao, J.; Liu, F.; Goodman, B.A.; Li, W.; Jia, Z.; Ran, W.; Zhang, R.; Shen, Q.; Yu, G. Contrasting effects of inorganic and organic fertilisation regimes on shifts in Fe redox bacterial communities in red soils. *Soil Biol. Biochem.* **2018**, *117*, 56–67. [CrossRef]
6. Urzedo, D.I.; Franco, M.P.; Pitombo, L.M.; Carmo, J.B. Effects of organic and inorganic fertilizers on greenhouse gas GHG emissions in tropical forestry. *For. Ecol. Manag.* **2013**, *310*, 37–44. [CrossRef]
7. Aguilera, E.; Lassaletta, L.; Sanz-Cobena, A.; Garnier, J.; Vallejo, A. The potential of organic fertilizers and water management to reduce N_2O emissions in Mediterranean climate cropping systems. A review. *Agric. Ecosyst. Environ.* **2013**, *164*, 32–52. [CrossRef]
8. Verdi, L.; Kuikman, P.J.; Orlandini, S.; Mancini, M.; Napoli, M.; Marta, A.D. Does the use of digestate to replace mineral fertilizers have less emissions of N_2O and NH_3? *Agric. For. Meteorol.* **2019**, *269–270*, 112–118. [CrossRef]

9. Ding, W.; Luo, J.; Li, J.; Yu, H.; Fan, J.; Liu, D. Effect of long-term compost and inorganic fertilizer application on background N_2O and fertilizer-induced N_2O emissions from an intensively cultivated soil. *Sci. Total Environ.* **2013**, *465*, 115–124. [CrossRef]
10. Agegnehu, G.; Bass, A.M.; Nelson, P.N.; Bird, M.I. Benefits of biochar, compost and biochar-compost for soil quality, maize yield and greenhouse gas emissions in a tropical agricultural soil. *Sci. Total Environ.* **2016**, *543*, 295–306. [CrossRef]
11. Case, S.D.C.; McNamara, N.P.; Reay, D.S.; Stott, A.W.; Grant, H.K.; Whitaker, J. Biochar suppresses N_2O emissions while maintaining N availability in a sandy loam soil. *Soil Biol. Biochem.* **2015**, *81*, 178–185. [CrossRef]
12. Sandrin, T.R.; Herman, D.C.; Maier, R.M. Physiological Methods. In *Environmental Microbiology*; Maier, R.M., Pepper, I.L., Gerba, C.P., Eds.; Academic Press: Cambridge, MA, USA, 2009; pp. 191–223.
13. Richarson, D.J. Bacterial respiration, a flexible process for a changing environment. *Microbiology* **2000**, *146*, 551–571.
14. Norton, J.M.; Alzerreca, J.J.; Suwa, Y.; Klotz, M.G. Diversity of ammonia monooxygenase operon in autotrophic ammonia-oxidizing bacteria. *Arch. Microbiol.* **2002**, *177*, 139–149. [CrossRef] [PubMed]
15. Jones, C.M.; Stres, B.; Rosenquist, M.; Hallin, S. Phylogenetic analysis of nitrite, nitric oxide, and nitrous oxide respiratory enzymes reveal a complex evolutionary history for denitrification. *Mol. Biol. Evol.* **2008**, *25*, 1955–1966. [CrossRef] [PubMed]
16. Levy-Booth, D.J.; Prescott, C.E.; Grayston, S.J. Microbial functional genes involved in nitrogen fixation, nitrification and denitrification in forest ecosystems. *Soil Biol. Biochem.* **2014**, *75*, 11–25. [CrossRef]
17. Yang, L.; Zhang, X.; Ju, X. Linkage between N_2O emission and functional gene abundance in an intensively managed calcareous fluvo-aquic soil. *Sci. Rep.* **2017**, *7*, 43283. [CrossRef]
18. Wang, C.; Li, H.; Li, Y.; Meng, Q.; Xie, F.; Xu, Y.; Wan, Z. Genetic characterization and fine mapping BrCER4 in involved cuticular wax formation in purple cai-tai Brassica rapa L. var purpurea. *Mol. Breed.* **2019**, *39*, 12. [CrossRef]
19. Li, C.; Zhang, Z.; Guo, L.; Cai, M.; Cao, C. Emissions of CH_4 and CO_2 from double rice cropping systems under varying tillage and seeding methods. *Atmos. Environ.* **2013**, *80*, 438–444. [CrossRef]
20. Yao, K.Y. Determination of soil microbial activities and character. In *Soil Microbial Ecology and Experimental Technology*; Yao, K.Y., Huang, C.Y., Eds.; Science Press: Beijing, China, 2006; pp. 187–192. (In Chinese)
21. Witt, C.; Gaunt, J.L.; Galicia, C.C.; Ottow, J.C.G.; Neue, H.U. A rapid chloroform-fumigation extraction method for measuring soil microbial biomass carbon and nitrogen in flooded rice soils. *Biol. Fertil. Soils* **2000**, *30*, 510–519. [CrossRef]
22. Allison, L.E. Organic carbon. In *Methods of Soil Analysis Part II*; Black, C.A., Ed.; American Society of Agronomy Publications: Madison, WI, USA, 1965; pp. 1367–1376.
23. Zaman, M.; Saggar, S.; Blennerhassett, J.D.; Singh, J. Effect of urease and nitrification inhibitors on N transformation, gaseous emissions of ammonia and nitrous oxide, pasture yield and N uptake in grazed pasture system. *Soil Biol. Biochem.* **2009**, *41*, 1270–1280. [CrossRef]
24. Evans, J.D. *Straight Forward Statistics for the Behavioral Sciences*; Brooks/Cole Publishing: Pacific Grove, CA, USA, 1996.
25. Pan, G.; Zhou, P.; Li, Z.; Smith, P.; Chen, X. Combined inorganic/organic fertilization enhances N efficiency and increases rice productivity through organic carbon accumulation in a rice paddy from the Tai Lake region, China. *Agric. Ecosyst. Environ.* **2009**, *131*, 274–280. [CrossRef]
26. Zhang, J.; Zhuang, M.; Shan, N.; Zhao, Q.; Li, H.; Wang, L. Substituting organic manure for compound fertilizer increases yield and decreases NH_3 and N_2O emissions in an intensive vegetable production systems. *Sci. Total. Environ.* **2019**, *670*, 1184–1189. [CrossRef] [PubMed]
27. Neumann, G. Root Exudates and Nutrient Cycling. In *Nutrient Cycling in Terrestrial Ecosystems. Soil Biology*; Marschner, P., Rengel, Z., Eds.; Springer: Berlin/Heidelberg, Germany, 2007; pp. 123–157.
28. Ali, M.A.; Sattar, M.A.; Islam, M.N.; Inubushi, K. Integrated effects of organic, inorganic and biological amendments on methane emission, soil quality and rice productivity in irrigated paddy ecosystem of Bangladesh, field study of two consecutive rice growing seasons. *Plant Soi.* **2014**, *378*, 239–252. [CrossRef]
29. Wang, F.; Chen, S.; Wang, Y.; Zhang, Y.; Hu, C.; Liu, B. Long-term nitrogen fertilization elevates the activity and abundance of nitrifying and denitrifying microbial communities in an upland soil, implications for nitrogen loss from intensive agricultural systems. *Front. Microbiol.* **2018**, *23*, 02424. [CrossRef]
30. Tian, X.; Hu, H.; Ding, Q.; Song, M.; Xu, X.; Zheng, Y.; Guo, L. Influence of nitrogen fertilization on soil ammonia oxidizer and denitrifier abundance, microbial biomass, and enzyme activities in an alpine meadow. *Biol. Fertil. Soils* **2013**, *50*, 703–713. [CrossRef]
31. Han, J.; Shi, J.; Zeng, L.; Xu, J.; Wu, L. Impacts of continuous excessive fertilization on soil potential nitrification activity and nitrifying microbial community dynamics in greenhouse system. *J. Soil Sediments* **2017**, *17*, 471–480. [CrossRef]
32. Grießmeier, V.; Leberecht, K.; Gescher, J. NO_3^- removal efficiency in field denitrification beds, key controlling factors and main implications. *Environ. Microb. Rep.* **2019**, *11*, 316–329. [CrossRef]
33. Kindaichi, T.; Ito, T.; Okabe, S. Ecophysiological interaction between nitrifying bacteria and heterotrophic bacteria in autotrophic nitrifying biofilms as determined by microautoradiography-fluorescence in situ hybridization. *Appl. Environ. Biol.* **2004**, *70*, 1641–1650. [CrossRef]
34. Martens, D.A. Denitrification. In *Encyclopedia of Soils in the Environment*; Hillel, D., Ed.; Academic Press: Cambridge, MA, USA, 2005; pp. 378–382.
35. Inglett, P.W.; Reddy, K.R.; Corstanje, R. Anaerobic soils. In *Encyclopedia of Soils in the Environment*; Hillel, D., Ed.; Academic Press: Cambridge, MA, USA, 2005; pp. 72–78.

36. Zhang, J.; Zhang, F.; Yang, J.; Wang, J.; Cai, M.; Li, C.; Cao, C. Emissions of N_2O and NH_3, and nitrogen leaching from direct seeded rice under different tillage practices in central China. *Agric. Ecosyst. Environ.* **2011**, *140*, 164–173.
37. Xu, Y.; Guo, L.; Xie, L.; Yun, A.; Li, Y.; Zhang, X.; Zhao, X.; Diao, T. Characteristics of background emissions and emission factors of N_2O from major upland fields in China. *Sci. Agric. Sin.* **2016**, *49*, 1729–1743.
38. Bateman, E.J.; Baggs, E.M. Contributions of nitrification and denitrification to N_2O emissions from soils at different water-filled pore space. *Biol. Fertil. Soils.* **2005**, *41*, 379–388. [CrossRef]
39. Ding, W.; Cai, Y.; Cai, Z.; Yagi, K.; Zheng, X. Nitrous oxide emissions from an intensively cultivated maize–wheat rotation soil in the north china plain. *Sci. Total. Environ.* **2007**, *373*, 501–511. [CrossRef] [PubMed]
40. Dobbie, K.; Smith, K. Nitrous oxide emission factors for agricultural soils in Great Britain, the impact of soil water-filled pore space and other controlling variables. *Glob. Chang. Biol.* **2003**, *9*, 204–218. [CrossRef]
41. Shcherbak, I.; Millar, N.; Robertson, G.P. Global metaanalysis of the nonlinear response of soil nitrous oxide N_2O emissions to fertilizer nitrogen. *Proc. Natl. Acad. Sci. USA* **2014**, *111*, 9199–9204. [CrossRef] [PubMed]
42. Nyamadzawo, G.; Wuta, M.; Nyamangara, J.; Smith, J.L.; Rees, R.M. Nitrous oxide and methane emissions from cultivated seasonal wetland dambo soils with inorganic, organic and integrated nutrient management. *Nutr. Cycl. Agroecosyst.* **2014**, *100*, 161–175. [CrossRef]

agriculture

Article

Side Deep Fertilizing of Machine-Transplanted Rice to Guarantee Rice Yield in Conservation Tillage

Qi-Xia Wu †, Bin Du †, Shuo-Chen Jiang, Hai-Wei Zhang and Jian-Qiang Zhu *

College of Agriculture, Yangtze University, Jingzhou 434025, China; qixiawu@yangtzeu.edu.cn (Q.-X.W.); xiaobin@stu.scau.edu.cn (B.D.); 202073052@yangtzeu.edu.cn (S.-C.J.); 202071644@yangtzeu.edu.cn (H.-W.Z.)
* Correspondence: 200572@yangtzeu.edu.cn
† These authors contributed equally to this work.

Abstract: Conservation tillage is an environmentally friendly and economical farming method, but its impact on rice yield is controversial. Artificially applied side deep fertilizing of machine-transplanted rice is when fertilizer is applied to the deep soil along with the machine transplantation of rice; this may improve the fertilizer utilization rate and rice yield and eliminate the possible negative effects of conservation tillage on rice yield. Using on machine-transplanted rice, this study aims to compare the effects of side deep fertilizing (SDF). We investigated the effects of artificially applying fertilizer (AAF) on rice growth and yield under conventional tillage (CT), reduced tillage (RT), and no tillage (NT). The rice root activity, root dry weight, leaf area index (LAI), net photosynthetic rate (Pn), chlorophyll content, panicle density, spikelets per panicle, and yield were all ranked as NT > RT > CT and SDF > AAF. The 1000-grain weight was also ranked as SDF > AAF. In addition, under NT conditions, the positive effect of SDF on rice growth and yield was higher than under RT and CT conditions. In general, conservation tillage combined with SDF saved costs and increased rice yield.

Keywords: reduced tillage; no tillage; side deep fertilizing of machine-transplanted rice; root function; photosynthesis

Citation: Wu, Q.-X.; Du, B.; Jiang, S.-C.; Zhang, H.-W.; Zhu, J.-Q. Side Deep Fertilizing of Machine-Transplanted Rice to Guarantee Rice Yield in Conservation Tillage. *Agriculture* **2022**, *12*, 528. https://doi.org/10.3390/agriculture12040528

Academic Editor: Manuel Ângelo Rosa Rodrigues

Received: 14 February 2022
Accepted: 3 April 2022
Published: 8 April 2022

Publisher's Note: MDPI stays neutral with regard to jurisdictional claims in published maps and institutional affiliations.

1. Introduction

Rice (*Oryza sativa* L.) is the main food source for more than half of the world's population and the most important food crop in Asia [1]. Tillage changes the distribution of soil nutrients and the stability of aggregates, reduces soil compactness, and allows air to enter deep soil [2]. Tillage can improve soil properties, promote rice growth, and increase rice yield [3]. However, the long-term use of conventional tillage (CT) has increased the degradation of organic matter, reduced the soil microbial content and enzyme activity, and caused a decline in soil function and soil erosion [4,5]. Since the 21st century, rural depopulation in China has not allowed the long-term development of traditional frequency tillage [6].

Conservation tillage such as reduced tillage (RT) and no tillage (NT) is a tillage technology for the protection of cultivated land. Hart et al. [7] reported that it has become apparent that the concomitant increase in losses of N and P from agricultural land is having a serious detrimental effect on water quality and the environment. Reportedly, conservation tillage promotes rice growth and increases rice yield by protecting soil aggregates [8] and by increasing soil organic matter contents [9], nutrient availability [10], microbial biomass [11,12], and enzyme activity [13]. In addition, conservation tillage has the effects of reducing the input of energy and labor [11] and reducing greenhouse gas emissions [14]. However, there are also reports that under conservation tillage, fertilizers are applied to the soil surface or the shallow soil layer, which increases fertilizer loss and reduces fertilizer use efficiency and rice yield [15].

Agricultural mechanization is an important process used to promote agricultural intensive management, adjust agricultural supply-side reforms, and accelerate the construction of modern agriculture, and it is also the basis for the development of smart agriculture in the future [16]. Machine-transplanted seedlings solve the disadvantages of traditional artificial direct-seeding rice, such as irregular growth, uneven row spacing, and susceptibility to lodging, reduces the labor input and production costs, and increases the planting speed, rice yield, and farmers' income [17]. Mechanical side deep fertilizing technology ensures that the fertilizer enters the soil cultivation layer, effectively reduces fertilizer loss, and improves fertilizer utilization and rice yield [18]. The side deep fertilizing (SDF) of machine-transplanted rice is an innovative approach, wherein rice seedlings are transplanted and granular fertilizers are applied simultaneously deep in the paddy soils [19]. This technique ensures close contact between nutrients and crop root systems, enhances nutrient absorption and utilization, and improves fertilizer use efficiency [20]. Zhong et al. [21] reported that SDF optimized agronomic traits and yield components, increased grain yield and economic return, and enhanced NPK fertilizer uptake but reduced their application rates. However, the application of SDF in rice production under conservation tillage has not yet been reported.

Aiming at the basal fertilizer application method for machine-transplanted rice, this study compared the effects of side deep fertilizing (SDF), artificially applying fertilizer (AAF), on rice growth and yield under three tillage management systems (CT, RT, and NT); thus, this study provides a reference for the sustainable development of conservation tillage and mechanized rice production.

2. Material and Methods

2.1. Experimental Site

Experiments were conducted during 2017–2019 at the Yangtze University farm, Jingzhou County (112°04′ N–112°05 N, 30°32′ E–30°33′ E), Hubei Province, China. The area belongs to the northern subtropical agricultural climate zone. The annual average temperature was 16.5 °C, the accumulated temperature $\geq$ 10 °C was 5094.9–5204.3 °C, the annual average precipitation was 1095 mm, and the sunshine time was 1718 h. In the experimental field, the soil texture was clay loam, and the index of the soil agrochemical properties were a pH of 6.02, 32.31 g kg^{-1} of organic matter, 243.05 mg kg^{-1} of available N, 11.03 mg kg^{-1} of available P, and 103.24 mg kg^{-1} of available K.

2.2. Experimental Design

The experiments were arranged in a split-plot design, with the fertilization methods as the main plot and the tillage managements as subplots, in three replications. The plot size was 5 × 10 m^2. Two fertilization methods (SDF and AAF) and three tillage management systems (RT, NT, and CT) were designed, which are briefly introduced as follows.

RT refers to only rotary tillage without plowing as the rice plants were planted; the field was irrigated to 3–4 cm deep after the wheat was harvested so as to prevent weeds (such as barnyard grass and tendon grass); at 3–5 days before rice transplanting, a paddy field stubble burying machine (HUHN RM320, Jining, Shandong, China) was used to press the rice stubble, and a rotary tiller (FENGYUAN 1GZL200, Huzhou, Zhejiang, China) was used to rotate (15 cm deep tillage) once, then field irrigation was performed and a water layer of 1–2 cm was kept in the field until transplanting.

NT means that the soil was not disturbed after the wheat harvest until rice planting. The residue of the wheat straw covered the ground until the rice harvest. At 3 weeks before planting, 60 mL ha^{-1} *Roundup* (Organophosphorus herbicides, Monsanto Company, St. Louis, MI, USA) aqua herbicide was sprayed on the field as the soil was drying, and the field had te be kept dry for 5–7 days in order to eliminate weeds; after this, the field was irrigated and kept with a water layer of 1–2 cm until transplanting.

CT is a local conventional tillage method, which is carried out as follows: the paddy field was first soaked with water to a depth of 4–5 cm and then plowed once (with a tillage depth of 25 cm) about 5 days later, and a rotary tillage was conducted (with a tillage depth of 15 cm) 5 days before transplanting; after this, the rice plants were transplanted into the field with a water layer of 1–2 cm.

SDF is completed by an integrated machine used for rice transplanting and fertilizing (Yameike RXA-60TK, Changzhou, Jiangsu, China), by which a basal fertilizer is put at a depth of 20 cm of the topsoil while the rice transplanting is completed. Transplanting and fertilizing are done almost at the same time.

Under the tillage management of CT and RT, AAF is used to apply the basal fertilizer on the day of the rotary tillage, and the basal fertilizer mixes with the soil by rotary tillage, then, the rice seedlings are transplanted by a rice transplanter (Yameike RXA-60TK, Changzhou, China). As NT was adopted, AAF spread the basal fertilizer on the field one day before the mechanical rice transplantation.

Rice was sown on 1 May and transplanted on 1 June in 2019 and 2020. Harvest was on 16 September 2019 and 21 September 2020. The planting density was 25 hills m^{-2}. The experimental plots in which the basal fertilizer was applied to the field using SDF and AAF had the same available nutrient dosage in each plot, with 120 kg ha^{-1} of N, 59 kg ha^{-1} of P_2O_5, and 120 kg ha^{-1} of K_2O, whereas 60% and 40% of N were applied, respectively, as the base fertilizer and tillering fertilizer. All the P_2O_5 and K_2O was applied as base fertilizers. N, P_2O_5, and K_2O were administered in the form of urea, disodium hydrogen phosphate, and potassium chloride, respectively. Rotary tillage was done twice before planting wheat in winter (HUHN RM320, Jining, Shandong, China).

2.3. *Measurements*

2.3.1. Rice Yield and Its Components

Grain yields and panicle density were measured at maturity by taking 5 m^2 plant samples at the center of each plot. The filled grains in each 5 m^2 plant sample were separated from the straws. The filled grains were dried in an oven at 70 °C to a stable weight and weighed, and the grain yield was calculated at a 14% moisture content. Rice plant samples plots were taken from 5 planting pits per for the determination of yield components (spikelets per panicle, grain filling rate, and 1000-grain weight).

2.3.2. Root Function

Rice plant samples were selected from 5 planting pits per plot to measure the root dry weight and root activity at the mid-tillering stage, heading stage, grain filling stage, and yellow ripe stage. For each root sample, a cube of soil (25 cm in length × 16 cm in width × 20 cm in depth) around each individual planting pit was removed using a sampling core, and such a cube contains about 95% of total root biomass [22]. The rice plants from 5 planting pits per plot formed a sample at each measurement, and the roots in each cube of soil were carefully rinsed with a hydropneumatic elutriation device (Gillison's Variety Fabrications, Benzonia, MI, USA). Portions of each root sample were used for the measurement of root activity, which was determined by measuring the oxidation of alpha-naphthylamine (α-NA) [23], whereas the other root samples were dried in an oven at 70 °C to stable weights and weighed. One gram of fresh roots was transferred into a 150 mL flask containing 50 mL of 20 ppm α-NA. The flasks were incubated for 2 h at room temperature in an end-over-end shaker. After this, the aliquots were filtered, and 2 mL of aliquot was mixed with 1 mL of 1.18 $mmol^{-1}$ NaNO and 1 mL of sulfanilic acid, and the resulting color was measured using a spectrophotometer.

2.3.3. Photosynthetic Properties

Rice plants were selected from 5 planting pits per plot to measure the leaf area index (LAI), net photosynthetic rate (Pn), and total chlorophyll content at the mid-tillering stage, heading stage, grain filling stage, and yellow ripe stage. The LAI of the top fully expanded leaves in the main stem was calculated as the measured leaf area divided by the ground surface area. The Pn of the top fully expanded leaves in the main stem was determined by a gas exchange analyzer (Li-6400, Li-COR Inc., Irving, TX, USA) between 9:30 and 11:00 a.m., when the photosynthetic active radiation above the canopy was 1200 mmol m^{-2} s^{-1}. After the determination of LAI and Pn, the measured leaves were cut, frozen immediately in liquid nitrogen, and stored at −80 °C for standby. The total chlorophyll content was extracted with about 0.2 g of fresh leaf disks using 25 mL of an alcohol and acetone mixture (v:v = 1:1) for 24 h in the dark at room temperature. The absorbance of the extract was measured at 663, 645, and 470 nm using a UV-VIS spectrophotometer (UV-2600, Shimadzu, Japan) to estimate the total chlorophyll content according to the method reported in [4].

2.4. Statistical Analysis

All the experimental data were collected in 2019 and 2020 and expressed as mean ± standard error (SE) of three replicates. The normal distribution and homogeneity variance of data were tested using Shapiro–Wilk's test and Levene's test on SPSS 21.0 (Statistical Product and Service Solutions, IBMLab) [24], respectively. The independent samples t-test was used to compare the differences in the relevant rice indicators between the two years (2019 and 2020). One-way analysis of variance was used to compare the differences between the relevant rice indicators among the tillage methods and fertilization methods, and two-factor analysis of variance was used to compare the impacts of the interaction of the tillage methods and fertilization methods on the rice indicators. In the statistical analysis, two significance levels were set at $p < 0.05$ and $p < 0.01$. The diagrams were drawn using the Origin 2017(*Origin* Lab) mapping software.

3. Results

3.1. Root Activity and Root Dry Weight

Figures 1 and 2 show the root activity and root dry weight under different fertilization and tillage systems. The results showed that there was no significant difference in root activity and root dry weight between the two years. With the growth of the rice, the root activity decreased, and the root dry weight increased first and then decreased. Different tillage and fertilization systems had significant effects on root activity and root dry weight. The root activity of SDF + NT was significantly higher than that of the other treatments in the mid-tillering stage, heading stage, full heading stage, and yellow maturity stage. Although AAF + NT was the second highest in the mid-tillering stage, SDF + RT was the second highest in the other growth stages, indicating that the SDF model was helpful in improving rice root activity. In the SDF model, the root activity in each growth stage was ranked as NT > RT > CT; in the AAF model, the root activity in each growth stage was ranked as NT > RT > CT; and the root activity of the NT treatment in the full heading stage and yellow mature stage was significantly higher than that of RT and CT. The results of the two fertilization models show that the NT treatment was helpful in improving the root activity of rice. The root dry weight of the SDF + NT model was significantly higher than that of the other treatments in the mid-tillering stage, heading stage, full heading stage, and yellow maturity stage. In the SDF and AAF models, the root dry weight at each growth stage was ranked as NT > RT > CT. It was concluded that the SDF model is better than the AAF model for rice root growth; the NT model is better than the RT and CT models for rice root growth, and the SDF + NT mode is better for rice root growth in the whole growth period.

3.2. Leaf Area Index (LAI), Net Photosynthetic Rate (Pn), and Total Chlorophyll Content

As shown in Figures 3–5, the interannual differences of LAI, PN, and chlorophyll contents were not significant. With the growth of rice, the photosynthetic rate and chlorophyll content showed a downward trend; the leaf area index first increased and then decreased, and was the highest at the heading stage. The LAI at each growth stage of the SDF and AAF models was ranked as NT > RT > CT, and SDF + NT was the highest. However, there was no significant difference between the mid-tillering stage, SDF + RT, and AAF + NT, and there was no significant difference between the yellow ripening stage and SDF + RT. SDF + NT in the other growth stages was significantly higher than that of the other treatments, indicating that the SDF + NT model was helpful in improving rice leaf growth. In the SDF model, the chlorophyll content in each growth stage was ranked as NT > RT > CT, and NT was significantly higher than RT in the middle growth stage. There was no significant difference between NT and RT in the other growth stages, but they were significantly higher than the CT model. In the AAF model, the chlorophyll content in each growth stage was ranked as NT > RT > CT, and the chlorophyll content in the SDF model was significantly higher than that in AAF under the same tillage model, among which SDF + NT was the highest. However, there was no significant difference between SDF + NT and SDF + RT in the mid-tillering stage and booting stage in 2020; SDF + NT in the other growth stages was significantly higher than that in the other treatments. In the SDF and AAF models, the PN in each growth stage was ranked as NT > RT > CT, and under the same tillage mode, the PN in the SDF model was significantly higher than that in the AAF model. It was concluded that, in the SDF and AAF models, the leaf area index, chlorophyll content, and PN in each growth stage were ranked as NT > RT > CT, and the SDF model is significantly higher than the AAF model, among which the SDF + NT model is the highest, indicating that the SDF + NT model is helpful in improving the rice leaf area and chlorophyll content, so as to improve rice photosynthesis.

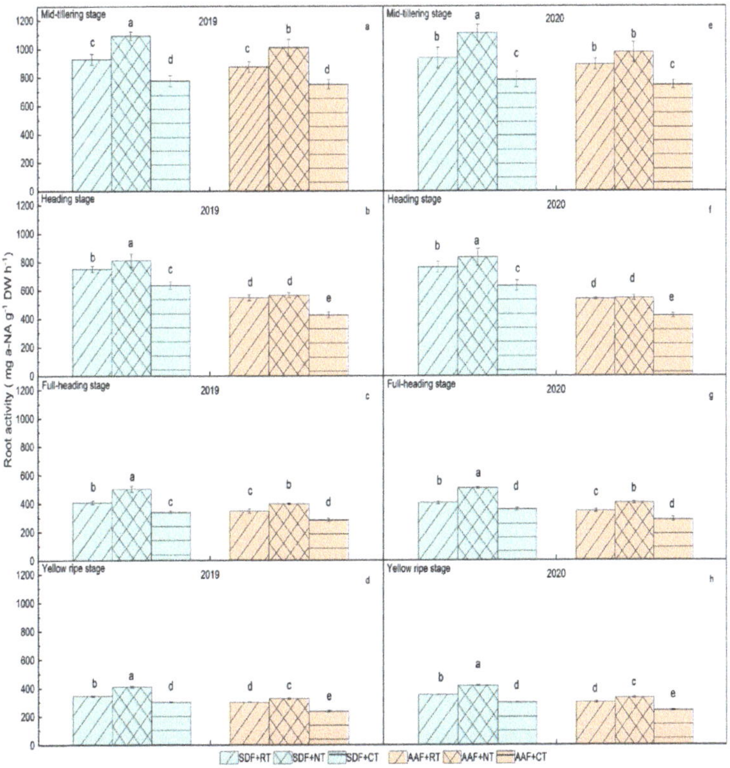

Figure 1. Root activity at different stages of rice under different tillage and fertilization treatments. (**a–d**) represent the root activity in mid-tillering stage, heading stage, full-heading stage and yellow ripe stage in 2019, respectively. (**e–h**) represent the root activity in mid-tillering stage, heading stage, full-heading stage, and yellow ripe stage in 2020, respectively. SDF + RT: Side Deep Fertilizing and Reduced Tillage. SDF + NT: Side Deep Fertilizing and No Tillage. SDF + CT: Side Deep Fertilizing and Conventional Tillage. AAF + RT: Artificially Applying Fertilizer and Reduced Tillage. AAF + NT: Artificially Applying Fertilizer and No Tillage. AAF + CT: Artificially Applying Fertilizer and Conventional Tillage. Different lowercase letters marked on the histogram mean significant differences among treatments at the 5% level according to Duncan's multiple-range test (0.05) (similarly hereinafter).

3.3. Rice Yield and Its Compositions

Table 1 shows the yield components of rice in 2019 and 2020 under different tillage and fertilization models, and the yield difference between the two years is not significant. The results of the variance analysis showed that the different tillage methods had significant or extremely significant effects on panicle density, grain number per panicle, and yield, and the different fertilization methods had significant effects on panicle density, grain number per panicle, 1000 grain weight, and yield. It can be seen from the data in the table that the yield under the SDF model was significantly higher than that under the AAF fertilization model. The NT yield under the SDF treatment was significantly higher than that of RT and CT, and there was no significant difference in yield among the three tillage methods of AAF treatment. It can be seen that the yield of SDF + NT is the highest, having increased by 14.03~15.17%, 22.06~30.22%, 26.99~30.22%, 19.35~34.43%, and 39.90~40.08%, respectively, compared with SDF + RT, SDF + CT, AAF + RT, and AAF + CT. In terms of yield components, there was no significant difference among the three tillage methods in the SDF model for the panicle density, grain number per panicle, seed setting rate, and 1000 grain weight, whereas NT in the SDF model was significantly higher than RT and CT, indicating that the main reason for the increase in SDF + NT yield was the increase in grain number per panicle.

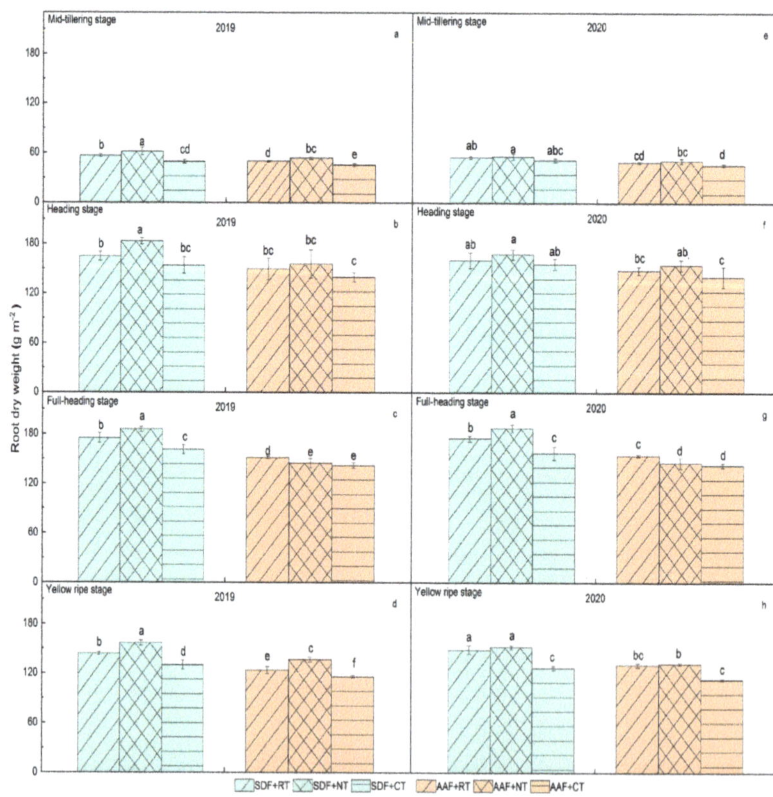

Figure 2. Root dry weight at different stages of rice under different tillage and fertilization treatments. (**a**–**d**) represent the root dry weight in mid-tillering stage, heading stage, and full-heading Scheme 2019, respectively. (**e**–**h**) represent the root dry weight in mid-tillering stage, heading stage, full-heading stage, and yellow ripe stage in 2020, respectively. Different lowercase letters marked on the histogram mean significant differences among treatments at the 5% level according to Duncan's multiple-range test (0.05) (similarly hereinafter).

Table 1. Rice yield and its components under different tillage and fertilization treatments.

Year	Treatments		Spike Density m^2	Spikelets per Panicle	Grain Filling Rate %	1000-Grain Weight g	Yield t·ha^{-1}
	Tillage	Fertilization					
2019	SDF	RT	268.19 ± 10.49 abc	177.49 ± 6.45 b	74.83 ± 0.76 bcd	26.28 ± 1.10 ab	9.56 ± 0.33 b
		NT	279.43 ± 6.10 a	194.42 ± 2.29 a	73.98 ± 2.00 bcd	27.30 ± 0.26 a	11.01 ± 0.22 a
		CT	274.69 ± 16.77 ab	172.40 ± 3.69 bc	76.06 ± 1.74 abc	26.23 ± 1.09 ab	9.02 ± 0.93 bc
	AAF	RT	272.58 ± 9.37 abc	168.77 ± 2.50 bc	75.76 ± 2.65 abcd	25.53 ± 0.88 bcd	8.67 ± 0.97 bcd
		NT	256.88 ± 2.2 bcd	166.03 ± 4.58 c	79.60 ± 2.81 a	26.06 ± 0.72 abc	8.19 ± 0.16 cde
		CT	254.47 ± 5.76 cd	168.12 ± 8.94 b c	71.69 ± 1.87 d	24.77 ± 1.50 bcd	7.86 ± 0.4 6 de
2020	SDF	RT	262.34 ± 11.52 bc	178.32 ± 11.57 b	75.4 ± 1.89 a	26.41 ± 0.85 ab	9.41 ± 0.71 b
		NT	291.16 ± 11.74 a	195.12 ± 9.77 a	78.40 ± 2.15 a	27.57 ± 0.83 a	10.73 ± 0.57 a
		CT	258.03 ± 13.7 bcd	168.28 ± 4.02 bc	77.80 ± 5.74 a	25.77 ± 1.56 abc	8.24 ± 0.53 cd
	AAF	RT	269.2 ± 6.07 b	171.13 ± 6.16 bc	72.81 ± 1.01 a	25.02 ± 1.18 bc	8.24 ± 0.50 cd
		NT	271.28 ± 12.52 b	171.40 ± 2.63 bc	75.46 ± 0.29 a	26.07 ± 1.29 abc	8.99 ± 0.74 bc
		CT	244.07 ± 6.72 def	165.59 ± 3.21 c	74.12 ± 1.88 a	25.40 ± 0.54 bc	7.67 ± 0.37 de
Tillage			*	**	ns	ns	**
Fertilization			**	**	ns	**	**
Tillage × Fertilization			*	**	**	ns	*

Within a column, different lowercase letters following numeric values mean significant differences among treatments at the 5% level according to Duncan's multiple-range test (0.05). * and ** indicate significant differences at the $p < 0.05$ and 0.01 levels, respectively. "ns" means not significant between a certain indicator of rice and tillage or fertilization treatment.

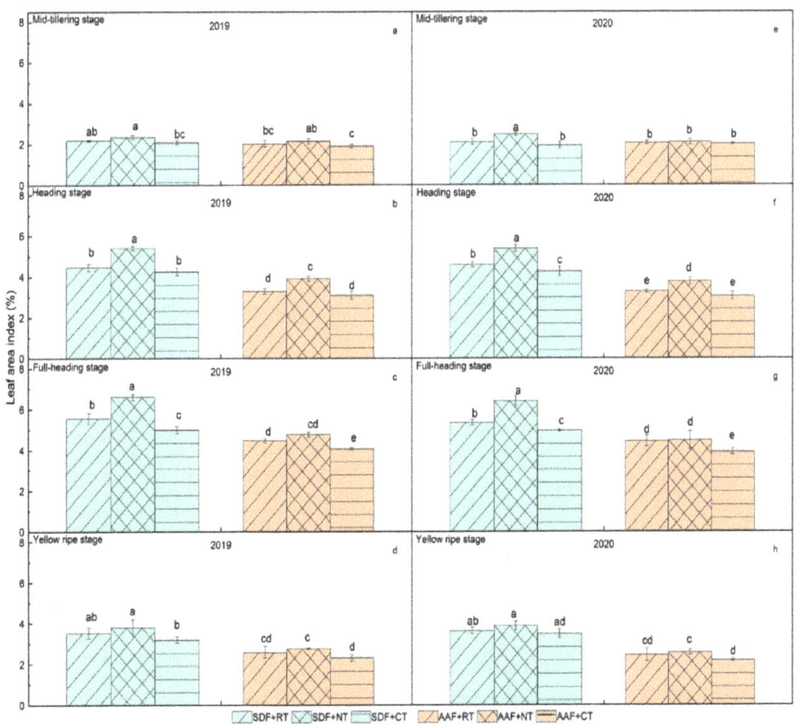

Figure 3. Leaf area index (LAI) at different stages of rice under different tillage and fertilization treatments. (**a–d**) represent the leaf area index in mid-tillering stage, heading stage, full-heading stage, and yellow ripe stage in 2019, respectively. (**e–h**) represent the leaf area index in mid-tillering stage, heading stage, full-heading stage, and yellow ripe stage in 2020, respectively. Different lowercase letters marked on the histogram mean significant differences among treatments at the 5% level according to Duncan's multiple-range test (0.05) (similarly hereinafter).

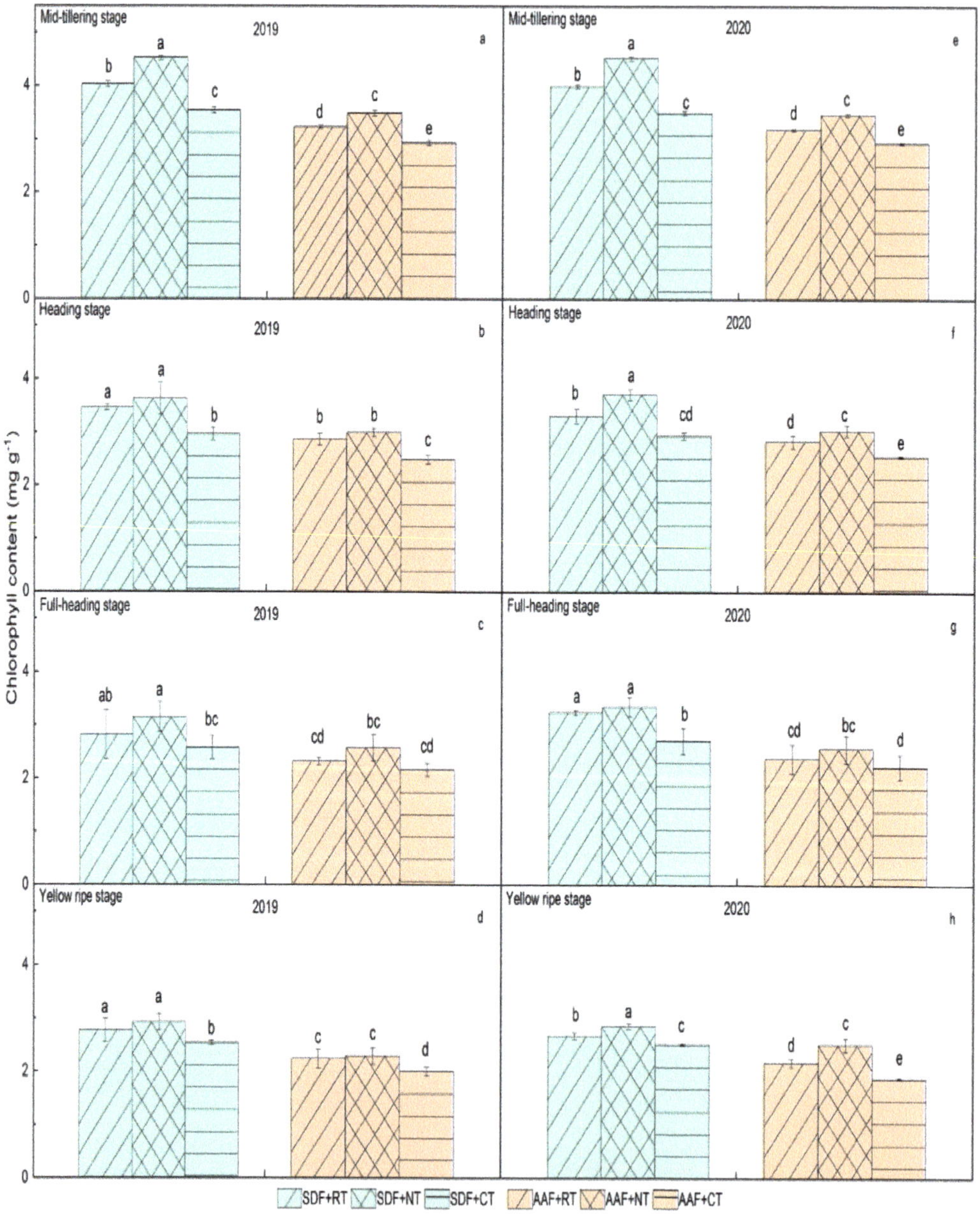

Figure 4. Chlorophyll content at different stages of rice under different tillage and fertilization treatments. (**a**–**d**) represent the chlorophyll content in mid-tillering stage, heading stage, full-heading stage, and yellow ripe stage in 2019, respectively. (**e**–**h**) represent the chlorophyll content in mid-tillering stage, heading stage, full-heading stage, and yellow ripe stage in 2020, respectively. Different lowercase letters marked on the histogram mean significant differences among treatments at the 5% level according to Duncan's multiple-range test (0.05) (similarly hereinafter).

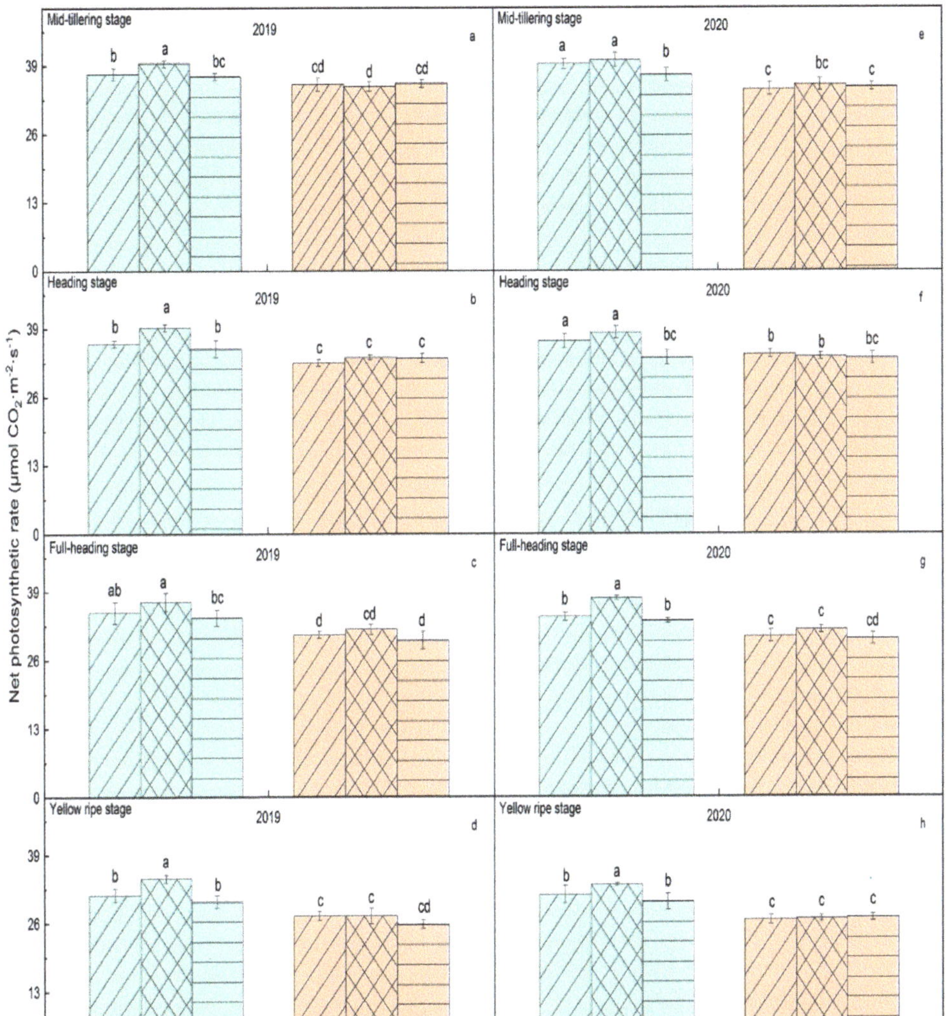

Figure 5. Net photosynthetic rate at different stages of rice under different tillage and fertilization treatments. (**a**–**d**) represent the net photosynthetic rate in mid-tillering stage, heading stage, full-heading stage, and yellow ripe stage in 2019, respectively. (**e**–**h**) represent the net photosynthetic rate in mid-tillering stage, heading stage, full-heading stage, and yellow ripe stage in 2020, respectively. Different lowercase letters marked on the histogram mean significant differences among treatments at the 5% level according to Duncan's multiple-range test (0.05) (similarly hereinafter).

4. Discussion

4.1. Effects of Tillage and Fertilization Treatments on Root Function

In this study, under different tillage management systems, the root function was ranked in the order of NT > RT > CT. Wang et al. [25] reported that compared with NT, both RT and CT mixed organic matter and fertilizers into deeper soils, which promoted the growth of rice roots in deeper soils. Wang et al. [26] reported that NT reduced the bulk density of the topsoil (0–5 cm) but increased the bulk density of deep soil (>5 cm), and a higher and inconsistent bulk density of deep soil limited root system growth. However, the results of Wang et al. [27] concern long-term (14 years) tillage effects on soil and rice. Long-term NT or RT could increase soil compactness, restrict air from entering deep soil, accumulate reducing substances in deep soil, and limit the growth of rice roots [28]. Das et al. [11] reported that short-term (4 years) NT increased soil organic carbon, microbial biological carbon, and dehydrogenase activity, and provided superior conditions for root growth. This was consistent with our findings, indicating that short-term NT or RT could improve soil conditions and promote root growth, as well as support the finding that the effects of NT were more pronounced than those of RT. The organic matter covering the topsoil in NT decomposes slowly, and the continuously decomposed organic matter also continuously provides nutrients for the rice [8]. In addition, organic matter (rice stalks and dead weeds) cover the topsoil in NT, which could help in maintaining a constant temperature and moisture and improve the growth of microorganisms and the synthesis of soil enzymes [11]. In addition, RT and CT make the organic matter fully integrated with the soil, which promotes the decomposition of organic matter into inorganic nutrients, and the inorganic nutrients unused by plants are lost to the environment [13].

In our study, under different fertilization treatments, the root function was ranked as SDF > AAF, which is consistent with previous studies [29,30]. N is essential to the growth and development of rice, and it participates in many metabolic processes, such as protein hydrolysis and amino acid metabolism [31]. The lack of N in the basal fertilizer limited the supply of nutrients for root growth. Min et al. [30] reported that SDF reduced chemical N fertilizer input without any reductions in yield, whereas it increased nitrogen use efficiency and reduced NH_3 volatilization and runoff N losses. In addition, the deep application of the fertilizer could make the fertilizer slowly dissolve in the soil and help retain it in the rice rhizosphere for a longer time such that the fertilizer can provide continuous nutrients for a longer period of time during rice growth [32]. SDF, as compared with AAF, had a better positive effect on root function under NT management, but such an effect did not occur under RT and CT; the reason may be that RT and CT brought organic matter into the deep soil. AAF causes the nutrients of the fertilizer to be retained in the topsoil under NT management, and side deep fertilization may bring fertilizer into the deep soil, which is advantageous for nutrient uptake by plant roots [27]. Therefore, side deep fertilization under NT management is very important for the growth of rice roots.

4.2. Effects of Tillage and Fertilization on Photosynthesis

Generally, under different tillage management systems, the photosynthesis in the four periods was ranked as NT > RT > CT. Our results agree with previous research stating that tillage could cause a decrease in photosynthetic capacity by limiting the nutrient uptake and growth of roots. In this study, under different fertilization treatments, the photosynthesis in the four periods was ranked as SDF > AAF, which is consistent with the findings of a previous study [33]. Leaves accumulated the most N in the plant, and as much as three quarters of leaf N was invested into the photosynthetic apparatus, which was the largest N sink in the plant [34]. It was reported that there was strong positive correlation between photosynthesis and leaf N content [35]. N application in the deep soil layer results in a higher NH_4^+-N concentration in the soil in the prime stage of rice growth and prolongs the availability of N for 2 months, which improves photosynthesis [36]. In addition, root function is closely related to photosynthesis [37], and higher root activity and root dry weight result in higher LAI, chlorophyll content, and Pn.

4.3. Effects of Tillage and Fertilization Treatments on Rice Yield and Its Compositions

In this study, the spike density, spikelets per panicle, and yield under different tillage management systems were ranked as NT > RT > CT. Conservation tillage, especially NT, may improve soil properties, enhance root function (Figures 1 and 2) and photosynthesis (Figures 3–5), increase the absorption of nutrients by the roots and the amount of carbohydrates assimilated by photosynthesis, and promote the formation of yield [38]. Conservation tillage also improves root function and photosynthesis at heading, full heading, and the yellow ripe stage, but only increases the rice "sink capacity" (that is, spikelet density) and does not increase the grain filling rate and 1000-grain weight. The reason may be the mutual restriction between the yield components [38]; the larger storage capacity formed in the early stage requires more carbohydrate filling. Therefore, on the basis of conservation tillage, increasing the nutrient supply in the later stage of rice growth could further increase the yield of rice. In our study, spike density, spikelets per panicle, 1000-grain weight, and yield under different fertilization treatments were ranked as SDF > AAF, which is consistent with the findings of a previous study [31]. The deep placement of N has a catalytic effect on roots, which provides more N in the deep root layer, ensures a longer availability of N, promotes plant N uptake, and increases crop yield [38].

5. Conclusions

From our findings, it can be concluded that under no tillage conditions, the positive effect of side deep fertilizing on rice growth and yield was higher than under reduced tillage and conventional tillage conditions. On the whole, side deep fertilizing under conservation tillage not only retains the advantages of conservation tillage for environmental protection, but also saves costs and maintains the high yield of rice, which is of value as a reference for the sustainable development of agriculture.

Author Contributions: Q.-X.W., B.D. and S.-C.J., conceptualization; Q.-X.W., B.D. and J.-Q.Z., methodology; B.D., S.-C.J. and Q.-X.W., formal analysis; Q.-X.W., B.D. and S.-C.J., investigation/writing—original draft/supervision; B.D., S.-C.J. and H.-W.Z., visualization; H.-W.Z., Q.-X.W. and S.-C.J., data curation; Q.-X.W., B.D. and S.-C.J., writing—review and editing; Q.-X.W., J.-Q.Z., B.D. and S.-C.J., project administration. All authors have read and agreed to the published version of the manuscript.

Funding: This research was funded by the National Natural Science Foundation (U21A2039) and National Key Research and Development Plan (2016YFD0300907).

Conflicts of Interest: The authors declare no conflict of interest.

References

1. Waqas, A.; Muhammad, Q.; Huang, J.; Dong, W.; Sun, G.; Liu, K.; Meng, Y.; Tang, A.; Sun, M.; Li, C.; et al. Tillage practices improve rice yield and soil phosphorus fractions in two typical paddy soils. *J. Soils Sedim.* **2020**, *20*, 850–861.
2. Lav, B.; Jagdish, K.L.; Raj, K.G.; Tirolccadre, S.S.A.; Saharawat, M.; Gathala, Y.S.; Pathak, H. Saving of Water and Labor in a Rice–Wheat System with No-Tillage and Direct Seeding Technologies. *Agron. J.* **2020**, *99*, 1288–1296.
3. Das, A.; Lal, R.; Patel, D.; Idapuganti, R.; Layek, J.; Ngachan, S.; Ghosh, P.; Bordoloi, J.; Kumar, M. Effects of tillage and biomass on soil quality and productivity of lowland rice cultivation by small scale farmers in North Eastern India. *Soil Tillage Res.* **2007**, *143*, 50–58. [CrossRef]
4. Denardin, L.G.D.O.; Carmona, F.C.; Veloso, M.G.; Martins, A.P.; Freitas, T.F.S.; Carlos, F.S. No-tillage increases irrigated rice yield through soil quality improvement along time. *Soil Tillage Res.* **2019**, *186*, 64–69. [CrossRef]
5. Ding, Z.; Kheir, A.S.; Ali, O.A.; Hafez, E.; Elshamey, E.A.; Zhou, Z.; Wang, B.; Lin, X.; Ge, Y.; Fahmy, A.F.; et al. Vermicompost and deep tillage system to improve saline-sodic soil quality and wheat productivity. *J. Environ. Manag.* **2020**, *277*, 111388. [CrossRef]
6. Du, B.; Luo, H.W.; He, L.X.; Zheng, A.X.; Chen, Y.L.; Zhang, T.T.; Wang, Z.M.; Hu, L.; Tang, X.R. Deep fertilizer placement improves rice growth and yield in zero tillage. *Appl. Ecol. Environ. Res.* **2019**, *16*, 8045–8054. [CrossRef]
7. Hart, M.R.; Quin, B.F.; Nguyen, M.L. Phosphorus runoff from agricultural land and direct fertilizer effects: A review. *J. Environ. Qual.* **2004**, *33*, 1954–1972. [CrossRef]
8. Du, B.; He, L.; Lai, R.; Luo, H.; Tang, X. Fragrant rice performances in response to continuous zero-tillage in ma-chine-transplanted double-cropped rice system. *Sci. Rep.* **2020**, *10*, 8326. [CrossRef]
9. Huang, M.; Shan, S.; Xie, X.; Cao, F.; Zou, Y. Why high grain yield can be achieved in single seedling ma-chine-transplanted hybrid rice under dense planting conditions? *J. Integrat. Agric.* **2019**, *17*, 1299–1306. [CrossRef]
10. Huang, M.; Zhou, X.; Cao, F.; Xia, B.; Zou, Y. No-tillage effect on rice yield in China: A meta-analysis. *Field Crops Res.* **2018**, *183*, 126–137. [CrossRef]
11. Huang, M.; Zou, Y.; Feng, Y.; Cheng, Z.; Mo, Y.; Ibrahim, M.; Xia, B.; Jiang, P. No-tillage and direct seeding for super hybrid rice production in rice–oilseed rape cropping system. *Eur. J. Agron.* **2011**, *34*, 278–286. [CrossRef]
12. Jiang, M.; Hu, X.; Chunga, J.; Lin, Z.; Fei, R. Does the popularization of agricultural mechanization improve energy-environment performance in China's agricultural sector? *J. Clean. Prod.* **2020**, *276*, 124210. [CrossRef]

13. Jiang, X.; Wright, A.; Wang, J.; Li, Z. Long-term tillage effects on the distribution patterns of microbial biomass and activities within soil aggregates. *CATENA* **2020**, *87*, 276–280. [CrossRef]
14. Lampayan, R.M.; Rejesus, R.M.; Singleton, G.; Bouman, B.A. Adoption and economics of alternate wetting and drying water management for irrigated lowland rice. *Field Crops Res.* **2015**, *170*, 95–108. [CrossRef]
15. Liang, X.; Zhang, H.; He, M.; Yuan, J.; Xu, L.; Tian, G. No-tillage effects on grain yield, N use efficiency, and nutrient runoff losses in paddy fields. *Environ. Sci. Pollut. Res.* **2016**, *23*, 21451–21459. [CrossRef]
16. Liu, X.; Wang, H.; Zhou, J.; Hu, F.; Liu, Y. Effect of N Fertilization Pattern on Rice Yield, N Use Efficiency and Fertilizer–N Fate in the Yangtze River Basin, China. *PLoS ONE* **2016**, *11*, e0166002. [CrossRef]
17. Xin, M.; Yang, Y.; Liu, Z.; Sun, Y.; Peng, X. Yield increasing effect of mechanical topdressing of polymer-coated urea mixed with compound fertilizer in cold area rice. *J. Plant Nutr. Fertil.* **2016**, *33*, 1095–1103.
18. Modather, M.; Azmi, Y.; Nor, M.; Ahmad, S.; Aimrun, W.; Suha, E. Rotary tillage effects on some selected physical properties of fine textured soil in wetland rice cultivation in Malaysia. *Soil Tillage Res.* **2017**, *194*, 104318.
19. Min, J.; Sun, H.; Wang, Y.; Pan, Y.; Kronzucker, H.J.; Zhao, D.; Shi, W. Mechanical side-deep fertilization mitigates am-monia volatilization and nitrogen runoff and increases profitability in rice production independent of fertilizer type and split ratio. *J. Clean. Prod.* **2019**, *316*, 128370. [CrossRef]
20. Pandey, D.; Agrawal, M.; Bohra, J.S. Effects of conventional tillage and no tillage permutations on extracellular soil enzyme activities and microbial biomass under rice cultivation. *Soil Tillage Res.* **2021**, *136*, 51–60. [CrossRef]
21. Ramasamy, S.; Ten Berge, H.F.M.; Purushothaman, S. Yield formation in rice in response to drainage and nitrogen application. *Field Crops Res.* **1997**, *51*, 65–82. [CrossRef]
22. Tang, H.; Li, C.; Shi, L.; Wen, L.; Cheng, K.; Li, W.; Xiao, X. Functional soil organic matter fraction in response to short-term tillage management under the double-cropping rice paddy field in southern of China. *Environ. Sci. Pollut. Res.* **2021**, *28*, 48438–48449. [CrossRef]
23. Thakur, A.; Mandal, K.G.; Mohanty, R.K.; Ambast, S.K. Rice root growth, photosynthesis, yield and water productivity improvements through modifying cultivation practices and water management. *Agric. Water Manag.* **2018**, *206*, 67–77. [CrossRef]
24. Chaikh, A.; Desgranges, C.; Balosso, J. Statistical methods to evaluate the correlation between measured and calculated dose using a quality assurance method in IMRT. *Int. J. Cancer Ther. Oncol.* **2015**, *3*, 3411–3419. [CrossRef]
25. Wang, X.; Qi, J.-Y.; Zhang, X.-Z.; Li, S.-S.; Virk, A.L.; Zhao, X.; Xiao, X.-P.; Zhang, H.-L. Effects of tillage and residue management on soil aggregates and associated carbon storage in a double paddy cropping system. *Soil Tillage Res.* **2019**, *194*, 104339. [CrossRef]
26. Wang, X.; Qi, J.; Liu, B.; Kan, Z.; Zhao, X.; Xiao, X.; Zhang, H. Strategic tillage effects on soil properties and agricultural productivity in the paddies of Southern China. *Land Degrad. Dev.* **2019**, *31*, 1277–1286. [CrossRef]
27. Yao, Y.; Zhang, M.; Tian, Y.; Zhao, M.; Zhang, B.; Zeng, K.; Zhao, M.; Yin, B. Urea deep placement in combination with Azolla for reducing nitrogen loss and improving fertilizer nitrogen recovery in rice field. *Field Crops Res.* **2018**, *218*, 141–149. [CrossRef]
28. Wang, X.; Wang, Y.; Zhang, Y.; Xiang, J.; Zhang, Y.; Zhu, D.; Chen, H. The nitrogen topdressing mode of indi-ca-japonica and indica hybrid rice are different after side-deep fertilization with machine transplanting. *Sci. Rep.* **2021**, *11*, 1–15.
29. Chu, G.; Chen, T.; Wang, Z.; Yang, J.; Zhang, J. Morphological and physiological traits of rice roots and their relationships to yield and nitrogen utilization as influenced by irrigation regime and nitrogen rate. *Agric. Water Manag.* **2021**, *203*, 385–394.
30. Xue, J.-F.; Liu, S.; Chen, Z.-D.; Chen, F.; Lal, R.; Tang, H.-M.; Zhang, H.-L. Assessment of carbon sustainability under different tillage systems in a double rice cropping system in Southern China. *Int. J. Life Cycle Assess.* **2014**, *19*, 1581–1592. [CrossRef]
31. Gulab, S.; Rattan, L.; Ram, S.; Subhash, B.; Anup, D.S.N.B.; Mrinmoy, D.; Jayanta, L.; Poulami, S. Conservation tillage and nutrient management effects on productivity and soil carbon sequestration under double cropping of rice in north eastern region of India. *Ecol. Indic.* **2019**, *105*, 303–315.
32. Yang, L.; Wang, Y.; Kobayashi, K.; Zhu, J.; Huang, J.; Yang, H.; Wang, Y.; Dong, G.; Liu, G.; Han, Y.; et al. Seasonal changes in the effects of free-air CO2enrichment (FACE) on growth, morphology and physiology of rice root at three levels of nitrogen fertilization. *Glob. Chang. Biol.* **2008**, *14*, 1844–1853. [CrossRef]
33. Zhao, Z.; Gao, S.; Lu, C.; Li, X.; Li, F.; Wang, T. Effects of different tillage and fertilization management practices on soil organic carbon and aggregates under the rice–wheat rotation system. *Soil Tillage Res.* **2021**, *212*, 105071. [CrossRef]
34. Zhong, X.; Peng, J.; Kang, X.; Wu, Y.; Luo, G.; Hu, W.; Zhou, X. Optimizing agronomic traits and increasing economic returns of machine-transplanted rice with side-deep fertilization of double-cropping rice system in southern China. *Field Crops Res.* **2021**, *270*, 108191. [CrossRef]
35. Zhong, X.; Zhou, X.; Fei, J.; Huang, Y.; Wang, G.; Kang, X.; Hu, W.; Zhang, H.; Rong, X.; Peng, J. Reducing ammonia vo-latilization and increasing nitrogen use efficiency in machine-transplanted rice with side-deep fertilization in a double-cropping rice system in Southern China. *Agric. Ecosyst. Environ.* **2021**, *306*, 107183. [CrossRef]
36. Zhu, C.; Xiang, J.; Zhang, Y.; Zhang, Y.; Zhu, D.; Chen, H. Mechanized transplanting with side deep fertilization increases yield and nitrogen use efficiency of rice in Eastern China. *Sci. Rep.* **2019**, *9*, 5653. [CrossRef]
37. Zhu, C.; Yang, Y.O.; Diao, Y.; Yu, J.; Luo, X.; Zheng, J.; Li, X. Effects of mechanized deep placement of nitrogen fertilizer rate and type on rice yield and nitrogen use efficiency in Chuanxi Plain, China. *J. Integr. Agric.* **2021**, *20*, 581–592. [CrossRef]
38. Jiang, S.; Wang, J.; Luo, H.; Xie, Y.; Feng, D.; Zhou, L.; Shi, L.; Chen, H.; Xu, Y.; Wang, M. Effect of meteorology and soil fertility on direct-seeded rice (*Oryza sativa* L.) performance in central China. *Appl. Ecol. Environ. Res.* **2019**, *17*, 12397–12406. [CrossRef]

agriculture

MDPI

Article

Reasonable Nitrogen Regime in the Main Crop Increased Grain Yields in Both Main and Ratoon Rice

Qiang Zhang [1,2], Xiangchen Liu [2], Guilong Yu [2], Bin Duan [2], Hao Wang [2], Haiying Zhao [2], Daqing Feng [2], Mengxuan Gu [2] and Lijun Liu [1,*]

1 Jiangsu Key Laboratory of Crop Genetics and Physiology/Jiangsu Co-Innovation Centre for Modern Production Technology of Grain Crops, Yangzhou University, Yangzhou 225009, China; yzdxzhangqiang@163.com
2 Xinyang Academy of Agricultural Sciences, Xinyang 464000, China; gsxlxc@163.com (X.L.); 15294809137@163.com (G.Y.); duanb@163.com (B.D.); wh1872575161@163.com (H.W.); zhy292@126.com (H.Z.); fdqkyk@126.com (D.F.); 15294808253@163.com (M.G.)
* Correspondence: ljliu@yzu.edu.cn; Tel.: +86-133-3886-7665

Abstract: Planting ratoon rice can realize one sowing and two harvests, which is of great significance for improving grain yield. However, the effects of nitrogen (N) regime in the main crop on the grain yield of ratoon rice and the associated physiological mechanisms are not clearly understood. The indica hybrid rice Liangyou 6326 was used, and three N fertilizer levels (100 kg ha^{-1} (low N, LN), 250 kg ha^{-1} (medium N, MN), and 400 kg ha^{-1} (high N, HN)) and four different ratios of basal tillering fertilizer to panicle fertilizer (7:3, 6:4, 5:5, and 4:6) applied to the main crop were designed to investigate their effects on the grain yields of the main and ratoon crops. The results showed that excessive N application rate and panicle N application rate in the main crop was not conducive to the improvement of yield and agronomic nitrogen use efficiency (ANUE) in both seasons. The increased yield in the ratoon crop was attributed to the increase in the regeneration rate. Appropriate increasing of the panicle N application rate was beneficial for increasing the ROA and NSC concentration in the main crop, resulting in an increase in the number, length, and fresh weight of regenerated buds, which caused an improvement in the regeneration rate. However, when excessive panicle N was applied in the main crop, the excessive germination of regenerated buds decreased the length and fresh weight of the regenerated bud and resulted in a decrease in the regeneration rate. These results suggest that in the production of ratoon rice, reasonable N regime in the main crop could increase the yield and ANUE in both seasons.

Keywords: nitrogen rate; nitrogen ratio; yield; regenerative ability

Citation: Zhang, Q.; Liu, X.; Yu, G.; Duan, B.; Wang, H.; Zhao, H.; Feng, D.; Gu, M.; Liu, L. Reasonable Nitrogen Regime in the Main Crop Increased Grain Yields in Both Main and Ratoon Rice. *Agriculture* **2022**, *12*, 527. https://doi.org/10.3390/agriculture12040527

Academic Editors: Chengfang Li and Lijin Guo

Received: 21 February 2022
Accepted: 6 April 2022
Published: 8 April 2022

Publisher's Note: MDPI stays neutral with regard to jurisdictional claims in published maps and institutional affiliations.

1. Introduction

The global population continues to grow and is expected to reach 8.5 billion by 2030 and 10.9 billion by 2100, according to UN's 2019 World Population Prospects. Global food production must be further increased to meet the needs of the growing population [1,2]. Rice (*Oryza sativa* L.) is an important food crop species worldwide, supporting more than half of the global population [3,4]. Given that global urbanization is increasing and water resources are becoming increasingly scarce, there are two main ways to increase global rice production: One is to increase the grain yield per unit area, and the other is to increase the multiple cropping index [5]. Because achieving major breakthroughs in rice yield per unit area is difficult, developing ratoon rice comprising one plant with two harvest periods was an effective way to increase rice yield [6,7]. Ratoon rice is rice planted in such a way that dormant buds in the stubble survive after the main crop is harvested, and subsequently, regenerate into panicles [8]. The use of ratoon rice has been adopted in many countries [9]. For example, in China, the planting area of ratoon rice reached nearly one million ha in 2019, and according to reports, there are more than 3.3 million ha of fields that are suitable for planting ratoon rice in southern China [10]. Therefore, increasing research on ratoon rice and further improving its yield are of great significance to ensure global food security.

The yield of ratoon rice is affected by many factors, such as the height of the stubble [11], water management [12], variety [13], and nitrogen (N) fertilizer management [14], among which N fertilizer management has a critical influence on the yield of ratoon rice. N fertilizer management of ratoon rice mainly involves the application of N fertilizer to the main crop and the ratoon crop (bud fertilizer and seedling fertilizer). Many scholars have researched the effects of N fertilizer management in ratoon season on the growth and development of ratoon crops, and most agree that the application of bud and seedling fertilizers can promote the growth of regenerated buds and improve the regenerative capacity and yield of ratoon crops [15,16]. However, there are relatively few studies about the effects of N fertilizer management in the main crop on the growth and development of ratoon crops, and the results have been inconsistent or even contradictory. For example, Huang et al. (2022) reported that increasing N application rate in the late growth stage of the main crop could improve the effective tillering percentage, leaf area index, canopy light interception rates, and transport rate of stem and sheath in the main crop, leading to an increase in yields of both seasons [17]. Properly postponed N application in the main crop was beneficial to increasing the root activity and promoting the growth of rice, resulting in increasing the yields of both seasons [18]. Liu et al. (2014) found that increasing the amount of panicle N fertilizer in the main crop was beneficial for increasing the yields of the main and ratoon crops [19]. However, Chen et al. (2010) suggested that a moderate amount of delayed N fertilizer could increase the yield of the main crop but had little effect on the yield of the ratoon crop [20]. Wang et al. (2019) indicated that the N application rate in the main crop had little effect on the yield of the ratoon crop [14].

However, there is a lack of in-depth research on the effects of N regime in the main crop on the growth and development of regenerated buds and the regeneration rate. In addition, the N regime in the main crop on the yield of the ratoon crop under different N application rate have not been reported so far. Since the N application rates vary greatly in different areas in the field production of ratoon rice, it is necessary to study the effects of N regime in the main crop on the yield of ratoon rice under different N application rates. In this study, N was applied at three different levels in conjunction with four panicle N ratios to the main crops to investigate the effects of the N regime in the main crop on the yield of the main and ratoon crops and to determine rational N management practices for the high-yield cultivation of ratoon rice.

2. Materials and Methods

2.1. Experimental Site and Materials

The field experiment was conducted in a farmer's field at Xinyang (32°07′ N, 114°05′ E), Henan Province, North China, in 2018 and repeated in 2019. The soil in the experimental field was a clay loam with contents of organic carbon of 11.4 g kg^{-1}, total N of 54.3 mg kg^{-1}, Olsen-phosphorus (P) of 9.7 mg kg^{-1}, and available potassium (K) of 79.8 mg kg^{-1}. Two years' weather information during the rice-growing periods (from March to November) is shown in Figure S1.

The tested rice variety was Liangyou 6326, an indica two-line hybrid rice variety that is a result of the combination of Xuan 69S and Zhongxian WH26. This variety was released by the Xuancheng Agricultural Science Research Institute and passed the National Crop Variety Approval in 2007 [National Approved Rice 2007013].

2.2. Experimental Design

The experiments were arranged in accordance with a split-plot design, with four replicates (plot size: 4 × 5 m). The main plots were divided according to the rate of total N applied to the main crop, and the three N application rates were low N (LN), medium N (MN), and high N (HN), corresponding to 100 kg ha^{-1}, 250 kg ha^{-1}, and 400 kg ha^{-1}, respectively. The N rates of MN and HN were based on the local standard for ratoon rice cultivation "Technique rule for planting ratoon rice in the south of Henan Province" (DB41/T 1564-2018) and "Theory and technology of rice precise and quantitative cultivation" [21]. Subplots corresponded to different ratios of basal tillering fertilizer to panicle fertilizer application to the main crop. The ratios were 7:3 (PN30), 6:4 (PN40), 5:5 (PN50), and 4:6 (PN60). The details of the N treatments are summarized in Table 1. In addition, 0 kg ha^{-1} N rate was set in both the main and ratoon crops to investigate the yield of N free. The seeds were sown in a greenhouse on March 4 in both 2018 and 2019. The seedlings were subsequently transplanted to the field on April 8 in both years, with a hill spacing of 0.33 × 0.15 m and 2 seedlings per hill. Before transplanting, all the plots were plowed and puddled. To prevent the flow of fertilizer between neighboring plots, the plots were separated by a 40 cm wide ridge created by a plastic film inserted into the soil to a depth of 30 cm. The technical drawing is shown in Figure S2.

Table 1. Nitrogen application rates (kg N ha^{-1}) in the main and ratoon crops.

N Rates in Main Crop			Treatment Code	N Rates in Ratoon Crop		
Total N Rate	**Basal and Tillering Fertilizer**	**Panicle Fertilizer**		**Total N Rate**	**Bud Fertilizer**	**Seedling Fertilizer**
100 (LN)	70 (70)	30 (30)	PN30	200	50	150
	60 (60)	40 (40)	PN40	200	50	150
	50 (50)	50 (50)	PN50	200	50	150
	40 (40)	60 (60)	PN60	200	50	150
250 (MN)	175 (70)	75 (30)	PN30	200	50	150
	150 (60)	100 (40)	PN40	200	50	150
	125 (50)	125 (50)	PN50	200	50	150
	100 (40)	150 (60)	PN60	200	50	150
400 (HN)	280 (70)	120 (30)	PN30	200	50	150
	240 (60)	160 (40)	PN40	200	50	150
	200 (50)	200 (50)	PN50	200	50	150
	160 (40)	240 (60)	PN60	200	50	150

Note: LN, low N rate (the total amount of N applied was 100 kg ha^{-1}); MN, medium N rate (the total amount of N applied was 250 kg ha^{-1}); HN, high N rate (the total amount of N applied was 400 kg ha^{-1}). PN30, PN40, PN50 and PN60, correspond to N application ratio of basal tillering fertilizer to panicle fertilizer of 7:3, 6:4, 5:5 and 4:6, respectively. The numbers in () indicate the percentage of N application to the total N application of the main crop.

The N fertilizer used was urea, with an N content of 46%. The ratio of basal fertilizer to tiller fertilizer was 7:3. Basal fertilizer was applied one day before transplantation, and tiller fertilizer was applied 5 days after transplantation. Panicle fertilizer was applied twice at panicle initiation and at the beginning of spikelet differentiation, accounting for 60% and 40%, respectively. The same amounts of P (as calcium superphosphate, 12% P_2O_5) and K (as potassium chloride, 60% K_2O) were applied in both years. All P and K were applied as basal fertilizer, and the amounts of P and K were 937.5 kg ha^{-1} and 187.5 kg ha^{-1}, respectively. At 15 days after heading of the main crop, 50 kg N ha^{-1} was applied to each plot as bud fertilizer. After the main crop was harvested, 150 kg ha^{-1} N was applied as a seedling fertilizer. Although the bud fertilizer was applied during the growth period of the main crop, it had a greater impact on the yield of the ratoon crop, so the bud fertilizer is usually included in the N management of ratoon season [22]. After transplanting, the field was kept flooded for 35 days, and then the water was drained. The field was reflooded at the jointing stage and drained again 7 days before the main crop was harvested. A water layer 1 to 3 cm above the soil surface was maintained during the entire ratoon crop growing season. Weeds were removed by hand. Diseases and insects were controlled by chemicals to avoid yield loss.

The main crop was harvested by hand on 12 August in both years with a stubble height of 0.35 m.

2.3. *Sampling and Measurements*

2.3.1. Root Oxidation Activity (ROA)

At full heading, at the 15th day after heading, and at maturity of the main crop, after the mean stem number in each plot was recorded, representative plants from 10 hills were selected to measure the ROA in each plot. A block of soil (20 × 20 × 20 cm) around each individual hill was removed, and after rinsing with running water (Figure S3), the root subsamples were taken to measure the ROA via oxidation of alpha-naphthylamine (α-NA) according to the methods of Ramasamy et al. [23].

2.3.2. Nonstructural Carbohydrate (NSC) Accumulation in the Stem and Leaf

At the same time as the procedures described above, after the mean stem number in each plot was recorded, representative plants from 10 hills (0.495 m^2) were selected in each plot and removed. The plants were separated into stems (culms + sheaths), leaves, and panicles. All the samples were first dried at 105 °C for 30 min and then dried to constant weight at 75 °C in an oven. The dried samples were subsequently crushed and passed through a 0.15 mm sieve to ultimately determine the NSC concentration [24]. Briefly, 100 mg dry sample was placed into 15 mL distilled water and boiled for 20 min. After filtration and constant volume, 1 mL of the content was taken, and 5 mL of anthrone reagent was added. Then spectrophotometer (UV-1800, Shimadzu, Tokyo, Japan) was used to measure the soluble sugar and starch content. According to the above method, at harvest of the main crop, 10 representative stubbles were taken from each plot to determine the NSC concentration.

2.3.3. Regenerated Buds

On the 15th day after heading, the 25th day after heading, maturity, and the 7th day after the main crop was harvested, 5 hills were selected from each plot according to the average level of seedlings in the whole field. The fresh weight and length of living regenerated buds at each internode were measured, and the number of living and dead regenerated buds per stem was investigated (buds larger than 1 cm were included) as described by Xu et al. [25] and Zhang et al. [26] (Figure S4).

2.3.4. Yield and Its Components

Plants from 1 m^2 in each plot were harvested to measure the spikelet number per panicle and the panicles number per square meter. The filled grains were separated by submerging all spikelets in tap water after threshing. In addition, the grain yields of the main and ratoon crops were measured by hand-harvesting rice plants from 5 m^2 in each plot (the moisture was adjusted to 14%).

2.3.5. Statistical Analysis

The regeneration rate was calculated as the number of panicles per square meter in the ratoon crop/the number of panicles per square meter in the main crop.

Agronomic nitrogen use efficiency (ANUE) = (Grain yield − Grain yield of 0 N)/N application rate.

The minimum sample size of each indicator in this experiment is 4. Analysis of variance (ANOVA) was performed using SPSS (version 17.0; SPSS, Inc., Chicago, IL, USA) to detect the effects of year and variety. The means were subjected to least significant difference (LSD) tests at $p < 0.05$ ($LSD_{0.05}$) and $p < 0.01$ ($LSD_{0.01}$). The number, length, and fresh weight of regenerated buds, regeneration rate, ROA and NSC content in stem are continuous variable variables. Therefore, this experiment uses Pearson's correlation and regression analysis to analyze their relationship.

3. Results

3.1. Differences in Experimental Factors

The yields of the main and ratoon crops and their components did not significantly differ between years but did significantly differ among N application rates and ratios (Table 2). The length, fresh weight, and number of regenerated buds per stem (unless otherwise specified, the following regenerated buds refer to living buds) at maturity of the main crop, ROA and NSC concentration in the stems and leaves on the 15th day after heading, NSC concentration in the stubble, and the regeneration rate of ratoon crop were not significantly different between years but were significantly different among N rates and ratios (Table 3). Since year was not a significant factor in any experiment, this paper mainly used the mean values of the two years for analysis purposes.

Table 2. Analysis-of-variance of F-values of the grain yield and yield components of the main and ratoon crops between/among years, N regimes.

Source of Variation		df	Panicle Number	Spikelet per Panicle	Filled Grain Rate	1000-Grain Weight	Grain Yield
Main crop	Y	1	NS	NS	NS	NS	NS
	N	2	159.72 **	150.20 **	131.57 **	NS	114.80 **
	R	3	24.62 **	24.63 **	NS	NS	53.75 **
	Y × N	2	NS	NS	NS	NS	NS
	Y × R	3	NS	NS	NS	NS	NS
	N × R	6	NS	NS	NS	NS	NS
	Y × N × R	6	NS	NS	NS	NS	NS
Ratoon crop	Y	1	NS	NS	NS	NS	NS
	N	2	359.22 **	229.83 **	NS	NS	175.72 **
	R	3	42.36 **	NS	NS	NS	7.10 **
	Y × N	2	NS	NS	NS	NS	NS
	Y × R	3	NS	NS	NS	NS	NS
	N × R	6	50.04 **	NS	NS	NS	7.10 **
	Y × N × R	6	NS	NS	NS	NS	NS

Note: Y, years. N, nitrogen rate. R, ratio of nitrogen application. ** represents a significant difference at the 1% level according to LSD tests, NS represents no significant difference at the 5% level according to LSD tests.

Table 3. Analysis-of-variance of F-values of key indices such as the growth and development of the main crop after full heading, growth of regenerative bud, and regenerative ability between/among years, N regime.

Source of Variation	df	Indices of Regenerated Bud at MS			ROA at 15DAH	NSC Concentration in Stem and Leaf at 15DAH	NSC Concentration in Stubble	Regeneration Rate
		Length	Fresh Weight	Number				
Y	1	NS	NS	NS	NS	NS	NS	NS
N	2	118.4 **	40.23 **	239.27 **	9.93 *	52.73 **	50.12 **	13.68 **
R	3	59.29 **	29.88 **	298.07 **	15.61 **	53.95 **	64.45 **	70.91 **
Y × N	2	NS	NS	NS	NS	NS	NS	NS
Y × R	3	NS	NS	NS	NS	NS	NS	NS
N × R	6	20.21 **	14.33	6.44 **	NS	NS	NS	29.25 **
Y × N × R	6	NS	NS	NS	NS	NS	NS	NS

Note: Y, years. N, rate nitrogen application. R, ratio of nitrogen application. NSC, nonstructural carbohydrate. ROA, root oxidation activity. DAH, days after heading of the main crop. MS, maturity stage of the main crop. ** represents a significant difference at the 1% level according to LSD test, * represents a significant difference at the 5% level according to LSD test, NS represents no significant difference at the 5% level according to LSD test.

3.2. Yield and Its Components in the Two Seasons

The yield of the main crop decreased in the order MN > HN > LN under different N levels. That is, a high N rate was not conducive to increasing the yield of the main crop. A significant decrease in the filled grain rate was the main reason for the decrease in yield in the main crop under HN (Figure 1A–C, and Table S1). Under the same N level, with increasing panicle N rate, the yield of the main crop decreased under LN, first increased and then decreased under MN (peaking at PN40) and increased under HN.

There was no significant difference in yield under MN and HN in the ratoon crops, but their yields were higher than that under LN. With the increasing proportion of panicle N fertilizer under the LN level, the yield increased. However, under MN and HN, the yield first increased and then decreased, and the highest yields were obtained at PN50 and PN40 (Figure 1D–F). According to the analysis of yield components of the ratoon crops, there were no significant differences in the number of spikes per panicle, filled grain rate, or 1000-grain weight among the different treatments, but there were significant differences in panicle numbers, which was the main reason for the changes in the yield of the ratoon crops (Table 4). Correlation analysis showed that there was a linear correlation between yield and panicle number of the ratoon crops and the coefficient of determination was high (Figure S5), indicating that an increased panicle number was the key to increasing the yield of the ratoon crops.

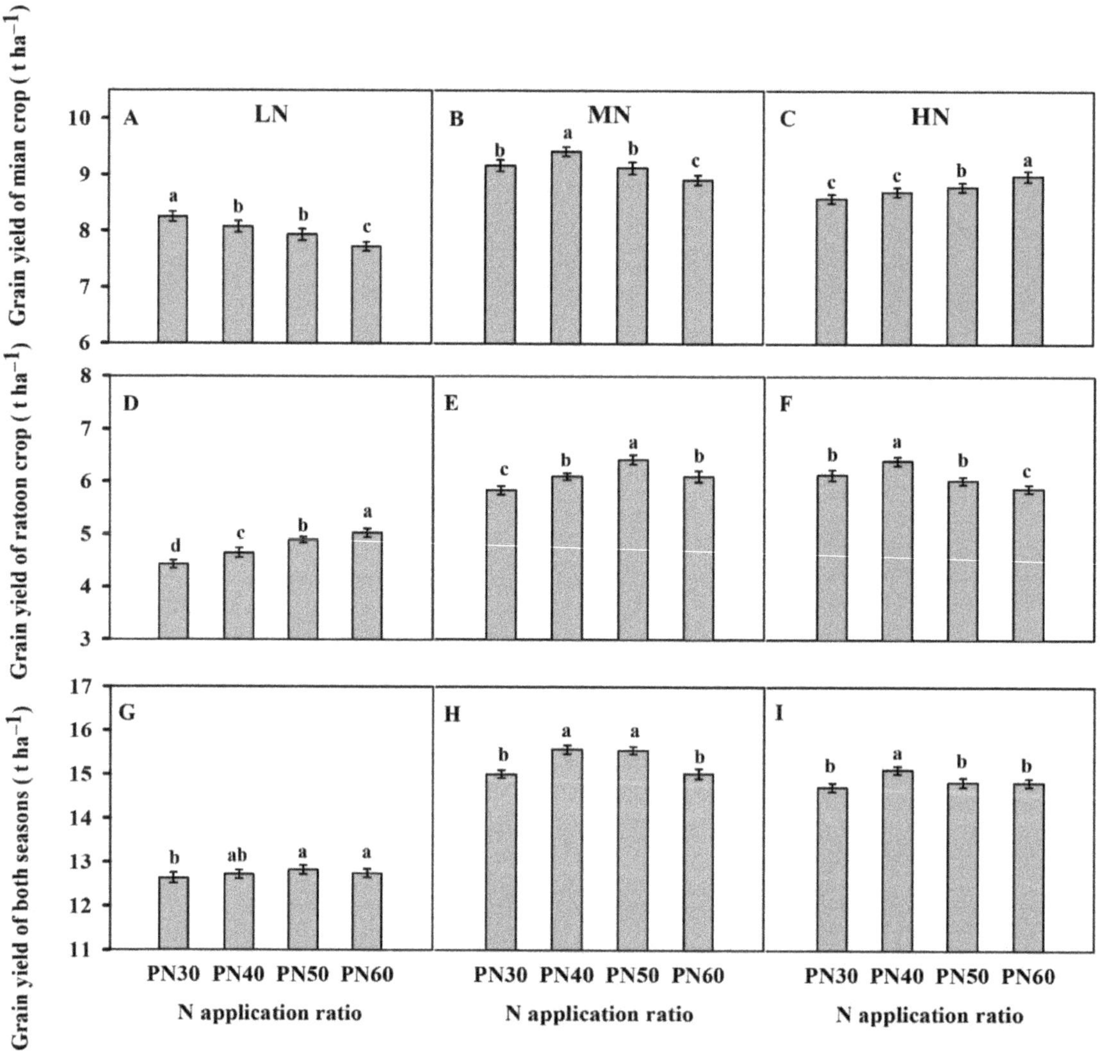

Figure 1. Effects of N regime in the main crop on the yield of the main crop (**A–C**), ratoon crop (**D–F**), and both seasons (**G–I**). Error bars are ±SE. Note: LN, low N rate (the total amount of N applied was 100 kg ha^{-1}); MN, medium N rate (the total amount of N applied was 250 kg ha^{-1}); HN, high N rate (the total amount of N applied was 400 kg ha^{-1}). PN30, PN40, PN50, and PN60 correspond to N application ratios of basal tillering fertilizer to panicle fertilizer of 7:3, 6:4, 5:5, and 4:6, respectively. The different lowercase letters at the same N application rate indicate significant differences at the 5% probability level according to LSD tests.

In terms of the total yield of the main crop and ratoon crop, the highest yields under the LN level were obtained at PN50 and PN60, both of which were 12.8 t ha^{-1}. The highest yield under the MN level was obtained at PN40 and PN50 (15.6 t ha^{-1}), and that under the HN level was obtained at PN40 (15.1 t ha^{-1}) (Figure 1G–I). These results indicate that N application to the main crop at a reasonable rate and ratio had important effects on the yield of the main crop and ratoon crop in both seasons.

Table 4. Effects of N regime in the main crop on the yield components of the ratoon crop.

Treatment		Panicles (10^4 ha^{-1})	Spikelet per Panicle	Filled Grain Rate (%)	1000-Grain Weight (g)
LN	PN30	296.8 g	58.2 b	88.1 a	29.0 a
	PN40	311.2 f	57.9 b	88.3 a	29.2 a
	PN50	328.6 e	58.1 b	87.7 a	29.2 a
	PN60	338.9 d	57.9 b	87.8 a	29.1 a
	Mean	318.8 B	58.0 B	88.0 A	29.1 A
MN	PN30	354.4 c	64.1 a	88.1 a	29.2 a
	PN40	371.9 b	64.1 a	88.2 a	29.2 a
	PN50	389.3 a	64.2 a	88.2 a	29.2 a
	PN60	368.3 b	64.3 a	88.1 a	29.3 a
	Mean	371.0 A	64.1 A	88.1 A	29.2 A
HN	PN30	371.0 b	65.7 a	87.8 a	28.8 a
	PN40	386.2 a	66.0 a	87.8 a	28.6 a
	PN50	365.9 b	65.6 a	87.7 a	28.7 a
	PN60	350.1 c	66.0 a	87.9 a	28.9 a
	Mean	368.3 A	65.8 A	87.8 A	28.7 A

Note: LN, low N rate (the total amount of N applied was 100 kg ha^{-1}); MN, medium N rate (the total amount of N applied was 250 kg ha^{-1}); HN, high N rate (the total amount of N applied was 400 kg ha^{-1}). PN30, PN40, PN50, and PN60, correspond to N application ratio of basal tillering fertilizer to panicle fertilizer of 7:3, 6:4, 5:5, and 4:6, respectively. Data in a column followed by different lower-case letters indicate significant differences at the 5% probability level according to the LSD test. Means followed by different upper-case letters indicate significant differences between the three nitrogen fertilizer levels at the 5% probability level according to the LSD test.

3.3. Regeneration Rate and Its Relationship with Panicle Number and Yield in the Ratoon Crop

The regeneration rate reflects the regeneration ability of ratoon rice, and this factor and the number of mother stems together determine the panicle number of the ratoon crop. The regeneration rate showed a trend of first increasing and then decreasing with increasing N fertilizer levels (Figure 2). As the panicle N application rate increased, the regeneration rate increased under the LN level, while under MN and HN, it first increased and then decreased. The change trend was similar to that of panicle number in the ratoon crop (Table 4), and the correlation analysis showed that compared with the number of mother stems, the regeneration rate had a greater correlation with the panicle number and yield of the ratoon crop (Table 5), indicating that the change in regeneration rate was the main reason for the changes in the panicle number and yield of the ratoon crop.

Table 5. Effects of N regime in the main crop on the yield components of the ratoon crop.

Parameters	N Rate	Panicle Number	Yield
Number of mother stem	LN	−0.9986 **	−0.9927 **
	MN	−0.4694	−0.5729
	HN	0.6810	0.5474
Regeneration rate	LN	0.9998 **	0.9947 **
	MN	0.8707	0.9220 *
	HN	0.7989	0.8374

Note: LN, low N rate (the total amount of N applied was 100 kg ha^{-1}); MN, medium N rate (the total amount of N applied was 250 kg ha^{-1}); HN, high N rate (the total amount of N applied was 400 kg ha^{-1}). ** represents the significant difference at the 1% level according to LSD test, * represents a significant difference at the 5% level according to LSD test.

3.4. Regenerated Buds and Its Relationship with Regeneration Rate

3.4.1. Number, Length, and Fresh Weight of Regenerated Buds

The total number of regenerated buds per unit area (including living and dead regenerated buds) increased from the 15th day after heading to the 7th day after harvesting of the main crop, but due to the increase in the number of dead regenerated buds, the number of living regenerated buds per unit area began to decrease after the main crop was harvested (Figure S6). During the growth of regenerative buds, due to the different growth rates of regenerative buds in different internodes, the regenerative buds that grow slowly are likely to die [9,27]. Compared with the dead bud, the living bud can better reflect the regeneration ability, so this paper mainly focuses on the growth of the living buds. Before the main crop was harvested, the number of regenerated buds per stem increased with increasing N application rate and proportion of panicle N fertilizer (Figure 3A–C). On the 7th day after harvesting, there was no significant difference in the number of regenerated buds per stem under MN and HN levels, and both of them were higher than that under LN level. With the increasing proportion of panicle N fertilizer, the number of regenerated buds per stem on the 7th day after harvest increased under the LN level but first increased and then decreased under the MN and HN levels, with the greatest values occurring at PN50 and PN40, respectively.

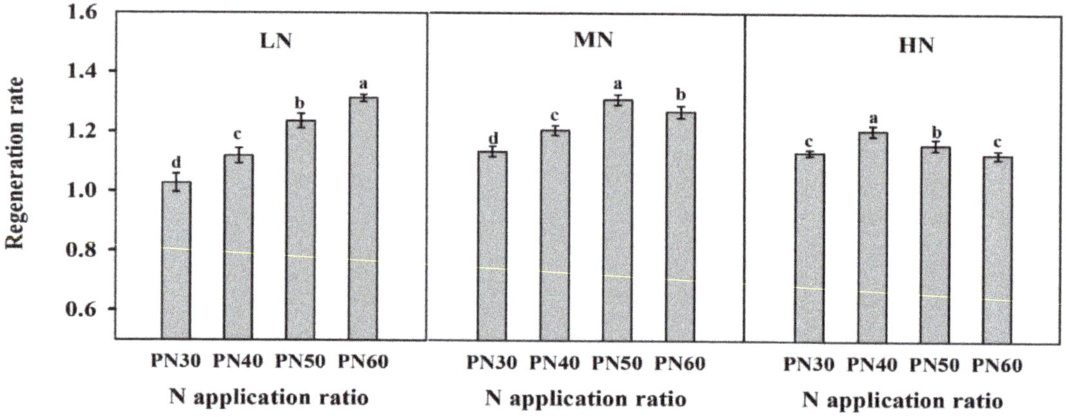

Figure 2. Effects of N rate and the ratio of the main crop on the regeneration rate. Error bars are ± SE. Note: LN, low N rate (the total amount of N applied was 100 kg ha^{-1}); MN, medium N rate (the total amount of N applied was 250 kg ha^{-1}); HN, high N rate (the total amount of N applied was 400 kg ha^{-1}). PN30, PN40, PN50, and PN60 correspond to N application ratios of basal tillering fertilizer to panicle fertilizer of 7:3, 6:4, 5:5, and 4:6, respectively. The different lowercase letters at the same growth stage indicate significant differences at the 5% probability level according to the LSD test.

Under different N levels, the length and fresh weight of regenerated buds followed the order MN > HN > LN (Figure 3D–I). With the increasing proportion of panicle N fertilizer, the length and fresh weight of regenerated buds increased under the LN level and first increased and then decreased under the MN and HN levels, with the greatest values occurring at PN50 and PN40, respectively. At maturity of the main crop, the number of regenerated buds per stem was nonlinearly positively correlated with the length and fresh weight of regenerated buds (Figure S7). The development level of regenerated buds gradually improved with an increasing number of regenerated buds per stem and decreased with excess germination of the regenerated buds.

3.4.2. Correlations between the Regeneration Rate and the Length, Fresh Weight, and Number of Regenerated Buds

The regeneration rate was linearly correlated with the length and fresh weight of regenerated buds (Figure 4B,C), but was nonlinearly correlated with the number of regenerated buds (Figure 4A). The regeneration rate first increases and then decreases with an increase in the number of regenerated buds per stem.

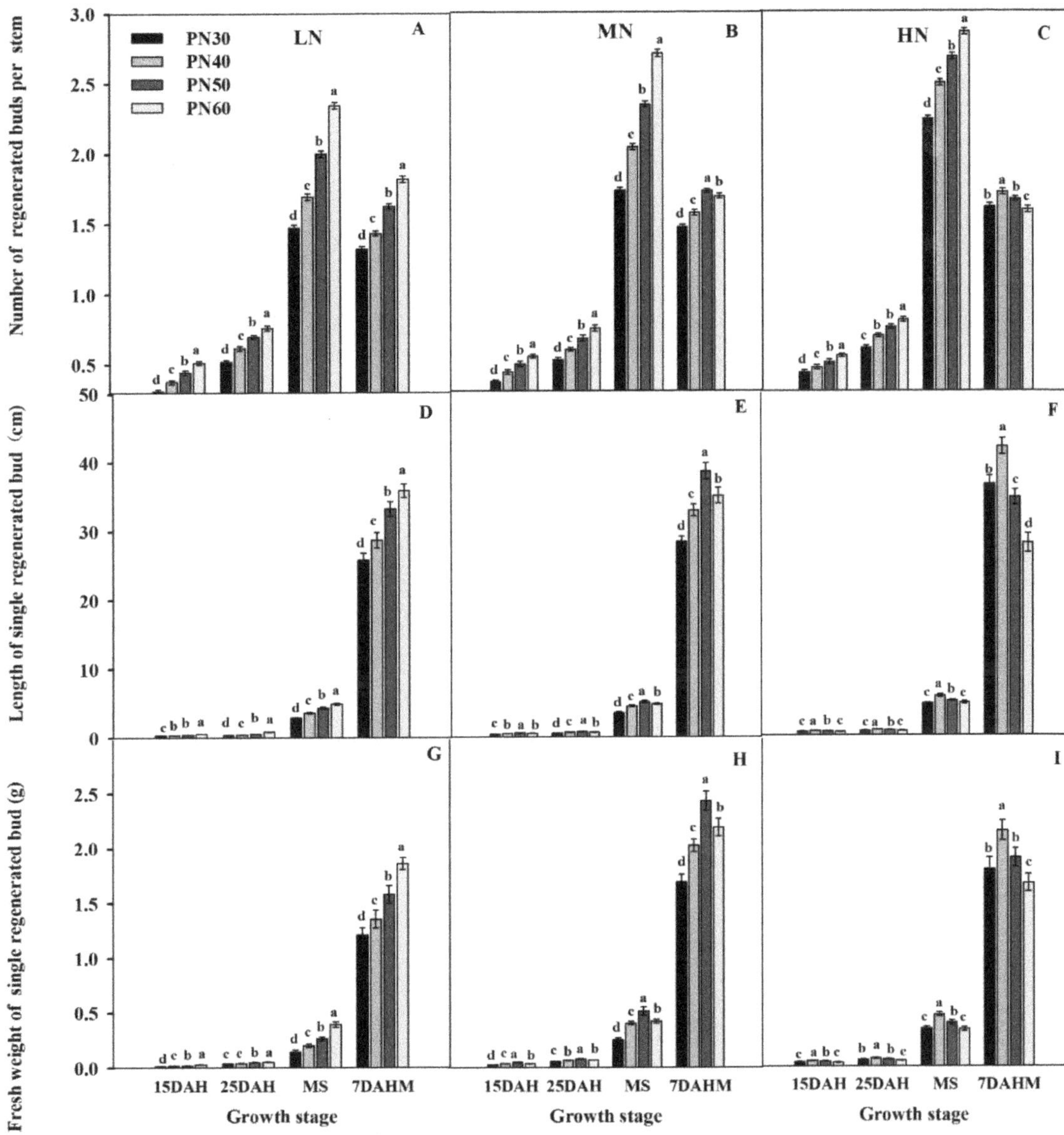

Figure 3. Effects of N rate and the ratio of the main crop on the number (**A–C**), length (**D–F**), and fresh weight (**G–I**) of regenerated buds. Error bars are ± SE. Note: LN, low N rate (the total amount of N applied was 100 kg ha^{-1}); MN, medium N rate (the total amount of N applied was 250 kg ha^{-1}); HN, high N rate (the total amount of N applied was 400 kg ha^{-1}). PN30, PN40, PN50, and PN60 correspond to N application ratios of basal tillering fertilizer to panicle fertilizer of 7:3, 6:4, 5:5, and 4:6, respectively. DAH, days after heading of the main crop. MS, maturity stage of the main crop. DAHM, days after harvest of the main crop. The different lowercase letters at the same growth stage indicate significant differences at the 5% probability level according to the LSD test.

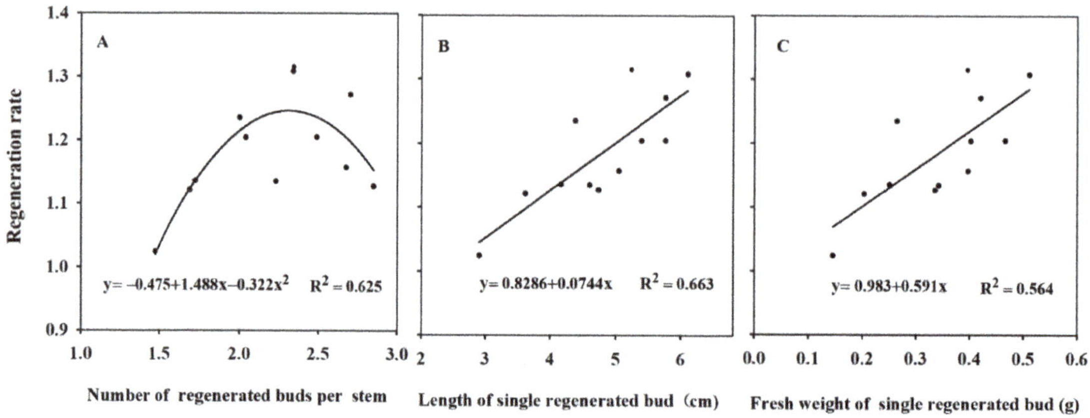

Figure 4. Relationships between the number (**A**), length (**B**), and fresh weight (**C**) of regenerated buds at maturity of the main crop and the regeneration rate.

3.5. Effects of N Regime on the ROA and NSC Concentration in the Stem and Leaf of the Main Crop

Under different N levels, the ROA at different growth stages decreased in the order MN > HN > LN (Figure 5A–C). Under the same N level, ROA at different growth stages increased with the increasing proportion of panicle N fertilizer, and this trend was most obvious at the maturity stage. At the maturity stage in the main crop, compared with that at PN30, the ROA at PN60 increased by 51.4%, 44.4%, and 43.1% under the LN, MN, and HN levels, respectively.

After full heading of the main crop, the NSC concentration in the stems and leaves decreased. The NSC concentration in stems and leaves and in stubble decreased in the order HN > MN > LN (Figure 5D–I). However, under the same N level, the concentration improved with an increasing proportion of panicle N fertilizer. At the maturity stage, compared with those at PN30, the stems and leaves NSC concentration at PN60 increased by 40.1%, 32.2%, and 33.8% under the LN, MN, and HN levels, respectively, and the stubble NSC concentrations increased by 55.1%, 35.2%, and 41.3%.

3.6. Relationships between the ROA and NSC Concentration in Stem and Leaf of the Main Crop and Regeneration Rate of the Ratoon Crop

The NSC concentration in the stems and leaves and the ROA in rice plants after heading of the main crop were linearly correlated with the maximum number of regenerated buds per stem but nonlinearly correlated with the length and fresh weight of regenerated buds and the regeneration rate (Figures 6 and 7). These results indicated that the number, length, fresh weight, and regeneration rate of regenerated buds increased with increasing NSC concentration in stems and leaves and increasing ROA. However, when the NSC concentration in stems and leaves and the ROA were too high, though the regenerated buds continued to increase, the development level of regenerated buds and regeneration rate decreased. Compared with ROA, the NSC concentration in stems and leaves had a higher coefficient of determination with the number, length, fresh weight, and regeneration rate of regenerated buds. Compared with the values at the full heading stage, the NSC concentration in stems and leaves and ROA on the 15th day after heading and at maturity were more strongly correlated with the number, length, fresh weight, and regeneration rate of regenerated buds.

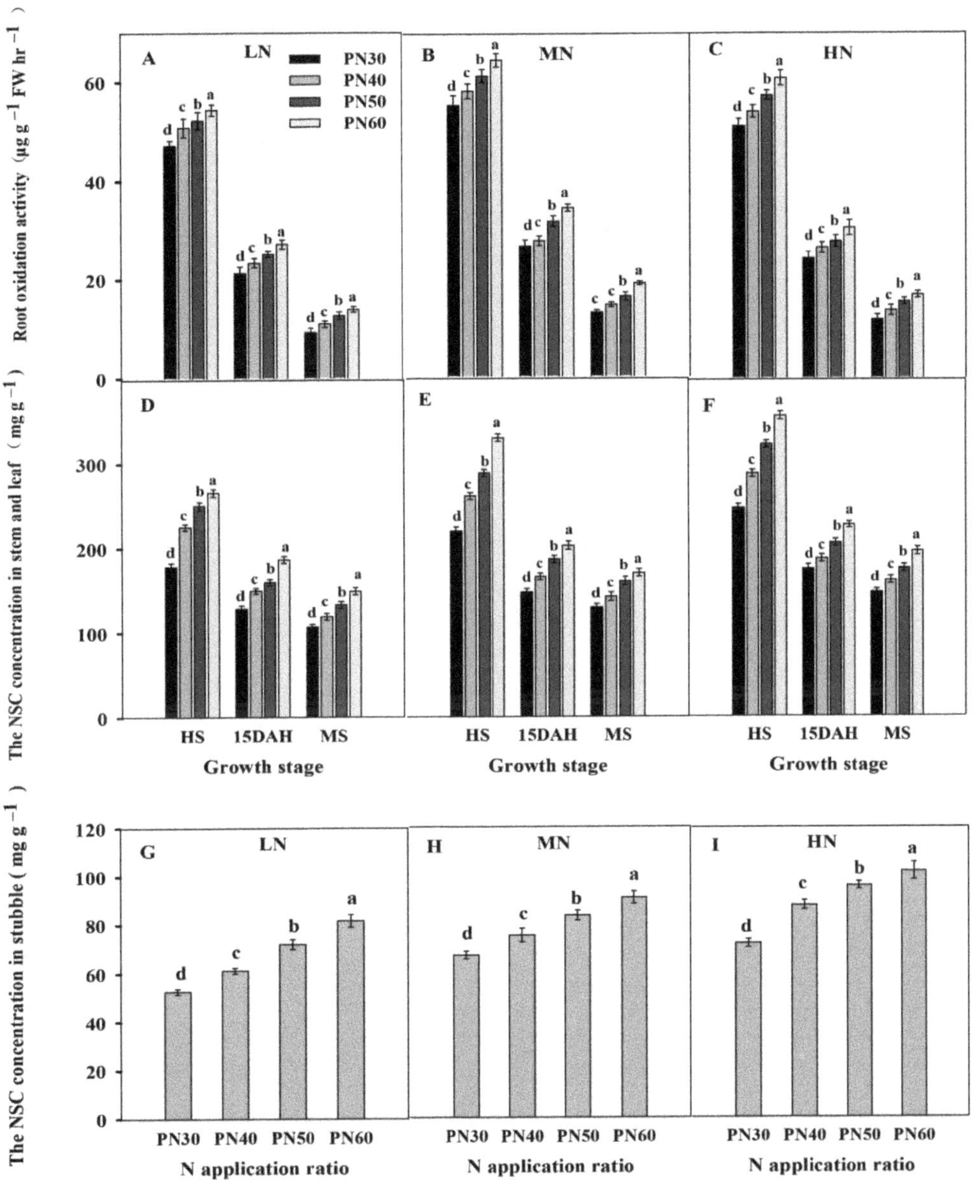

Figure 5. Effects of N rate and the ratio of the main crop on root oxidation activity (**A–C**), NSC concentration in the stem and leaf (**D–F**) after heading of the main crop and in the stubble (**G–I**). Error bars are ± SE. Note: LN, low N rate (the total amount of N applied was 100 kg ha^{-1}); MN, medium N rate (the total amount of N applied was 250 kg ha^{-1}); HN, high N rate (the total amount of N applied was 400 kg ha^{-1}). PN30, PN40, PN50, and PN60 correspond to N application ratios of basal tillering fertilizer to panicle fertilizer of 7:3, 6:4, 5:5, and 4:6, respectively. HS, full heading stage of the main crop. DAH, days after heading of the main crop. MS, maturity stage of the main crop. The different lowercase letters at the same growth stage indicate significant differences at the 5% probability level according to the LSD test.

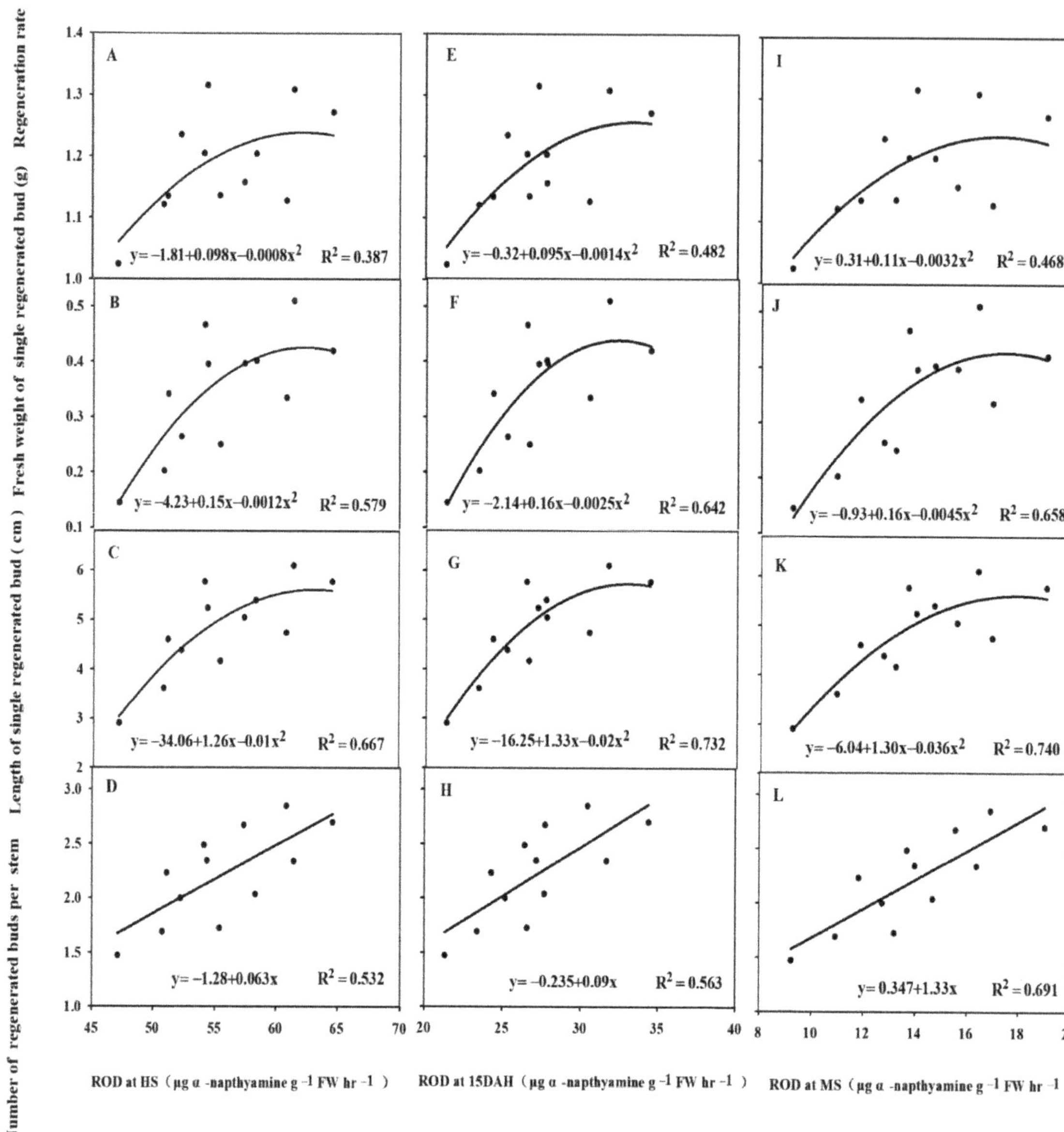

Figure 6. Relationships between root oxidation activity in the main crop and number (**D**,**H**,**L**), length (**C**,**G**,**K**), and fresh weight of regenerated buds (**B**,**F**,**J**) at maturity of main crop and regeneration rate (**A**,**E**,**I**). Note: HS, full heading stage of the main crop. DAH, days after heading of the main crop, MS, maturity stage of the main crop. ROA, root oxidation activity.

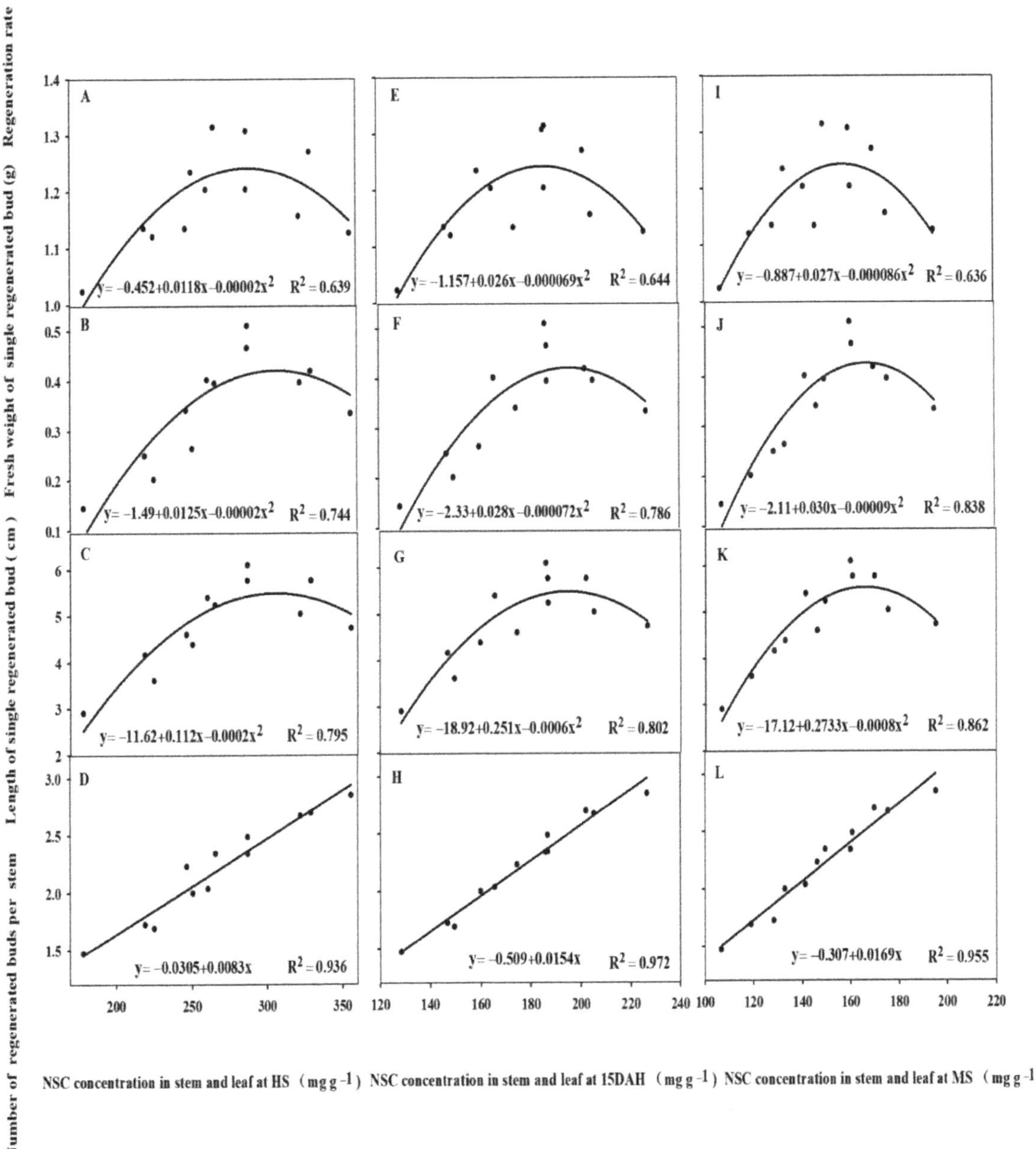

Figure 7. Relationships between NSC concentration of stem and leaf in the main crop and number (**D**,**H**,**L**), length (**C**,**G**,**K**), and fresh weight of regenerated buds (**B**,**F**,**J**) at maturity of main crop and regeneration rate (**A**,**E**,**I**). Note: HS, full heading stage of the main crop. DAH, days after heading of the main crop, MS, maturity stage of the main crop. ROA, root oxidation activity.

The stubble NSC concentration was nonlinearly correlated with the regeneration rate. With increasing stubble NSC concentration, the regeneration rate first increased and then decreased (Figure 8).

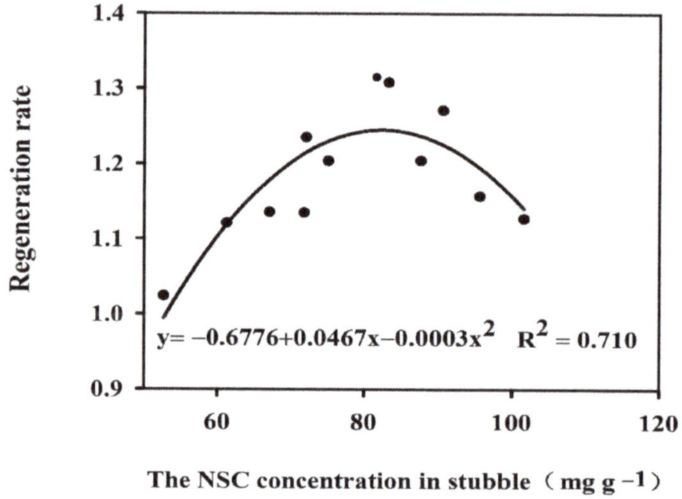

Figure 8. Relationships between stubble NSC concentration and regeneration rate.

3.7. Effects of N Regime in the Main Crop on the Nitrogen Use Efficiency in Main and Ratoon Corp

The ANUE in the main crop decreased with the N application rate in the main crop increased. As the panicle N application rate increased in the main crop, the ANUE in the main crop decreased under LN, increased under HN, first increased and then decreased under MN. In the ratoon crop, there was no difference in ANUE between HN and MN, but they were all higher than LN. And the ANUE increased under LN, first increased and then decreased under MN and HN with the panicle N application rate increased in the main crop. In terms of the ANUE in both seasons, the ANUE showed the order MN > LN > HN. And as the panicle N application rate increased in the main crop, the ANUE first increased and then decreased under LN, MN, and HN. The results showed that excessive N application rate and panicle N application rate in the main crop were not conducive to the increase of ANUE in both seasons.

4. Discussion

4.1. Reasonable N Regime in the Main Crop Improved the Growth and Development of Regenerated Buds

There are few reports on the effect of N regime in the main crop on the growth and development of regenerated bud. The growth and development of regenerated buds are usually reflected by their number, length, and fresh weight. The number, length, and fresh weight were closely related to the NSC content in stems and ROA in plants after full heading in the main crop [28–30]. Before the main crop was harvested, the growth of regenerated buds depended on the nutrients stored in the stems and leaves of the main crop [31]. Therefore, under a high stubble cutting height, the NSC concentration in stems and leaves after heading in the main crop was positively correlated with the number and length of regenerated buds [30]. Zhang et al. (2005) reported that increasing the ROA in plants after full heading in the main crop could promote the growth of regenerated buds [27]. In this paper, the NSC concentration in the stems and leaves and ROA in the main crop were positively correlated with the number, length, and fresh weight of regenerated buds (Figures 6 and 7).

Previous studies reported that appropriately increasing the panicle N application rate reduced the number of ineffective tillers and increased the percentage of productive tillers, leading to an increase in the dry matter weight per stem [32–34]. The canopy transmittance and root activity were improved by applying more N in the late stage of rice [17,35]. Our result is consistent with those of our predecessors. Increasing the panicle N application rate increased the NSC concentration in stems and leaves and ROA after heading of the main crop (Figure 5). Therefore, increasing panicle N application rate and N application rate could improve the number, length, and fresh weight of regenerated bud by improving NSC concentration in stems and leaves and ROA (Figures 6 and 7). However, our study also showed that the length and fresh weight of regenerated buds were nonlinearly correlated with the ROA and NSC in the main crop (Figures 6 and 7). With the increase of panicle N application rate and N application rate, the length and fresh weight first increased and then decreased. The decrease in length and fresh weight of regenerated buds was related to the number of regenerated buds. The length and fresh weight of regenerated buds first increased and then decreased with the increasing number of regenerated buds (Figure S7). This may have occurred because nutrient competition among regenerated buds intensified after excessive germination, leading to low development in regenerated buds.

4.2. Reasonable N Regime in the Main Crop Improved the Regeneration Rate

The regeneration rate is the ratio of the panicle number in the ratoon crop to that in the main crop and can directly reflect the regeneration ability of individual plants. It was closely related to the growth and development of regenerated buds. Xu et al. (2021) found that the fresh weight and length of regenerated buds were significantly and positively correlated with the regeneration rate [31]. Increasing the length and fresh weight of regenerated buds could reduce the mortality of regenerated buds and increase the regeneration rate [36]. Our result also showed that the regeneration rate was linearly correlated with the length and fresh weight of regenerated buds (Figure 4B,C). Since the ROA and NSC in the stem and leaf of the main crop were nonlinearly correlated with the length and fresh weight of regenerated buds, the regeneration rate had a similar trend with ROA and NSC in the stem and leaf of the main crop (Figures 6 and 7). However, there are few studies on the relationship between the number of regenerated buds and regeneration rate. This research showed that the regeneration rate was nonlinearly correlated with the number of regenerated buds (Figure 4A). This may be due to the nonlinear correlation between the number of regenerated buds and the length and fresh weight of regenerated buds (Figure S7). The length and fresh weight of regenerated buds decreased after regenerated bud excessive germination, and the regenerated buds with slow growth tended to die at the later stage of the ratoon crop. Therefore, although excessive N application rate and panicle N application rate were beneficial to increase the number of regenerated buds, it would reduce the length and fresh weight of regenerated buds, leading to a decrease in regeneration rate. We can conclude that a reasonable N regime in the main crop could simultaneously increase the number of regenerated buds and the development of regenerated buds, improving the regeneration rate and the yield of the ratoon crop.

4.3. Effects of N Regime in the Main Crop on the Yield of the Main and Ratoon Crops

The N regime strongly affected rice yields. Zhu et al. (2017) and Liang et al. (2021) revealed that rice yield first increased and then decreased with increasing N application rate [37,38], and we reached a similar conclusion (Figure 1A–C). Ye et al. (2019) indicated that N application at a later stage of rice reduced the panicle number and was not conducive to increasing grain yields [39]. However, Li et al. (2018) reported that increasing the proportion of panicle N fertilizer improved N use efficiency and increased grain yield [40]. Our results are different from those of previous studies. With increased panicle N application, we found that the yield of the main crop decreased under the LN level, increased under the HN level, and first increased and then decreased under the MN level (Figure 1A–C). These results indicated that the yield was not only related to the N application rate but also closely related to the application time and N ratio.

There have been few studies on the effects of the N regime in the main crop on the yield of ratoon crops, and the results have been inconsistent. Zhang et al. (2019) revealed that N application rate and N application method in the main crop had a great effect on the yield of the ratoon crop [12]. Some scholars reported that increasing N application rate in the late growth stage of the main crop was beneficial to promoting the growth and development of the aboveground and underground organs of rice, which could improve the yield of the main and ratoon crops [17–19]. Nakano et al. (2009) indicated that N application rate of 22.5 g N m^{-2} in the main crop had a higher dry matter yield for the ratoon crop than with 15.0 g N m^{-2}, and more N was applied early in the main crop could obtain higher dry matter yield [41]. However, Chen et al. (2010) and Wang et al. (2019) showed that an increase in total N application or panicle N fertilizer in the main crop had little effect on the yield of the ratoon crops [14,20]. Our research showed that, the effect of increasing panicle N application on the yield of the ratoon crops was different under different N application rates. With increased panicle N application, the yield in the ratoon crop increased under LN level, while it first increased and then decreased under MN and HN levels, with the greatest values occurring at PN50 and PN40, respectively (Figure 1D–F). High N application rate in the main crop was not conducive to increasing yield in the main and ratoon crops (Figure 1). Further analysis showed that the N regime in the main crop mainly affected the yield of the ratoon crop by changing the regeneration rate in the ratoon crop (Table 5). In this experiment, the yield of the ratoon crop was mainly affected by the panicle number in the ratoon crop (Figure S5). The panicle number in the ratoon crop is usually determined by the number of mother stem and regeneration rate. Although the number of mother stems and the regeneration rate were both changed by different N regime, the correlation analysis revealed that compared with the number of mother stems, the regeneration rate had a greater correlation with panicle number in the ratoon crop (Table 5). Especially under the LN and MN levels, when the number of mother stems was reduced by increased panicle N application, the improvement in regeneration rate still led to a higher yield in PN60 and PN50 (Figure 1 and Table 4). Therefore, the appropriate way to obtain a high yield in the ratoon crop should be to continuously improve the regeneration rate through breeding methods and cultivation techniques.

We can conclude that an appropriate proportion of basal tillering fertilizer to panicle fertilizer under different N rates could improve the regeneration rate and yield of the ratoon crop and simultaneously increase the yield of the main crop, achieving the purpose of high yield in both seasons.

4.4. Effects of the N Regime in the Main Crop on the N Use Efficiency of the Main and Ratoon Crops

N use efficiency (NUE) is also an important factor in determining whether the application of N fertilization is reasonable. N uptake rate and NUE are closely related to the N application rate. Xu et al. (2015) reported that appropriately increasing N application rate was beneficial to improving N uptake and NUE, leading to an increase in rice yield [42]. However, under high N application rate, the N recovery efficiency decreased, the ANUE first increased and then decreased, and rice yield did not increase. Zhang et al. (2019). revealed that with the increase of N application rate, the N uptake rate and ANUE first increased and then decreased [43]. Our result showed that with the N application rate increased, ANUE decreased in the main crop, there was no increasing trend in HN in the ratoon crop, and it first increased then decreased in both seasons (Table 6). Excessive N application reduces the number and physiological activity of roots [44] and reduces the absorption of inorganic N in paddy soil [43], resulting in less N absorbed by plants and a lower N uptake rate and NUE.

The N application ratio also has a great influence on the ANUE. Because the seedlings had less root system and less N was absorbed in the early stage of rice growth, appropriately reducing basal N and appropriately increasing panicle and tiller N could improve the NUE [45]. Our result is consistent with that (Table 6). In addition, Xu et al. (2011) reported that N uptake and N use efficiency were not only related to N regime, but also to soil characteristics [46]. Under high N supply of soil, reducing the N application rate in basal was beneficial to improve NUE and rice yield. However, under low N supply of soil, reducing the N application rate in basal would result in decrease in NUE and rice yield [47]. Liu et al. (2005) indicated that compared with low N supply of soil, the effect of N rate on rice yield was reduced, and NUE was also decreased under high N supply of soil [48]. In China, the yield of rice is usually 5–6 t ha^{-1} in N-free areas of paddy soils [49]. In this experiment, the yield of N free area was 5.28 t ha^{-1} in the main crop and 8.43 t ha^{-1} in the total yield of two seasons, belonging to the area with medium and upper fertility. Therefore, during the ratoon rice production of the area, appropriately reducing the application rate of base tillering fertilizer and increasing the application rate of panicle fertilizer is beneficial to improve the NUE and the yield of both seasons.

Table 6. Effects of N regime in the main crop on the nitrogen use efficiency in the main and ratoon crop.

Treatment		ANUE in Main Crop	ANUE in Ratoon Crop	ANUE in Both Seasons
LN	PN30	29.7 a	6.3 f	14.1 b
	PN40	27.9 b	7.4 e	14.2 b
	PN50	26.6 c	8.6 d	14.6 b
	PN60	24.5 d	9.3 d	14.4 b
	Mean	27.2 A	7.9 B	14.3 B
MN	PN30	15.5 f	13.4 c	14.6 b
	PN40	16.6 e	14.9 b	15.8 a
	PN50	15.3 f	16.4 a	15.8 a
	PN60	14.6 g	14.7	14.7 b
	Mean	15.5 B	14.8 A	15.2 A
HN	PN30	8.3 i	15.0 b	10.5 d
	PN40	8.6 i	16.2 a	11.1 c
	PN50	8.8 i	14.5 b	10.7 d
	PN60	9.3 h	13.6 c	10.7 d
	Mean	8.7 C	14.8 A	10.8 C

Note: ANUE, agronomic nitrogen use efficiency. LN, low N rate (the total amount of N applied was 100 kg ha^{-1}); MN, medium N rate (the total amount of N applied was 250 kg ha^{-1}); HN, high N rate (the total amount of N applied was 400 kg ha^{-1}). PN30, PN40, PN50, and PN60, correspond to N application ratio of basal tillering fertilizer to panicle fertilizer of 7:3, 6:4, 5:5, and 4:6, respectively. Data in a column followed by different lower-case letters indicate significant differences at the 5% probability level according to the LSD test. Means followed by different upper-case letters indicate significant differences between the three nitrogen fertilizer levels at the 5% probability level according to the LSD test. ANUE in the ratoon cop = (Grain yield of ratoon crop − Grain yield of 0 N in ratoon crop)/N rate in ratoon crop. ANUE in both seasons = (Grain yield of both seasons − Grain yield of 0 N in both seasons)/N rate in both seasons. The grain yield of 0 N in the main and ratoon crops was 5.28 t ha^{-1} and 3.15 t ha^{-1}, respectively.

5. Conclusions

The rate and ratio of N application in the main crop had an important influence on the rice yield and ANUE of the main and ratoon crops. Excessive total and panicle N application rate in the main crop were not conducive to improving the grain yield and ANUE in both seasons. Under different N application rate, appropriate increasing the panicle N application rate was beneficial for increasing the ROA and NSC concentration in stem and leaf in the main crop, resulting in improving the growth and development of regenerated buds, which caused an increase in the regeneration rate and yield of the ratoon crop. A total N rate of 250 kg ha^{-1} and a ratio of basal tillering fertilizer to panicle fertilizer of 5:5 in the main crop could increase grain yields and ANUE in both seasons.

Supplementary Materials: The following supporting information can be downloaded at: https://www.mdpi.com/article/10.3390/agriculture12040527/s1, Figure S1: Mean temperature (A), sunshine hours (B) and rainfall (C) monthly during rice growth in 2018 and 2019. Figure S2: Technical drawings for the experimental program. Figure S3: The process of obtaining root samples from the field. Figure S4: Field investigation on the growth and development of regenerated buds. Figure S5: Relationships between the yield and panicle number in the ratoon crop. Figure S6: Effects of N regime in the main crop on the number of living (A–C) and dead (D–F) regenerated buds per square meter. Figure S7. Relationships between number, fresh weight, and length of regenerated buds at maturity of the main crop. Table S1: Effects of N regime in the main crop on the yield components of the main crop.

Author Contributions: Conceptualization, L.L. and X.L.; Writing-original draft, Q.Z.; Writing—review and editing, Q.Z. and L.L.; Project administration, X.L.; Data curation, G.Y. and H.W.; Investigation, D.F. and H.Z.; Formal analysis, B.D. and M.G.; Supervision, L.L. All authors have read and agreed to the published version of the manuscript.

Funding: This study was supported by Modern Agricultural Industrial Technology System of Henan Province (Z2012-04-G01), the National Natural Science Foundation of China (32071947,31871557), the Priority Academic Program Development of Jiangsu Higher Education Institutions (PAPD).

Institutional Review Board Statement: Not applicable.

Informed Consent Statement: Not applicable.

Data Availability Statement: Not applicable.

Conflicts of Interest: The authors declare no conflict of interest.

References

1. Sanderine, N. Global food supply and the impacts of increased use of biofuels. *Energy* **2011**, *37*, 115–121. [CrossRef]
2. Dong, C.F.; Xu, N.X.; Ding, C.L.; Gu, H.G.; Zhang, W.J.; Sun, L. Developing ratoon rice as forage in subtropical and temperate areas. *Field Crops Res.* **2020**, *245*, 1–7. [CrossRef]
3. Zhou, Y.J.; Li, X.X.; Cao, J.; Li, Y.; Huang, J.L.; Peng, S.B. High nitrogen input reduces yield loss from low temperature during the seedling stage in early-season rice. *Field Crops Res.* **2018**, *228*, 68–75. [CrossRef]
4. Xu, J.M.; Amelia, H.; Nese, S. Rice yield formation under high day and night temperatures—A prerequisite to ensure future food security. *Plant Cell Environ.* **2020**, *43*, 1595–1680. [CrossRef]
5. Shen, X.; Zhang, L.; Zhang, J.B. Ratoon rice production in central China: Environmental sustainability and food production. *Sci. Total Environ.* **2020**, *764*, 142850. [CrossRef]
6. He, A.B.; Wang, W.Q.; Jiang, G.L.; Sun, H.J.; Jiang, M.; Man, J.G.; Cui, K.H.; Huang, J.L.; Peng, S.B.; Nie, N.X. Source-sink regulation and its effects on the regeneration ability of ratoon rice. *Field Crops Res.* **2019**, *236*, 155–164. [CrossRef]
7. Golam, F.; Rosna, T.; Zakaria, P. Rice ratoon crop: A sustainable rice production system for tropical hill agriculture. *Sustainability* **2014**, *6*, 5785–5800. [CrossRef]
8. Zhang, Q.; Liu, X.C.; Yu, G.L.; Wang, H.; Feng, D.Q.; Zhao, H.Y.; Liu, L.J. Agronomic and physiological characteristics of high-yielding ratoon rice varieties. *Agron. J.* **2021**, *113*, 5063–5075. [CrossRef]
9. Wang, W.Q.; He, A.B.; Jiang, G.L.; Sun, H.J.; Jiang, M.; Man, J.G.; Ling, X.X.; Cui, K.H.; Huang, J.L.; Peng, S.B.; et al. Ratoon rice technology: A green and resource-efficient way for rice production. *Adv. Agron.* **2020**, *159*, 135–167. [CrossRef]
10. Huang, J.W.; Pan, Y.P.; Chen, H.F.; Zhang, Z.X.; Fang, C.X.; Shao, C.H.; Amjad, H.; Lin, W.W.; Lin, W.X. Physiochemical mechanisms involved in the improvement of grain-filling, rice quality mediated by related enzyme activities in the ratoon cultivation system. *Field Crops Res.* **2020**, *258*, 107962. [CrossRef]
11. Dustin, L.H.; Jason, A.B.; Sterling, B. Evaluation of main-crop stubble height on ratoon rice growth and development. *Field Crops Res.* **2009**, *114*, 396–403. [CrossRef]
12. Zhang, L.; Jiang, P.; Gou, X.Y.; Zhou, X.B.; Zhu, Y.C.; Liu, M.; Xiong, H.; Xu, F.X. Integrated water and nitrogen management practices to enhance yield and environmental goals in rice–ratoon rice systems. *Agron. J.* **2019**, *111*, 2821–2831. [CrossRef]
13. Chen, Q.; He, A.B.; Wang, W.Q.; Peng, S.B.; Huang, J.L.; Cui, K.H.; Nie, L.X. Comparisons of regeneration rate and yields performance between inbred and hybrid rice cultivars in a direct seeding rice-ratoon rice system in central China. *Field Crops Res.* **2018**, *223*, 164–170. [CrossRef]
14. Wang, Y.C.; Zheng, C.; Xiao, S.; Sun, Y.T.; Huang, J.L.; Peng, S.B. Agronomic responses of ratoon rice to nitrogen management in central China. *Field Crops Res.* **2019**, *241*, 107569. [CrossRef]
15. Gribaldi, G.; Nurlaili, N.; Firnawati, S.; Nurmala, D.; Ardi, A. Strategy of nitrogen fertilizer application to increase growth and yield of rice in ratoon system at tidal swampland. *Aust. J. Crop Sci.* **2020**, *16*, 1004–1010. [CrossRef]
16. Jason, A.B.; Patrick, K.B. Ratoon rice response to nitrogen fertilizer. *Crop Manag.* **2006**, *5*, 1–5. [CrossRef]
17. Huang, J.W.; Wu, J.Y.; Chen, H.F.; Zhang, Z.X.; Fang, C.X.; Shao, C.H.; Lin, W.W.; Weng, P.Y.; Muhammad, U.K.; Lin, W.X. Optimal management of nitrogen fertilizer in the main rice crop and its carrying-over effect on ratoon rice under mechanized cultivation in Southeast China. *J. Integr. Agric.* **2022**, *21*, 351–364. [CrossRef]
18. Huang, J.W.; Wu, J.Y.; Chen, H.F.; Zhang, Z.X.; Fang, C.X.; Shao, C.H.; Lin, W.W.; Weng, P.Y.; Lin, W.X. Nitrogen fertilizer management for main crop rice and its carrying-over effect on rhizosphere function and yield of ratoon rice. *Chin. J. Rice Sci.* **2021**, *35*, 383–395. [CrossRef]
19. Liu, M.X.; Zhang, L.; Fan, Q.H.; Xiong, H.; Zhou, X.B.; Zhu, Y.C.; Jiang, P.; Liu, M.; Guo, X.Y.; Xu, F.X. Effects of shift part of nitrogen fertilizer from basal-tillering to panicle initiation on grain yield of mid-season hybrid rice and ratooning rice. *Chin. Rice* **2014**, *20*, 48–50. [CrossRef]

20. Chen, H.F.; Yang, D.; Liang, Y.Y.; Zhang, Z.X.; Liang, K.J.; Lin, W.X. Effect of nitrogen application strategy in the first cropping rice on dry matter accumulation, grain yield and nitrogen utilization efficiency of the first cropping rice and its ratoon rice crop. *Chin. J. Eco-Agric.* **2010**, *18*, 50–56. [CrossRef]
21. Ling, Q.H. Theory and technology of rice precise and quantitative cultivation. *Hybrid Rice* **2010**, *25*, 27–34. [CrossRef]
22. Yang, D.S.; Peng, S.B.; Zheng, C.; Xiang, H.S.; Huang, J.L.; Cui, K.H.; Wang, F. Effects of nitrogen fertilization for bud initiation and tiller growth on yield and quality of rice ratoon crop in central China. *Field Crops Res.* **2021**, *272*, 108286. [CrossRef]
23. Ramasamy, S.; Berge, H.F.M.; Purushothaman, S. Yield formation in rice in response to drainage and nitrogen application. *Field Crops Res.* **1997**, *51*, 65–82. [CrossRef]
24. Pan, J.F.; Cui, K.H.; Wei, D.; Huang, J.L.; Xiang, J.; Nie, L.X. Relationships of non-structural carbohydrates accumulation and translocation with yield formation in rice recombinant inbred lines under two nitrogen levels. *Physiol. Plant* **2011**, *141*, 321–331. [CrossRef] [PubMed]
25. Xu, F.X.; Xiong, H.; Hong, S. Relationship between axillary bud growth matter accumulation of stem-sheath after heading of main crop hybrid rice. *Chin. J. Rice Sci.* **1997**, *11*, 160–164. [CrossRef]
26. Zhang, L.; Xiong, H.; Xu, F.X.; Zhu, Y.C.; Guo, X.Y.; Zhou, X.B.; Liu, M. Relationship between living rate of bud and emergence rate of ratoon rice and characteristics of the first cropping mid-season hybrid rice. *Agric. Sci. Technol.* **2012**, *13*, 1873–1876. [CrossRef]
27. Zhang, G.L.; Tu, N.M.; Zhang, S.T. Ratooning properties of axillary buds in hybrid rice. *Chin. J. Rice Sci.* **2005**, *19*, 323–327. [CrossRef]
28. Turner, F.T.; Jund, M.F. Rice ratoon crop yield linked to main crop stem carbohydrates. *Crop Sci.* **1993**, *33*, 150–153. [CrossRef]
29. Zheng, J.S.; Lin, W.; Zhuo, C.Y.; Fang, X.J.; Lin, W.X. The correlation of dry biomass and activity of root system with grain yield in ratoon rice. *Chin. J. Eco-Agric.* **2004**, *22*, 112–115.
30. Nakano, H.; Tanaka, R.; Wada, H.; Okami, M.; Nakagomi, K.; Hakata, M. Breaking rice yield barrier with the ratooning method under changing climatic conditions: A paradigm shift in rice-cropping systems in southwestern Japan. *Agron. J.* **2020**, *112*, 3975–3992. [CrossRef]
31. Xu, F.X.; Zhang, L.; Zhou, X.B.; Guo, X.Y.; Zhu, Y.C.; Liu, M.; Xiong, H. The ratoon rice system with high yield and high efficiency in China: Progress, trend of theory and technology. *Field Crops Res.* **2021**, *272*, 108282. [CrossRef]
32. Yu, Y.; Peng, X.L.; Liu, Y.Y.; Zhang, H.; Cheng, L.N. Effects of N application at later stage on absorbability of rice root in cold area. *Soils* **2011**, *43*, 548–553. [CrossRef]
33. Hu, Q.; Xia, M.; Zhang, H.C.; Cao, L.Q.; Guo, B.W.; Wei, H.Y.; Chen, H.C.; Dai, Q.G.; Huo, Z.Y.; Xu, K.; et al. Effect of nitrogen application regime on yield, nitrogen absorption and utilization of mechanical pot-seedling transplanting rice with good taste quality. *Acta Agron. Sin.* **2016**, *42*, 1666–1676. [CrossRef]
34. Wang, X.D.; Wang, Y.L.; Zhang, Y.P.; Xiang, J.; Zhang, Y.K.; Zhu, D.F.; Chen, H.Z. The nitrogen topdressing mode of indica-japonica and indica hybrid rice are different after side-deep fertilization with machine transplanting. *Sci. Rep.* **2021**, *11*, 1494. [CrossRef]
35. Fu, J.; Wang, Y.T.; Yin, H.Q.; Wang, S.X.; Wang, F.H.; Chen, X.G.; Wang, Y.; Yang, W.B.; Bai, T. Effect of nitrogen application rate on root morphological and physiological characteristics and yield of japonica rice in region along the yellow river. *Henan Agric. Sci.* **2017**, *46*, 18–25. [CrossRef]
36. Xu, F.X.; Xiong, H.; Zhao, G.L.; Hong, S. A Study on the death mechanism of the axillary buds before harvest of the hybrid midseason rice and its improvement. *Sci. Agric. Sin.* **2000**, *33*, 31–37. [CrossRef]
37. Zhu, D.W.; Zhang, H.C.; Guo, B.W.; Xu, K.; Dai, Q.G.; Wei, H.Y.; Gao, H.; Hu, Y.J.; Cui, P.Y.; Huo, Z.Y. Effects of nitrogen level on yield and quality of japonica soft super rice. *J. Integr. Agric.* **2017**, *16*, 1018–1027. [CrossRef]
38. Liang, H.L.; Gao, S.Y.; Ma, J.X.; Zhang, T.; Wang, T.Y.; Zhang, S.; Wu, Z.X. Effect of nitrogen application rates on the nitrogen utilization, yield and quality of rice. *Food Nutr. Sci.* **2021**, *12*, 13–27. [CrossRef]
39. Ye, C.; Huang, X.; Chu, G.; Chen, S.; Xu, C.M.; Zhang, X.F.; Wang, D.Y. Effects of postponing topdressing-N on the yield of different types of japonica rice and its relationship with soil fertility. *Agronomy* **2019**, *9*, 868. [CrossRef]
40. Li, G.H.; Lin, J.J.; Xue, L.H.; Ding, Y.F.; Wang, S.H.; Yang, L.Z. Fate of basal N under split fertilization in rice with (15)N isotope tracer. *Pedosphere* **2018**, *28*, 135–143. [CrossRef]
41. Nakano, H.; Morita, S. Effects of planting time and nitrogen application on dry matter yield of the forage rice cultivar Tachiaoba in southwestern Japan. *Plant Prod. Sci.* **2009**, *12*, 351–358. [CrossRef]
42. Xu, X.P.; Zhou, W.; Liang, G.Q.; Sun, J.W.; Wang, X.B.; He, P.; Xu, F.S.; Yu, X.C. Effects of nitrogen and density interactions on grain yield and nitrogen use efficiency of double-rice systems. *J. Plant Nutr. Fertil.* **2015**, *21*, 763–772. [CrossRef]
43. Zhang, B.; Wei, W.W. Effects of different nitrogen supply levels on soil inorganic nitrogen residue, nitrogen balance and yield of rice. *J. Shandong Agric. Univ.* **2019**, *50*, 566–570. [CrossRef]
44. Dong, G.C.; Chen, C.; Yuan, Q.M.; Yang, B.; Zhu, Z.K.; Cao, W.Y.; Zhong, J.; Zhou, J.; Luo, G.; Wang, Y.; et al. The effect of nitrogen fertilizer treatments on root traits and nitrogen use efficiency in indica rice varieties with high nitrogen absorption efficiency. *Acta Ecol. Sin.* **2016**, *36*, 642–651. [CrossRef]
45. Cassman, K.G.; Peng, S.; Olk, D.C.; Ladha, J.K.; Reichardt, W.; Dobermann, A.; Singh, U. Opportunities for increased nitrogen-use efficiency from improved resource management in irrigated rice systems. *Field Crops Res.* **1998**, *56*, 7–39. [CrossRef]

46. Xu, F.X.; Xiong, H.; Zhang, L.; Guo, X.Y.; Zhu, Y.C.; Zhou, X.B.; Liu, M. Variation of nitrogen uptake and utilization efficiency of mid-season hybrid rice at different ecological sites under different nitrogen application levels. *Agric. Sci. Technol.* **2011**, *12*, 1001–1012. [CrossRef]
47. Zhao, J.Y.; Yu, Z.W. Effects of nitrogen rate on nitrogen fertilizer use of winter wheat and content of soil nitrate-N under different fertility condition. *Acta Ecol. Sin.* **2006**, *26*, 815–822. [CrossRef]
48. Liu, L.J.; Xu, W.; Tang, C.; Wang, Z.Q.; Yang, J.C. Effect of indigenous nitrogen supply of soil on the grain yield and fertilizer-N use efficiency in rice. *Chin. J. Rice Sci.* **2005**, *12*, 267–274. [CrossRef]
49. Peng, S.B.; Buresh, R.J.; Huang, J.L.; Yang, J.C.; Zou, Y.B.; Zhong, X.H.; Wang, G.H.; Zhang, F.S. Research strategy in impoving fertilizer-nitrogen use efficiency of irrigated rice in China. *Sci. Agric. Sin.* **2002**, *35*, 1095–1103. [CrossRef]

agriculture

MDPI

Article

Residue Mulching Alleviates Coastal Salt Accumulation and Stimulates Post-Fallow Crop Biomass under a Fallow–Maize (*Zea mays* L.) Rotation System

Yifu Zhang [1,2,*], Wei Yuan [1] and Lianjie Han [1]

[1] School of Mechanical Engineering, Yangzhou University, Yangzhou 225127, China; mx120210538@yzu.edu.cn (W.Y.); mx120210537@stu.yzu.edu.cn (L.H.)

[2] Jiangsu Engineering Center for Modern Agricultural Machinery and Agronomy Technology, Yangzhou 225127, China

* Correspondence: zyfu@yzu.edu.cn

Abstract: Fallow, a field where living plants are unplanted for a period, is continually implemented to accumulate moisture for the upcoming cultivation. However, there are less studies on the fallow strategies in one-crop-per-annum cropping system for coastal saline soils. In this study, 2-year "fallow + maize (*Zea mays* L.)" rotation experiments were carried out from 2016 to 2018 to assess how the mulching determine post-fallow soil moisture, salt distribution, and crop performance. Three treatments were designed, i.e., traditional cultivation without residue retention (TT), traditional tillage with total straw mulching during fallow (TT + SM), and no-till cultivation combined fallow mulching (NT + SM). After 2 years of fallow mulching with maize rotation, TT + SM reduced soil electrical conductivity (EC) and total salt of the upper 30 cm soil profile by 22.9% and 25.4% (p = 0.05), respectively, compared with the TT treatment. The results also indicate an improvement in volumetric soil water content (SWC) by 10.3%, soil organic matter (SOM) by 17.8%, and ultimately grain yield by 11.3% (p = 0.05) under the TT + SM treatment. Fallow mulching is recommended as an acceptable way to protect soil health in coastal fresh-starved or rain-fed farming practice.

Keywords: coastal salt-affected soil; one-crop-per-annum cropping; fallow mulching; salt accumulation; crop growing

Citation: Zhang, Y.; Yuan, W.; Han, L. Residue Mulching Alleviates Coastal Salt Accumulation and Stimulates Post-Fallow Crop Biomass under a Fallow–Maize (*Zea mays* L.) Rotation System. *Agriculture* **2022**, *12*, 509. https://doi.org/10.3390/agriculture12040509

Academic Editors: Chengfang Li and Lijin Guo

Received: 26 February 2022
Accepted: 31 March 2022
Published: 3 April 2022

1. Introduction

Soil salinization is a process during which the salt in the deep soil and groundwater rises to the surface via evaporation, and then accumulates in the topsoil. Salt accumulation has been one of the most severe ecological environmental problems that restricts the agricultural sustainable development in arid and semi-arid areas [1–4]. Currently in China, principally distributed in the northeast, northwest, and coastal areas, more than 36 million hectare farmlands are suffering from salinity, accounting for approximately 4.9% of the whole available lands [5–7].

Stimulating grain yield in coastal farmlands is a vital part of ensuring food security. Focusing on the special climate and hydrological conditions, soil desalination for coastal areas was conducted mainly by following three aspects, i.e., salt leaching, capillary water blocking, and biological desalination [8–10]. However, it is of great importance to introduce different ways to minimize salt constraints and expanding agricultural output. Generally, coastal croplands are vulnerable to anthropogenic activities and climatic changes. Especially in the coastline of Bohai bay, east China, the fluctuation in sea level and the excessive consumption of groundwater will inevitably induce the invasion from seawater. Reasons for the severe salt stress primarily come from the following three aspects [11,12]. Firstly, year-round intrusion by seawater leads to excessively groundwater salinity concentration. Secondly, mainly concentrated in summer, the precipitation in this region is uneven throughout the four seasons as affected by the oceanic climate. Thirdly, from autumn to next spring, salt accumulates upward into the topsoil via evaporation, a result from the monsoon and drought. Therefore, "fallow + maize (*Zea mays* L.)" rotation cultivation has been applied for decades in this region, and the local farmers tend to cultivate summer maize due to the scarce fresh water and fallow in the post season.

Fallow, a period during which no living plants are grown, is frequently utilized to collect moisture and nutrition for subsequent cultivation [13]. Fallow is feasible to solve environmental deterioration through the self-recovery of the barren or low-yield lands. It was reported that fallow practice had advantages in improving soil hardening, desertification, enhancing biodiversity, and thereby increasing grain production and ensuring food security [14,15]. Overall, fallow is a feasible practice in arid and part of subhumid regions, when the accessible rainfall during cropping season is less than that required for expected yield [16,17]. However, fallow inevitably incurs shrinking output due to the extended time without cash crop cover.

Straw mulching, soil cover with crop residues, has been well confirmed to have positive influences on soil water and heat redistribution, soil physicochemical properties and nutrient, and ultimately facilitate crop performance [18–21]. In addition, straw mulching also produced the expected effect in salt-affected soils. Wang et al. [22] found that water evaporation decreased significantly, while the saline soil was covered. Moreover, cotton straw was demonstrated to have a positive effect on soil fertility and crop yields [23]. Based on the previous studies, straw, as a by-product from farmland crops, has great potential in improving the soil environment, especially because it is available and easy to apply. More importantly, Yang [24] indicated that fallow rotation under straw mulching was beneficial to improve soil structure, which provides a feasible reference for the exploration on the "fallow + summer maize" rotation.

Therefore, this study aims to achieve a fallow–maize rotation crop system with severe salt accumulation and insufficient freshwater, and we attempt to introduce reasonable fallow managements, as well as to survey appropriate agronomic solutions for coastal salt-affected croplands. We hypothesized that, if maize straw mulching is beneficial to reduce the upward salt accumulation during fallow, the subsequent cropping season would obtain better water and salt conditions. In this study, a 2-year fallow–maize rotation cultivation was carried out, and during the fallow period, the field was covered with maize straw after harvest to ascertain how they impact the salt movement and crop growth. Before each cropping season, soil electrical conductivity (EC), total salt, soil organic matter (SOM), bulk density, and volumetric soil water content (SWC) were measured, as well as the relevant grain yield, to investigate the comprehensive response in comparison to traditional management. Accordingly, the objective of this study was to assess the response of soil moisture distribution and crop performance to fallow straw mulching, and to provide a reproducible approach to cultivate coastal saline croplands.

2. Materials and Methods

2.1. Experimental Site and Soil

From May 2016 to September 2018, fallow combined with summer maize rotation experiments were conducted in the Binhai district, Tianjin city, China (38°46′ N, 117°13′ E, Figure 1). The climate is semi-humid and monsoons with 211 frost-free days and 12.3 °C of annual average temperature. The 570 mm annual precipitation is fluctuant and imbalanced, mainly (>70%) concentrated from June to September. The annual evaporation is about 1800 mm, and the evaporation–precipitation ratio exceeds 3:1.

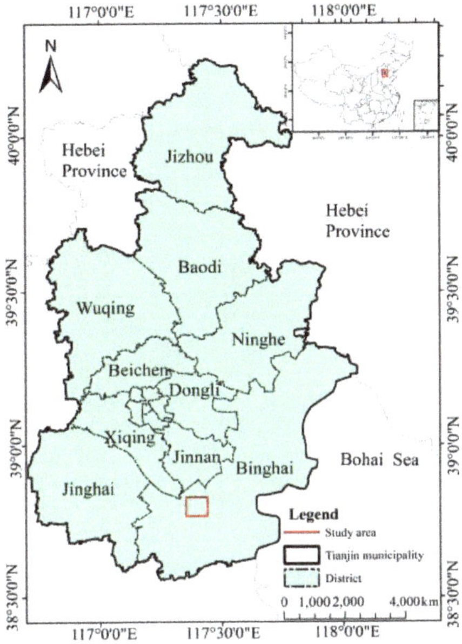

Figure 1. Location of the experimental site.

Before the initiation of field experiments, the soil was defined as solonchak, with 2.67 mg kg^{-1} of sodium, 7.08 g kg^{-1} of salt content, 10.4 g kg^{-1} of SOM, 64.5 mg kg^{-1} of alkaline-hydrolysable nitrogen, 31.4 mg kg^{-1} of available phosphorus (by Olsen), and 63.2 mg kg^{-1} of available potassium. The soil texture was silty clay loam, according to the USDA classification. The physical properties of the 0–30 cm profile prior to experiments are shown in Table 1.

Table 1. Soil physical properties of the 0–30 cm profile before experiments.

Items	Mean Value	Unit
Sand (0.05~2 mm)	10.6	%
Silt (0.002~0.05 mm)	61.0	%
Clay (<0.002 mm)	28.4	%
Soil bulk density	1.39	g cm^{-3}
Field capacity (by weight)	28.4	%
Total porosity	46.5	%

2.2. Experimental Design

Prior to 2016, the site was farmed traditionally for decades with summer maize cultivation. In this study, the whole sites were ploughed to eliminate the existing plough pan in early June. Then, the seedbed was renovated by rotary harrowing to a depth of 0.15 m and smoothing before planting. In late September, maize harvest and total stubble removing were executed manually, according to traditional tillage management. This research was performed after maize harvest, and three treatments were designed in this study: (1) TT, traditional tillage for maize cultivation with no straw being returned to the field after maize harvest, as the control group; (2) TT + SM, traditional tillage for maize cultivation with 100% fallow straw mulching from harvest to planting in the next year; and (3) NT + SM, 100% fallow straw mulching without tillage. The three treatments were applied to a randomized design with three replicates. The plots were ridged against cross contamination, and each was 36 m long and 15 m wide.

Table 2 describes the cultivated details of the fallow–maize rotation. In late September 2016, dry maize stubble was equally supplied in each mulch-treated plot at 4000 kg ha^{-1} for the TT + SM and NT + SM treatments. In late September 2017, the residues harvested from the previous cropping season were all retained (i.e., 100% straw retention) in each site by mulching. For the sowing procedure of the NT + SM treatment, the maize was planted directly without seedbed renovation.

Table 2. Cropping schedules for the fallow–maize rotation system. The precipitation was calculated during the cropping season or fallow period, i.e., from maize planting to harvest and from maize harvest to the next year planting, respectively).

Rotation Design	Start or Planting Date	End or Harvest Date	Precipitation (mm)
Prior to experiment		Late September 2016	
Fallow	Late September 2016	Early June 2017	105.1
Maize cropping	Early June 2017	Late September 2017	350.3
Fallow	Late September 2017	Early June 2018	162.9
Maize cropping	Early June 2018	Late September 2018	433.3

During each in-season cultivation, a coincident cropping method was applied in accordance with the local farming practice. A no-till maize seeder was applied to execute sowing and fertilizing simultaneously. In detail, the experimental cultivar was Zhengdan958 with a row spacing of 60 cm and 28 cm seed spacing in a row, which was sown on 9 June 2017 and harvested on 29 September and again sown on 6 June 2018 and harvested on 27 September. The fertilizer was incorporated at a rate of 45 kg hm^{-2} of N, 45 kg hm^{-2} of P_2O_5, and 40 kg hm^{-2} of K_2O, while sowing. In addition, 40 kg hm^{-2} of N was supplied as topdressing at the jointing stage. Plant protection, such as weeds, insect pests, and diseases. was performed when needed in accordance with the local agronomic specifications.

2.3. Sampling and Measurement

Soil samples were collected at the end of fallow period, i.e., early June of 2017 and 2018, before the seedbed renovation for maize sowing. The disturbed samples were air dried, then pulverized and screened for chemical properties measurements. Soil EC was measured using the soil water suspension (1:5, w/v) by an EC meter. SOM was determined under the dichromate oxidation method by Liu et al. [25]. Total salt storage to the calculated soil profile was measured as the mass per unit area, as described by [26]:

$$\text{Total salt} = 10\sum \rho_{bi}\, s_i\, z_i, \tag{1}$$

where ρ_{bi} is the soil bulk density of the i soil layer; in g cm^{-3}; s_i is the soil salt content of the i soil layer, in g kg^{-1}; and z_i is the thickness of the i soil layer, in cm.

The undisturbed soil cores were taken before the seedbed renovation using the constant volume cutting ring. The volumetric SWC investigation was performed using the oven drying method, as described by He et al. [27]:

$$b_v = b_m \times (\rho_b/\rho_w), \tag{2}$$

where b_v expresses the volumetric SWC, in cm^3 cm^{-3}; b_m is the gravimetric SWC, in g g^{-1}; ρ_b and ρ_w are the soil bulk density and water density, respectively, in g cm^{-3}.

Plant samples were collected at maturity stage from five randomly selected plants in each plot. Root samples were collected within a 0.15 × 0.15 m square, and to a depth of 0.40 m. Adhered soil particles and unrelated impurities were removed by running tap water, and then the roots were air dried and oven dried at 70 °C unto constant weight to provide the root biomass. Three 5 m long rows were randomly selected under different treatment to determine maize yields. The grain yield was adjusted to 12.0% moisture content.

2.4. Statistical Analyses

The mean values were calculated for each measurement, and analysis of variance (ANOVA) was performed to evaluate the effect of different treatments on the variables with SPSS software (International Business Machines Corporation, New York, NY, USA). Normality and homoscedasticity were tested for original data before the ANOVA test. If the homogeneity did not show, the original data were classified to conform to the requirement. Multiple comparisons were conducted based on the least significant difference test (LSD) at a 5% level of probability ($p = 0.05$).

3. Results

3.1. Soil EC

Figure 2 compared mean soil EC in the top 30 cm soil layer at the end of the fallow period, i.e., before the seedbed renovation for maize sowing. In the entire 2-year observation, the TT treatment showed the highest values in soil EC in comparison with the TT + SM and NT + SM treatments ($p = 0.05$). The mean EC under the TT + SM treatment appeared to be the lowest, which showed a reduction by 9.8% in 2017 and 22.9% in 2018, in comparison to the TT treatment ($p = 0.05$). Additionally, despite the lack of a significant difference in 2017 (by 7.9%), the NT + SM treatment had a 12.6% significant improvement in EC, as compared to TT ($p = 0.05$).

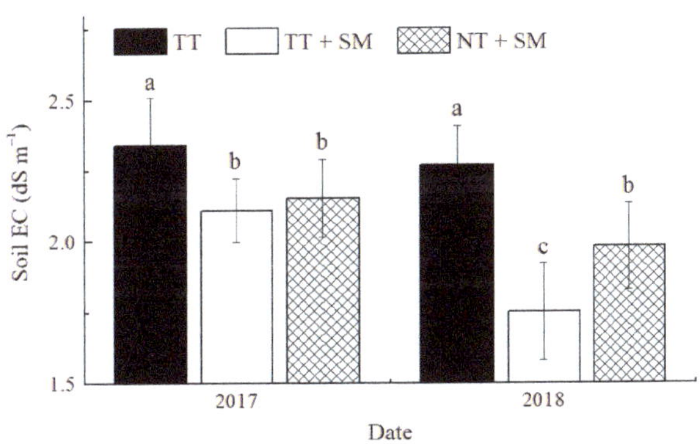

Figure 2. Mean soil electrical conductivity (EC) in the upper 30 cm profile for the TT, TT + SM, and NT + SM treatments. TT: traditional tillage for maize cultivation without straw return after maize harvest; TT + SM: traditional tillage with 100% straw mulching after harvest; NT + SM: no-till cultivation combined with 100% straw mulching. Data were measured at the end of fallow period, i.e., early June of 2017 and 2018 before maize sowing. Means in the same year followed by a different letter are significantly different ($p = 0.05$).

3.2. Total Salt

The total salt in the top 30 cm soil layer was calculated and is shown in Table 3, which reveals its similar tendency with that of soil EC. Compared with the TT treatment, total salt under the TT + SM treatment significantly decreased by 11.3% in 2017 and 25.4% in 2018, respectively ($p = 0.05$). The NT + SM treatment showed an 8.9–13.2% decrease in total salt ($p = 0.05$). Additionally, the total salt under the TT + SM treatment tended to be lower, and a significant decrease was observed in 2018 by 14.0%, when compared with the NT + SM treatment ($p = 0.05$).

Table 3. Total salt and mean soil organic matter (SOM) in the top 30 cm soil layer at the end of the fallow period, i.e., early June of 2017 and 2018 before maize sowing, for the TT, TT + SM, and NT + SM treatments. TT: traditional tillage for maize cultivation without straw return after maize harvest; TT + SM: traditional tillage with 100% straw mulching after harvest; NT + SM: no-till cultivation combined with 100% straw mulching. Means in the same year followed by a different letter are significantly different ($p = 0.05$).

Treatment Identifiers		Total Salt (g/m^2)	SOM (g/kg)
Year	Treatments		
2017	TT	3551.9 a	16.04 a
	TT + SM	3149.5 b	17.08 a
	NT + SM	3236.2 b	16.61 a
2018	TT	3424.0 a	16.65 b
	TT + SM	2553.7 c	19.61 a
	NT + SM	2969.6 b	17.29 b

3.3. SOM

Table 3 describes the pre-planting mean SOM in the top 30 cm soil layer at the end of the fallow period. Generally, the mean SOM tended to be highest under the TT + SM treatment, while TT had the lowest values, i.e., TT + SM > NT + SM > TT. Before the first cropping season of early June 2017, the three treatments had no significant difference in SOM. However, in 2018, TT + SM accelerated SOM significantly by 17.8% and 13.4%, in comparison to the TT and NT + SM treatments, respectively ($p = 0.05$).

3.4. Soil Bulk Density

Mean soil bulk density in the top 30 cm profile of the soil is shown in Figure 3; the means were measured prior to maize sowing. Generally, prior to the first cropping season (early June 2017), no significant difference in soil bulk density was observed between the TT, TT + SM, and NT + SM treatments. However, a significant improvement was observed in 2018 under the TT + SM treatment, which decreased the soil bulk density by 3.4% and by 2.7%, respectively ($p = 0.05$), in comparison to the TT and NT + SM treatments. In addition, the difference in 2018 between TT and NT + SM was not significant.

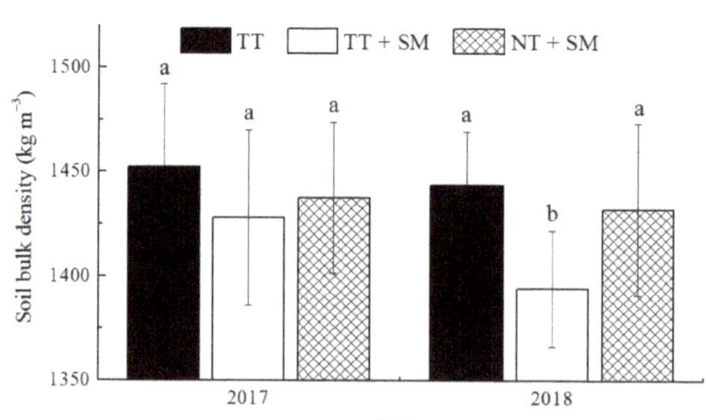

Figure 3. Mean soil bulk density to the depth of 30 cm for traditional tillage without straw return (TT), traditional tillage with straw mulching (TT + SM), and no-till cultivation combined with straw mulching (NT + SM) treatments. Data were measured before maize sowing. Mean values within the same year followed by a different letter are significantly different ($p = 0.05$).

3.5. SWC

Figure 4 shows the mean volumetric SWC in the upper 30 cm soil profile after each fallow period. Volumetric SWC under treatments with straw mulching tended to be higher throughout the 2-year experiments. Particularly, prior to the second cropping season in 2018, TT + SM significantly accumulated more volumetric SWC by 10.3% (p = 0.05), in comparison to the TT treatment. Furthermore, the mean values in volumetric SWC under the NT + SM treatment tended to be medium, but no significant variation was observed, both compared with the TT and TT + SM treatments.

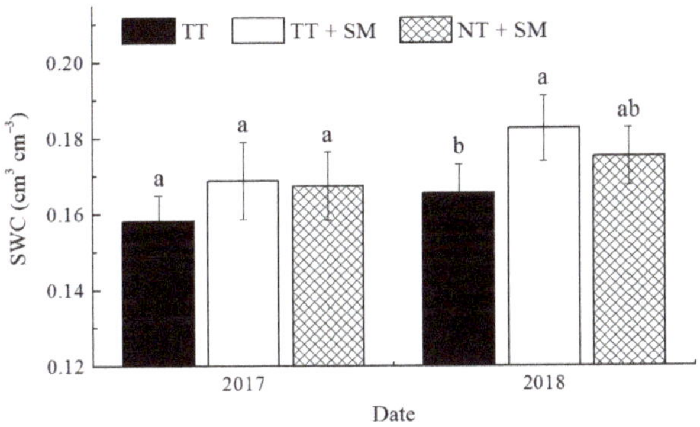

Figure 4. Mean volumetric soil water content (SWC) in the top 30 cm profile at the end of each fallow period, for the TT, TT + SM, and NT + SM treatments. TT: traditional tillage for maize cultivation without straw return after maize harvest; TT + SM: traditional tillage with 100% straw mulching after harvest; NT + SM: no-till cultivation combined with 100% straw mulching. Means in the same year followed by a different letter are significantly different (p = 0.05).

3.6. Crop Performance

At the end of the fallow period in early June, in-season maize cultivation was conducted, and the crop growth at maturity stage is listed in Table 4. No significant difference was observed regarding plant height among various treatments. Moreover, for root dry weight, means under treatments with straw mulching tended to be higher. In detail, TT + SM increased root dry weight by 16.2–18.7%, as compared with the TT treatment (p = 0.05). Moreover, compared with the TT treatment, no significant promotion was observed in root dry weight under the NT + SM treatment.

Table 4. Plant height and root dry weight at maturity stage, as well as the final grain yield for maize cultivation under the traditional tillage without straw return (TT), traditional tillage with straw mulching (TT + SM), and no-till cultivation combined with straw mulching (NT + SM) treatments. Mean values within a column in the same year followed by a different letter are significantly different (p = 0.05).

Treatment Identifiers		Plant Height (cm)	Root Dry Weight (g/plant)	Grain Yield (kg/hm^2)
Year	Treatments			
2017	TT	164.8 a	43.9 b	4655.3 b
	TT + SM	173.7 a	52.1 a	5009.7 a
	NT + SM	170.3 a	46.2 b	4741.3 b
2018	TT	172.5 a	47.2 b	4789.3 b
	TT + SM	181.2 a	54.8 a	5331.0 a
	NT + SM	179.3 a	51.4 a b	4914.7 b

Grain yield for each crop season followed a similar trend with that of root dry weight (Table 4). Treatments with straw mulching tended to harvest more grain, while the yield in TT was lower, i.e., TT + SM > NT + SM > TT. In detail, TT + SM increased grain yield by 7.6% in 2017 and 11.3% in 2018, when compared with the TT treatment ($p = 0.05$). Moreover, compared with the TT treatment, no significant promotion was observed for grain yield under the NT + SM treatment.

4. Discussion

It was reported that, under the one-crop-per-annum system of coastal regions, rather than the rhizosphere nutritional conditions within the cropping season, farmers must pay attention to salt fluctuations during fallow [28]. Due to monsoon and tidal activities, the farmland environment in coastal areas is difficult to predict and utilize. Particularly in the domestic Tianjin Binhai district of the west Bohai Gulf (the experimental plot in this paper), local farmers prefer to conduct maize cultivation resorting to the rainfall leaching in summer. However, salt accumulation in the topsoil during the fallow periods is less reported. Therefore, this study was extremely different from previous demonstrations.

Firstly, in response to such a "fallow + summer maize" rotation cropping system, we attempted to optimize fallow management to provide an acceptable rhizosphere environment for subsequent sowing. Prior to the fallow, we covered soil surface with the maize residues, and at the end of the fallow, positive information was obtained. After 2 years of "fallow + summer maize" rotation cultivation, the TT + SM treatment reduced EC and total salt in the upper 30 cm soil profile by 22.9% and 25.4% ($p = 0.05$) compared with TT. The results confirmed that fallow mulching was conducive to minimize the upward salt accumulation, which was consistent with Deng et al. [29] within an adjacent experimental region. This may be attributed to the following three reasons: first, the solar radiation on the surface is shielded by maize straw, thereby reducing the temperature of the topsoil; second, the exposed area was reduced; third, straw mulching is also conducive to the prevention of wind, which may result in the weakened soil evaporation and reduced the upward movement of water. Hence, we infer that fallow mulching is conducive to diminishing salt accumulation through inhibiting bottom salt rising to the topsoil via water evaporation, which was also reported by Yusefi et al. [30].

Secondly, we also focused on the physicochemical properties of the top 30 cm soil profile after fallow mulching. Beneficial results were observed in volumetric SWC, bulk density, and SOM before the second cropping season, as affected by mulching treatment. From farmland measurements, TT + SM appeared to increase volumetric SWC by 10.3%, accelerate SOM by 17.8%, and decrease bulk density by 3.4% ($p = 0.05$) in comparison to the TT treatment. Adequate soil water storage is a requisite for crop germination, growth, and thereby gaining higher grain yield in fresh-starved farming [31]. The results showed that pre-seeding volumetric SWC had a significantly positive correlation relationship with grain yield, with a correlation coefficient of 0.827 ($p = 0.05$, Figure 5). Similar to our findings, Choudhary and Kumar [32] reported that a higher soil moisture while sowing contributes to a better crop performance with mulching under maize-based cropping practice. Importantly, maize grain yield under different treatments was significantly related to the applied mulching practices ($p = 0.05$, Tables 4 and 5); the crop performance showed a trend of TT + SM > NT + SM > TT. The post-fallow SOM and grain yield with straw mulching were significantly accelerated than those without mulching ($p = 0.05$), which was consistent with Zhao et al. [33] and Xue et al. [34]. This could be explained by the alleviation of salt accumulation and the improvement of nutrients in the topsoil treated by fallow mulching.

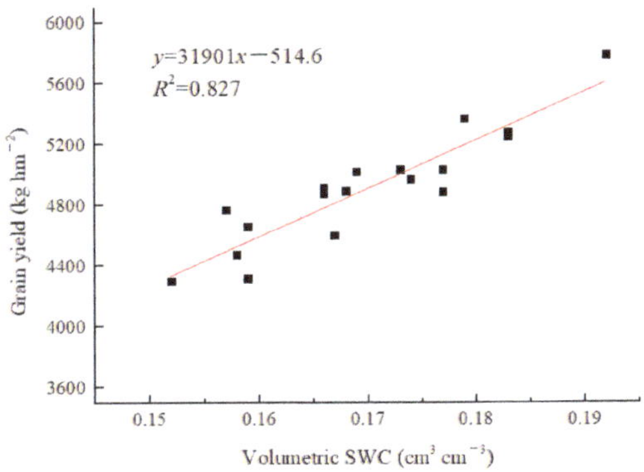

Figure 5. Relationship between pre-seeding volumetric soil water content (SWC) in the top 30 cm profile and in-season maize grain yield.

Table 5. ANOVA of the maize grain yield in line with the diverse treatments from 2016 to 2018. Y: year; T: treatments; Y × T: interaction influence of treatment and year. ** indicates significant difference at $p = 0.01$ level.

Variation Source	Degree of Freedom	Mean Square	*F*-Value	*p*-Value
T	2	329,061.5556	45.5246 **	0.0001
Y	1	197,610.3889	27.3388 **	0.0008
Y × T	2	14,636.2222	2.0249	0.1943
Error	8	7228.2222		
Total variation	17			

In terms of seedbed preparation (i.e., tillage mainly), we compared the effects of rotary tillage and no-till on crop growth in coastal salinized farmlands. The results showed that, under fallow mulching conditions, the rotary tillage seemed to be more favorable for matter accumulation (between the TT + SM and NT + SM treatments). This was due to better root development, which ultimately resulted in a higher maize yield [35], because rotary harrowing could loosen soil particles, cut off soil capillaries, and thereby slow down water evaporation, which is conducive to providing a better seedbed environment for sowing [36,37]. However, in coastal farmlands, soils are highly argillaceous with poor permeability, and in undisturbed soil, it is difficult to have positive impacts on root growing under the no-till treatment.

Meanwhile, this study was an adaptability exploration of conservation agriculture (CA) in coastal salt-affected soils. Despite the positive effects from the fallow mulching treatment, the no-till treatment did not achieve an optimal ecological environment and grain accumulation. In fact, we also conducted a no-till treatment alone (no-till seeding without fallow mulching), but only seldom was emergence monitored (results not shown). Pittelkow et al. [38] found that the no-till application alone had a negative impact on crop yield, while the negative effects of no-till could be minimized when other principles of CA (i.e., mulching or crop rotation) were applied. In dry or hydropenic climates, the yield profits with no-till combined with mulching may be due to improved soil moisture [39]. In this study, NT + SM gained an advantage in the post-fallow volumetric SWC by 5.8% and maize yield by 2.6% over the TT treatment after a 2-year cultivation, which was consistent with previous ones.

In the one-crop-per-annum cropping system, grain yield is only one of the diverse components that reflect soil productivity, and there is an urgent need for farmers and researchers to ameliorate farming management, among other socio-economic and ecological indicators. Especially in "fallow + summer maize" rotation systems, we are required to focus on both of seasonal cultivation and fallow management, rather than crop cultivation alone. The findings confirm our hypothesis that rational fallow management can reach a lower salt stress and higher water conditions for subsequent maize sowing. This will help to increase post-fallow crop yields.

However, there are several deficiencies in this research. Firstly, despite the improved soil physicochemical properties before sowing, soil evaporation during fallow was not monitored. In fact, the upward salt accumulation via capillaries is closely related to soil evaporation. Secondly, the principles of CA (i.e., no-till, soil cover, and crop rotation) were discussed in this study, but their positive influence under the no-till treatment was limited in response. A deeper interpretation of the farming patterns in coastal areas is required.

5. Conclusions

Compared with traditional tillage, fallow mulching showed an advantage in reducing the total salt of topsoil, increasing water storage, and enhancing maize growth. After 2 years of the "fallow + maize" rotation system cultivation, TT + SM reduced soil EC and the total salt of the upper 30 cm soil profile by 22.9% and 25.4% ($p = 0.05$), respectively, compared with the TT treatment. The results also indicate an improvement in volumetric SWC by 10.3%, SOM by 17.8%, and ultimately, grain yield by 11.3% ($p = 0.05$) under the TT + SM treatment.

Based on the above findings, this study could provide some guidance for scholars. Firstly, as a by-product from croplands after harvest, residues provide a method (retention by mulching) for solving agriculture-related social and economic problems, such as straw burning and biomass recycling. Secondly, aiming at coastal salinized soils, fallow mulching combined with crop rotation can also be extended to inland dry agricultural areas that require fallow to preserve soil moisture. Of course, agricultural production is a complicated, multifaceted collaborative system, and fallow cover is not an immediate management. In the coastal farming practice, it is recommended to carry out long-term fallow mulching to maintain an acceptable water and salt environment in the rhizosphere. Future studies will focus on the long-term impact of fallow mulching on rotation farming and multifaceted analyses in terms of transpiration, microorganism, soil structure, etc., will be introduced. In the meantime, this study will be applied in other soil environments, such as dryland or fresh-starved farming systems.

Author Contributions: Conceptualization, Y.Z.; methodology, L.H.; software, W.Y. and L.H.; formal analysis, Y.Z.; investigation, Y.Z.; data curation, L.H.; writing—original draft preparation, W.Y.; writing—review and editing, Y.Z.; funding acquisition, Y.Z. All authors have read and agreed to the published version of the manuscript.

Funding: This research was funded by the Natural Science Foundation of the Jiangsu Higher Education Institutions (20KJB416008), the Jiangsu Modern Agricultural Machinery Equipment and Technology Demonstration and Promotion Project (NJ2021-16), the Yangzhou University Interdisciplinary Research Foundation for Crop Science Discipline of Targeted Support (yzuxk202007).

Institutional Review Board Statement: Not applicable.

Informed Consent Statement: Not applicable.

Data Availability Statement: Not applicable.

Conflicts of Interest: The authors declare no conflict of interest.

References

1. Kumawat, K.C.; Nagpal, S.; Sharma, P. Potential of plant growth-promoting rhizobacteria-plant interactions in mitigating salt stress for sustainable agriculture: A review. *Pedosphere* **2022**, *32*, 223–245. [CrossRef]
2. Marsack, J.M.; Connolly, B.M. Generalist herbivore response to volatile chemical induction varies along a gradient in soil salinization. *Sci. Rep.* **2022**, *12*, 1689. [CrossRef] [PubMed]
3. Li, Y.Q.; Chai, Y.H.; Wang, X.S.; Huang, L.Y.; Luo, X.M.; Qiu, C.; Liu, Q.H.; Guan, X.Y. Bacterial community in saline farmland soil on the Tibetan plateau: Responding to salinization while resisting extreme environments. *BMC Microbiol* **2021**, *21*, 119. [CrossRef] [PubMed]
4. Cho, K.H.; Beon, M.; Jeong, J. Dynamics of soil salinity and vegetation in a reclaimed area in Saemangeum, Republic of Korea. *Geoderma* **2018**, *321*, 42–51. [CrossRef]

5. Lv, Z.Z.; Liu, G.M.; Yang, J.S.; Zhang, M.M.; He, L.D.; Shao, H.B.; Yu, S.P. Spatial variability of soil salinity in Bohai Sea coastal wetlands, China: Partition into four management zones. *Plant Biosyst. Int. J. Deal. All Asp. Plant Biol.* **2013**, *147*, 1201–1210. [CrossRef]
6. Liu, S.; Hou, X.; Yang, M.; Cheng, F.; Coxixo, A.; Wu, X.; Zhang, Y. Factors driving the relationships between vegetation and soil properties in the Yellow River Delta, China. *Catena* **2018**, *165*, 279–285. [CrossRef]
7. Yu, P.; Liu, S.; Yang, H.; Fan, G.; Zhou, D. Short-term land use conversions influence the profile distribution of soil salinity and sodicity in northeastern China. *Ecol. Indic.* **2018**, *88*, 79–87. [CrossRef]
8. Wang, X.; Li, Z.; Xing, Y. Effects of mulching and nitrogen on soil temperature, water content, nitrate-N content and maize yield in the Loess Plateau of China. *Agric. Water Manag.* **2015**, *161*, 53–64. [CrossRef]
9. She, D.; Liu, D.; Liu, Y.; Liu, Y.; Xu, C.; Qu, X.; Chen, F. Profile characteristics of temporal stability of soil water storage in two land uses. *Arab. J. Geosci.* **2014**, *7*, 21–34. [CrossRef]
10. Gao, L.; Shao, M.; Peng, X.; She, D. Spatio-temporal variability and temporal stability of water contents distributed within soil profiles at a hillslope scale. *Catena* **2015**, *132*, 29–36. [CrossRef]
11. Xie, X.; Pu, L.; Wang, Q.; Zhu, M.; Xu, Y.; Zhang, M. Response of soil physicochemical properties and enzyme activities to long-term reclamation of coastal saline soil, Eastern China. *Sci. Total Environ.* **2017**, *607*, 1419–1427. [CrossRef] [PubMed]
12. Shi, S.; Tian, L.; Nasir, F.; Bahadur, A.; Batool, A.; Luo, S.; Yang, F.; Wang, Z.; Tian, C. Response of microbial communities and enzyme activities to amendments in saline-alkaline soils. *Appl. Soil. Ecol.* **2019**, *135*, 16–24. [CrossRef]
13. Williams, A.; Kay, P.; Stirling, G.; Weng, X.; Bell, L. Impacts of reducing fallow periods on indicators of soil function in subtropical dryland farming systems. *Agric. Ecosyst. Environ.* **2022**, *324*, 107727. [CrossRef]
14. Ti, J.; Yang, Y.; Pu, L.; Wen, X.; Yin, X.; Chen, F. Ecological compensation for winter wheat fallow and impact assessment of winter fallow on water sustainability and food security on the North China Plain. *J Clean Prod* **2021**, *328*, 129431. [CrossRef]
15. Li, G.; Zhang, M.; Wu, C. Short-term fallow practices drive soil bacterial community changes: A case study from China. *Appl. Soil. Ecol.* **2021**, *165*, 103988. [CrossRef]
16. Passioura, J.B.; Angus, J.F. Improving Productivity of Crops in Water-Limited Environments. *Adv. Agron.* **2010**, *106*, 37–75. [CrossRef]
17. Tanwar, S.P.S.; Rao, S.S.; Regar, P.L.; Datt, S.; Praveen-Kumar; Jodha, B.S.; Santra, P.; Kumar, R.; Ram, R. Improving water and land use efficiency of fallow-wheat system in shallow Lithic Calciorthid soils of arid region: Introduction of bed planting and rainy season sorghum-legume intercropping. *Soil Tillage Res.* **2014**, *138*, 44–55. [CrossRef]
18. Guan, S.; Liu, S.; Liu, R.; Zhang, J.; Ren, J.; Cai, H.; Lin, X. Soil organic carbon associated with aggregate-size and density fractions in a Mollisol amended with charred and uncharred maize straw. *J. Integr. Agric.* **2019**, *18*, 1496–1507. [CrossRef]
19. Zhang, P.; Chen, X.; Wei, T.; Yang, Z.; Jia, Z.; Yang, B.; Han, Q.; Ren, X. Effects of straw incorporation on the soil nutrient contents, enzyme activities, and crop yield in a semiarid region of China. *Soil Tillage Res.* **2016**, *160*, 65–72. [CrossRef]
20. Stagnari, F.; Galieni, A.; Speca, S.; Cafiero, G.; Pisante, M. Effects of straw mulch on growth and yield of durum wheat during transition to Conservation Agriculture in Mediterranean environment. *Field Crops Res.* **2014**, *167*, 51–63. [CrossRef]
21. Akhtar, K.; Wang, W.; Ren, G.; Khan, A.; Feng, Y.; Yang, G. Changes in soil enzymes, soil properties, and maize crop productivity under wheat straw mulching in Guanzhong, China. *Soil Tillage Res.* **2018**, *182*, 94–102. [CrossRef]
22. Wang, M.; Chen, W.; Song, X.; Li, X.; Hu, Q.; Deng, C. Preliminary Study on Effect of Straw Mulching and Incorporation on Water and Salt Movement in Salinized Soil. *Acta Pedol. Sin.* **2017**, *54*, 1395–1403. (In Chinese)
23. Mao, L.; Guo, W.; Yuan, Y.; Qin, D.; Wang, S.; Nie, J.; Zhao, N.; Song, X.; Sun, X. Cotton stubble effects on yield and nutrient assimilation in coastal saline soil. *Field Crops Res.* **2019**, *239*, 71–81. [CrossRef]
24. Yang, Y. *Effects of Fallow Rotation on Soil Aggregates and Its Nutrients in Maize Land under Straw Returning to Field*; University of Chinese Academy of Sciences: Beijing, China, 2019. (In Chinese)
25. Liu, E.; Yan, C.; Mei, X.; He, W.; Bing, S.H.; Ding, L.; Liu, Q.; Liu, S.; Fan, T. Long-term effect of chemical fertilizer, straw, and manure on soil chemical and biological properties in northwest China. *Geoderma* **2010**, *158*, 173–180. [CrossRef]
26. Zhang, Y.; Zhang, R.; Zhang, B.; Xi, X. Artificial Macropores with Sandy Fillings Enhance Desalinization and Increase Plant Biomass in Two Contrasting Salt-Affected Soils. *Appl. Sci.* **2021**, *11*, 3037. [CrossRef]
27. He, J.; Li, H.; McHugh, A.D.; Wang, Q.; Lu, Z.; Li, W.; Zhang, Y. Permanent raised beds improved crop performance and water use on the North China Plain. *J. Soil. Water Conserv.* **2015**, *70*, 54–62. [CrossRef]
28. Zhang, Y.; Wang, W.; Yuan, W.; Zhang, R.; Xi, X. Cattle Manure Application and Combined Straw Mulching Enhance Maize (*Zea mays* L.) Growth and Water Use for Rain-Fed Cropping System of Coastal Saline Soils. *Agriculture* **2021**, *11*, 745. [CrossRef]
29. Deng, L.; Wei, W.; Hu, J.; Jiang, X.; Yang, J.; Ge, Y.; Dong, J.; Zhang, Y.; Wu, Q. Effects of Straw Mulching on Water and Salt Movement in Coastal Saline-alkali Soil. *J. Agric.* **2017**, *7*, 23–26. (In Chinese)
30. Yusefi, A.; Farrokhian Firouzi, A.; Aminzadeh, M. The effects of shallow saline groundwater on evaporation, soil moisture, and temperature distribution in the presence of straw mulch. *Hydrol. Res.* **2020**, *51*, 720–738. [CrossRef]
31. He, G.; Wang, Z.; Li, F.; Dai, J.; Li, Q.; Xue, C.; Cao, H.; Wang, S.; Malhi, S.S. Soil water storage and winter wheat productivity affected by soil surface management and precipitation in dryland of the Loess Plateau, China. *Agric. Water Manag.* **2016**, *171*, 1–9. [CrossRef]
32. Choudhary, V.; Kumar, S. Influence of mulching on productivity, root growth and weed dynamics of maize (*Zea mays*) based cropping systems. *Indian J. Agron.* **2014**, *59*, 364–370.

33. Zhao, H.; Mao, A.; Yang, H.; Wang, T.; Dou, Y.; Wang, Z.; Malhi, S. Summer fallow straw mulching and reducing nitrogen fertilization: A promising practice to alleviate environmental risk while increasing yield and economic profits of dryland wheat production. *Eur. J. Agron.* **2022**, *133*, 126440. [CrossRef]
34. Xue, J.-F.; Yuan, Y.-Q.; Zhang, H.-L.; Ren, A.-X.; Lin, W.; Sun, M.; Gao, Z.-Q.; Sun, D.-S. Carbon footprint of dryland winter wheat under film mulching during summer-fallow season and sowing method on the Loess Plateau. *Ecol. Indic.* **2018**, *95*, 12–20. [CrossRef]
35. Choudhary, V.K.; Meena, R.S. Assessment of diverse tillage system with mulching for water-cum-energy efficiency and soil carbon stabilization in maize (*Zea mays* L.)-rapeseed (*Brassica campestris* L.) system. *Soil Tillage Res.* **2022**, *219*, 105326. [CrossRef]
36. Zhang, Y.; Tan, C.; Wang, R.; Li, J.; Wang, X. Conservation tillage rotation enhanced soil structure and soil nutrients in long-term dryland agriculture. *Eur. J. Agron.* **2021**, *131*, 126379. [CrossRef]
37. Mijangos, I.; Albizu, I.; Garbisu, C. Beneficial Effects of Organic Fertilization and No-Tillage on Fine-Textured Soil Properties Under Two Different Forage Crop Rotations. *Soil Sci.* **2010**, *175*, 173–185. [CrossRef]
38. Pittelkow, C.M.; Liang, X.; Linquist, B.A.; van Groenigen, K.J.; Lee, J.; Lundy, M.E.; van Gestel, N.; Six, J.; Venterea, R.T.; van Kessel, C. Productivity limits and potentials of the principles of conservation agriculture. *Nature* **2015**, *517*, 365–482. [CrossRef] [PubMed]
39. Serraj, R.; Siddique, K.H.M. Conservation agriculture in dry areas. *Field Crops Res.* **2012**, *132*, 1–6. [CrossRef]

agriculture

MDPI

Article

Fenlong-Ridging Promotes Microbial Activity in Sugarcane: A Soil and Root Metabarcoding Survey

Mingzheng Duan [1,2], Yanyan Long [1,2,3], Hongzeng Fan [1,2], Li Ma [1,2], Shijian Han [1,2], Suli Li [1,2], Benhui Wei [3] and Lingqiang Wang [1,2,*]

1 Guangxi Key Laboratory of Sugarcane Biology, College of Agriculture, Guangxi University, Nanning 530004, China; duanmingzheng@gxu.edu.cn (M.D.); longyanyan@gxaas.net (Y.L.); 2017301007@st.gxu.edu.cn (H.F.); 2017301029@st.gxu.edu.cn (L.M.); hsjly@gxu.edu.cn (S.H.); lisuli@gxu.edu.cn (S.L.)

2 State Key Laboratory for Conservation and Utilization of Subtropical Agro—Bioresources, College of Agriculture, Guangxi University, Nanning 530004, China

3 Guangxi Academy of Agricultural Sciences, 174 Daxue East Rd., Nanning 530007, China; jzs@gxnky.cn

* Correspondence: lqwang@gxu.edu.cn

Citation: Duan, M.; Long, Y.; Fan, H.; Ma, L.; Han, S.; Li, S.; Wei, B.; Wang, L. Fenlong-Ridging Promotes Microbial Activity in Sugarcane: A Soil and Root Metabarcoding Survey. *Agriculture* **2022**, *12*, 244. https://doi.org/10.3390/agriculture12020244

Academic Editors: Chengfang Li and Lijin Guo

Received: 30 November 2021
Accepted: 1 February 2022
Published: 8 February 2022

Abstract: Fenlong-ridging (FL) is a recently proposed conservation tillage technology which has dramatic differences to traditional ones. Previous studies have demonstrated in many crops that FL has yield-increasing effects without additional inputs. However, little is known about the role that microbes play in mediating the growth-promoting effects of FL, which restricts its further application and improvement. Here, we characterized variation in the soil and root microbial diversity of sugarcane (GT44) under FL and traditional turn-over plough tillage (CK) by conducting 16S rRNA and ITS metabarcoding surveys. We also measured several phenotypic traits to determine sugarcane yields and analyzed the chemical properties of soil. We found that: (i) plant height (PH) and total biomass weight (TW) of sugarcane plants were 9.1% and 21.7% greater under FL than those under CK, indicating\increased biomass yield of the sugarcane in FL operation; (ii) contents of organic matter, total nitrogen, available phosphorus, and available potassium were lower in soil under FL than those under CK, which indicates the utilization of soil nutrients was greater in FL soil; (iii) FL promoted the activity of endophytic microbes in the roots, and these diverse microbial taxa might have an effect on sugarcane yield and soil chemical properties; and (iv) *Sphingomonas*, *Rhizobium*, and *Paraburkholderia* and *Talaromyces*, *Didymella*, and *Fusarium* were the top three most abundant genera of bacteria and fungi, respectively, in soil and root samples. In addition, strains from *Rhizobium* and *Talaromyces* were isolated to verify the results of the metabarcoding survey. Overall, our study provides new insights into the role of microbes in mediating the growth-promoting effects of FL. These findings could be used to further improve applications of this novel conservation tillage technology.

Keywords: conservation tillage; metabarcoding; smash ridging; soil chemical properties; soil microbial diversity; sugarcane

1. Introduction

The sustainable production of food is being increasingly challenged by human population growth and climate changes [1]. Conservation tillage is primarily used to protect soils from erosion and compaction, conserve soil moisture, and reduce production costs [2]. Soil and root microbial diversity and community composition are important for sustainable agriculture and conservation tillage because microbes mediate the processes supporting agricultural production [3–5]. However, many of these agriculturally important soil and root microbial taxa, and the impacts of different tillage practices on their abundances are largely unknown [6]. More studies are required to identify the soil and root microbial taxa under different types of tillage operations [7].

Fenlong-ridging (i.e., smash ridging, FL) is an advanced conservation tillage technology that has recently been proposed to increase the yields of many crops, such as rice and sugarcane, without requiring increases in fertilizer application [8,9]. Whereas traditional tillage methods involve plowing the soil, FL is a deep tillage technology (up to 40 cm in depth) that works by horizontally crushing the soil in situ. It maintains soil nutrients and moisture and increases soil air permeability, thereby enhancing the growth of crop roots [8]. This sustainable tillage method has become increasingly used in China in recent years and has helped contribute to achieving China's carbon neutrality target [10,11]. Several studies have been tried to reveal the agronomic and/or physiological mechanism underlying the yield differences under Fenlong-ridging processing [9,12–17] but little work has been done on the alteration of the physicochemical properties of the soils surrounding plants root systems. As we know, the soil and endophytic microbial diversity have substantial effects on crop yield, play an important role in regulating the supply of nutrients for crops, and mediate resistance to plant diseases and insect pests [7]. Plants and the associated microbiota form a "holobiont" [7]. When plants are facing biotic stress, they may combat stress by altering root exudates to recruit beneficial microbes from the soil, and also can improve soil chemical properties condition by the same approach [7,18]. We can speculate that the FL should causes many differences in root micro-ecological environments. Understanding the role of microbes in FL will be benefit to the application and improvement of this technology.

Sequencing technology is generally considered one of the most effective approaches for characterizing the diversity of soil microbes [19]. Many previous studies have used various sequencing technologies to study bacterial communities, and these studies have provided key insights into the diverse ways in which microbes can affect plants. For example, Wang et al. [20] studied the response of the sugarcane rhizosphere bacterial community to drought stress, Achouak et al. [21] examined the control of microbial denitrification activity by plant hosts; and Guyonnet et al. [22] found that plant nutrient resource use strategies shape active rhizosphere microbiota through root exudation using metabarcoding sequencing.

Here, we studied the role of microbes in mediating the growth-promoting effects of FL in sugarcane (*Saccharum officinarum* L.), which is the world's largest sugar-yielding crop and the second largest source of biofuel globally [23]. Specifically, we measured phenotypic indicators of yield, the chemical properties of soil, and the diversity of fungi and bacteria in the roots and rhizosphere of sugarcane through metabarcoding under FL tillage and conventional tillage. The results of the metabarcoding survey were verified by a culture-omics experiment.

2. Materials and Methods

2.1. Materials

2.1.1. Experimental Design

The experiment was conducted on the campus of Guangxi University, Nanning City, China. The experimental sugarcane field was surrounded by other fields of crops, including rice, corn, and multiple fruit trees covering 380 m^2 (Figure 1a). Two tillage methods were used before planting sugarcane, FL and conventional tillage (CK), each of which were applied every other row (i.e., tillage methods were alternated among rows). For FL, the soil layers were crushed and loosened to a depth of 40 cm. CK was conducted by turnover plowing with a mini-tiller, and the soil was tilled to a depth of 20 cm. Our tillage methods were based on the procedures described by Zhang et al. [8]. To minimize the effect of sampling on sugarcane phenotype data, we established protection rows and designated specific areas from which phenotype data and soil and root samples were collected (Figure 1a).

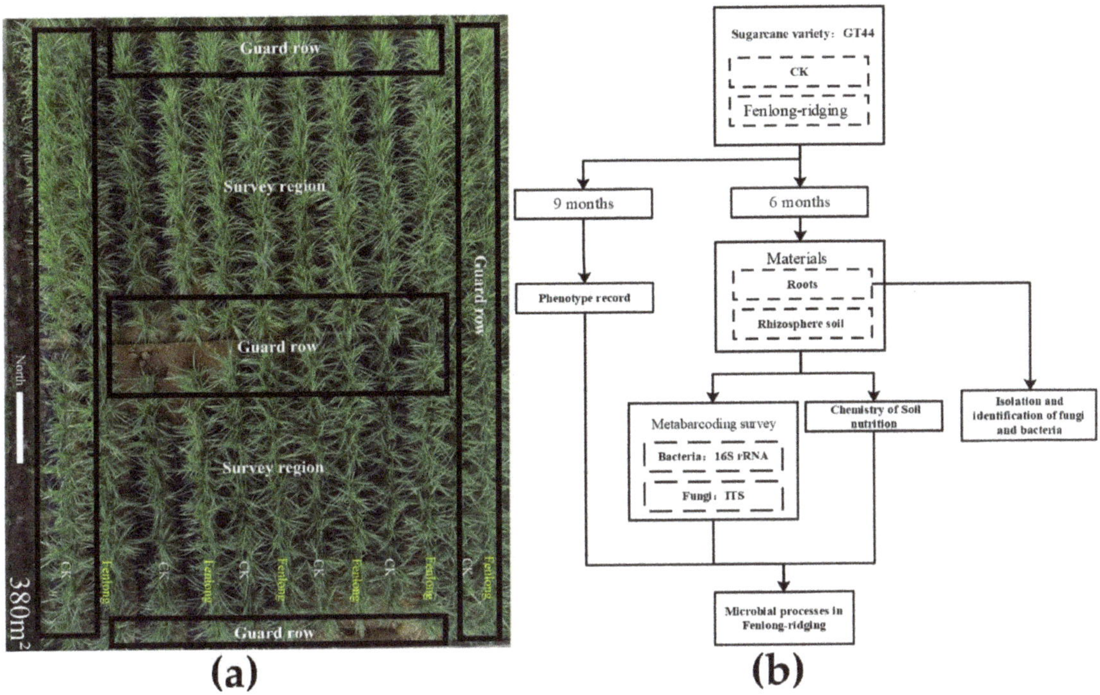

Figure 1. Design of the plots in field experiment (**a**) and the procedure of the sampling and analysis (**b**). "Fenlong" indicates Fenlong-ridging while "CK" is conventional tillage. At the sixth month after planting, the samples of sugarcane roots and rhizosphere soil were collected for metabarcoding sequencing and testing of the soil chemistry properties. Meanwhile, artificial isolation of endophytic bacteria and fungi from roots were conducted. At the ninth month, the sugarcane yield traits were investigated. Based on the data obtained from above processes, the microbial process of Fenlong-ridging in sugarcane was evaluated and analyzed.

2.1.2. Soil and Root Sampling

Soil and roots were sampled after six months of growth (Figures 1b and 2b). We randomly selected six sugarcane plants in CK and FL rows from the soil and root sample collection areas (Figure 1a) for sampling. First, we extracted entire sugarcane plants, removed the soil directly under the root system, crushed the soil, and then sifted it through a 0.6-mm sieve to obtain soil samples. The taproots were then cut and washed three times with sterile water, three times with 75% ethanol, and finally three times with sterile water (cleaning with residual ethanol) to obtain root samples.

2.2. Methods

This study was conducted per the procedures shown in Figure 1b.

2.2.1. Estimation of Sugarcane Yield

We evaluated sugarcane yields using two phenotypic traits, including total biomass weight (TW) and plant height (PH). A violin plot was created in R using the ggplot2 package (version 3.3.5; http://CRAN.R-project.org/package=ggplot2; accessed on 1 May 2021).

2.2.2. Analysis of Soil Chemical Properties

The mixed soil samples from FL and CK rows were used to determine chemical indicators, including organic matter (OM), total nitrogen (TN), available phosphorus

(AP), and available potassium (APO), which were measured at the Center of Agricultural Analysis, Testing and Research, Guangxi University, Nanning City, China.

Figure 2. Improved agronomic performance and phenotypes of sugarcane plants and the soil nutrients alteration in Fenglong compared to traditional tillage. (**a**) Statistical analysis of agronomic traits of sugarcanes under conventional tillage (CK) and Fenlong (FL) at six months. The violin-shaped columns indicate the distributions of the data. The curves of the violin-shaped columns represent the probability curve of the data distribution. The number of data points at a particular value is positively correlated with the width of the probability curve. The upper and lower ends of each violin-shaped column indicate the maximum and minimum values of non-outlier data, respectively. The upper and lower edges of the vertical line in each violin-shaped column indicate the 75th and 25th percentiles of the data, respectively; and the central dot indicates the median. (**b**) Phenotypes of the sugarcane plants at six months. (**c**) Soil nutrient traits of the soils under CK and FL conditions. The box-plot shows the maximum (top whisker), minimum (bottom whisker), median (line inside the box), upper quartile (top margin of the box), and lower quartile (lower margin of the box).

2.2.3. Metabarcoding Sequencing

Microbial DNA was extracted using HiPure Soil DNA Kits (Magen, Guangzhou, China) and DNA Isolation Kits (Sangon, No. B518231, China) per the manufacturer's protocols. The 16S rRNA V5–V7 and ITS 1–2 regions of the metabarcoding biomarkers were amplified by PCR with the primers 799F: AACMGGATTAGATACCCKG and 1193R: ACGTCATCCCCACCTTCC [24] for bacteria and the primers ITS1-F: CTTGGTCATTTAGAGGAAGTAA and ITS2: GCTGCGTTCTTCATCGATGC [25] for fungi. The purified amplicons were pooled in equimolar ratios and paired-end sequenced (PE250) on an Illumina platform (Novaseq 6000 sequencing) following standard protocols.

2.2.4. Statistical Analysis

Representative operational taxonomic unit (OTU) sequences were classified by a naïve Bayesian model using an RDP classifier [26] (version 2.2) based on the SILVA database (for 16S rRNA metabarcoding data) [27] (version 132) and UNITE database (for ITS metabarcoding data) [28] (version 8.0), with a confidence threshold value of 0.8. All figures were made using R projects. Venn analysis was used to show OTU differences among different groups and was performed in R using the VennDiagram package (version 1.6.16); [29] (version 1.6.16); Sob (to assess species richness level), Shannon and Simpson (to comprehensively assess richness and evenness of species), and Good's coverage (to assess sequencing saturation of samples). Indices were calculated in QIIME [30] (version 1.9.1).

Principal component analysis (PCA, to assess sample composition relation) and Tukey's honestly significant difference test (HSD, to assess genera significance of differences in abundance between groups) were performed in R using the vegan package (version 2.5.3; http://CRAN.R-project.org/package=vegan; accessed on 4 March 2021). Circular layout representations of species abundance were graphed using Circos [31] (version 0.69-3). All the above data were based on quantitative statistics of OTU numbers without any model transformation before analysis.

2.2.5. Isolation and Identification of Bacterial and Fungal Strains

First, clean sugarcane taproots from FL rows were collected, cut into pieces, and coated in medium (Fungi: PDA, which consisted of 200 g of potatoes, 20 g of glucose, and 16 g of agar per liter; bacteria: NB, which consisted of nutrient broth, 10 g of peptone, 3 g of beef extract powder, and 5 g of NaCl per liter) for culture at 25 °C (fungi) and 37 °C (bacteria). After 12 to 72 h, single colonies were selected for culture and preserved. We used two pairs of primers of ITS 16S rRNA as the DNA barcoding markers to identify the isolated strains, ITS1: TCCGTAGGTGAACCTGCGG and ITS4: TCCTCCGCTTATTGATATGC [32] (fungi) and 27f: AGAGTTTGATCATGGCTCAG and 1492r: ACGGTTACCTTGTTACGACTT [33] (bacteria). We identify the taxon of isolated strains by comparing with reference sequences in the database via phylogenetic trees. The reference sequences used in this study were downloaded from Genbank, and all DNA barcoding sequences together with reference sequences were aligned using Clustal X (1.83). Phylogenetic analysis based on the neighbor-joining method with 1000 bootstrap replications was conducted using MEGA v.4.0.

3. Results

3.1. The Improved Agronomic Performace and the Altered Soil Properties Were Found in Fenlong Compared with the CK

We evaluated sugarcane yield using two agricultural traits: total biomass weight (TW) and plant height (PH). The mean values of TW for FL and CK were 4.6 kg and 3.6 kg per plant (21.7% increase in FL), and the mean values of PH were 1.32 m and 1.2 m per plant (9.1% increase in FL), respectively (Figure 2a). Generally, TW and PH were increased under FL compared with CK. The growth-promoting effects of FL were apparent at six months into the experiment, as the mean PH was approximately 9% higher under FL at this point compared with CK (CK: 2.18 m; FL: 2.39 m, Figure 2b). Overall, sugarcane yield was higher under FL than under CK.

We evaluated soil chemical properties by measuring five soil nutrient parameters. For FL and CK, the mean values of OM were 19.13 and 20.96 g/kg; the mean values of TN were 1.07 and 1.13 g/kg; the mean values of AP were 121.87 and 134.41 mg/kg; and the mean value of APO were 154.66 and 158 mg/kg, respectively (Figure 2c). The mean values of OM, TN, AP, and APO were lower in the FL group than in the CK group, which suggests that sugarcane cultivated by FL utilized soil nutrients more effectively than when it was cultivated by CK.

3.2. Metabarcoding Survey of Soil and Root Microbes

3.2.1. Sequencing Analysis Revealed the Greater Diversity in Fenlong Samples than in CK Samples for Fungi and/or Bacteria

A total of 5,613,900 metabarcoding tags were obtained from the sequencing data. The clustering analyses of the soil and root samples revealed 1618 and 648 operational taxonomic units (OTUs) on average for bacteria (based on 16S rRNA) and fungi (based on ITS), respectively (Table 1). A Venn diagram is a picture showing sets of things that have a shared quality as circles that cross over each other, to show which qualities the different sets have in common. It was revealed that 36.5% (859/2356) of bacterial OTUs and 23.3% (309/1325) of fungal OTUs were shared among the four groups of samples (ROOT-FL, SOIL-FL, ROOT-CK, and SOIL-CK) (Figure 3a,b). Alpha diversity was analyzed by Tukey's HSD to assessing species diversity. The mean values of the observed species (Sob) index of soil samples from the CK and FL groups were 2058 and 2057.5 (bacteria) and 854.16 and 823.66 (fungi), respectively. The mean values of the Sob index of root samples from the CK and FL groups were 1141.83 and 1215.66 (bacteria) and 470.16 and 444.16 (fungi), respectively. The mean values of the Shannon index of soil samples from the CK and FL groups were 8.54 and 8.55 (bacteria) and 5.75 and 5.47 (fungi), respectively. The mean values of the Shannon index of root samples from the CK and FL groups were 7.03 and 6.95 (bacteria) and 3.98 and 3.0 (fungi), respectively. The mean values of the Simpson index of soil samples from the CK and FL groups were 0.99 and 0.99 (bacteria) and 0.94 and 0.92 (fungi), respectively. The mean values of the Simpson index of root samples from the CK and FL groups were 0.98 and 0.97 (bacteria) and 0.85 and 0.68 (fungi), respectively. The mean values of the Good's coverage index were all under 0.99, indicating that the level of sequencing was adequate for elucidating microbial diversity (Figure 3c,d). To further assess sample composition relation, we performed principal components analysis (PCA). It was shown that the PC1 alone could divided the OTU of bacteria into the soil group and the root group, while the PC2 further distinguish the differences existed within root group (Figure 3e). However, the differences in OTU of fungi between soil and root was not obviously (Figure 3f). It was revealed that greater variation in FL samples than in CK samples for fungi in both root samples and soil samples (Figure 3f), while the greater variation in FL samples than in CK samples for bacteria in root samples but not in soil samples (Figure 3e). Overall, the diversity of endophytic bacteria and fungi in roots was generally lower than that of soil bacteria and fungi, and there was no significant difference in the diversity of OTUs between FL and CK soil and root samples according to the Sob index (Figure 3c,d). However, significant range variation in the Shannon and Simpson indices was observed among FL and CK soil and root samples. For example, Simpson index values ranged from 0.78 to 0.90 in CK root samples but ranged from 0.47 to 0.90 in FL root samples (Figure 3c,d).

Table 1. Statistics of the metabarcoding sequencing data for soil and root samples. FL stands for Fenlong-ridging, and CK stands for conventional tillage.

Sample ID	16S rRNA			ITS		
	Tags	N90 (bp)	OTUs	Tags	N90 (bp)	OTUs
SOIL-CK-1	118718	409	2098	123409	301	823
SOIL-CK-2	103827	408	2041	115703	297	878
SOIL-CK-3	113775	409	2031	119342	301	833
SOIL-CK-4	105755	409	2027	127228	301	891
SOIL-CK-5	112379	408	2044	127989	300	865
SOIL-CK-6	112899	409	2107	128203	300	835
SOIL-FL-1	116253	409	2069	119658	302	725
SOIL-FL-2	103995	406	2019	131919	302	899
SOIL-FL-3	112407	409	2075	120641	302	788
SOIL-FL-4	110659	403	2070	117866	301	821
SOIL-FL-5	110378	409	1998	114546	301	829
SOIL-FL-6	111498	408	2114	125921	301	880
ROOT-CK-1	119801	409	1234	127742	310	491
ROOT-CK-2	103773	409	1095	113294	285	457
ROOT-CK-3	109902	409	1244	120016	301	451
ROOT-CK-4	112555	409	1111	116902	320	451
ROOT-CK-5	114501	409	1045	127469	302	531
ROOT-FL-1	109488	409	1122	128731	297	440
ROOT-FL-1	109488	409	1257	126253	302	401
ROOT-FL-2	111747	410	1098	124065	339	438
ROOT-FL-3	107628	409	1202	128470	320	444
ROOT-FL-4	111854	410	1047	113670	318	449
ROOT-FL-5	116919	409	1367	124252	296	478
ROOT-FL-6	108549	409	1323	121863	301	455

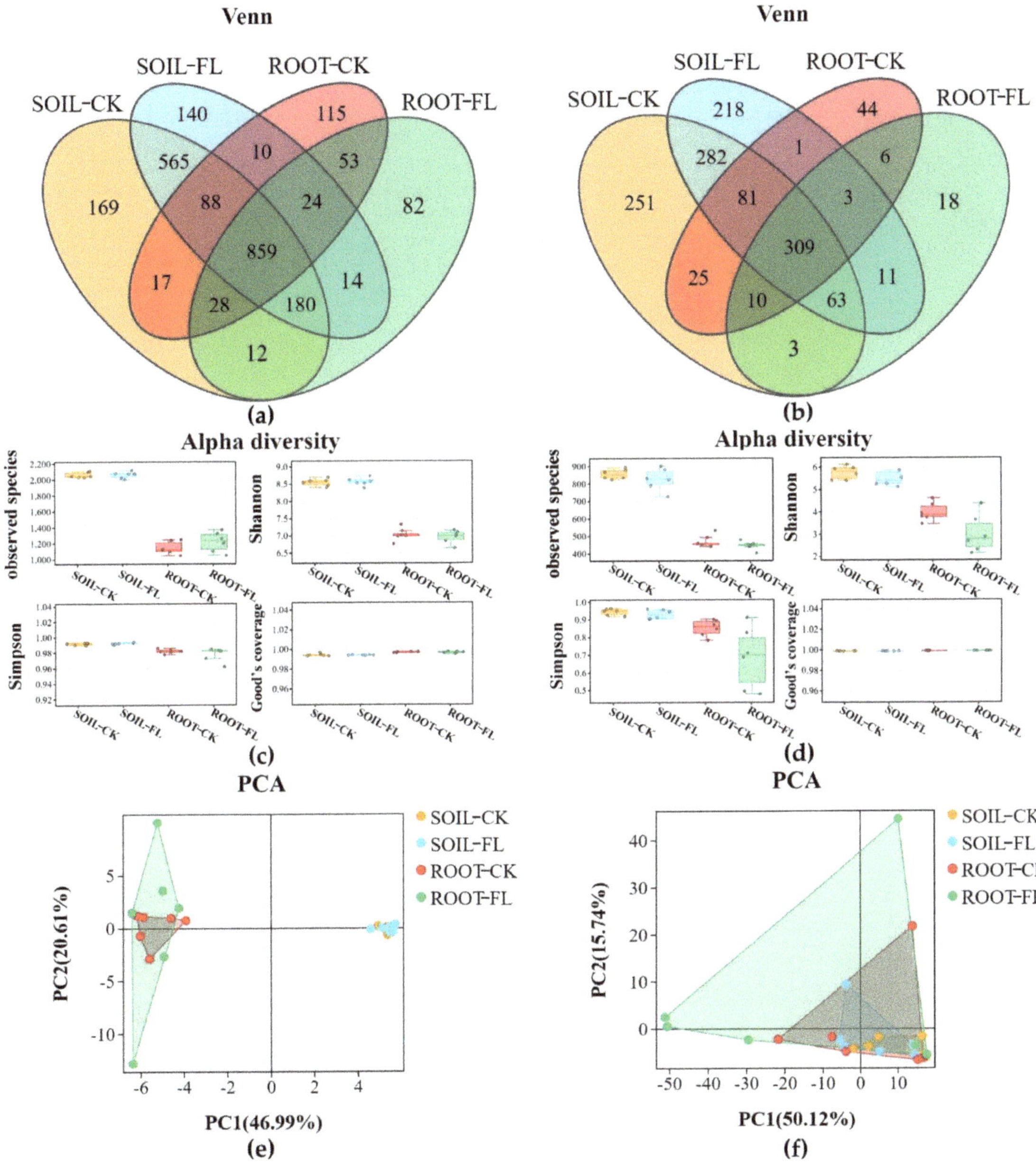

Figure 3. Venn analysis, alpha diversity analysis, and principal component analysis based on the recovered OTUs. (**a**,**b**) Venn analysis of bacteria and fungi, respectively; (**c**,**d**) alpha diversity analysis for bacteria and fungi using Tukey's HSD; (**e**,**f**) principal component analysis of the OUT of the bacteria and fungi from soil and roots under FL and CK. The colored dots in the figures correspond to the different sample groups.

3.2.2. The Predominant Microbial Genera Identified in Fenlong Operation

We analyzed differences in the community compositions of bacteria and fungi in soil and root samples from the CK and FL groups based on the SILVA and UNITE databases. The main microbial genera detected are shown in Figure 4a,b. After low-abundance taxa and unmatched OTUs were removed, the top 10 most abundant bacteria were *Sphingomonas* (24.57, 21.38, 25.17%, and 28.89 in SOIL-CK, SOIL-FL, ROOT-CK, and ROOT-FL, respectively), *Rhizobium* (6.22, 6.3, 33.76, and 53.72%), *Paraburkholderia* (13.16, 12.03, 49.18, and 25.63%), *Bradyrhizobium* (11.17, 12.21, 33.63, and 42.98%), *Dyella* (10.74, 10.95, 42.26, and 36.06%), *Amycolatopsis* (5.89, 5.05, 35.85, and 53.21%), *Pseudolabrys* (26.41, 27.14, 24.16, and 22.29%), *Nocardioides* (36.18, 32.21, 13.63, and 18.09%), *Devosia* (9.49, 9.39, 48.73, and 32.39%), and *Haliangium* (17.4, 16.61, 46.01, and 19.98%). The top 10 most abundant fungi were *Talaromyces* (14.83, 19.79, 27.92, and 34.46%), *Didymella* (48.08, 43.3, 4.48, and 4.15%), *Fusarium* (34.51, 36.69, 15.89, and 12.91%), *Corynascella* (6.78, 7.42, 56.12, and 29.68%), *Ramichloridium* (39.63, 53.22, 3.61, and 3.54%), *Rhizoctonia* (50.3, 17.41, 29.33, and 6.65%), *Penicillium* (47.23, 27.39, 19.96, and 5.42%), *Cladosporium* (41.86, 50.69, 3.38, and 4.08%), *Curvularia* (42.44, 36.53, 11.8, and 9.24%), and *Zopfiella* (20.64, 7.96, 63.45, and 7.95%).

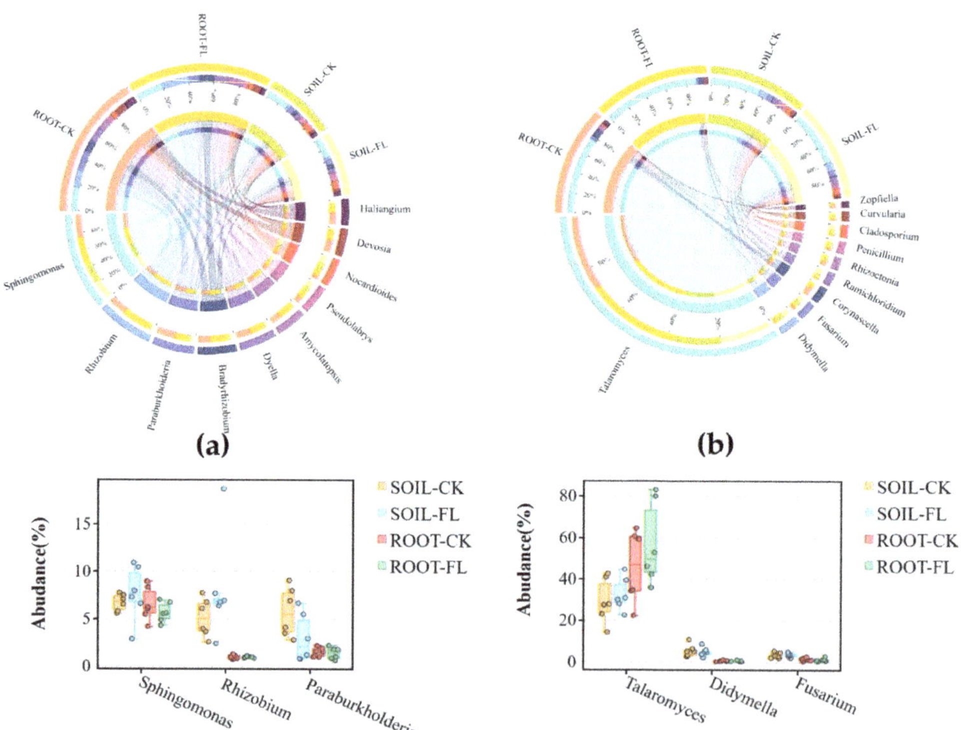

Figure 4. Microbial community composition and taxa (genera) of the top three biomarker species of bacteria and fungi at the genus level. (**a**,**b**) Top 10 abundant bacterial and fungal genera in soil and roots in the CK and FL groups. The colors of the upper half of the circle indicate the different sample groups, and the color of the lower half of the circle indicates the main genera. The colors of the outermost ring of the lower half of the circle indicate the genera, and the innermost ring of the circle indicates the abundance of the genera in the different groups. The thickness of the lines connecting genera to samples indicates the abundance of the genera in particular samples. (**c**,**d**) Biomarker genus abundance analysis for bacteria and fungi by Tukey's HSD.

We characterized differences in the distribution of the top three abundant genera between all groups (including low-abundance taxa and unmatched OTUs). The mean total relative abundances of the top three bacterial genera *Sphingomonas*, *Rhizobium*, and *Paraburkholderia* were 6.62, 3.77, and 2.9%, respectively (Figure 4c). The mean total relative abundances of the top three fungal genus *Talaromyces*, *Didymella*, and *Fusarium* were 40.04, 2.88%, and 2.4%, respectively (Figure 4d). No significant differences in the relative abundances of fungal genera in soil and root samples in the CK and FL groups were observed.

Although no statistically significant differences between CK and FL samples were detected, two bacterial and fungal genera, *Rhizobium* and *Talaromyces*, were more common in the ROOT-FL group than in the ROOT-CK group. Specifically, the abundance of *Rhizobium* was 33.76 and 53.72% in the ROOT-CK and ROOT-FL groups, respectively, and the abundance of *Talaromyces* was 27.92 and 34.46% in the ROOT-CK and ROOT-FL groups, respectively (Figure 4a,b).

3.3. Isolation and Classification of the Specific Endophytic Root Bacteria and Fungi from Sugarcane Rhizosphere

To verify the above findings, we performed a culture-omics experiment on sugarcane samples from the ROOT-FL group. A total of 100 bacterial strains and 50 fungal strains were isolated, and 14 bacterial strains and 11 fungal strains could be resolved by DNA barcoding sequencing (16S rRNA 27f/1492r was used for bacteria, and ITS 1/4 was used for fungi). The sequences of related species downloaded from Genbank (Table 2) were used to construct phylogenetic trees of fungi and bacteria to identify the isolated strains. A total of 13 of the 14 bacterial strains clustered with sequences from Genbank (Figure 5a; Table 3). R1 was not closely clustered with sequences of type species, but instead was most closely clustered with *Rhizobium* species (Figure 5a; Table 3). R3, Lx2.2, Rx11, and R2 were most closely clustered with *Bacillus aryabhattai*; Rx4 and Lx2.1 were most closely clustered with *Bacillus aerius*; Rx12 and Rx18 were most closely clustered with *Bacillus safensis*; and Rx1, Rx16, Rx5, Rx13, and R5 were most closely clustered with *Ralstonia* sp. (Figure 5a; Table 3). Among fungi, T16 and T13 were most closely clustered with *Penicillium ludwigii*; RT8 was most closely clustered with *Penicillium raperi*; T5 was most closely clustered with *Penicillium refeldin*; T24 was most closely clustered with *Penicillium* sp.; T3 was most closely clustered with *Aspergillus terreus*; T19 was most closely clustered with *Talaromyces* sp.; RT4, T8, and R3 were most closely clustered with *Talaromyces argentin*; and T18 was most closely clustered with *Curvularia petersoni* (Figure 5b; Table 3). We thus successfully isolated species from the high-abundant genera *Rhizobium* and *Talaromyces* from sugarcane roots.

Table 2. Sequence information (Genbank ID) used in this study.

Bacteria		Fungi	
Genbank ID	**Taxon**	**Genbank ID**	**Taxon**
NR026067	*Rhizobium tropici*	NR138339	*Penicillium ludwigii*
NR115466	*Rhizobium tropici*	NR121230	*Penicillium raperi*
NR118084	*Rhizobium tropici*	NR138263	*Penicillium brefeldi*
NR113739	*Rhizobium tropici*	NR158825	*Penicillium panissan*
NR044063	*Rhizobium miluonense*	NR131276	*Aspergillus terreus*
NR109703	*Rhizobium mayense*	NR077153	*Penicillium crustosu*
EF061096	*Rhizobium miluonense*	NR135337	*Aspergillus glaucus*
AY738130	*Rhizobium lusitanum*	NR103665	*Talaromyces calidica*
NR118139	*Rhizobium mesoameric*	NR147413	*Talaromyces flavus*
FN908229	*Rhizobium mesoamerica*	NR170732	*Talaromyces annesoph*
NR117203	*Rhizobium nepotum*	NR172395	*Talaromyces coprophi*
NR115953	*Bacillus aryabhattai*	NR165525	*Talaromyces argentin*
NR118439	*Bacillus aerius*	NR138223	*Curvularia lunata*
NR041794	*Bacillus safensis*	NR158448	*Curvularia petersoni*
NR113945	*Bacillus safensis*		
NR114126	*Ralstonia* sp.		

Note: 16S rRNA 27f/1492r was used to identify bacteria, and ITS1/4 was used to identify fungi.

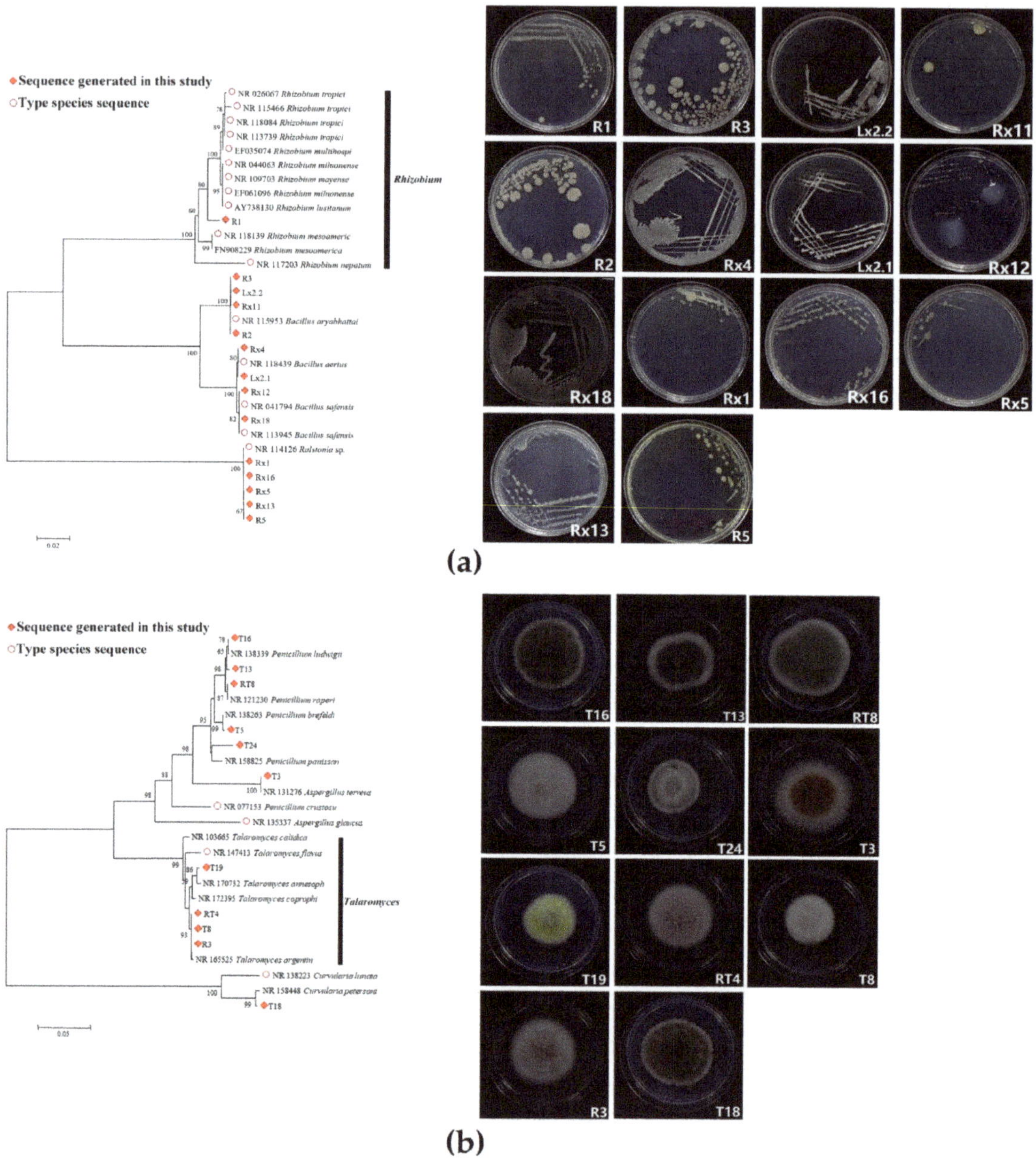

Figure 5. Phylogenetic tree of bacteria (**a**) and fungi (**b**). The similarity distance scale is provided in the lower left corner. Values on the nodes of the phylogenetic tree are bootstrap values.

Table 3. Endophytic strains of fungi and bacteria isolated in this study.

Bacteria		**Fungi**	
Strain ID	**Clustering of Specie**	**Strain ID**	**Clustered Species**
R1	*Rhizobium* sp.	T16	*Penicillium ludwigii*
R3		T13	
Lx2.2	*Bacillus aryabhattai*	RT8	*Penicillium raperi*
Rx11		T5	*Penicillium brefeldi*
R2		T24	*Penicillium* sp.
Rx4	*Bacillus aerius*	T3	*Aspergillus terreus*
Lx2.1		T19	*Talaromyces* sp.
Rx12	*Bacillus safensis*	RT4	
Rx18		T8	*Talaromyces argentin*
Rx1		R3	
Rx16		T18	*Curvularia petersoni*
Rx5	*Ralstonia* sp.		
Rx13			
R5			

4. Discussion

The development of sustainable systems of tillage with reduced effort and reduced expenditure is important for agriculture [34,35]. Fenlong (FL) is an advanced tillage operation newly developed that has been shown to significantly increase the yield of many crops, including sugarcane, without extra inputs [9,12,13,17]; however, the mechanism by which FL promotes crop growth has not been far from enough explored to date. We identified the bacteria and fungi in both soil and roots of sugarcane under FL and CK to provide insight into how soil and root microbiota mediate the growth-promoting effects of FL.

Some previous work reported that FL significantly increased sugarcane yield up to 20% [9,12]. Plant height of the sugarcane was the most robust indicator of crop yield in our data set (Figure 2a,b). Similar increases in yield have been reported in rice [8]. Our results were basically consistent with these previous studies. In addition, we also found that FL increased the yield of sugarcane by increasing the efficiency with which soil nutrients could be utilized by plants (Figure 2). The effects of tillage practices on the chemical properties of soil as well as crop growth and yields vary [36]. In FL, the soil can be deeply plowed with minimal disturbance [8]. Thus, FL provides the advantages of deep tillage, including the stability of tilled soil, which promotes the development of crop roots. There was no significant difference in the available potassium content of soil in the FL and CK groups. Potassium is key for the synthesis and translocation of sucrose [37]. This finding suggests that FL does not affect Brix value of sugarcane. Overall, our findings confirmed the efficacy of FL for increasing crop yields.

Our microbial metabarcoding survey revealed that FL promoted the activity of endophytic microbes in sugarcane roots. Although FL affected the Sob index slightly in sugarcane soil and roots, analysis of alpha indices revealed significant differences in the abundance of specific OTUs in the ROOT-FL group relative to the other three groups (Figure 3c,d), indicating that the abundance of endophytic bacteria and fungi varied greatly after FL. In addition, principal components analysis revealed that FL could increase differences in the abundances of OTUs among root samples (Figure 3d–f). These findings indicate that FL increases the diversity of the root environment. We supposed FL may enhance soil-root interaction due to the soil being smashed while the main soil layer that makes the contact area between the roots and the soil is not disturbed. This may increase intensity of competition among microbial taxa. Competition between microbial taxa might also result in the appearance of additional metabolic processes [18], and this might contribute to explaining the sugarcane yield-promoting effects of FL.

Among the top three most abundant bacterial genera, *Rhizobium* was particularly noteworthy because the abundance of this genus varied greatly among all groups (Figure 4c), and was most abundant in the FL group (Figure 4a). Rhizobia species are plant growth-promoting bacteria that provide nitrogen to hosts by binding to plant roots [38]. Rhizobia populations have been previously studied in the soil and roots of sugarcane [20,39]. We also isolated a strain (R1) from the roots of sugarcane under FL that was most closely clustered with *Rhizobium*, and the phylogenetic tree suggested that this isolate might represent a new species (Figure 5a). Other strains of soil and root bacteria that were isolated or identified included: *Sphingomonas*, which is a common genus that has been widely isolated from soil [40]; *Paraburkholderia*, which plays a role in promoting soil metabolism [41]; *Bacillus* spp., which produce various compounds that contribute to the biocontrol of plant pathogens and promote plant growth [42]; and *Bacillus aryabhattai*, which plays a role in soil bioremediation [43] (Figure 5a); *Ralstonia* sp., which has been reported produce volatile compounds that promote plant growth [44], and that may related to the growth-promoting properties of FL. Among the top abundant fungal genera detected and isolated strains, *Talaromyces* was dominant in both soil and root samples. *Talaromyces* is known to be able to carry out phosphate solubilization [45]. The abundance of *Talaromyces* was higher in root and soil samples from the FL group (Figure 4d). Thus, *Talaromyces* might affect PLT and TW traits; however, this hypothesis requires further testing. Besides, with respect to the other two fungal top genera and isolated strains (genera of *Didymella*, *Fusarium*, *Penicillium*, *Aspergillus* and *Curvularia*), their relative abundance was low which implies their association with FL may not significant.

In summary, we revealed differences in the diversity of microbial taxa in the soil and roots of sugarcane under FL and CK. Our findings provide new insights that could be used to enhance sugarcane yields. The results of this research will also aid further improvement and application of FL.

Author Contributions: Conceptualization, L.W. and B.W.; methodology, M.D. and S.H.; software, M.D.; investigation, M.D., H.F., L.M. and Y.L.; resources, L.W. and Y.L.; data curation, M.D.; writing—original draft preparation, M.D.; writing—review and editing, L.W. and Y.L.; visualization, M.D.; supervision, L.W.; project administration, L.W. and S.L.; funding acquisition, L.W. All authors have read and agreed to the published version of the manuscript.

Funding: This research was funded by China Postdoctoral Science Foundation (2020M683619XB); Guangxi innovation driven development project (No. AA20302020-3); Natural Science Foundation of Guangxi (2020GXNSFDA238027).

Institutional Review Board Statement: Not applicable.

Informed Consent Statement: Not applicable.

Data Availability Statement: The raw amplicon sequencing dataset is available in the NCBI Sequence Read Archive under BioSample accession PRJNA783171.

Acknowledgments: We thank TopEdit (www.topeditsci.com, accessed on 11 November 2021) for linguistic assistance during the preparation of this manuscript.

Conflicts of Interest: The authors declare no conflict of interest.

References

1. Zhang, H.; Li, Y.; Zhu, J.K. Developing naturally stress-resistant crops for a sustainable agriculture. *Nat. Plants* **2018**, *4*, 989–996. [CrossRef] [PubMed]
2. Holland, J.M. The environmental consequences of adopting conservation tillage in Europe: Reviewing the evidence. *Agric. Ecosyst. Environ.* **2004**, *103*, 1–25. [CrossRef]
3. Wang, Z.; Li, Y.; Li, T.; Zhao, D.; Liao, Y. Conservation tillage decreases selection pressure on community assembly in the rhizosphere of arbuscular mycorrhizal fungi. *Sci. Total Environ.* **2020**, *710*, 136326. [CrossRef] [PubMed]
4. Wipf, H.M.; Xu, L.; Gao, C.; Spinner, H.B.; Taylor, J.; Lemaux, P.; Mitchell, J.; Coleman-Derr, D. Agricultural Soil Management Practices Differentially Shape the Bacterial and Fungal Microbiome of Sorghum bicolor. *Appl. Environ. Microbiol.* **2020**, *87*, e02345-20. [CrossRef]
5. Li, Y.; Li, T.; Zhao, D.; Wang, Z.; Liao, Y. Different tillage practices change assembly, composition, and co-occurrence patterns of wheat rhizosphere diazotrophs. *Sci. Total Environ.* **2021**, *767*, 144252. [CrossRef]

6. Fierer, N. Embracing the unknown: Disentangling the complexities of the soil microbiome. *Nat. Rev. Microbiol.* **2017**, *15*, 579–590. [CrossRef]
7. Trivedi, P.; Leach, J.E.; Tringe, S.G.; Sa, T.; Singh, B.K. Plant–microbiome interactions: From community assembly to plant health. *Nat. Rev. Microbiol.* **2020**, *18*, 607–621. [CrossRef]
8. Zhang, J.; Li, F.; Liao, P.; Khan, A.; Hussain, I.; Iqbal, A.; Ali, I.; Wei, B.; Jiang, L. Smash ridge tillage strongly influence soil functionality, physiology and rice yield. *Saudi J. Biol. Sci.* **2021**, *28*, 1297–1307. [CrossRef]
9. Suli, L.; Jinling, H.; Benhui, W.; Xiaoying, L.; Ruijie, L.; Lingqiang, W.; Zhigang, L. Effects of Fenlong Tillage on Photosynthetic and Physiological Characteristics, Yield and Quality of Sugarcane (*Saccharumofficinarum*). *Chin. J. Trop. Crops* **2021**, *42*, 726–731.
10. Bin, W.; Yu-E, L.; An-Dong, C.; Shuo, L.; Tian-Jing, R.; Jia-Qi, Z. Global policies in agricultural greenhouse gas reduction and carbon sequestration and their enlightenment to China in the view of carbon neutrality. *Clim. Chang. Res.* **2022**, *18*, 1–13.
11. Zhenke, Z.; Mouliang, X.; Liang, W.; Shuang, W.; Jina, D.; Jianping, C.; Tida, G. The key biogeochemical processes of carbon sequestration in paddy soil and its countermeasures for carbon neutralization. *Chin. J. Eco-Agric.* **2021**, 1–11. Available online: http://kns.cnki.net/kcms/detail/13.1432.S.20211224.1121.001.html (accessed on 29 November 2021).
12. Ben-hui, W.; Xiu-qin, G.; Zhang-you, S.; Xiu-cheng, N.; Liu-ying, L.; Guang-po, W.; Yan-ying, L.; Po, H.; Bin, L.; Yan-yong, W. Yield Increase of Smash-Ridging Cultivation of Sugarcane. *Sci. Agric. Sin.* **2011**, *44*, 4544–4550. [CrossRef]
13. Benhui, W.; Zhangyou, S.; Jia, Z.; Lingzhi, Z.; Po, H.; Xian, Z. Study on Effect and Mechanism of Improving Saline-alkali Soil by Fenlong Tillage. *Soils* **2020**, *52*, 699–703. [CrossRef]
14. Jianghan, L.; Wenshou, H. Effects of Smash-ridging Technology on Soil Properties and Potato Yield. *J. Northeast Agric. Sci.* **2020**, *45*, 20–25. [CrossRef]
15. Shijia, W.; Daihua, J.; Wenguo, Z.; Rongrong, Z.; Junwei, L.; Benhui, W. Effect of Deep Vertical Rotary Tillage on Aggregate Structure in Farmland of Lateritic Red Soil. *Acta Pedol. Sin.* **2020**, *57*, 326–335. [CrossRef]
16. Xiaohua, D.; Xinyue, W.; Hongwu, Y.; Yongjun, L.; Yongsheng, D.; Miliang, Z.; Mingfa, Z.; Jiongping, Z.; Qi, L.; Weimin, W.; et al. Effects of Smashing Ridge Tillage on Growth, Dry Matter Accumulation, Output and Quality of Flue-cured Tobacco. *Chin. Tob. Sci.* **2020**, *41*, 28–35. [CrossRef]
17. Hao, L.; Jin-ling, H.; Zhi-gang, L.; Ben-hui, W.; Xiao-ru, C.; Shi-jian, H.; Xiao-ying, L.; Su-li, L. Fenlong tillage increase soil nutrient availability, and benefit vascular tissue structure and nutrient absorption of sugarcane. *J. Plant Nutr. Fertil.* **2021**, *27*, 204–214. [CrossRef]
18. Rolfe, S.A.; Griffiths, J.; Ton, J. Crying out for help with root exudates: Adaptive mechanisms by which stressed plants assemble health-promoting soil microbiomes. *Curr. Opin. Microbiol.* **2019**, *49*, 73–82. [CrossRef]
19. Nesme, J.; Achouak, W.; Agathos, S.N.; Bailey, M.; Baldrian, P.; Brunel, D.; Frostegård, Å.; Heulin, T.; Jansson, J.K.; Jurkevitch, E.; et al. Back to the Future of Soil Metagenomics. *Front. Microbiol.* **2016**, *7*, 73. [CrossRef]
20. Liu, Q.; Zhao, X.; Liu, Y.; Xie, S.; Xing, Y.; Dao, J.; Wei, B.; Peng, Y.; Duan, W.; Wang, Z. Response of Sugarcane Rhizosphere Bacterial Community to Drought Stress. *Front. Microbiol.* **2021**, *12*, 716196. [CrossRef]
21. Achouak, W.; Abrouk, D.; Guyonnet, J.; Barakat, M.; Ortet, P.; Simon, L.; Lerondelle, C.; Heulin, T.; Haichar, F.E.Z. Plant hosts control microbial denitrification activity. *FEMS Microbiol. Ecol.* **2019**, *95*, fiz021. [CrossRef] [PubMed]
22. Guyonnet, J.P.; Guillemet, M.; Dubost, A.; Simon, L.; Ortet, P.; Barakat, M.; Heulin, T.; Achouak, W.; Haichar, F.E.Z. Plant Nutrient Resource Use Strategies Shape Active Rhizosphere Microbiota Through Root Exudation. *Front. Plant Sci.* **2018**, *9*, 1662. [CrossRef] [PubMed]
23. Singh, R.K.; Singh, P.; Li, H.B.; Song, Q.Q.; Guo, D.J.; Solanki, M.K.; Verma, K.K.; Malviya, M.K.; Song, X.P.; Lakshmanan, P.; et al. Diversity of nitrogen-fixing rhizobacteria associated with sugarcane: A comprehensive study of plant-microbe interactions for growth enhancement in *Saccharum* spp. *BMC Plant Biol.* **2020**, *20*, 220. [CrossRef] [PubMed]
24. Beckers, B.; De Beeck, M.O.; Thijs, S.; Truyens, S.; Weyens, N.; Boerjan, W.; Vangronsveld, J. Performance of 16s rDNA Primer Pairs in the Study of Rhizosphere and Endosphere Bacterial Microbiomes in Metabarcoding Studies. *Front. Microbiol.* **2016**, *7*, 650. [CrossRef]
25. Toju, H.; Tanabe, A.S.; Yamamoto, S.; Sato, H. High-coverage ITS primers for the DNA-based identification of ascomycetes and basidiomycetes in environmental samples. *PLoS ONE* **2012**, *7*, e40863. [CrossRef]
26. Wang, Q.; Garrity, G.M.; Tiedje, J.M.; Cole, J.R. Naive Bayesian classifier for rapid assignment of rRNA sequences into the new bacterial taxonomy. *Appl. Environ. Microbiol.* **2007**, *73*, 5261–5267. [CrossRef]
27. Pruesse, E.; Quast, C.; Knittel, K.; Fuchs, B.M.; Ludwig, W.; Peplies, J.; Glöckner, F.O. SILVA: A comprehensive online resource for quality checked and aligned ribosomal RNA sequence data compatible with ARB. *Nucleic Acids Res.* **2007**, *35*, 7188–7196. [CrossRef]
28. Nilsson, R.H.; Larsson, K.H.; Taylor, A.F.S.; Bengtsson-Palme, J.; Jeppesen, T.S.; Schigel, D.; Kennedy, P.; Picard, K.; Glöckner, F.O.; Tedersoo, L.; et al. The UNITE database for molecular identification of fungi: Handling dark taxa and parallel taxonomic classifications. *Nucleic Acids Res.* **2019**, *47*, D259–D264. [CrossRef]
29. Chen, H.; Boutros, P.C. VennDiagram: A package for the generation of highly-customizable Venn and Euler diagrams in R. *BMC Bioinform.* **2011**, *12*, 35. [CrossRef]
30. Caporaso, J.G.; Kuczynski, J.; Stombaugh, J.; Bittinger, K.; Bushman, F.D.; Costello, E.K.; Fierer, N.; Peña, A.G.; Goodrich, J.K.; Gordon, J.I.; et al. QIIME allows analysis of high-throughput community sequencing data. *Nat. Methods* **2010**, *7*, 335–336. [CrossRef]

31. Krzywinski, M.; Schein, J.; Birol, I.; Connors, J.; Gascoyne, R.; Horsman, D.; Jones, S.J.; Marra, M.A. Circos: An information aesthetic for comparative genomics. *Genome Res.* **2009**, *19*, 1639–1645. [CrossRef] [PubMed]
32. White, T.J.; Bruns, T.; Lee, S.; Taylor, J. 38—Amplification and Direct Sequencing of Fungal Ribosomal RNA Genes for Phylogenetics. In *PCR Protocols*; Innis, M.A., Gelfand, D.H., Sninsky, J.J., White, T.J., Eds.; Academic Press: San Diego, CA, USA, 1990; pp. 315–322. [CrossRef]
33. Avşar, C.; Aras, E.S. Community structures and comparison of nosZ and 16S rRNA genes from culturable denitrifying bacteria. *Folia Microbiol.* **2020**, *65*, 497–510. [CrossRef]
34. Busari, M.A.; Kukal, S.S.; Kaur, A.; Bhatt, R.; Dulazi, A.A. Conservation tillage impacts on soil, crop and the environment. *Int. Soil Water Conserv. Res.* **2015**, *3*, 119–129. [CrossRef]
35. Selman, P. Learning to Love the Landscapes of Carbon-Neutrality. *Landsc. Res.* **2010**, *35*, 157–171. [CrossRef]
36. Li, J.; Wang, Y.-K.; Guo, Z.; Li, J.-B.; Tian, C.; Hua, D.-W.; Wang, H.-Y.; Han, J.-C.; Xu, Y. Effects of conservation tillage on soil physicochemical properties and crop yield in an arid Loess Plateau, China. *Sci. Rep.* **2020**, *10*, 4716. [CrossRef] [PubMed]
37. Kwong, K. The effects of potassium on growth, development, yield and quality of sugarcane. In *Pasricha and Bansal. Potassium for Sustainable Crop Production: Potash Research Institute of India and International Potash Institute*; Springer: Horgen, Switzerland, 2002; pp. 430–440. Available online: https://www.ipipotash.org/uploads/udocs/Potassium%20in%20sugarcane.pdf (accessed on 29 November 2021).
38. Poole, P.; Ramachandran, V.; Terpolilli, J. Rhizobia: From saprophytes to endosymbionts. *Nat. Rev. Microbiol.* **2018**, *16*, 291–303. [CrossRef]
39. Júnior, I.D.A.M.; de Matos, G.F.; de Freitas, K.M.; da Conceição Jesus, E.; Rouws, L.F.M. Occurrence of diverse Bradyrhizobium spp. in roots and rhizospheres of two commercial Brazilian sugarcane cultivars. *Braz. J. Microbiol.* **2019**, *50*, 759–767. [CrossRef]
40. An, D.S.; Liu, Q.M.; Lee, H.G.; Jung, M.S.; Kim, S.C.; Lee, S.T.; Im, W.T. *Sphingomonas ginsengisoli* sp. nov. and *Sphingomonas sediminicola* sp. nov. *Int. J. Syst. Evol. Microbiol.* **2013**, *63*, 496–501. [CrossRef]
41. Cyle, K.T.; Klein, A.R.; Aristilde, L.; Martínez, C.E. Ecophysiological Study of *Paraburkholderia* sp. Strain 1N under Soil Solution Conditions: Dynamic Substrate Preferences and Characterization of Carbon Use Efficiency. *Appl. Environ. Microbiol.* **2020**, *86*, e01851-20. [CrossRef]
42. Miljaković, D.; Marinković, J.; Balešević-Tubić, S. The Significance of *Bacillus* spp. in Disease Suppression and Growth Promotion of Field and Vegetable Crops. *Microorganisms* **2020**, *8*, 1037. [CrossRef]
43. Inthama, P.; Pumas, P.; Pekkoh, J.; Pathom-Aree, W.; Pumas, C. Plant Growth and Drought Tolerance-Promoting Bacterium for Bioremediation of Paraquat Pesticide Residues in Agriculture Soils. *Front. Microbiol.* **2021**, *12*, 604662. [CrossRef] [PubMed]
44. Tahir, H.A.S.; Gu, Q.; Wu, H.; Raza, W.; Safdar, A.; Huang, Z.; Rajer, F.U.; Gao, X. Effect of volatile compounds produced by Ralstonia solanacearum on plant growth promoting and systemic resistance inducing potential of Bacillus volatiles. *BMC Plant Biol.* **2017**, *17*, 133. [CrossRef] [PubMed]
45. Rubio, P.J.S.; Godoy, M.S.; Mónica, I.F.D.; Pettinari, M.J.; Godeas, A.M.; Scervino, J.M. Carbon and Nitrogen Sources Influence Tricalcium Phosphate Solubilization and Extracellular Phosphatase Activity by Talaromyces flavus. *Curr. Microbiol.* **2016**, *72*, 41–47. [CrossRef] [PubMed]

agriculture

Article

Phytoremediation of Secondary Salinity in Greenhouse Soil with *Astragalus sinicus*, *Spinacea oleracea* and *Lolium perenne*

Shumei Cai, Sixin Xu, Deshan Zhang, Zishi Fu, Hanlin Zhang * and Haitao Zhu

Institute of Eco-Environment and Plant Protection, Shanghai Academy of Agricultural Sciences, Shanghai 201403, China; caishumei@saas.sh.cn (S.C.); xsxofsaas@163.com (S.X.); zds234@163.com (D.Z.); fzs@foxmail.com (Z.F.); htzhu123@163.com (H.Z.)
* Correspondence: zhanghanlinchick@163.com

Abstract: Phytoremediation is an effective and ecological method used to control soil secondary salinization in greenhouses. However, the plant–soil interactions for phytoremediation have not been studied sufficiently. In this study, three crop species (*Astragalus sinicus* (CM), *Spinacea oleracea* (SP) and *Lolium perenne* (RY)) were compared in a greenhouse experiment. The results showed that all three crops increased the soil microbial biomass, the abundance of beneficial microorganisms, available phosphorus and soil pH, and reduced the soil salt content. The crop nutrient accumulation was positively correlated with the relative abundance of bacterial 16S rRNA sequences in the soil. CM and RY respectively increased the relative abundances of *norank_f_Gemmatimonadaceae* and *norank_f_Anaerolineaceae* within the soil bacterial community, while SP increased the relative abundances of *Gibellulopsis* within the fungal community. Correlation analysis revealed that pH and total dissolved salts were the vital factors affecting soil microbial communities in the secondary salinized soil. Our results suggest that phytoremediation could effectively alleviate secondary salinization by regulating the balance of soil microbial community composition and promoting crop nutrient accumulation.

Keywords: phytoremediation; secondary salinization; salt tolerance; microbial diversity; nutrient accumulation

Citation: Cai, S.; Xu, S.; Zhang, D.; Fu, Z.; Zhang, H.; Zhu, H. Phytoremediation of Secondary Salinity in Greenhouse Soil with *Astragalus sinicus*, *Spinacea oleracea* and *Lolium perenne*. *Agriculture* **2022**, *12*, 212. https://doi.org/10.3390/agriculture12020212

Academic Editors: Chengfang Li and Lijin Guo

Received: 21 December 2021
Accepted: 28 January 2022
Published: 1 February 2022

Publisher's Note: MDPI stays neutral with regard to jurisdictional claims in published maps and institutional affiliations.

1. Introduction

Secondary soil salinization, which is mainly caused by intensified anthropogenic agricultural production, has been recognized as an extensive form of land degradation [1–3]. High inputs of agrochemicals, high evaporation, and mineral leaching in the intensified production system significantly intensify secondary soil salinization, as well as high Na^+ accumulation in surface soil, which restricts agricultural productivity worldwide [4,5]. The salinization causes soil compaction and an imbalance in nutrient supply, which directly harms the normal growth of crops. Furthermore, the salinization alters the status of soil microorganisms, thereby indirectly affecting the entire ecological environment, and thus hindering the sustainable development of agricultural production [6].

Phytoremediation can alleviate secondary salinization in facility cultivation soils and reduce the dependence on mineral fertilizers [7]. In previous studies, it has been suggested that soil microbes are susceptible to farming practices, and that selecting an effective crop has positive effects on microbial communities and functions [8]. For example, applying a green manure crop has been shown to significantly change the soil microbial community composition and function [9]. Fungi, bacteria, and actinomycetes have been found to be active in the rhizosphere of Italian ryegrass [10]. The symbioses of these microorganisms accelerated nutrient cycling processes [11]. Therefore, understanding the structure of the soil microbial community and its responses to applications of different types of crops is necessary to elucidate the effects of microbes on secondary salinization.

In this study, it was hypothesized that the saline soil biochemical properties and microbial communities would change consistently with the type of crop species planted. Further, it was postulated that there would be considerable linkage between crop nutrient accumulation, soil salinization indicators and soil biochemical properties during the process of phytoremediation. The objectives of this study were to clarify the impact of different crop species on soil biochemical properties and microbial communities in cultivation facility soil, and to explore the linkage between crop nutrient accumulation and the composition of soil microbial communities.

2. Materials and Methods

2.1. Field Site Setup, Management and Sampling

A greenhouse experiment was performed at the Zhuanghang Comprehensive Experimental Station of the Shanghai Academy of Agricultural Sciences, Fengxian, Shanghai, China (30°53′20.0″ N, 121°23′06.4″ E). The study site was flat and the soil type was calcareous alluvium. Three replicates of four treatments were arranged in a randomized block design using 30 m × 2 m plots constructed in January 2015 (Figure 1). Nylon screen fabric was erected around every plot to avoid runoff effects, and it was buried beneath the soil surface with a height of 40 cm. Four treatments were set up, including the fallow control (CK), Chinese milk vetch (*Astragalus sinicus* L., CM), Spinach (*Spinacia oleracea* L., SP) and Ryegrass (*Lolium perenne* L., RY). CM, SP and RY were selected because they are the major native winter cover crop species that are easily accessible and widely applied to ameliorate soil salinization. Seeds of CM, SP and RY were obtained from Shanghai Nongle Planting Co. Ltd. (Shanghai, China). After 3 years of continuous planting, soil samples were collected on 30 January, 2018. In each separated plot, soil samples from the 0–20 cm surface layer were collected from 8 points to form a mixed composite soil sample, which was then divided into two parts, with one part air dried prior to the determination of basic physicochemical properties, and the other stored at −80 °C prior to the DNA extraction.

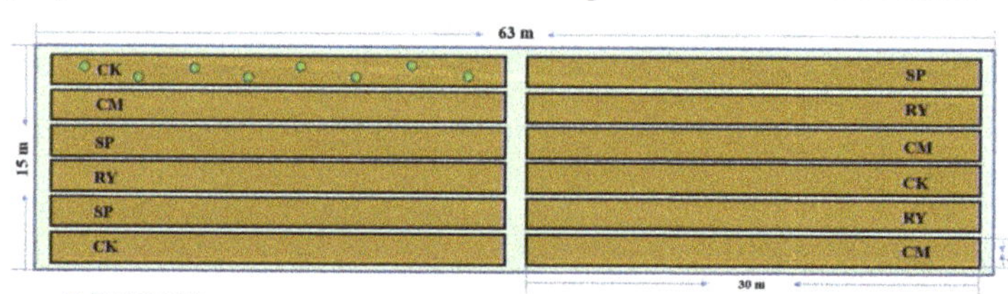

Figure 1. The schematic for field site setup, management and sampling. CK, control; CM, *Astragalus sinicus*; SP, *Spinacia oleracea*; RY, *Lolium perenne*.

2.2. Crop Yield and Nutrient Accumulation

The experimental crops were planted by sowing the equivalent of 75 kg ha^{-1} of seed per plot in early October, and the crops were grown and harvested until the end of January annually. Equal amounts of irrigation water were supplied to each plot and no fertilization was used during the period of phytoremediation. The former crop for all treatments was pakchoi (*Brassica chinensis* L.), and urea (N 46%); compound fertilizer (17-17-17) and potassium sulphate (K_2O 52%) were applied as the N, P and K fertilizer, with application rates of N 120 kg ha^{-1}, P_2O_5 45 kg ha^{-1} and K_2O 90 kg ha^{-1}. Crops were harvested at the same time as the soil samples were taken. The selected uniformly growing plants were taken to the laboratory immediately and were dried to a constant weight in preparation for nutrient determination. The total nitrogen (N), phosphorus (P) and potassium (K) contents of the dry matter were quantified using the Kjeldahl nitrogen determination method, the vanadium-molybdenum-yellow photometric method, and the flame photometry approach, respectively [12].

2.3. Determination of Soil Physicochemical Properties

Soil chemical properties, including the pH, total N, soil organic matter (SOM), alkali-hydrolyzable nitrogen (AMN), available P (AP), available K (AK), and total dissolved salts (TDS), were tested according to the methods of Bao (2000). The soil pH was tested using a soil-to-water ratio of 1:2.5. The soil total N was determined via the Kjeldahl method and SOM was determined via the potassium dichromate oxidation method. The soil AMN content was measured using the alkaline hydrolysis diffusion method. The AP and AK were measured using the molybdenum blue colorimetric method and the flame photometry method, respectively. The TDS in the soil were determined using the gravimetric method. The soil microbial biomass C (MBC) and N (MBN) were measured using the chloroform fumigation method [13].

2.4. Soil DNA Extraction and Microbial Community Analysis

Bacterial and fungal DNA were extracted as three replicates from each soil sample using a FastDNA Spin Kit for Soil and were stored at −80 °C. The bacterial V3–V4 region of the 16S rRNA gene was amplified using the primers 338F and 806R [6]. The internal transcribed spacer (ITS) region of the fungal rRNA gene was amplified using ITS1F and ITS2R [14]. All PCR reactions were performed according to the methods described by Cai et al. [15]. Pyrosequencing was carried out by Majorbio Bio-Pharm Biotechnology Co., Ltd., Shanghai, China, using the Illumina Miseq PE250 platform. After high-throughput sequencing and optimization, 566,160 bacterial 16S rRNA sequences with 235,957,809 bp were obtained from the four treatments (N = 12), and the average sequence length was 416.8 bp. Meanwhile, 727,233 fungal ITS sequences with 173,088,882 bp were obtained, and the average sequence length was 238.0 bp. Sequence data were deposited in the National Center for Biotechnology Information (NCBI) Sequence Read Archive (SRA) under the accession number SRP273207.

2.5. Real-Time Quantitative Polymerase Chain Reaction (qPCR)

qPCR was used to examine the effects of bioremediation on soil microbial abundance. Standard reactions were performed for all samples in triplicate with an ABI7500 Real-time PCR System (Applied Biosystems INC, Foster City, CA, USA) using the SYBR green qPCR method. The standard curves and amplification curves are shown in Figures S1–S4. The qPCR mixture (20 μL) contained 10 μL of Maxima SYBR green/ROX qPCR Master Mix (Thermo Fisher Scientific, Waltham, MA, USA), 0.8 μL of each primer, 1.0 μL of template DNA and 7.4 μL of dd H_2O. The amplification conditions of 16S rRNA comprised pre-denaturation for 5 min at 95 °C, followed by 30 amplification cycles of denaturation at 95 °C for 30 s, annealing for 30 s at 55 °C and extension at 72 °C for 1 min. The amplification conditions of ITS were nearly the same despite the difference of annealing temperature at 62 °C. The gene abundances of each reaction were calculated based on the constructed standard curves and then converted to copies per gram of soil, assuming 100% DNA extraction efficiency.

2.6. Data Analysis

The effects of the different crop treatments on the physicochemical soil properties and crop biomass values were tested using one-way ANOVA in SPSS 17.0 (SPSS Inc., Chicago, IL, USA). QIIME (1.7.0) software was used to calculate the alpha and beta diversities of the soil bacterial and fungal communities. The OTUs were used to characterize the alpha diversity. The Chao1, ACE, Shannon and Simpson indices were calculated. Principal coordinates analysis (PCoA) of the unweighted UniFrac distances between the samples was used to characterize the similarities (beta diversity) in the bacterial and fungal communities among the treatments [16]. The vegan data package in R was used for redundancy analysis (RDA), which was used to identify factors that affected microbial community structure.

3. Results

3.1. Response of Soil Biochemical Properties to Green Manure Crops

The application of green manure in the form of the three different crops improved the biochemical properties associated with soil fertility (Table 1). Compared to CK, all cultivation treatments displayed lower TDS contents, but higher soil pH, MBC and MBN contents ($p < 0.05$). The AP and AK contents were also significantly higher than in the control, irrespective of the type of crop applied ($p < 0.05$). The AMN and MBN contents were significantly higher in the CM treatment than in the other treatments ($p < 0.05$). The AP and AK contents were significantly higher in RY than in the other treatments ($p < 0.05$).

Table 1. Biochemical properties of the soil in each bioremediation treatment.

Treatments	pH	TDS (g kg^{-1})	SOC (g kg^{-1})	AMN (mg kg^{-1})	AP (mg kg^{-1})	AK (mg kg^{-1})	MBC (mg kg^{-1})	MBN (mg kg^{-1})
CK	4.98 c	3.25 a	11.70 b	120.58 b	45.56 c	205.20 a	108.45 c	38.70 c
CM	5.29 b	2.81 b	12.58 ab	139.33 a	90.47 b	336.92 c	192.94 ab	62.60 a
SP	5.33 b	2.74 b	11.79 b	104.06 c	92.44 b	282.98 b	188.36 b	52.56 b
RY	5.52 a	2.52 b	13.09 a	95.89 c	113.60 a	426.08 d	194.84 a	54.92 b

Note: TDS, total dissolved salts; SOC, soil organic carbon; AMN, alkali-hydrolyzable nitrogen; AP, available phosphorus; AK, available potassium; MBC, microbial biomass carbon; MBN, microbial biomass nitrogen. CK, control; CM, *Astragalus sinicus*; SP, *Spinacia oleracea*; RY, *Lolium perenne*. Different letters in the same column indicate a significant difference between treatments at the 0.05 level ($n = 3$).

3.2. The Yield and Nutrient Uptake and Accumulation of the Different Crop Species

Crop yield significantly differed both in terms of shoot biomass and root biomass (Figure 2). RY had a significantly higher yield than SP and CM ($p < 0.01$), with the whole fresh biomass of RY reaching 83.8 kg ha^{-1}. As shown in Figure 3, the three crops displayed the highest cumulative uptake for K, followed by N, and then P. The cumulative K uptake by RY was significantly greater than that exhibited by SP and CM ($p < 0.01$). It was also observed that the root-to-shoot ratios of dry weight for SP and CM were significantly higher than the ratio of RY (Table S1). Further, significant differences were observed in the nutrient contents of the three crops (Figure S5). The N content of RP was significantly lower than that of CM and SP. The P content was higher in SP than in CM and RY. The average K content of the three crops was in the following order: SP > RY > CM.

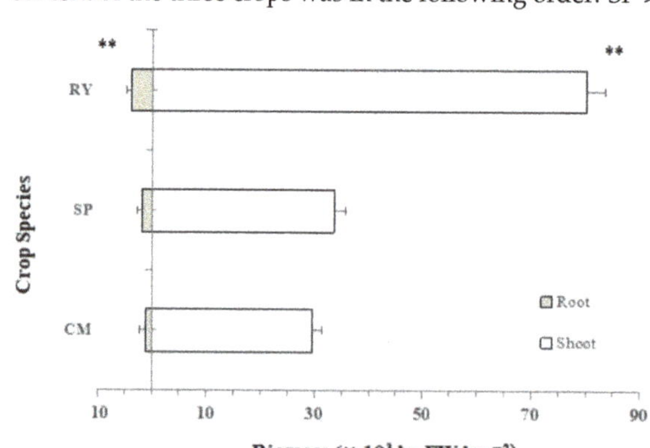

Figure 2. Shoot and root biomass (fresh weight [FW]) of *Lolium perenne* (RY), *Spinacia oleracea* (SP) and *Astragalus sinicus* (CM). Vertical bars represent the standard error of the mean. ** denotes statistically significant differences between crop varieties ($p < 0.01$, Duncan's test).

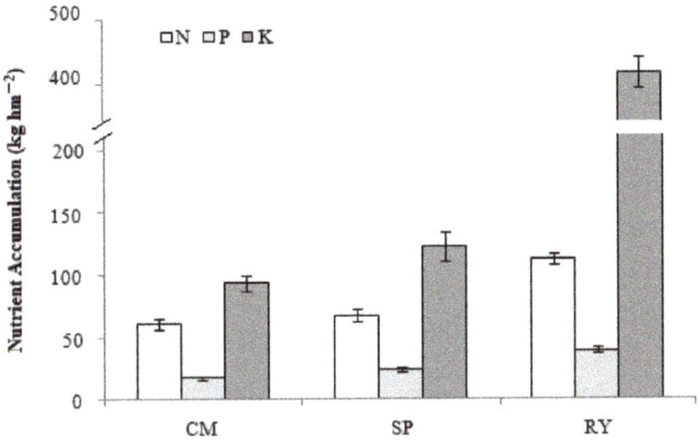

Figure 3. Nutrient accumulation of the three different crop species, *Astragalus sinicus* (CM), *Spinacia oleracea* (SP) and *Lolium perenne* (RY). Vertical bars represent the standard error of the mean.

3.3. Abundance and Diversity of Doil Microbial Communities

Compared to the CK samples, significantly more 16S rRNA sequences but fewer ITS copies ($p < 0.05$) were found in the CM and SP samples (Table 2). However, RY had no significant effects on the number of ITS sequences in the soil compared with CK. Quantitative PCR (qPCR) data also indicated that the bacteria-to-fungi (B/F) ratio declined in the following order: SP > CM > RY > CK.

Table 2. Bacterial (16S rRNA) and fungal (ITS) gene copy numbers in soil samples.

Treatments	16S Gene Copy Numbers (Copies × 10^{10})	ITS Gene Copy Numbers (Copies × 10^8)	B/F (Bacteria/Fungi 10^3)
CK	2.50 d	1.44 a	0.17 c
CM	4.21 b	0.82 b	0.51 b
SP	2.83 c	0.26 c	1.11 a
RY	6.69 a	1.47 a	0.46 b

Note: CK, control; CM, *Astragalus sinicus*; SP, *Spinacia oleracea*; RY, *Lolium perenne*. Different letters in the same column indicate a significant difference between treatments at the 0.05 level ($n = 3$).

The α-diversity analysis showed that the bacterial and fungal community richness (Chao1 and ACE) and diversity (Shannon and Simpson) indices varied markedly among the treatments (Table S2). Crop application increased the bacterial and fungal richness indices ($p < 0.05$) and the bacterial Shannon indices ($p < 0.05$) when compared to the control.

The crop treatments were related to an increase in the relative abundance of Proteobacteria and Bacteroidetes, and a decrease in the relative abundance of Actinobacteria for soil bacteria at the phylum level (Figure 4). The fungal community at the phylum level was comparable among all soil samples except for SP samples, which had a high abundance of Basidiomycota and Unclassified_k_Fungi (Figure 5). With respect to the bacterial community at the genus level, CM increased the relative abundances of *norank_f_Gemmatimonadaceae*. With respect to the fungal community, CM and RY both increased the relative abundances of *Chaetomium* and *Humicola*. A higher relative abundance of *Gibellulopsis* was observed in the SP samples.

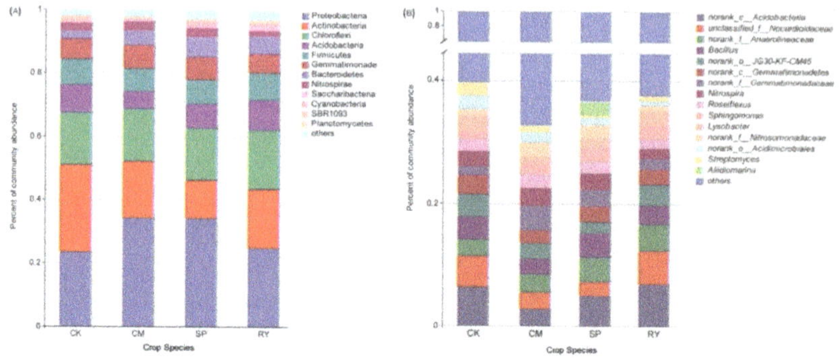

Figure 4. Relative abundance maps of the dominant bacterial taxa in the soils of the different crop treatments based on 16S rRNA sequences at the phylum (**A**) and genus (**B**) levels. CK, control; CM, *Astragalus sinicus*; SP, *Spinacia oleracea*; RY, *Lolium perenne*.

3.4. Relationships between the Relative Abundance of Soil Microorganisms and Crop Nutrient Accumulation

The relative abundances of 16S rRNA and ITS sequences in the soils were significantly correlated with crop nutrient accumulation. The accumulation of all the nutrients by the crops was positively correlated with the relative abundance of soil bacterial and fungal sequences (Figure S6).

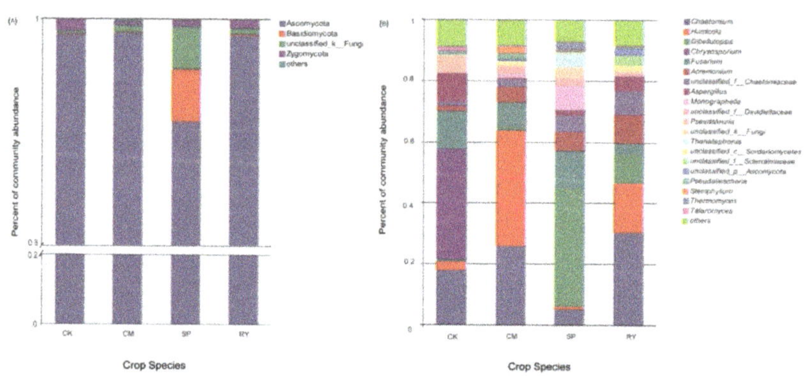

Figure 5. Relative abundance maps of the dominant fungal taxa in the soils of the different crop treatments based on internal transcribed spacer sequences at the phylum (**A**) and genus (**B**) levels. CK, control; CM, *Astragalus sinicus*; SP, *Spinacia oleracea*; RY, *Lolium perenne*.

RDA based on the soil biochemical properties explained 85.13% of the variation in the first two components of the 16S rRNA community diversity (Figure 6A). The first component (RDA1) represented 58.31% of the variability and was dominated by pH and MBN. The second component (RDA2) represented 26.82% of the variability and was dominated by AK and TDS. With respect to the ITS community diversity, the first two trait axes of the RDA accounted for 50.86% and 43.03% of the total variation, respectively; AMN scored high on the first axis, and MBN, SOC, AP, TDS, pH and MBC scored high on the second axis (Figure 6B). For the bacterial 16S rRNA community, most bacterial genera were clustered and scattered in the directions of pH, TDS and AK. Meanwhile, most fungal genera were clustered and scattered in the directions of TDS, AMN, pH and AK.

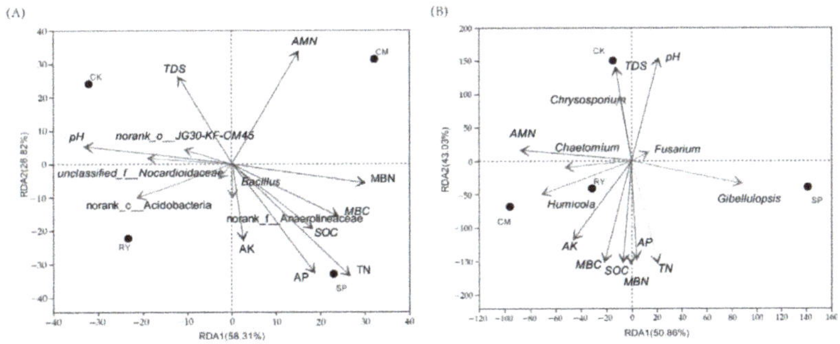

Figure 6. Redundancy analysis (RDA) of soil microbial community diversity and soil biochemical properties using the five most dominant genera according to bacterial 16S rRNA (**A**) and fungal internal transcribed spacer (**B**) sequences. TN, total nitrogen; SOC, soil organic carbon; AMN, alkali-hydrolyzable nitrogen; AP, available phosphorus; AK, available potassium; TDS, total dissolved salts; MBC, microbial biomass carbon; MBN, microbial biomass nitrogen. CK, control; CM, *Astragalus sinicus*; SP, *Spinacia oleracea*; RY, *Lolium perenne*.

4. Discussion and Conclusions

4.1. Crop Biomass Accumulation and Nutrient Absorption in Secondary Salinized Soil

Depending on the ability of a crop to adapt to the stress of secondary salinization, crop growth and nutrient absorption differs. In this study, consistent with the yields of the crops, the order of the total amount of nutrient absorption and accumulation of the three crops was as follows: RY > SP > CM (Figures 1 and 2). Previous studies have shown that when the saline conditions of the soil are aggravated, the growth of legumes is inhibited, and the amount of biomass and nutrient absorption and accumulation decreases [17]. CM, which is a legume used as green manure, can obtain the nutrients required for crop growth through biological nitrogen fixation even in soils with low fertility. However, milk vetch is sensitive to the soil pH and salt content, which restricts its growth and nitrogen-fixing ability in the face of saline-alkali adversity [9,18]. In the present study, the fresh biomass of CM was 30.8 kg ha^{-1}, which was only 1/3 of the average yield of RY.

In this study, it was confirmed that crop yield and nutrient content together determine the amount of nutrient accumulation. For example, the yield of RY was significantly higher than that of the other crops (Figure 1), meanwhile it also performed better with respect to nutrient accumulation. Overall, compared to CK, all cultivation treatments increased soil available nutrients except for soil AMN in RY and SP (Table 1), which may be due to high N accumulation and low N fertilizer application. The results also showed that the root-to-shoot ratio was another key factor affecting crop nutrient accumulation. Previous studies on soil salinization in cultivation facilities have noted that aboveground plant parts were less sensitive to environmental changes than belowground parts [19,20]. Similar variation trends were observed for the root-to-shoot ratios of dry weight and the aboveground N contents of the three crops in the present study. This suggests that the variation in root-to-shoot ratio of dry weight could be a good indicator of aboveground plant N uptake status in cultivation facility soils subjected to secondary salinization.

4.2. Plant–Soil Feedback in Soil Subjected to Secondary Salinization

The mechanisms underlying plant–soil feedback in agrosystems are complex. Previous studies have reported that when exposed to salt stress, plants actively change their strategy for the absorption of inorganic ions, and synthesize proline and other substances to osmotically adjust the cytoplasmic microenvironment [21,22]. Through these changes, the plants can resist the damage caused by saline-alkali stress. The results of the present study show that when crops are grown on secondary salinized soil, the absorption of K^+ by crops increases. In addition, the content of AP in the soil solution increased, which was possibly related to the pH value and K^+ saturation in the soil solution.

Furthermore, the results of the present study confirmed that phytoremediation increased the B/F ratio in the secondary salinized soil of the facility, especially for SP (Table 2). Previous studies have reported that SP had strong salt tolerant and antifungal ability, compared with other vegetable species on the aspects of phytoremediation and food safety [23,24]. It was found in the present study that leguminous green manure (CM) increased the relative abundances of the bacterial groups *norank_f_Gemmatimonadaceae*, and of the fungal genera *Chaetomium* and *Humicola*. *Gemmatimonadaceae* has been reported for the capacity of accumulating polyphosphate [25]. *Chaetomium* and *Humicola* have been found to be major groups of biological control agents, which not only reduce the incidence of soil-borne pathogens and plant disease, but also degrade a wide range of recalcitrant compounds [26]. These fungi possess a variety of genes that produce high-value enzymes, including chitinase and glucanase. The present study revealed that the nutrient accumulation of the crops was positively correlated with soil microbial communities, and soil pH, MBN, AK and TDS play important roles in maintaining microbial flora balance. However, we recommend future studies using dependency analysis of accuracy methods to create a holistic view of soil microbial succession and crop nutrient accumulation in the cultivation facility soils subjected to secondary salinization.

The utilization of phytoremediation in the form of planting salt-tolerant crops can alleviate the secondary salinization of soils in cultivation facilities. Such bioremediation can optimize the structure of the soil microbial community by increasing the soil microbial biomass, AP, AK and soil pH, and by reducing the soil salt content. The bacterial and fungal community compositions differed among the soils planted with the different salt-tolerant crop species. This study stresses the importance of phytoremediation for soils subjected to secondary salinization, and confirms that the crop species influences the correlations between crop nutrient accumulation and soil microbial community compositions.

Supplementary Materials: The following are available online at https://www.mdpi.com/article/10.3390/agriculture12020212/s1, Figure S1: qPCR standard curve for the 16S rRNA gene; Figure S2: qPCR amplification curve for the 16S rRNA gene; Figure S3: qPCR standard curve for the ITS gene; Figure S4: qPCR amplification curve for the ITS gene; Figure S5: Shoot and root nutrient contents of the three different crop species; Figure S6: Correlations (Pearson's *P*-value) between crop nutrient accumulation and soil microbial abundance for Chinese milk vetch (A,B); spinach (C,D) and ryegrass (E,F); Table S1: Biomass accumulation and root-to-shoot ratio of the bioremediation crops; Table S2: Bacterial (16S rRNA) and fungal (ITS) α-diversity in soil samples.

Author Contributions: Data curation and writing-original draft preparation, S.C.; investigation, S.X. and Z.F.; formal analysis, D.Z.; visualization and writing—review and editing, H.Z. (Hanlin Zhang); supervision, H.Z. (Haitao Zhu). All authors have read and agreed to the published version of the manuscript.

Funding: This work was supported by the Domestic Cooperation Program of Shanghai Science and Technology Commission [Grant No. 20025800500], the Agricultural Achievement Transformation and Demonstration Application Program of Shanghai Science and Technology Commission [Grant No. 20392003400] and the Outstanding Team Program of Shanghai Academy of Agricultural Science [Grant No. Hu-Nong-Ke-Zhuo 2022 (008)].

Institutional Review Board Statement: Not applicable.

Conflicts of Interest: The authors declare no conflict of interest in this manuscript.

References

1. Chu, S.H.; Zhang, D.; Wang, D.X.; Zhi, Y.; Zhou, P. Heterologous expression and biochemical characterization of assimilatory nitrate and nitrite reductase reveals adaption and potential of *Bacillus megaterium* NCT-2 in secondary salinization soil. *Int. J. Biol. Macromol.* **2017**, *101*, 1019–1028. [CrossRef] [PubMed]
2. Hu, S.; Liu, L.J.; Zuo, S.F.; Ali, M.; Wang, Z.L. Soil salinity control and cauliflower quality promotion by intercropping with five turfgrass species. *J. Clean. Prod.* **2020**, *266*, 121991. [CrossRef]

3. Shen, W.S.; Ni, Y.Y.; Gao, N.; Bian, B.Y.; Zheng, S.N.; Lin, X.G.; Chu, H.Y. Bacterial community composition is shaped by soil secondary salinization and acidification brought on by high nitrogen fertilization rates. *Appl. Soil Ecol.* **2016**, *108*, 76–83. [CrossRef]
4. Fan, Y.N.; Zhang, Y.X.; Wan, M.X.; Hu, W.Y.; Chen, Z.K.; Huang, B. Plastic shed production intensified secondary soil salinization in perennial fruit production systems. *Agr. Ecosyst. Environ.* **2021**, *316*, 107469. [CrossRef]
5. Gavrichkova, O.; Brykova, R.A.; Brugnoli, E.; Calfapietra, C.; Cheng, Z.Q.; Kuzyakov, Y.; Liberati, D.; Moscatelli, M.C.; Pallozzi, E.; Vasenev, V.I. Secondary soil salinization in urban lawns: Microbial functioning, vegetation state, and implications for carbon balance. *Land Degrad. Dev.* **2020**, *31*, 2591–2604. [CrossRef]
6. Wu, R.Y.; Sun, H.W.; Xue, J.; Yan, D.; Liu, Y.; Gui, D.W.; Wang, X.G.; Yang, J.Z. Acceleration of soil salinity accumulation and soil degradation due to greenhouse cultivation: A survey of farmers' practices in China. *Environ. Monitor. Assess.* **2020**, *192*, 399. [CrossRef]
7. Jesus, J.M.; Danko, A.S.; Fiúza, A.; Borges, M.T. Comparison of phytoremediation and chemical amendments for non-calcareous highly saline-sodic soil remediation. *Water Air Soil Pollut.* **2018**, *229*, 1–10. [CrossRef]
8. Zhang, J.; Shi, H.T.; Wang, Y.J.; Li, S.; Cao, Z.; Ji, S.; He, Y.; Zhang, H. Effect of dietary forage to concentrate ratios on dynamic profile changes and interactions of ruminal microbiota and metabolites in holstein heifers. *Front. Microbiol.* **2017**, *8*, 2206. [CrossRef]
9. Zhou, Q.; Chen, J.; Xing, Y.X.; Xie, X.Y.; Wang, L.C. Influence of intercropping Chinese milk vetch on the soil microbial community in rhizosphere of rape. *Plant Soil* **2019**, *440*, 85–96. [CrossRef]
10. He, H.B.; Li, W.X.; Zhang, Y.W.; Cheng, J.K.; Jia, X.; Li, S.; Yang, H.; Chen, B.; Xin, G.R. Effects of Italian ryegrass residues as green manure on soil properties and bacterial communities under an Italian ryegrass (*Lolium multiflorum* L.)-rice (*Oryza sativa L.*) rotation. *Soil Till. Res.* **2020**, *196*, 104487. [CrossRef]
11. Guo, J.J.; Liu, W.B.; Zhu, C.; Luo, G.W.; Kong, Y.L.; Ling, N.; Wang, M.; Dai, J.Y.; Shen, Q.R.; Guo, S.W. Bacterial rather than fungal community composition is associated with microbial activities and nutrient-use efficiencies in a paddy soil with short-term organic amendments. *Plant Soil* **2017**, *424*, 335–349. [CrossRef]
12. Lu, R.K. *Soil Argrochemistry Analysis Protocoes*; China Agriculture Science Press: Beijing, China, 1999.
13. Moore, J.M.; Klose, S.; Tabatabai, M.A. Soil microbial biomass carbon and nitrogen as affected by cropping systems. *Biol. Fertil. Soils* **2020**, *31*, 200–210. [CrossRef]
14. Li, Z.Q.; Zhang, X.; Xu, J.X.; Cao, K.; Wang, J.H.; Xu, C.X.; Cao, W.D. Green manure incorporation with reductions in chemical fertilizer inputs improves rice yield and soil organic matter accumulation. *J. Soil Sediment* **2020**, *20*, 2784–2793. [CrossRef]
15. Cai, S.M.; Lv, W.G.; Zhu, H.T.; Zhang, D.S.; Fu, Z.S.; Zhang, H.L.; Xu, S.X. Effect of nitrogen application rate on soil fungi community structure in a rice-fish mutualistic system. *Sci. Rep.* **2019**, *9*, 16188. [CrossRef] [PubMed]
16. Li, T.T.; Long, M.; Gatesoupe, F.J.; Zhang, Q.; Li, A.; Gong, X. Comparative analysis of the intestinal bacterial communities in different species of carp by pyrosequencing. *Microb. Ecol.* **2014**, *69*, 25–36. [CrossRef]
17. Kotula, L.; Kwa, H.Y.; Nichols, P.G.; Colmer, T.D. Tolerance and recovery of the annual pasture legumes *Melilotus siculus*, *Trifolium michelianum* and *Medicago polymorpha* to soil salinity, soil waterlogging and the combination of these stresses. *Plant Soil* **2019**, *444*, 267–280. [CrossRef]
18. Jeromela, A.M.; Mikic, A.M.; Vujic, S.; Cupina, B.; Krstic, D.; Dimitrijevic, A.; Vasiljevic, S.; Mihailovic, V.; Cvejic, S.; Miladinovic, D. Potential of legume-brassica intercrops for forage production and green manure: Encouragements from a temperate southeast european environment. *Front. Plant Sci.* **2017**, *8*, 312. [CrossRef]
19. Chen, Q.; Wang, Y.; Zou, C.B.; Wang, Z.L. Aboveground biomass invariance masks significant belowground productivity changes in response to salinization and nitrogen loading in reed marshes. *Wetlands* **2017**, *37*, 985–995. [CrossRef]
20. Stagg, C.L.; Schoolmaster, D.R.; Piazza, S.C.; Snedden, G.; Steyer, G.D.; Fischenich, C.J.; McComas, R.W. A landscape-scale assessment of above- and belowground primary production in coastal wetlands: Implications for climate change-induced community shifts. *Estuar. Coast* **2016**, *40*, 856–879. [CrossRef]
21. Rane, N.R.; Tapase, S.; Kanojia, A.; Watharkar, A.; Salama, E.S.; Jang, M.; Yadav, K.K.; Amin, M.A.; Cabral-Pinto, M.M.; Jadhav, J.P.; et al. Molecular insights into plant–microbe interactions for sustainable remediation of contaminated environment. *Bioresour. Technol.* **2022**, *344*, 126246. [CrossRef]
22. Thakur, M.P.; Putten, W.; Wilschut, R.A.; Veen, C.; Bezemer, T.M. Plant-soil feedbacks and temporal dynamics of plant diversity-productivity relationships. *Trends Ecol. Evol.* **2021**, *36*, 651–661. [CrossRef] [PubMed]
23. Ferreira, J.; Sandhu, D.; Liu, X.; Halvorson, J. Spinach (*Spinacea oleracea* L.) response to salinity: Nutritional value, physiological parameters, antioxidant capacity, and gene expression. *Agriculture* **2018**, *8*, 163. [CrossRef]
24. Singh, P.; Pallavi; Negi, R.; Rani, A.; Parul. Fungal isolation and characterization from spoiled vegetables *Lycopersicon esculentum*, *Brassica oleracea*, *Spinacia oleracea*. *Indo Am. J. P. Sc.* **2016**, *3*, 1271–1275.
25. Zhang, X.Y.; Zha, L.N.; Jiang, P.Y.; Wang, X.Y.; Lu, K.W.; He, S.B.; Huang, J.C.; Zhou, W.L. Comparative study on nitrogen removal and functional genes response between surface flow constructed wetland and floating treatment wetland planted with *Iris pseudacorus*. *Environ. Sci. Pollut. R.* **2019**, *26*, 23696–23706. [CrossRef]
26. Hung, P.M.; Wattanachai, P.; Kasem, S.; Poeaim, S. Efficacy of chaetomium species as biological control agents against phytophthora nicotianae root rot in citrus. *Mycobiology* **2015**, *43*, 288–296. [CrossRef]

agriculture

MDPI

Article

Root System Architecture Differences of Maize Cultivars Affect Yield and Nitrogen Accumulation in Southwest China

Song Guo [1], Zhigang Liu [2], Zijun Zhou [1], Tingqi Lu [3], Shanghong Chen [1], Mingjiang He [1], Xiangzhong Zeng [1], Kun Chen [1], Hua Yu [1], Yuxian Shangguan [1], Yujiao Dong [1], Fanjun Chen [2], Yonghong Liu [4] and Yusheng Qin [1,5,*]

1 Institute of Agricultural Resources and Environment, Sichuan Academy of Agricultural Sciences, Chengdu 610066, China; guosong1999@163.com (S.G.); zhouzijun1007@163.com (Z.Z.); 13908040304@163.com (S.C.); hemj9331@163.com (M.H.); xzhzeng@163.com (X.Z.); chenkun410@163.com (K.C.); yuhua353@163.com (H.Y.); 1987329002@sohu.com (Y.S.); echo-215@163.com (Y.D.)

2 College of Resources and Environmental Sciences, China Agricultural University, Beijing 100193, China; lzglzglhy@163.com (Z.L.); caucfj@cau.edu.cn (F.C.)

3 Mianyang Academy of Agricultural Sciences, Mianyang 621023, China; lutingqi0822@126.com

4 Crop Research Institute, Sichuan Academy of Agricultural Sciences, Chengdu 610066, China; 13908189593@163.com

5 Monitoring and Experimental Station of Plant Nutrition and Agro-Environment for Sloping Land in South Region, Ministry of Agriculture and Rural Affairs, Chengdu 610066, China

* Correspondence: shengyuq@126.com; Tel.: +86-028-84504285

Abstract: Root system architecture (RSA) plays a critical role in the acquisition of water and mineral nutrients. In order to understand the root characteristics that contribute to enhanced crop yield and N accumulation high-yielding and N efficient cultivars under N-stressed conditions. Here, grain yield, N accumulation and RSA traits of six dominant maize cultivars (CD30, ZH311, ZHg505, CD189, QY9 and RY1210) grown in the Southwestern part of China were investigated in field experiment under three different N regimes in 2019–2020; N300 (300 kg N ha^{-1}), N150 (150 kg N ha^{-1}) and N0 (no N supplied). Using Root Estimator for Shovelomics Traits (REST) for the quantitative analysis of maize root image obtained in the field, RSA traits including total root length (RL), root surface area (RA), root angle opening (RO), and root maximal width (RMW) were quantified in this study. The results showed that Yield, N accumulation and RSA were significantly affected by N rates, cultivars and their interactions. Grain yield, N accumulation and root weight showed a similar trend under N300 and N150 conditions compared to N0 conditions. With the input of N fertilizer, the root length, surface area, and angle increase, but root width does not increase. Under the N300 and N150 condition, RL, RA, RO and RMW increased by 17.96%, 17.74%, 18.27%, 9.22%, and 20.39%, 18.58%, 19.92%, 16.79%, respectively, compared to N0 condition. CD30, ZH505 and RY1210 have similar RO and RMW, larger than other cultivars. However, ZH505 and RY1210 have 13.22% and 19.99% longer RL, and 11.41% and 5.17% larger RA than CD30. Additionally, the grain yield of ZH505 and RY1210 is 17.57% and 13.97% higher compared with CD30. The N accumulation of ZH505 and RY1210 also shows 4.55% and 9.60% higher than CD30. Correlation analysis shows that RL, RA, RO and RMW have a significant positive correlation with grain yield while RO and RMW have a significant positive correlation with N accumulation. Linear plus plateau model analysis revealed that when the RO reaches 99.53°, and the RMW reaches 15.18 cm, the N accumulation reaches its maximum value under 0–300 kg N ha^{-1} conditions. Therefore, selecting maize cultivars with efficient RSA suitable for different soil N inputs can achieve higher grain yield and N use efficiency.

Keywords: maize; root system architecture; nitrogen rates; cultivars; yield

Citation: Guo, S.; Liu, Z.; Zhou, Z.; Lu, T.; Chen, S.; He, M.; Zeng, X.; Chen, K.; Yu, H.; Shangguan, Y.; et al. Root System Architecture Differences of Maize Cultivars Affect Yield and Nitrogen Accumulation in Southwest China. *Agriculture* **2022**, *12*, 209. https://doi.org/10.3390/agriculture12020209

Academic Editors: Chengfang Li and Lijin Guo

Received: 26 December 2021
Accepted: 30 January 2022
Published: 1 February 2022

1. Introduction

The root is an essential organ in plants, and it plays an important role in nutrient uptake, growth and yield formation [1,2]. Root system architecture (RSA), including root length, root numbers, root surface area, root angle and root width, is an important trait in crops for the acquisition of underground resources [3,4]. The RSA of maize is influenced by genotype and environmental factors such as water, nutrients and temperature [5,6]. Nitrogen (N) is the key limiting nutrient in crop production. At the same time, Ninputinagricultural systems are also an important factor affecting environmental degradation and climate change [7]. Enhancing N use efficiency (NUE) is one of the most effective ways in sustainable agriculture to meet the 2050 global food demand projected [7,8]. Understanding the relationship between N uptake and utilization efficiency and RSA in maize is an important step towards improving maize productivity. Breeding new varieties based on RSA differences will improve N use efficiency (NUE) in maize production [3,9].

Modern varieties with deeper root distribution can increase yield under low N conditions [10]. A strong root system is an important factor for high yields [11]. There is a positive correlation between root weight and above-ground biomass and ultimately yield [12]. Under high planting density, a medium root system with more root distribution is more likely to result in high yield [6]. The interaction between RSA and soil N absorption determines yield largely [6,13]. Lynch considered that steep, cheap and deep are ideal RSA for obtaining N fertilizer and water in a low N input system in maize [5]. The maize root length and surface of different eras showed an increasing trend followed by a decreasing trend in China [14]. It has been reported that the effect of root horizontal distribution on grain yield is greater than that of root horizontal distribution [15]. Under low N conditions, the root horizontal expansion decreased [16]. The RSA of modern maize varieties in China is characterized by "horizontal contraction and vertical extension", which is more suitable for planting at a higher N level [17]. Response to N fertilizer varies with genotype, the yield and root biomass of maize varieties with high N efficiency are higher than those with low N efficiency [10,18].

High temperature, drought, poor soil fertility and nutrient leaching are persistent agronomic challenges in spring maize production in the central Sichuan Basin [19]. Thus, it is crucial to identify maize varieties with an ideal RSA suitable for this environment. Root Estimator for Shovelomics Traits (REST) is a simple, rapid and effective method for the quantitative analysis of plant root images obtained in the field [20–22]. In this study, we used this high-throughput root phenotype analysis method to study the genotypic differences and N response of RSA among maize varieties grown in this area.

2. Material and Methods

2.1. Plant Materials

A field experiment was conducted at Sichuan Agricultural Research Institute Modern Agriculture Experimental Station, Deyang City, Sichuan Province (30°36.784′ N, 105°01.322′ E) in 2019–2020. A total of 6 maize varieties were tested (Table 1), all of which are currently dominant high-yield spring maize cultivars in the hilly region of central Sichuan, namely Chengdan 30 (CD30), Zhenghong505 (ZH505), Zhenghong 311 (ZH311), Chuandan 189 (CD189), Quanyu 9 (QY9) and Rongyu 1210 (RY1210).

Table 1. Characteristics of cultivars used in this study.

Cultivar	Year of Release	Parents	Breeding Institution
CD30	2004	Chengzi2142 × Zhengzi205-22	Sichuan Academy of Agricultural Sciences
ZH311	2006	K236 × 21-ES	Zhenghong Seeds Co., Ltd. (Chengdu, China)
ZH505	2008	K305 × K389	Zhenghong Seeds Co., Ltd. (Chengdu, China)
CD189	2009	SCML203 × SCML1950	Sichuan Agricultural University
QY9	2011	Y3052 × 18-599	Sichuan Academy of Agricultural Sciences
RY1210	2015	SCML202 × LH8012	Sichuan Agricultural University

2.2. Experimental Design

The experimental design was a split-plot with three replicates, with N fertilizer treatments in the main plots and the cultivars in the subplots. The variable between plots was three application rates of N fertilizer, namely 300 kg N ha^{-1} (N300), 150 kg N ha^{-1} (N150), and no N supplied (N0). The plots were fertilized with 90 kg ha^{-1} P_2O_5 and 150 kg ha^{-1} K_2O. Phosphorus and potassium fertilizers were applied before sowing, 50% of the N fertilizer was applied as a base dressing, and the remaining 50% was applied at the stem jointing stage in N300 and N150 treatments. The subplots area was 20 m^2 (5-m-long × 4-m-width). Maize was over-seeded on 1 April in 2019 and 7 April in 2020. At the V3 stage, seedlings were thinned to a final density of 50,000 plants ha^{-1}, which is the optimum population for maize cultivars grown in local areas. Cultivars were harvested on 11 August in 2019 and 15 August in 2020. The field was irrigated before sowing. Plots were kept free of weeds, insects and diseases with chemicals based on standard practices. No irrigation was applied during the growing season. Rainfall throughout the growing season was as shown in Figure 1.

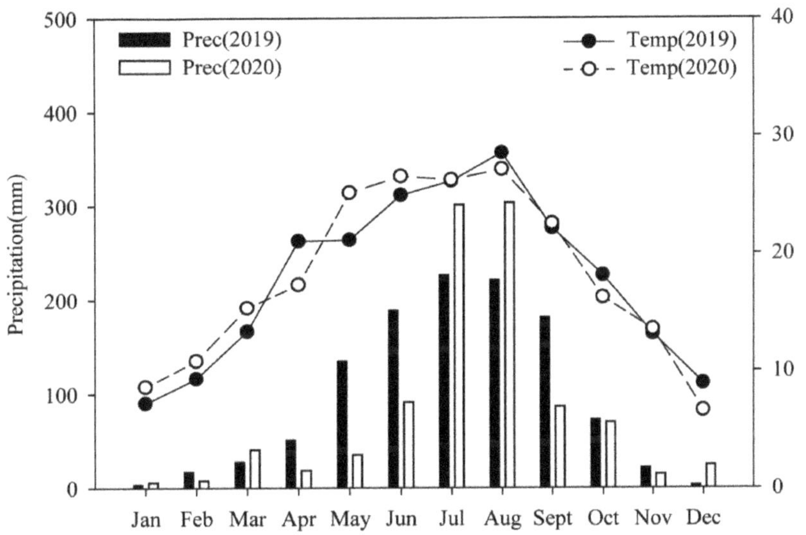

Figure 1. Temperature and precipitation of Zhongjiang County during the study period in 2019 and 2020. Note: Prec, Total monthly precipitation (mm); Temp, Average monthly temperature (°C).

Soil physical and chemical characteristics were evaluated at the beginning of the experiment for each treatment by analyzing three soil samples. The topsoil layer (0–30 cm) contained organic matter 11.5 g kg^{-1}, total N 0.99 g kg^{-1}, alkali-hydrolyzable N 74.0 mg kg^{-1}, available phosphorus (Olsen-P) 14.5 mg kg^{-1}, ammonium acetate extractable potassium (K) 172 mg kg^{-1} and pH 7.94 (1:1.25 g/v) in N300 treatment. The chemical characteristics in N150 treatment were as follows: organic matter 10.8 g kg^{-1}, total N 0.81 g kg^{-1}, alkali-hydrolyzable N 32.5 mg kg^{-1}, Olsen-P 17.5 mg kg^{-1}, ammonium acetate extractable potassium (K) 161.8 mg kg^{-1} and pH 7.98. In No treatment, chemical characteristics were organic matter 9.6 g kg^{-1}, total N 0.69 g kg^{-1}, alkali-hydrolyzable N 8.7 mg kg^{-1}, Olsen-P 11.9 mg kg^{-1}, ammonium acetate extractable potassium (K) 143.7 mg kg^{-1} and pH 7.85. The soil type is brown and is classified as Cambisols with sandy loam according to the FAO classification system (IUSS Working Group WRB, 2015).

2.3. Agronomic Trait Measurements

At silking and physiological maturity stage, three uniform plants from each plot were cut at the soil surface and separated into leaves, stem and grain (only at maturity). At the silking stage, roots were excavated within a soil volume of 30 cm (length) × 30 cm (width) × 25 cm (depth) for each plant and were then shaken off a large fraction of the soil adhering to the root system. Afterward, the roots were washed under low pressure using a water hose and nozzle. Root imaging and processing were as described by Colombi et al. [21]. Briefly, root images were captured in the imaging tent using a digital camera (Canon EOS 400D, Canon, Tokyo, Japan) with a 35 mm fixed focal lens (Canon EF 35 mm f/2.0, Canon, Tokyo, Japan). The image size was 35 × 52.5 cm resulting in a pixel size of 0.13 mm. Root images analyses were processed using REST software (Figure 2). RSA traits, including total root length, surface area, angle opening, and maximal width, were quantified in this study.

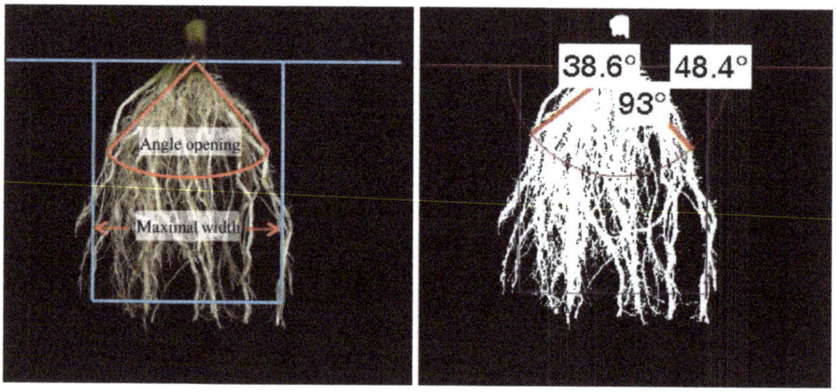

Figure 2. Root image processing with Root Estimator for Shovelomics Traits (REST).

At both sampling dates, all samples were heat-treated at 105 °C for 30 min, dried at 65 °C to constant weight. After obtaining dry matter weight, the samples were ground into fine powder for N measurement. N concentration at silking (leaves and stem) and maturity (leaves, stem, and grain) were determined by the semi-micro Kjeldahl procedure [23]. At maturity, the ears in each plot were harvested. Grain was oven-dry to determine grain moisture content. The grain yield was determined and then standardized to 14% moisture.

2.4. Statistical Analysis

Statistical analysis was performed using SPSS Statistics 17.0 (SPSS, Inc., Chicago, IL, USA). Three-way analysis of variance (ANOVA) was used to test for significant differences among N treatment, cultivar, year and N treatment × cultivar × year interaction. N level and cultivar were treated as a fixed effect. The least significant difference test (LSD) was used to evaluate significant differences between means when a significant effect was detected by ANOVA. Means for each cultivar were used for correlation analysis and linear platform model fitting.

3. Results

3.1. Grain Yield and N Accumulation Properties

Across the two years, significant differences in grain yield and N accumulation were found among N treatments (Tables 2 and 3). ANOVA showed significant effects of N levels (0, 150 and 300 kg ha^{-1}) (N), cultivar (C), and years (Y) on grain yield, ED, KPR and N accumulation at silking and maturity. The interaction of N × C had a significant effect on N accumulation at the silking stage, and grain yield and N accumulation at maturity. Due to the less rain from April to June 2020 (Figure 1), the grain yield and N accumulation were significantly lower than those in 2019.

The yield in N300 was similar to that of N150, with both of them having higher yields than N0, except that there was no difference in 2019. N accumulation showed a similar trend under N300 and N150 conditions compared to N0 conditions. The grain yield and N accumulation among cultivars were significantly different. The grain yield per plant ranged from 126.16 to 148.33 g, and the maximum value was 17.58% in ZH505, higher than the cultivar with the minimum value, CD30 (Table 2). There was no significant difference in grain yield betweenZH505, ZH311, CD189 and RY1210, although they were significantly higher than CD30. The grain weight of RY1210 between 2019 and 2020 had no difference at the N300 level. At the silking stage, the difference of N accumulation between varieties was mainly on the leaf. ZH311 showed the highest value in N accumulation, and 10.95% higher than CD30 which has the minimum value. However, the difference in N accumulation between varieties was mainly on leaf and grain at maturity. The minimum value of total N accumulation was observed in QY9 and 10.57% lower when compared with RY1210 (Table 3).

Table 2. Analysis of variance in GW, HKW, EL, ED, RPE and KPR on maize of six cultivars under three N conditions.

Treatment	GW (g Plant^{-1})	HKW (g)	EL (cm)	ED (cm)	RPE	KPR
Nitrogen (N)						
N0	127.12 b	28.52 b	15.81 a	47.72 b	16.44 a	31.847 b
N150	149.02 a	32.57 a	16.54 a	49.65 a	16.71 a	35.667 a
N300	149.11 a	32.17 a	17.93 a	48.26 ab	16.35 a	35.639 a
Cultivar (C)						
CD30	126.16 c	29.55 c	16.08 a	46.07 c	17.08 b	34.03 b
ZH311	146.03 ab	33.61 a	18.06 a	50.37 a	16.42 b	31.31 c
ZH505	148.33 a	26.21 d	17.78 a	49.84 a	19.17 a	34.64 ab
CD189	145.74 ab	32.41 ab	15.75 a	48.02 b	14.86 c	36.06 a
QY9	140.46 b	31.0 bc	16.47 a	49.39 a	16.25 b	35.83 ab
RY1210	143.79 ab	33.74 a	16.42 a	47.59 b	15.22 c	34.44 ab
Year (Y)						
2019	157.72 a	31.50 a	17.43 a	50.08 a	16.89 a	38.02 a
2020	125.78 b	30.68 a	16.09 a	47.01 b	16.11 b	30.75 b
Source of variation						
N	**	**	ns	**	ns	**
C	**	**	ns	**	**	**
Y	**	ns	ns	**	**	**
N × C	*	ns	ns	**	ns	*
N × Y	**	ns	ns	**	ns	ns
C × Y	*	ns	ns	*	ns	*
N × C × Y	ns	ns	ns	*	ns	ns

Within N or cultivar or year, numbers followed by different letters indicate significant difference ($p < 0.05$). * significant at $p < 0.05$, ** significant at $p < 0.01$, ns: not significant ($p > 0.05$). GW, Grain weight; HKW, hundred-kernel weight; EL, Ear length; BHL, Bald head length; ED, Ear diameter; RPE, Rows per ear; KPR, Kernels per row.

3.2. Root System Architecture Traits Evaluation

Variance analysis of root traits showed that N treatments had a significant effect on root traits (Table 4). The interaction of the N × cultivar had a significant effect on all root traits. Therefore, the N × cultivar was further analyzed. Root dry weight, total length, surface area, and angle opening in N150 were similar to that of N300, with both of them having higher values than the N0 condition. In addition, the maximal width in N150 was higher than that of N300 and N0 treatments. Under N150 condition, the root angle, root width, root length and root surface area increased by 19.93%, 16.79%, 20.39% and18.58% compared with no fertilizer treatment. The RW among cultivars was significantly different. The maximum value of RW was 20.36 g in RY1210 higher than the cultivar with the minimum value, QY9. The maximum value of total length was 1651.29 cm in RY1210 higher than the cultivar with the minimum value, CD30. The maximum value of the surface area was 2851.67 cm^2 in ZH505 higher than the cultivar with the minimum value, CD189. The root angle opening of RY1210 and CD30 were larger than others. The root maximal width of ZH 505, RY1210 and CD30 were larger than others.

Table 3. Evaluation of N accumulation traits of six cultivars under three nitrogen (N) conditions.

Treatment	N Accumulation at Silking (g plant^{-1})			N Accumulation at Maturity (g plant^{-1})			
	Stem	Leaf	Total	Stem	Leaf	Grain	Total
Nitrogen (N)							
N0	0.50 b	0.57 b	1.07 b	0.37 b	0.20 c	0.87 b	1.44 b
N150	0.66 a	0.90 a	1.56 a	0.62 a	0.34 b	1.35 a	2.31 a
N300	0.65 a	0.92 a	1.57 a	0.59 a	0.36 a	1.38 a	2.33 a
Cultivar (C)							
CD30	0.56 a	0.76 bc	1.32 b	0.51 a	0.34 a	1.14 c	1.98 b
ZH311	0.62 a	0.84 a	1.46 a	0.54 a	0.29 bc	1.14 c	1.97 b
ZH505	0.62 a	0.78 bc	1.39 ab	0.51 a	0.30 bc	1.26 ab	2.07 ab
CD189	0.62 a	0.82 ab	1.43 ab	0.57 a	0.31 ab	1.15 c	2.03 ab
QY9	0.64 a	0.73 c	1.37 ab	0.49 a	0.28 c	1.17 bc	1.94 b
RY1210	0.58 a	0.84 a	1.42 ab	0.53 a	0.28 c	1.35 a	2.17 a
Year (Y)							
2019	0.62 a	0.88 a	1.50 a	0.64 a	0.35 a	1.22 a	2.21 a
2020	0.59 a	0.71 b	1.30 b	0.41 b	0.25 b	1.18 a	1.84 b
Source of variation							
N	ns	**	**	**	**	**	**
C	**	**	*	ns	**	**	*
Y	ns	**	**	**	**	ns	**
N × C	ns	**	*	*	*	ns	*
N × Y	ns	ns	ns	ns	**	ns	*
C × Y	ns	**	*	ns	ns	**	ns
N × C × Y	ns	*	ns	ns	**	*	ns

Within N or cultivar or year, numbers followed by different letters indicate significant difference ($p < 0.05$). * significant at $p < 0.05$, ** significant at $p < 0.01$, ns: not significant ($p > 0.05$).

Table 4. Evaluation of root system traits of six cultivars under three nitrogen (N) conditions.

Treatment	Root Weight (g Plant^{-1})	Total Length (cm)	Surface Area (cm^2)	Angle Opening (°)	Maximal Width (cm)
Nitrogen (N)					
N0	12.21 b	1325.30 b	228.56 b	88.08 b	15.08 c
N150	19.81 a	1595.54 a	271.03 a	105.63 a	17.61 a
N300	19.39 a	1563.32 a	269.11 a	104.17 a	16.47 b
Cultivar (C)					
CD30	17.30 b	1376.23 c	252.83 bc	104.42 a	17.63 a
ZH311	17.09 b	1551.27 b	255.06 bc	97.19 b	15.67 bc
ZH505	17.28 b	1558.24 b	281.67 a	97.62 b	17.08 a
CD189	16.56 b	1353.67 c	249.44 c	97.40 b	16.02 b
QY9	14.22 c	1477.61 b	232.50 d	93.74 b	14.79 c
RY1210	20.36 a	1651.29 a	265.89 b	105.41 a	17.11 a
Year (Y)					
2019	22.76 a	1863.40 a	284.17 a	101.05 a	16.45 a
2020	11.52 b	1126.04 b	228.30 b	97.55 b	16.32 a
Source of variation					
N	**	**	**	**	**
C	**	**	**	**	**
Y	**	**	**	*	ns
N × C	**	*	**	*	**
N × Y	**	ns	ns	ns	ns
C × Y	*	**	ns	ns	ns
N × C × Y	**	*	**	*	**

Within N or cultivar or year, numbers followed by different letters indicate significant difference ($p < 0.05$). * significant at $p < 0.05$, ** significant at $p < 0.01$, ns: not significant ($p > 0.05$).

The root dry weights under N150 and N300 were increased by 62.24% and 58.80%, respectively, compared with N0 (Figure 3). It indicated that N application might increase maize root weight, while excessive N will reduce it (Figure 4). The root dry weight was also significantly different among cultivars. Under N0 treatment, RY1210 showed higher root weights, while QY9 had the lowest values in two years. Under N150 treatment, RY1210, and CD189 had higher root weights, while QY9 had the lowest values. The root weight under N300 treatment, of which CD189 had the least root weight compared to other varieties. CD189 were very sensitive to N stress, and the root weight decreased significantly under N deficiency or excess.

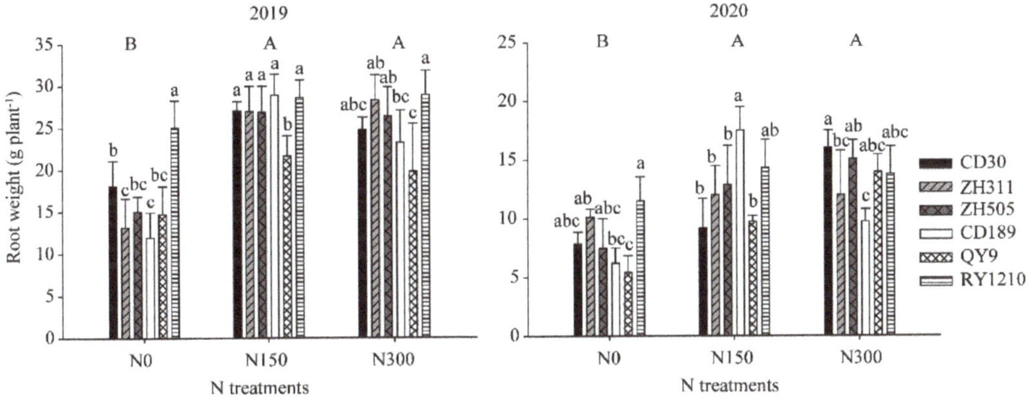

Figure 3. Root biomass of six maize cultivars under three N treatments in 2019 and 2020. Bars indicate standard error. Different lower case letters indicate significant differences at $p < 0.05$ among the cultivars in the same N treatment. Different capital letters indicate significant differences among the N treatments.

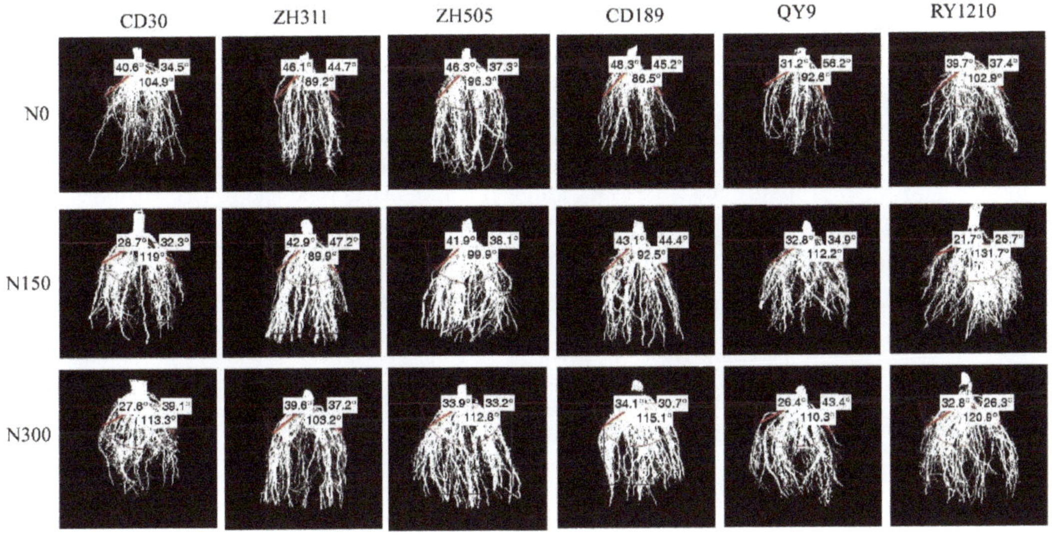

Figure 4. The distribution of the root system of six maize cultivars in three N treatments.

The total root length and root surface area were regulated by N fertilizer (Figure 5). There was no significant difference in root total length and surface area between N150 and

N300, which were higher than that of N0. Compared with N0, the root length and root surface area of N150 were increased by 20.39% and 18.58%, respectively. The root length and root surface area of N300 were increased by 17.96% and 17.74%, respectively (Table 4). This shows that N can promote the growth of roots. However, excessive application of N fertilizer inhibits root elongation and root surface area increase. The root length and root surface area of most maize varieties showed a trend of increasing at first and then decreasing at the three N levels, while RY1210 and CD189 showed an increasing trend all the time, indicating that RY1210 and CD189 were not sensitive to high N. The root length of RY1210 under N0 treatment is higher than that under N300 treatment. Root surface area under N0 is similar to that of N300. It indicates that this genotype is not sensitive to low N stress and belongs to low N tolerance varieties.

The root angle and width are regulated by N as well (Figure 5); they increased at first and then decreased with the increase in N supply, while the difference between N150 and N300 was significant in root width. The root angles of N150 and N300 were increased by 19.92% and 18.27% compared to the N0 treatment, respectively. The root maximal width of N150 and N300 was increased by 16.79% and 9.22% compared with the N0 treatment, respectively. These results indicate that N might promote the growth of maize roots, and the angle and width of the root system are significantly increased. However, excessive N may inhibit the increase in root angle and width.

The root angle opening and width between genotypes were significantly different. Under three N treatments, CD30 and RY1210 had larger root angles, CD189 and QY9 showed smaller ones, while ZH505 had a larger root width at N150 and N300, respectively. However, the change in root width was not completely consistent with the root angle. Under three N treatments, CD30, ZH505 and RY1210 had larger root widths, while QY9 had a smaller root width.

3.3. Relationship between RSA, Grain Yield, and N Accumulation

Significant correlations were found between RSA, grain yield and N accumulation (Figure 6). The grain yield increased logarithmically with increasing root length and surface area, and the regression multiple R^2 values were 0.84 and 0.69 ($p < 0.01$). Furthermore, N accumulation increased logarithmically with increasing root length and surface area ($p < 0.01$). With the increase of root length of maize roots, the grain yield and N accumulation continued to increase, while after reaching a certain point, the yield and N accumulation did not continue to increase, but showed a downward trend (Figure 7).

Yield and N accumulation are significantly related to root angle and root width (Figure 8). With the increase of root angle and root width, the yield and N accumulation continue to increase, while their continuous increase cannot further increase yield. On the contrary, too large root width will reduce maize yield and N accumulation. The trends in yield and N accumulation at the three N levels are consistent with the linear + plateau model ($p < 0.05$). Results from the model showed that when the root angle reaches 99.53° and 97.39°, the N uptake will reach the plateau value of 2.56 g plant^{-1} and 2.11 g plant^{-1} in 2019 and 2020, respectively; when the root width reaches 15.18 cm and 14.83 cm, the y N accumulation plateau value will be 2.34 g plant^{-1} and 1.90 g plant^{-1} in 2019 and 2020, respectively. Therefore, when the root angle of cultivars reaches 99.53°, and the root width reaches 15.18 cm, higher yield and N accumulation can be obtained.

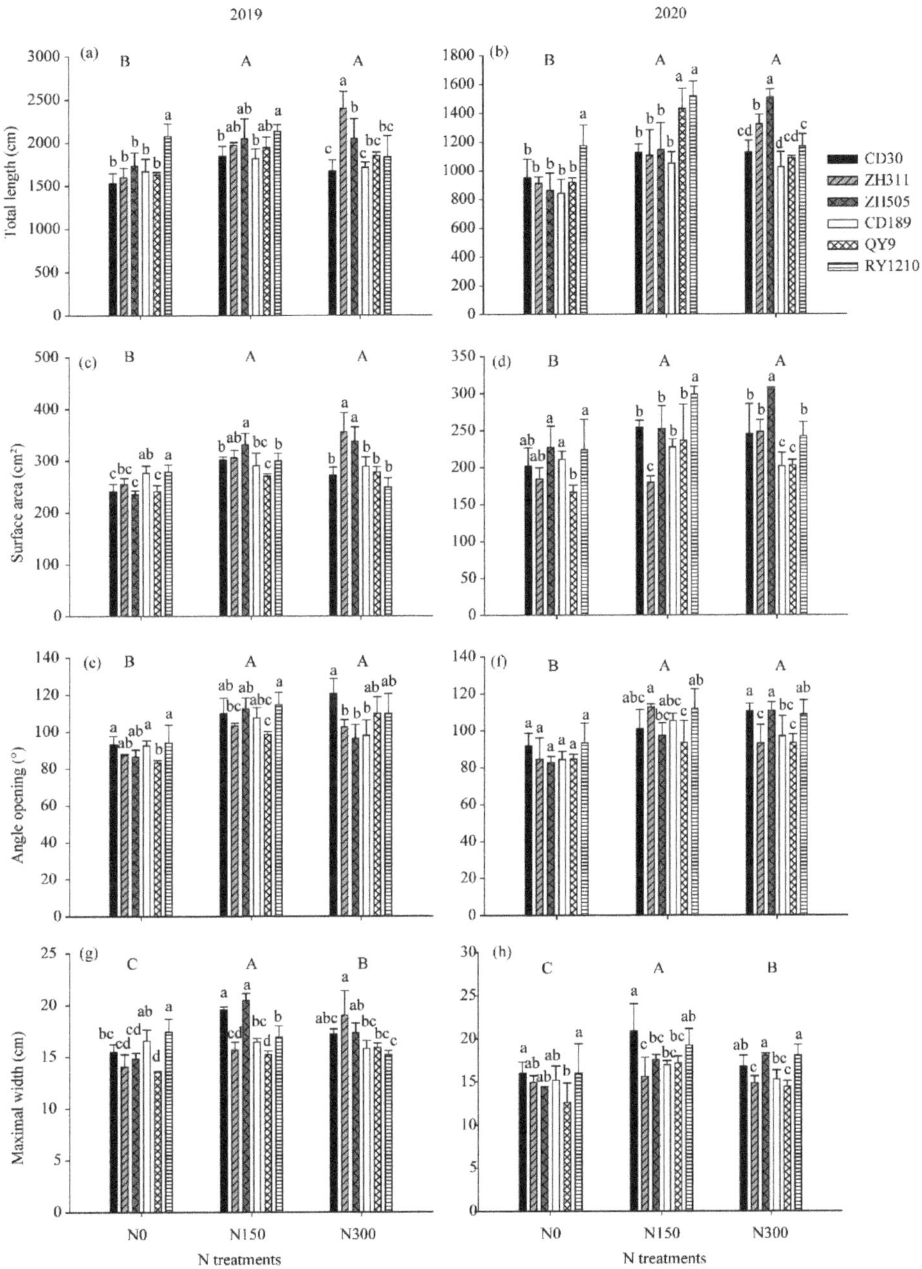

Figure 5. Root total length, surface area, angel opening, and maximal width of six maize cultivars under three N treatments in 2019 and 2020: (**a**) root total length in 2019, (**b**) root total length in 2020, (**c**) root surface area in 2019, (**d**) root surface area in 2020, (**e**) root angel opening in 2019, (**f**) root angel opening in 2020, (**g**) root maximal width in 2019, (**h**) root maximal width in 2020. Bars indicate standard error. Different lower case letters indicate significant differences at $p < 0.05$ among the cultivars in the same N treatment. Different capital letters indicate significant differences among the N treatments.

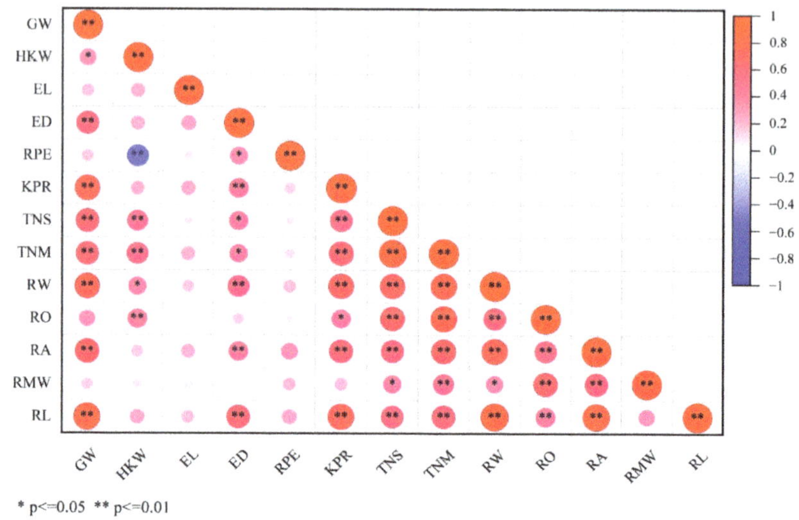

Figure 6. Correlation coefficients between grain yield, N accumulation, and root system traits of six cultivars under three nitrogen (N) conditions. * significant at the 0.05 probability level; ** significant at the 0.01 probability level. GW, Grain weight; HKW, hundred-kernel weight; EL, Ear length; BHL, Bald head length; ED, Ear diameter; RPE, Rows per ear; KPR, Kernels per row; TNS, total N accumulation at silking; TNM, total N accumulation at maturity; RW, root weight; RO, root angle opening; RA, root surface area; RMW, root maximal width; RL, root length.

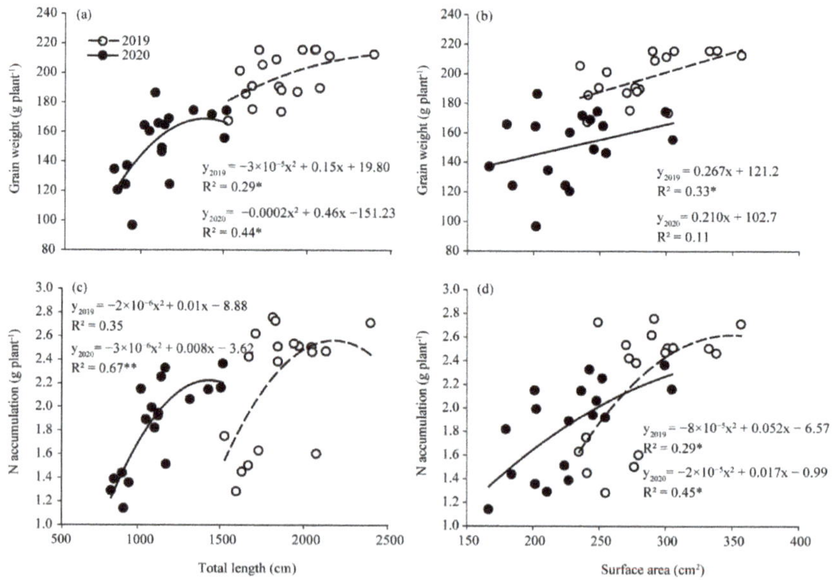

Figure 7. Relationship among total root length, surface area, grain yield, and N accumulation in six maize cultivars under three different N treatments. (**a**) Relationship between total root length and grain yield, (**b**) Relationship between root surface area and grain yield, (**c**) Relationship between total root length and N accumulation, (**d**) Relationship between root surface area and N accumulation. * significant at $p < 0.05$, ** significant at $p < 0.01$.

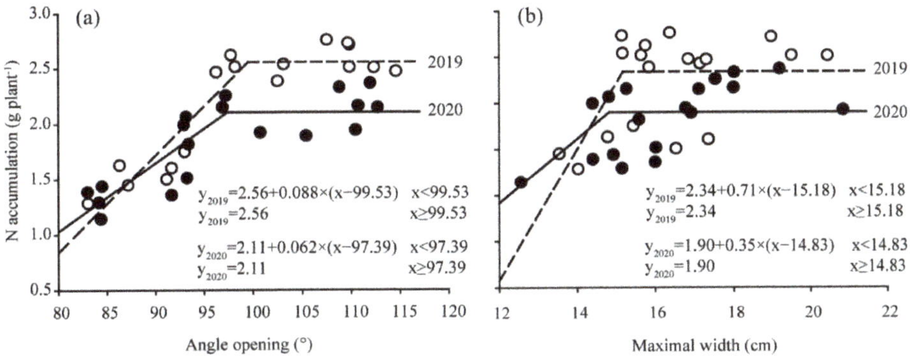

Figure 8. Relationship among angle opening, maximal width, and N accumulation in six maize cultivars under three different N treatments. (**a**) Relationship between root angle opening and N accumulation, (**b**) Relationship between root maximal width and N accumulation. The white and black cicles represent the 2019 and 2020 values, respectively.

4. Discussion

4.1. Influence of RSA Traits on Grain Yield and N Accumulation

Under field conditions, 80–90% of maize roots are distributed in 0–30 cm soil layers [24]. The root system of maize continues to grow and develop during the growth period, reaching the maximum at the silking stage and then begins to senesce [25]. Therefore, the silking stage is a critical period for analyzing root traits in fields [6]. The growth and development of roots are affected by genetics and the environment [26]. In this study, ANOVA showed significant effects of N levels, cultivar and years on grain yield, N accumulation and RSA (Tables 2–4). Their interactions (N × cultivar) suggested the existence of strong environmental effects on grain yield, NUE and RSA of the different maize cultivars. The differences in root biomass, root total length, surface area, root angle, and width among maize cultivars were significant. These root differences are the key factors that cause differences in N uptake and grain yield among these cultivars (Figures 6 and 7).

The function of the root system depends on the biomass and spatial distribution of the root system. Larger root systems are often closely related to higher yield, biomass and N uptake [27,28]. In this study, ZH505 and RY1210 have a larger root system and higher yield and N accumulation (Tables 2 and 3). The root length of RY1210 is longer, and it has a higher N uptake during the silking period because the root length of maize is directly related to the uptake and utilization of nitrate-nitrogen in the soil [29]. The root surface area of ZH505 is larger, and it increases the ability to obtain N in the soil through the root surface by diffusion in the rhizosphere [30]. The root system of QY9 is small, its yield and N uptake are low because the smaller root system will restrict the aboveground access to water and nutrients and will also lead to a decline in yield [5]. The angle between the maize root system and the ground is 10–80°, and the angle of the root system affects the depth of root penetration [5]. The root expansion width can reflect the horizontal distribution of the root system, ultimately affecting nutrient uptake, and gradually decreasing as the root system extends downward. In this study, RY1210 had a larger root angle and width, which can expand the root growth space and promote the uptake of nutrients and water, while the root angle of ZH505 is smaller, but the N uptake is higher, which may be because the deeper root can get more N in deep soil [6]. Although CD30 had a larger root angle and width, smaller root length and root surface area affected N absorption, while modern varieties need to have higher N acquisition capacity in deep soil [18].

Adverse weather conditions, such as droughts, elevated temperatures can reduce maize yields. In this study, the grain weight of RY1210 between 2019 and 2020 was similar at the N300 level. RY1210 has larger root width and root opening angle, which may be important for yield in arid environments (Figure 5). Drought stress inhibited the growth of shoots and reduced the number of lateral roots, but the growth of roots usually continued [3]. The deeper the root angle, the greater the root growth angle has better water and N absorption capacity [31]. Under high planting density, competition between N and water among the plants, medium root systems with more root distribution are more likely to result in high yield [32]. RY1210 and ZH505 have the same RMW, but RO is smaller, and RA is larger in ZH505, making it more suitable for high-density planting (Table 4).

4.2. Correlation between RSA, Grain Yield and N Accumulation

N fertilizer application affects maize yield and N uptake by affecting root growth and distribution in the soil [33] (Tables 2–4). The growth of roots is usually improved by increasing N application, but excessive N application can inhibit the growth of lateral roots and the elongation of roots, while N deficiency can promote the increase in root biomass [16,34]. In this study, root architecture significantly correlated with yield and N accumulation, while there was no significant correlation between root angle and root width and yield (Figure 6). It may be because, under excessive N, the yield has nothing to do with root angle and root width. Studies have shown that on the clay soil of northeast China, the total root length of maize continues to increase with the increase of N application, and the total root length reaches the maximum at 168–240 kg N ha^{-1}, then decreases with the increase of N application [10]. In this study, excessive N application not only led to a decrease in yield but also reduced the root width (Figure 5, Table 4). Under low N conditions, the root angle and width are significantly reduced (Figure 5). This is because N deficiency forces the root system to obtain N fertilizer in deep soil, and the growth becomes steeper [35].

A robust root system is an essential feature of maize N-efficient varieties. Studies have shown that N efficient varieties have higher root length and root biomass under low and medium N conditions [3]. In this study, QY9 and CD189 had significantly reduced root weight under N deficiency, while CD30, ZH505 and RY1210 had higher root biomass (Table 3). We found that applying an appropriate amount of N fertilizer (150 kg N ha^{-1}) can significantly increase root angle by 19.93%, root width by 16.79%, root length by 20.39%, and root surface area by 18.58% compared with no fertilizer treatment. Therefore, the rational application of N fertilizer can promote the development of the maize root system and improve the absorption of nutrients in the soil, resulting in an efficient N absorption and high crop yield. The study found that biochar application can increase root angle by about 14%, root width about 20%, and root surface area about 54%, thereby increasing yield by 45%. The linear + plateau model revealed that when the RO reaches 97.39°, and the RMW reaches 15.18 cm, the N uptake will reach a plateau (Figure 8). However, an excessively large angle will affect the depth of root penetration, which is not efficient in the uptake and utilization of nutrients and water from deep soil and reduces the plant's ability to resist adverse environmental conditions [3]. Under high-density conditions, horizontally distributed roots reduce competition among plant roots and improve yield than vertically distributed roots [15]. Therefore, it is more suitable to plant maize varieties with medium root size at high density, because too large a root system will lead to competition between root systems, resulting in decreased yield [6].

5. Conclusions

Root architecture can be an important index to evaluate high-yielding and N efficient cultivars. In this study, the RSA of RL and RS displayed a significant positive correlation with grain yield and N accumulation. The RSA of RO and RMW also showed a significant positive correlation with N accumulation, no grain yield. Oversize RO and RMW cannot further improve the N accumulation. Therefore, although CD30 has larger RO and RMW, smaller RL and RS result in lower yield and N accumulation. The root system of the cultivar ZH505 and RY1210 are moderate in size, reducing the excessive carbon consumption and penetrating down to absorb N in deeper soil. The root architecture of RY1210 has good adaptability to the arid climate. In comparison, ZH505 is suitable for high-density planting. It can be expected, selecting maize cultivars with an ideal root system architecture and applying the appropriate N fertilizer inputs in the hilly area of Southwest China can improve N efficiency and crop yield in a sustainable way.

Author Contributions: S.G., X.Z. and Y.Q. conceived of and designed the study; S.G., Z.L., M.H. and Z.Z. analyzed the data; S.G. wrote the manuscript; S.G., K.C., H.Y., Y.S. and Y.D. carried out the field measurements and soil analysis; F.C. and Y.L. assisted with manuscript writing and editing; T.L. and S.C. provided the material. All authors have read and agreed to the published version of the manuscript.

Funding: This work was supported by the National Natural Science Foundation of China (41907080), the National Key R&D Program of China (2018YFD0200700), the Sichuan Science and Technology Program (2020YFH0171), and the Research Foundation of Sichuan Academy of Agricultural Sciences (2019QYXK022, 2020BJRC005).

Institutional Review Board Statement: Not applicable.

Informed Consent Statement: Not applicable.

Data Availability Statement: Data can be achieved by contacting with the first author and responsdense author.

Acknowledgments: We sincerely thank the reviewers for their helpful comments and supplementary proposal. We thank Yuanzhong Bianji for editing the English text of a draft of this manuscript.

Conflicts of Interest: The authors declare no conflict of interest.

References

1. Lynch, J.P. Root architecture and plant productivity. *Plant Physiol.* **1995**, *109*, 7. [CrossRef] [PubMed]
2. Hochholdinger, F. Untapping root system architecture for crop improvement. *J. Exp. Bot.* **2016**, *67*, 4431. [CrossRef] [PubMed]
3. Mi, G.H.; Chen, F.J.; Yuan, L.X.; Zhang, F.S. Ideotype root system architecture for maize to achieve high yield and resource use efficiency in intensive cropping systems. *Adv. Agron.* **2016**, *139*, 73–97.
4. Li, J.P.; Chen, F.J.; Li, Y.Q.; Li, P.C.; Wang, Y.Q.; Mi, G.H.; Yuan, L.X. ZmRAP2. 7, an AP2 Transcription Factor, Is Involved in Maize Brace Roots Development. *Front. Plant Sci.* **2019**, *10*, 820. [CrossRef] [PubMed]
5. Lynch, J.P. Steep, cheap and deep: An ideotype to optimize water and N acquisition by maize root systems. *Ann. Bot.* **2013**, *112*, 347–357. [CrossRef] [PubMed]
6. Shao, H.; Shi, D.F.; Shi, W.J.; Ban, X.B.; Chen, Y.C.; Ren, W.; Chen, F.J.; Mi, G.H. Genotypic difference in the plasticity of root system architecture of field-grown maize in response to plant density. *Plant Soil* **2019**, *439*, 201–217. [CrossRef]
7. Zhang, X.; Davidson, E.A.; Mauzerall, D.L.; Searchinger, T.D.; Dumas, P.; Shen, Y. Managing nitrogen for sustainable development. *Nature* **2015**, *528*, 51–59. [CrossRef]
8. Kheir, A.M.S.; Alkharabsheh, H.M.; Seleiman, M.F.; Al-Saif, A.M.; Ammar, K.A.; Attia, A.; Zoghdan, M.G.; Shabana, M.M.A.; Aboelsoud, H.; Schillaci, C. Calibration and Validation of AQUACROP and APSIM Modelsto Optimize Wheat Yield and Water Saving in Arid Regions. *Land* **2021**, *10*, 1375. [CrossRef]
9. Rogers, E.D.; Benfey, P.N. Regulation of plant root system architecture: Implications for crop advancement. *Curr. Opin. Biotechnol.* **2015**, *32*, 93–98. [CrossRef]
10. Su, W.N.; Kamran, M.; Xie, J.; Meng, X.P.; Han, Q.F.; Liu, T.N.; Han, J. Shoot and root traits of summer maize hybrid varieties with higher grain yields and higher nitrogen use efficiency at low nitrogen application rates. *PeerJ* **2019**, *7*, e7294. [CrossRef]
11. Zhang, L.; Meng, Y.; Li, S.Q.; Yue, S.C. High-yield characteristics and root support of rain-fed maize under film mulching. *Agron. J.* **2020**, *112*, 2115–2131. [CrossRef]
12. Chen, X.Y.; Liu, P.; Zhao, B.; Zhang, J.W.; Ren, B.Z.; Li, Z.; Wang, Z.Q. Root physiological adaptations that enhance the grain yield and nutrient use efficiency of maize (*Zea mays* L.) and their dependency on phosphorus placement depth. *Field Crops Res.* **2022**, *276*, 108378. [CrossRef]
13. Wu, Q.; Pages, L.; Wu, J. Relationships between root diameter, root length and root branching along lateral roots in adult, field-grown maize. *Ann. Bot.* **2016**, *117*, 379–390. [CrossRef]
14. Zhang, F.L.; Niu, X.K.; Zhang, Y.M.; Xie, R.Z.; Liu, X.; Li, S.K.; Gao, S.J. Studies on the root characteristics of maize varieties of different eras. *J. Integr. Agric.* **2013**, *12*, 426–435. [CrossRef]
15. Gao, J.; Lei, M.; Yang, L.J.; Wang, P.; Tao, H.B.; Huang, S.B. Reduced row spacing improved yield by optimizing root distribution in maize. *Eur. J. Agron.* **2021**, *127*, 126291. [CrossRef]

16. Le Marie, C.A.; York, L.M.; Strigens, A.; Malosetti, M.; Camp, K.H.; Giuliani, S.; Lynch, J.P.; Hund, A. Shovelomics root traits assessed on the EURoot maize panel are highly heritable across environments but show low genotype-by-nitrogen interaction. *Euphytica* **2019**, *215*, 173. [CrossRef]
17. Chen, X.C.; Chen, F.J.; Chen, Y.L.; Gao, Q.; Yang, X.L.; Yuan, L.X.; Zhang, F.S.; Mi, G.H. Modern maize hybrids in Northeast C hina exhibit increased yield potential and resource use efficiency despite adverse climate change. *Glob. Chang. Biol.* **2013**, *19*, 923–936. [CrossRef]
18. Chen, X.C.; Zhang, J.; Chen, Y.L.; Li, Q.; Chen, F.J.; Yuan, L.; Mi, G.H. Changes in root size and distribution in relation to nitrogen accumulation during maize breeding in China. *Plant Soil* **2014**, *374*, 121–130. [CrossRef]
19. Zha, L.; Xie, M.L.; Zhu, M.; Dou, P.; Cheng, Q.B.; Wang, X.L.; Yuan, J.C.; Kong, F.L. Effects of ridge-cultivation and plastic film mulching on root distribution and yield of spring maize in hilly area of central Sichuan basin, China. *J. Appl. Ecol.* **2016**, *27*, 855–862.
20. Abiven, S.; Hund, A.; Martinsen, V.; Cornelissen, G. Biochar amendment increases maize root surface areas and branching: A shovelomics study in Zambia. *Plant Soil* **2015**, *395*, 45–55. [CrossRef]
21. Colombi, T.; Kirchgessner, N.; Le Marié, C.A.; York, L.M.; Lynch, J.P.; Hund, A. Next generation shovelomics: Set up a tent and REST. *Plant Soil* **2015**, *388*, 1–20. [CrossRef]
22. Xu, H.; Vandecasteele, B.; Maenhout, P.; Pannecoucque, J.; Neve, S.D.; Sleutel, S. Maize root biomass and architecture depend on site but not on variety: Consequences for prediction of C inputs and spread in topsoil based on root-to-shoot ratios. *Eur. J. Agron.* **2020**, *119*, 126121. [CrossRef]
23. Bremner, J. Nitrogen-total. In *Methods of Soil Analysis: Part 3 Chemical Methods*; Soil Science Society of America: Madison, WI, USA, 1996; Volume 5, pp. 1085–1121.
24. Peng, Y.F.; Niu, J.F.; Peng, Z.P.; Zhang, F.S.; Li, C.J. Shoot growth potential drives N uptake in maize plants and correlates with root growth in the soil. *Field Crops Res.* **2010**, *115*, 85–93. [CrossRef]
25. Shao, H.; Xia, T.T.; Wu, D.L.; Chen, F.J.; Mi, G.H. Root growth and root system architecture of field-grown maize in response to high planting density. *Plant Soil* **2018**, *430*, 395–411. [CrossRef]
26. Zhai, L.C.; Zhang, L.K.; Yao, H.P.; Zheng, M.J.; Ming, B.; Xie, R.Z.; Zhang, J.T.; Jia, X.L.; Ji, J.J. The Optimal Cultivar—Sowing Date—Plant Density for Grain Yield and Resource Use Efficiency of Summer Maize in the Northern Huang–Huai–Hai Plain of China. *Agriculture* **2022**, *12*, 7. [CrossRef]
27. Mi, G.H.; Chen, F.J.; Wu, Q.P.; Lai, N.W.; Yuan, L.X.; Zhang, F.S. Ideotype root architecture for efficient nitrogen acquisition by maize in intensive cropping systems. *Sci. China Life Sci.* **2010**, *53*, 1369–1373. [CrossRef] [PubMed]
28. Liu, Z.G.; Zhao, Y.; Guo, S.; Cheng, S.; Guan, Y.; Cai, H.; Yuan, L.X.; Chen, F.J. Enhanced crown root number and length confers potential for yield improvement and fertilizer reduction in nitrogen-efficient maize cultivars. *Field Crops Res.* **2019**, *241*, 107562. [CrossRef]
29. Garnett, T.; Conn, V.; Kaiser, B.N. Root based approaches to improving nitrogen use efficiency in plants. *Plant Cell Environ.* **2009**, *32*, 1272–1283. [CrossRef] [PubMed]
30. Ning, P.; Li, S.; White, P.J.; Li, C.J. Maize varieties released in different eras have similar root length density distributions in the soil, which are negatively correlated with local concentrations of soil mineral nitrogen. *PLoS ONE* **2015**, *10*, e0121892. [CrossRef]
31. Lynch, J.P. Harnessing root architecture to address global challenges. *Plant J.* **2021**, *109*, 415–431. [CrossRef] [PubMed]
32. Chen, F.J.; Liu, J.C.; Liu, Z.G.; Chen, Z.; Ren, W.; Gong, X.P.; Wang, L.F.; Cai, H.G.; Pan, Q.C.; Yuan, L.X.; et al. Breeding for high-yield and nitrogen use efficiency in maize: Lessons from comparison between Chinese and US cultivars. *Adv. Agron.* **2021**, *166*, 251–275.
33. Chun, L.; Mi, G.H.; Li, J.S.; Chen, F.J.; Zhang, F.S. Genetic analysis of maize root characteristics in response to low nitrogen stress. *Plant Soil* **2005**, *276*, 369–382. [CrossRef]
34. Feng, G.Z.; Zhang, Y.J.; Chen, Y.L.; Li, Q.; Chen, F.J.; Gao, Q.; Mi, G.H. Effects of nitrogen application on root length and grain yield of rain-fed maize under different soil types. *Agron. J.* **2016**, *108*, 1656–1665. [CrossRef]
35. Trachsel, S.; Kaeppler, S.; Brown, K.M.; Lynch, J.P. Maize root growth angles become steeper under low N conditions. *Field Crops Res.* **2013**, *140*, 18–31. [CrossRef]

agriculture

MDPI

Article

Intercropping of Rice and Water Mimosa (*Neptunia oleracea* Lour.): A Novel Model to Control Pests and Diseases and Improve Yield and Grain Quality while Reducing N Fertilizer Application

Zewen Hei [1,2], Huimin Xiang [1,2,3,4], Jiaen Zhang [1,2,3,4,*], Kaiming Liang [5], Jiawen Zhong [1,2], Meijuan Li [1,2] and Xiaoqiao Ren [1,2]

1. Guangdong Provincial Key Laboratory of Eco-Circular Agriculture, South China Agricultural University, Guangzhou 510642, China; zewenhei@163.com (Z.H.); hmxiang@scau.edu.cn (H.X.); zhongjiawen2021@163.com (J.Z.); limeijuan028@163.com (M.L.); hahaeihaha@163.com (X.R.)
2. Department of Ecology, College of Natural Resources and Environment, South China Agricultural University, Guangzhou 510642, China
3. Guangdong Provincial Engineering Technology Research Center of Modern Eco-Agriculture and Circular Agriculture, Guangzhou 510642, China
4. Key Laboratory of Agro-Environment in the Tropics, Ministry of Agriculture and Rural Affairs, South China Agricultural University, Guangzhou 510642, China
5. The Rice Research Institute of Guangdong Academy of Agricultural Sciences, Guangdong Key Laboratory of New Technology for Rice Breeding, Guangzhou 510640, China; kaiming-liang@163.com

* Correspondence: jeanzh@scau.edu.cn

Citation: Hei, Z.; Xiang, H.; Zhang, J.; Liang, K.; Zhong, J.; Li, M.; Ren, X. Intercropping of Rice and Water Mimosa (*Neptunia oleracea* Lour.): A Novel Model to Control Pests and Diseases and Improve Yield and Grain Quality while Reducing N Fertilizer Application. *Agriculture* **2022**, *12*, 13. https://doi.org/10.3390/agriculture12010013

Academic Editor: Chengfang Li

Received: 16 October 2021
Accepted: 5 December 2021
Published: 23 December 2021

Abstract: Cereal/legume intercropping is an effective agricultural practice for pest and disease control and crop production. However, global research on rice and aquatic legume intercropping is relatively rare. A field experiment during two seasons (2018 late season and 2019 early season) was conducted to explore the effects of rice and water mimosa intercropping on rice canopy microclimate, pest and disease, yield, grain quality, and economic income. Two cultivation patterns including rice/water mimosa intercropping and rice monocropping were employed, and three nitrogen (N) fertilizer application levels, including zero N (ZN, 0 kg ha^{-1} N), reduced N (RN, 140 kg ha^{-1} N), and conventional N (CN, 180 kg ha^{-1} N) levels, were applied for the above two cultivation patterns. The results showed that rice/water mimosa intercropping formed a canopy microclimate of rice with higher temperature and lower relative humidity and dew point temperature. In addition, there was a significant reduction in the occurrences of rice leaf blast by 15.05%~35.49%, leaf folders by 25.32%~43.40%, and sheath blight by 16.35%~41.91% in the intercropping treatments. Moreover, rice/water mimosa intercropping increased rice per unit yield by 43.00%~53.10% in the late season of 2018 and 21.40%~26.18% in the early season of 2019. Furthermore, rice grain quality was totally improved, among which brown and head rice rates increased but rice chalky rate and chalkiness degree decreased in the intercropping system. We suggest that combining rice/water mimosa intercropping and N fertilizer reduction can be used as an environmentally friendly eco-farming technique because it can decrease N fertilizer application by approximately 40 kg·ha^{-1}. This combination would not only mitigate nonpoint source pollution but also obtain advantages for controlling rice pests and diseases that would alleviate pesticide usage and improve rice yield and grain quality, which can be extended for green rice production to increase income for producers.

Keywords: rice; intercropping; water mimosa; pest and disease; microclimate; grain quality; yield

1. Introduction

Globally, rice (*Oryza sativa* L.) is one of the oldest and most important staple foods and is consumed by approximately 50% of the world's population and 60% of China's population [1–5]. In addition, modern, intensive agriculture has resulted in a series of problems, such as soil degradation, biodiversity loss, and nonpoint source pollution [6,7]. In many large, modern farms, crops are cultivated as monocultures that reduce the margins [8]. Therefore, achieving plant diversity is extremely important for the development of sustainable agroecological systems [9]. Intercropping is a vital, traditional agriculture practice. In intercropping systems, no less than two crops are simultaneously cultivated in the same field [9]. The intercropping system can efficiently utilize light, heat, water, nutrients, and other environmental resources [1]. Intercropping usually brings plenty of benefits, such as increasing crop yield, land equivalent ratio, and economic income, reducing soil degradation, and controlling weeds, pests, and diseases [10].

The influence of structural diversity of plants on microclimatic parameters has become increasingly clear [11]. In the intercropping system, the aboveground and belowground microclimates were changed through horizontal and vertical plant growth [12]. For instance, coffee (*Coffea arabica* L.) and macauba (*Acrocomia aculeata*. (Jacq.) Lood. ex Mart) intercropping decreased the air temperature of the coffee canopy because the coffee was shaded by macauba [13]. Another study showed that soybeans (*Glycine max* L.) had smaller roots and shoots than the 26-year-old hybrid poplar (*Populus deltoides* X nigra.), so the active radiation, soil water content, and ambient temperature decreased, but the relative humidity increased in their intercropping system [14].

Besides modifying the microclimate, intercropping is an environmentally friendly approach to control pests and disease. Rice blast (*Pyricularia oryzae* Cav.), sheath blight (*Rhizoctonia* solani), and leaf folder (*Cnaphalocrocis medinalis* Guenee) are common pests and diseases of rice. In addition, the disease incidence is related to the initial fungi amount, field microclimate, and plant traits. High temperature and humidity are the important causes of the disease, while aeration and permeability can reduce pest growth [1]. In China, rice blast, sheath blight, and leaf folders are the main diseases and pests that reduce rice yield [15]. The pesticide sales for rice were approximately 538 million US dollars in 2006 [16]. Using intercropping methods to control rice diseases and pests is an optimal choice in paddy fields. The mechanisms of intercropping to reduce the incidence of pests and diseases are possibly attributed to microclimate change, physical barrier, dilution effect, non-lodged effect, allelopathy, etc. [1]. Compared with rice monocultures, rice/water spinach (*Ipomoea aquatica* Forssk.) intercropping showed great control of rice sheath blight, leaf folders, and leaf blast [1,4]. In addition, rice/water chestnut (*Eleocharis dulcis* (Burm. f.) Trin.) intercropping suppressed rice sheath blight and blast [3]. Generally, intercropping rice and other crops would be a sustainable agricultural strategy to reduce the occurrence of pests and diseases in paddy fields.

In all kinds of intercropping systems, the cereal/legume intercropping system shows the advantages of increasing crop yield and improving grain quality. In a dry land rice/legume intercropping system, legumes fix N_2 from the atmosphere and transfer N to rice [17]. This biofertilizer supplied by legumes promoted crop growth and increased yield and grain quality. For instance, maize/lentil (*Lens culinaris* L.) intercropping promoted plant growth and increased the yield and grain protein [18]. In addition, compared with maize monoculture, maize/lablab (*Lablab purpureus* (Linn.) Sweet) intercropping increased the yield and grain quality of maize with higher crude protein, acid, and neutral detergent fiber contents [19].

In paddy fields, rice can be intercropped with some aquatic plants, such as water chestnut, water spinach, alligator flags (*Thalia dealbata* Fraser), and Azolla (*Azolla imbricate* (Roxb.) Nakai) [3–6]. However, few studies have focused on the intercropping of rice and aquatic leguminous plants in paddy fields. Water mimosa (*Neptunia oleracea* Lour.), an aquatic legume vegetable, is consumed by many Asian countries due to its high nutritional value [20,21]. Previous studies showed that rice intercropping with water mimosa increased the rice yield [22,23]. N fertilizer also promoted rice growth and increased the yield and grain quality of rice [24,25]. However, there are limited references studying the effect of rice/water mimosa intercropping and N fertilizer on rice pest and disease control, yield, grain quality, and economic income in paddy fields. In addition, previous studies on the microclimate in rice/water mimosa intercropping systems are relatively rare.

Therefore, the present study was conducted to explore whether the microclimate of the rice canopy would be changed by rice/water mimosa intercropping and whether this intercropping can control pests and diseases and increase rice yield and grain quality while reducing N fertilizer application. Our hypotheses were as follows: (1) Rice/water mimosa intercropping can modify the rice canopy microclimate; and (2) rice/water mimosa intercropping can control diseases and pests and then increase yield and grain quality while reducing N fertilizer application.

2. Materials and Methods

2.1. Study Site and Materials

A field experiment was performed at Zengcheng Teaching and Research Farm (23°14′ N, 113°38′ E), South China Agricultural University, Guangzhou, Guangdong Province, China. The area has a subtropical monsoon climate with a warm winter and hot summer (Figure S1). In addition, rice (*Oryza sativa* L., Huanghuazhan), a typical local crop that is widely cultivated in southern China for its good taste, grain quality, strong lodging resistance, and steady productivity, and water mimosa (*Neptunia oleracea* Lour.), an aquatic leguminous plant, were selected for the experiment. The basic physical and chemical properties of the soil were as follows: sandy loam, pH 4.88, containing 15.80 g kg^{-1} organic matter, 2.27 g kg^{-1} total N, 0.51 g kg^{-1} total phosphorous, 10.91 g kg^{-1} total potassium, 9.06 mg kg^{-1} ammonium nitrogen (N), 4.69 mg kg^{-1} nitrate N, 43.82 mg kg^{-1} available phosphorus, and 47.56 mg kg^{-1} available potassium.

2.2. Cultivating Experimental Design

A field experiment was conducted during 2018 and 2019 consisting of two rice growing seasons. The 2018 late growing season was from August to November, and the 2019 early growing season was from April to July. Six treatments with four replicates were applied to the experiment: rice monocropping (0, 140, and 180 $kg{\cdot}ha^{-1}$ N) and rice/water mimosa intercropping (0, 140, and 180 $kg{\cdot}ha^{-1}$ N). A completely randomized design was used in the field experiment. Each plot area was 35 m^2 (5 m × 7 m) and irrigated and drained independently. The seedlings' density of rice was 250,000 holes/ha and 166,667 holes/ha in monocropping and intercropping treatments. The density of water mimosa was 83,333 plants/ha in intercropping treatments. The cultivation standard per strip in the intercropping treatments was performed every four rows of rice and then connected to three rows of water mimosa. Within the rice monocropping treatments, rice row spacing was 0.2 m. In intercropping treatments, the intrarow spacing of rice and water mimosa was 0.2 m and 0.15 m, respectively, and the row distance between rice and water mimosa was 0.25 m (Figure 1). The area ratio of rice and water mimosa was 2:1 in the intercropping treatment. There were 2/3 and 1/3 areas of each plot occupied by rice and water mimosa in intercropping treatments, respectively. Urea ($CO(NH_2)_2$) was applied as N fertilizer before transplanting and at the rice tillering, heading, and filling stages with proportions of 40%, 20%, 30%, and 10%, respectively (Figure 1). In addition, the application method of phosphate and potassium fertilizer was the same in each treatment. Calcium superphosphate (P_2O_5 12%), as the phosphate fertilizer, was applied only as the base fertilizer at 45 kg ha^{-1}. Potassium chloride (KCl 60%) was used as potassium fertilizer at 135 kg ha^{-1}; one half was applied as the base, and the other half was applied as heading stage fertilizer. Seeds were soaked in water for 24 h at room temperature and then germinated under moisture conditions for seedling preparation. Germinated seeds were sown on 25 July (late growing season of 2018) and on 10 March (early growing season of 2019) for rising nurseries. In addition, 15 kg ha^{-1} urea fertilizer was applied during the rice seedling stage. Water mimosa (length of 0.3 m) and rice seedlings (three-leaf-and-one-leaflet stage) were then concurrently transplanted to the paddy field. Fertilization management methods were the same for the two seasons. The drainage and irrigation managements were to keep the water layer at 6~8 cm throughout the rice-growing season in the field because water mimosa grows better in water flooded condition, but irrigation was stopped 1 week before the rice harvest. We did not apply any pesticides, herbicides, or weed control practices for any treatments.

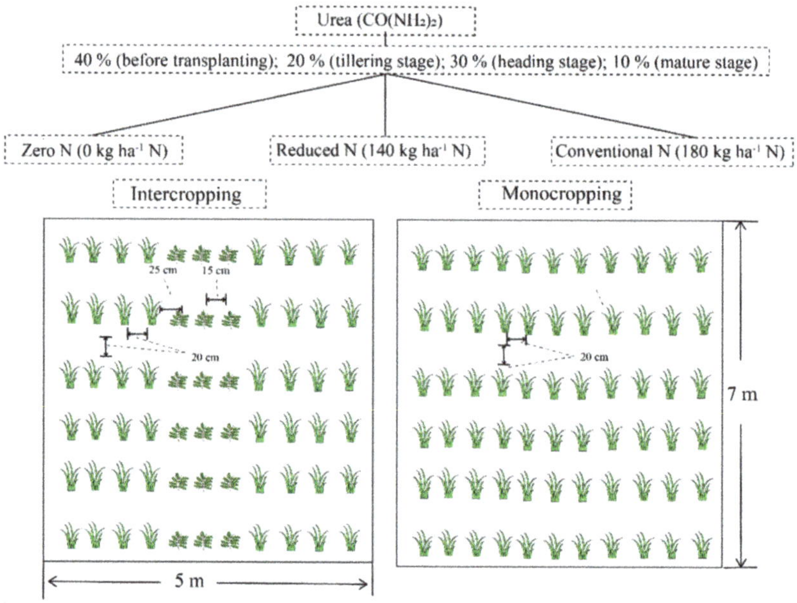

Figure 1. Experimental design of two cropping systems under three different levels of N fertilizer application treatments.

2.3. *Sampling and Data Collection*

2.3.1. Canopy Microclimate

The canopy microclimate of rice in the monocropping and intercropping treatments under the reduced N fertilizer level was measured by a small, portable climate instrument UA-002064 (Onset, Bourne, MA, USA). On the 50th, 55th, 60th, 65th, 70th, and 75th days after transplanting, daily changes of dew point temperature, air temperature, and relative humidity (RH) in the rice canopy were measured for the four plots in each treatment. A small, portable climate instrument was placed among the rows of rice. The instrument was tied on a pole at the height of the rice canopy, and its height was changed according to the height of the rice canopy. The data were recorded every 20 min, and then the dew point temperature, air temperature, and RH value were calculated to obtain an average value per hour.

2.3.2. Pest and Disease Survey

Rice leaf blast, leaf folders, and sheath blight were investigated on the 50th, 55th, 60th, 65th, 70th, and 75th days after transplanting in each season. Three holes of rice in different rows were selected randomly as the initiation points, and then five holes were extended in each row when using a parallel jumping sampling method. The calculation methods of rice leaf blast, leaf folders, and sheath blight were as follows.

The damaged leaf number caused by rice leaf blast and leaf folders and the total leaf number of the 15-hole rice were counted. The incidences of rice leaf blast and leaf folders were calculated by using the following Equation (1) [1]:

$$\text{The incidence of rice leaf blast (leaf folders) (\%)} = \text{damaged leaf number}/\text{total leaf number} \times 100\% \quad (1)$$

Rice sheath blight was surveyed, according to Wu et al. (2012) [26], and the disease categories of rice sheath blight were as follows:

0—No lesion;
1—Lesions on any leaf except the top three leaves;
2—Lesions up to the third topmost leaf;
3—Lesions up to the second topmost leaf;
4—Lesions up to the flag leaf or panicle.

The numbers of infected stems in different disease categories and total stems were counted. The disease index was calculated according to Equation (2) [1,26]:

$$\text{Disease index of rice sheath blight (\%)} = [\Sigma\ (\text{disease category} \times \text{infected stems number in this disease category}) / (\text{total stems number} \times \text{the highest disease category})] \times 100\% \quad (2)$$

2.3.3. Grain Yield, Yield Components, Grain Quality, and Economic Income

Rice and water mimosa yields were measured from four plots with an area of 1 m × 1 m for each replicate, and each sample contained insiders and outsiders of the plants. In addition, rice was threshed manually and sun-dried (adjusted to 14% moisture content) to obtain the grain yield, and the water mimosa yield was determined by harvesting the top tender part. Per unit yield indicated the rice yield which was calculated in the 3 × 1 m^2 plots, and the actual yield of rice and water mimosa represented the yield considering the area ratio of rice and water mimosa in the intercropping treatment. Yield components were measured according to Li et al. (2019) [27] and determined from five hills for each replicate. Effective panicle numbers per hill were counted and calculated per unit effective panicle numbers and actual panicle numbers. Then, the panicles were threshed manually and divided into filled grains and unfilled grains. Five subsamples of filled and unfilled grains were used to estimate the grain numbers per panicle and seed-setting rate. The 1000-grain weight was also calculated from the sampled grains.

Rice grain quality was measured according to Li et al. (2019) [27] and determined from five samples for each replicate. A 100-g sample of rice grain was passed through a dehusker for polishing and then divided into broken and unbroken grains. The brown, milled, and head rice rates were calculated as the percentages of the total (100 g) rice grains. Amylose content, soluble protein content, and alkali value were measured using an Infratec-1241 grain analyzer (FOSS-TECATOR, Hilleroed, Denmark). The chalky rice rate, chalkiness degree, and length/width were scanned with Plant Mirror Image Analysis (MICROTEK, Hsinchu, Taiwan, China), and then the resulting images were processed with SC-E software (HangzhouWanshen Detection Technology Co., Ltd., Hangzhou, Zhejiang, China).

Net income for each treatment was estimated using Equation (3) [28]:

$$\text{Net income} = \text{Gross income} - \text{Total cost of cultivation} \quad (3)$$

Here, the total cost of cultivation comprised the costs of inputs and labors. The costs of inputs (seeds and fertilizers) were based on the local market prices. Water mimosa is a perennial herb that can be reproduced in the field so that its plant prices are free [20,21]. The costs of labors were calculated by the cultural activities (land preparation, seedling, transplanting, applying fertilizer, and harvesting) and paid at the rate of 120 yuan (Chinese yuan) person-day^{-1} of 8 h. Gross income was calculated as the total value of economic yield (water mimosa and rice) per treatment. The market prices of rice and water mimosa were 4 yuan kg^{-1} and 7 yuan kg^{-1}, respectively.

2.4. Data Analysis

Data are expressed as the mean value ± standard error and subjected to two-way ANOVA in the two cultivation patterns with three different N fertilizer levels ($p < 0.05$). In addition, treatment differences in the same cropping system were statistically assessed by using Duncan's method of one-way ANOVA for multiple comparisons when the data met the normality and homoscedasticity hypotheses; otherwise, the data were evaluated through a Games–Howell method ($p < 0.05$). The differences between intercropping and monocropping were statistically evaluated by independent T-tests ($p < 0.05$). All statistical analyses were performed using SPSS 17.0. The data were tested by the Pearson correlation analysis using the "corrplot" package in R [29].

3. Results

3.1. Rice Canopy Microclimate

On the 50th, 55th, 60th, and 75th days in the late season of 2018 and on the 55th, 65th, and 75th days in the early season of 2019, the air temperatures were significantly higher under intercropping treatments than under the monocropping treatments, by 0.07~2.30 °C (Figure S2a,b). In contrast, in the late season of 2018, in the middle of the day, the dew point temperature and RH in the intercropping treatments were generally significantly lower, by 0.07~1.77 °C and 0.88~11.85% compared with those in the rice monocropping treatments, respectively (Figure S2c,e).

3.2. Rice Pest and Disease

During the whole investigation, intercropping treatments reduced the occurrence of rice pests and diseases. For instance, the incidences of rice leaf blast in intercropping treatments were generally significantly lower than those in monocropping treatments except on the 50th day after transplanting in the late season of 2018 (Figure 2a,b, Table S1). Intercropping treatments also significantly reduced the occurrence of rice leaf folders compared with the monocropping treatments on the 60th~75th days after transplanting in both seasons (Figure 2c,d, Table S3). Moreover, there was a significant reduction in the incidence of rice sheath blight in the intercropping treatment on the 55th and 70th days after transplanting in the 2018 late season and on the 60th, 65th, and 75th days after transplanting in the 2019 early season (Figure 2e,f, Table S5). Additionally, the incidences of rice leaf blast, leaf folders, or sheath blight in 2018 late season generally were significantly higher than those in 2019 early season after the 70th day of transplanting (Tables S2, S4 and S6).

More importantly, the incidences of rice leaf blast, leaf folders, and sheath blight in the intercropping with reduced N treatments were significantly lower than those in the monocropping with zero, reduced, and conventional N treatments (Tables S1, S3 and S5). No significant difference was found between the N fertilizer treatments in pests and diseases.

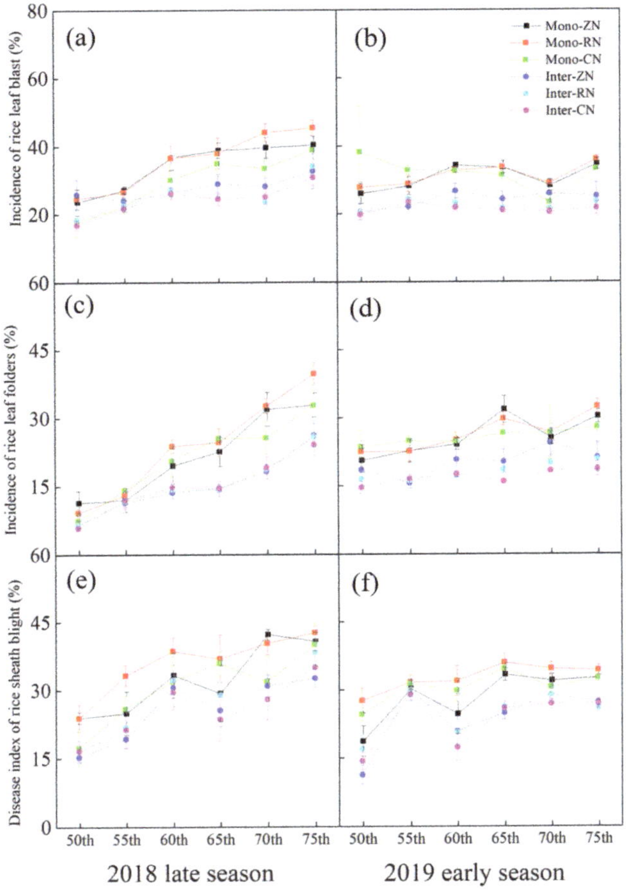

Figure 2. Pest and disease of rice on the 50th, 55th, 60th, 65th, 70th, and 75th days after transplanting. "Mono", monocropping; "Inter", Intercropping. ZN, RN, and CN indicate the 0, 140, and 180 kg ha^{-1} N fertilizer levels, respectively. (**a**,**c**,**e**) and (**b**,**d**,**f**) denotes 2018 late season and 2019 early season, respectively.

3.3. Rice Yield and Yield Components

There was a significant interactive effect between cultivation pattern and N fertilizer on rice yield and per unit effective panicle number (Table S7). The per unit yield, actual yield, and per unit effective and actual panicle number of rice and the yield of water mimosa in the CN and RN treatments were significantly higher than those in the ZN treatments. In addition, intercropping treatments significantly increased the rice's per unit yield and per unit effective panicle numbers by 43.00%~53.10% and 22.15%~41.41% in the late season of 2018 and 21.40%~26.18% and 15.72%~41.54% in the early season of 2019, respectively. However, no significant difference was found in grain numbers per panicle, seed-setting rate, or 1000-grain weight (Table 1). Additionally, rice yield and yield components in 2019 early season were generally significantly higher than those in 2018 late season except the grains per panicle (Table S8).

Furthermore, compared with rice monocropping with conventional N application treatments, intercropping with reduced N treatments significantly increased the rice's per unit yield by 35.64% in the 2018 late season and increased the per unit effective panicle numbers by 17.01% in the 2019 early season (Table 1).

3.4. Rice Grain Quality

The interactions between the two factors (cultivation pattern and N fertilizer) had significant effect on rice chalky rice rate and chalkiness degree (Table S9). Compared with the ZN treatments, the RN and CN treatments significantly increased the brown rice rate, protein content, length/width, and alkali value of rice in the early season of 2019 (Table 2). As our prediction, brown and head rice rates in intercropping treatments were slightly higher than those in monocropping treatments in the late season of 2018. There was a significant reduction in chalky rice rate and chalkiness degree in intercropping treatments compared with monocropping treatments during the two seasons (Table 2). No significant difference was found in the milled rice rate or amylose content (Table 2). Additionally, rice grain quality in 2019 early season was generally significantly higher than that in 2018 late season except the chalky rice rate, chalkiness degree, and length/width (Table S10).

The chalky rice rate in the intercropping with reduced N treatment was significantly lower than that in the monocropping with the conventional N treatment (Table 2). In addition, the length/width and alkali value in the intercropping with zero N treatments were significantly lower than those in the monocropping with the conventional N treatment in the early season of 2019 (Table 2).

3.5. Correlation Analysis

In the late season of 2018, the incidences of rice blast, leaf folders, and sheath blight generally had significant negative correlations with the air temperature, rice per unit yield, per unit effective panicle number, and head rice rate, but had significant positive correlations with the dew point temperature, RH, chalky rice rate, and chalkiness degree (Figure S3a). Although the incidences of pathogens did not have obvious correlations with microclimatic parameters, the incidences of pathogens had negative correlations with rice per unit yield, effective number, and head rice rate and had significant positive correlations with chalky rice rate and chalkiness degree in the early season of 2019 (Figure S3b).

3.6. Economic Analysis

In the two seasons, intercropping with reduced N treatments had the highest economic income and accounted for 18,234 and 18,830 yuan ha^{-1} in the 2018 late season and 2019 early season, respectively (Table 3). Thus, combining rice/water mimosa intercropping and N reduction could obtain the maximum benefit for farmers and producers.

The prices of rice and water mimosa are 4 and 7 yuan kg^{-1}, respectively. ZN, RN and CN indicate the 0, 140 and 180 kg ha^{-1} N fertilizer levels, respectively. "Mono"and "Inter" denote the mono-cropping and intercropping treatments.

Table 1. Yield and yield components in monocropping and intercropping treatments with the three different N fertilizer application levels.

Season	N Fertilizer	Pattern	Rice							Water Mimosa
			Per Unit Yield (t·ha^{-1})	Actual Yield (t·ha^{-1})	Per Unit Effective Panicle (10^4 ha^{-1})	Actual Effective Panicle (10^4 ha^{-1})	Grains per Panicle	Seed-Setting Rate (%)	1000-Grain Weight (g)	Yield (t·ha^{-1})
2018 late season	ZN	Mono	3.22 ± 0.16 Aa	-	252.75 ± 16.75 Bb	-	118.76 ± 4.93 Aa	54.15 ± 1.85 Aa	19.11 ± 0.31 Aa	-
		Inter	3.48 ± 0.08 Ab	2.32 ± 0.05 b	357.50 ± 12.75 Aa	238.33 ± 8.50 a	121.39 ± 5.03 Aa	55.72 ± 2.55 Aa	20.10 ± 0.06 Aa	1.64 ± 0.07 b
	RN	Mono	3.20 ± 0.26 Ba	-	331.50 ± 16.5 Ba	-	108.18 ± 2.66 Aa	56.77 ± 3.74 Aa	19.05 ± 0.32 Aa	-
		Inter	4.90 ± 0.16 Aa *	3.27 ± 0.10 a	405.00 ± 13.75 Aa	270.00 ± 9.17 a	116.95 ± 1.86 Aa	53.92 ± 3.99 Aa	20.14 ± 0.41 Aa	2.70 ± 0.19 a
	CN	Mono	3.61 ± 0.26 Ba *	-	382.75 ± 24.25 Aa	-	117.95 ± 7.94 Aa	54.66 ± 5.45 Aa	18.60 ± 0.63 Aa	-
		Inter	5.17 ± 0.12 Aa	3.45 ± 0.08 a	382.50 ± 16.50 Aa	255.00 ± 11.00 a	125.44 ± 11.32 Aa	57.01 ± 3.14 Aa	19.60 ± 0.18 Aa	2.36 ± 0.18 a
2019 early season	ZN	Mono	4.72 ± 0.15 Bb	-	325.00 ± 12.50 Ba	-	94.34 ± 2.56 Aa	79.09 ± 1.07 Aa	21.14 ± 0.15 Aa	-
		Inter	5.96 ± 0.21 Aa	3.97 ± 0.14 a	460.00 ± 15.25 Ab	306.67 ± 10.17 b	91.69 ± 2.83 Aa	77.85 ± 1.70 Aa	21.2 ± 0.26 Aa	1.38 ± 0.12 b
	RN	Mono	5.63 ± 0.11 Aa	-	397.50 ± 9.25 Ba	-	89.19 ± 4.04 Aa	76.36 ± 1.89 Aa	20.68 ± 0.17 Aa	-
		Inter	5.97 ± 0.19 Aa	3.98 ± 0.13 a	460.00 ± 10.75 Aa *	306.67 ± 7.17 b	97.96 ± 4.16 Aa	77.82 ± 2.49 Aa	20.81 ± 0.14 Aa	2.48 ± 0.15 a
	CN	Mono	5.34 ± 0.06 Ba	-	393.25 ± 12.25 Aa *	-	89.63 ± 3.57 Aa	76.59 ± 0.71 Aa	20.85 ± 0.23 Aa	-
		Inter	6.48 ± 0.23 Aa	4.32 ± 0.15 a	448.75 ± 16.75 Aa	299.17 ± 11.17 b	93.69 ± 6.05 Aa	78.20 ± 2.42 Aa	20.90 ± 0.30 Aa	2.31 ± 0.17 a

All the data presented are the means of four replicates ± standard errors. Different capital letters indicate significant differences between cultivation patterns under the same N fertilizer level ($p < 0.05$). Different lowercase letters indicate significant differences between N fertilizer levels in the same cultivation pattern ($p < 0.05$). Actual yield and effective panicle denote the indexes according to the area ratio of the two crops in the intercropping treatment. ZN, RN and CN indicate the 0, 140 and 180 kg ha^{-1} N fertilizer levels, respectively. "Mono"and "Inter" denote the monocrop-ping and intercropping treatments. The "*" represents a significant difference between mono-cropping with the conventional N treatments and intercropping with zero and reduced N treat-ments ($p < 0.05$).

Table 2. Grain quality of rice in monocropping and intercropping treatments with the three different N fertilizer application levels.

Season	N Fertilizer	Pattern	Brown Rice Rate (%)	Milled Rice Rate (%)	Head Rice Rate (%)	Length/ Width	Chalky Rice Rate (%)	Chalkiness Degree (%)	Amylose Content (%)	Protein Content (%)	Alkali Value
2018 late season	ZN	Mono	75.88 ± 0.31 Ba	62.13 ± 0.64 Aa	51.68 ± 1.77 Aa	3.42 ± 0.03 Aa	3.14 ± 0.41 Aa	0.75 ± 0.17 Aa	17.50 ± 0.16 Aa	12.68 ± 1.36 Aa	6.75 ± 0.06 Aa
		Inter	76.83 ± 0.10Ab	63.00 ± 0.63 Aa	52.93 ± 1.04 Aa	3.43 ± 0.06 Aa	2.28 ± 0.35 Aa	0.29 ± 0.05 Ba	17.30 ± 0.10 Aa	14.13 ± 0.03 Aa	6.75 ± 0.05 Aa *
	RN	Mono	77.15 ± 0.60 Aa	62.03 ± 0.69 Aa	50.15 ± 0.42 Ba	3.47 ± 0.03 Aa	3.25 ± 0.23 Aa	0.61 ± 0.07 Aa	17.83 ± 0.23 Aa	13.18 ± 0.83 Aa	7.03 ± 0.05 Aa
		Inter	76.13 ± 0.59 Aab	61.23 ± 1.06 Aa	52.20 ± 0.30 Aa	3.42 ± 0.05 Aa	1.72 ± 0.25 Ba *	0.24 ± 0.06 Ba	17.63 ± 0.11 Aa	13.15 ± 0.93 Aa	7.05 ± 0.05 Aa
	CN	Mono	76.65 ± 0.85 Aa	62.18 ± 1.14 Aa	50.08 ± 1.53 Aa	3.49 ± 0.05 Aa	2.77 ± 0.35 Aa *	0.44 ± 0.07 Aa	17.75 ± 0.09 Aa	13.05 ± 0.85 Aa	7.08 ± 0.09 Aa *
		Inter	78.05 ± 0.22 Aa	63.20 ± 0.16 Aa	51.40 ± 0.36 Aa	3.48 ± 0.06 Aa	1.59 ± 0.31 Aa	0.43 ± 0.07 Aa	17.48 ± 0.11 Aa	13.23 ± 0.81 Aa	7.08 ± 0.05 Aa
2019 early season	ZN	Mono	73.38 ± 0.33 Ab	58.36 ± 1.07 Aa	38.35 ± 2.32 Aa	3.37 ± 0.02 Ab	2.88 ± 0.43 Aa	0.46 ± 0.04 Aa	17.28 ± 0.23 Aa	8.38 ± 0.31 Ab	6.65 ± 0.02 Ab
		Inter	73.45 ± 0.31 Aa	57.43 ± 1.81 Aa	39.35 ± 0.65 Aa	3.38 ± 0.02 Aa *	1.98 ± 0.41 Aa	0.28 ± 0.038 Ba	17.30 ± 0.39 Aa	8.50 ± 0.31 Ab *	6.73 ± 0.06 Aa *
	RN	Mono	74.13 ± 0.25 Aab	57.74 ± 0.23 Aa	39.12 ± 1.69 Aa	3.42 ± 0.02 Aab	2.98 ± 0.26 Aa	0.35 ± 0.02 Aa	16.90 ± 0.15 Aa	9.95 ± 0.12 Aa	6.88 ± 0.03 Aa
		Inter	73.73 ± 0.37 Aa	60.66 ± 1.58 Aa	41.01 ± 1.28 Aa	3.42 ± 0.01 Aa	1.53 ± 0.17 Ba *	0.25 ± 0.02 Ba	17.08 ± 0.29 Aa	9.83 ± 0.23 Aa	6.88 ± 0.03 Aa
	CN	Mono	74.55 ± 0.23 Aa	60.05 ± 1.31 Aa	39.43 ± 0.65 Aa	3.46 ± 0.01 Aa *	2.47 ± 0.27 Aa *	0.34 ± 0.07 Aa	16.85 ± 0.13 Aa	10.00 ± 0.26 Aa *	6.98 ± 0.05 Aa *
		Inter	74.20 ± 0.42 Aa	59.20 ± 0.51 Aa	37.48 ± 1.86 Aa	3.43 ± 0.01 Aa	1.59 ± 0.12 Ba	0.30 ± 0.02 Aa	17.15 ± 0.44 Aa	9.83 ± 0.28 Aa	6.88 ± 0.85 Aa

All the data presented are the means of four replicates ± standard errors. Different capital letters indicate significant differences between cultivation patterns under the same N fertilizer level ($p < 0.05$). Different lowercase letters indicate significant differences between N fertilizer levels in the same cultivation pattern ($p < 0.05$). ZN, RN and CN indicate the 0, 140 and 180 kg ha^{-1} N fertilizer levels, respectively. "Mono"and "Inter" denote the monocropping and intercropping treatments. The "*" represents a significant difference between monocropping with the conventional N treatments and intercropping with zero and reduced N treatments ($p < 0.05$).

Table 3. Economic analysis (yuan ha^{-1}) of monocropping and intercropping treatments with the three different N fertilizer application levels.

Season	N Fertilizer	Pattern	Rice Seed	Water Mimosa	Fertilizer	Labor	Rice Value	Water Mimosa Value	Net Income
2018 late season	ZN	Mono	600	-	731	7500	12,880	-	4049
		Inter	300	0	731	8250	6960	11,495	9174
	RN	Mono	600	-	1949	7500	12,800	-	2751
		Inter	300	0	1949	8250	9800	18,933	18,234
	CN	Mono	600	-	2291	7500	14,440	-	4049
		Inter	300	0	2291	8250	10,340	16,522	16,021
2019 early season	ZN	Mono	600	-	731	7500	18,880	-	10,049
		Inter	300	0	731	8250	11,920	9694	12,333
	RN	Mono	600	-	1949	7500	22,520	-	12,471
		Inter	300	0	1949	8250	11,940	17,389	18,830
	CN	Mono	600	-	2291	7500	21,360	-	10,969
		Inter	300	0	2291	8250	12,960	16,183	18,302

4. Discussion

4.1. Canopy Microclimate Modifying

Intercropping is a simple and effective way to modify the microclimate of rice canopies [30–32]. In our two-season field experiments, intercropping treatments increased the canopy air temperature compared with the rice monocropping treatments (Figure S2). These results were similar to those of a previous study that showed that canopy temperature increased in maize/cowpea intercropping systems relative to sole maize cropping [33,34]. Additionally, rice/water mimosa intercropping reduced the dew point temperature and relative humidity of rice canopy (Figure S2). Similarly, many studies have demonstrated that intercropping reduces the canopy relative humidity and dew point temperature [30,31]. In the rice/water mimosa intercropping system, water mimosa prostrated growth, rice erected growth, and the height of rice was much higher than that of water mimosa [21]. It is known that higher plants have stronger competition for light in intercropping systems, and high irradiation is usually combined with high temperature and low humidity [14,32]. In addition, the height difference between the two crops exhibited better air circulation. Moreover, this intercropping system had a higher air temperature and lower relative humidity and dew point temperature and, thus, exhibited a special canopy microclimate of rice as a result (Figures 3 and S2).

4.2. Pest and Disease Control

It is believed that intercropping usually reduces the occurrence of harmful organisms due to the high biodiversity and stability of the ecosystem [1]. In the present study, rice/water mimosa intercropping reduced the incidences of rice leaf blast, leaf folders, and sheath blight (Figure 2, Tables S1, S3 and S5). A similar study on rice and water spinach intercropping in paddy fields also showed that pests and diseases were substantially lower in intercropping systems [4]. In addition, our study also showed that the occurrence of pathogens had a positive correlation with dew point temperature and relative humidity and a negative correlation with air temperature in the late season of 2018 (Figure S3). High air temperature inhibited rice leaf roller egg hatching and reduced the longevity of this insect [1]. Hence, the high air temperature in this intercropping system might contribute to reducing the breeding of rice folders. Universally, dew is formed at night, and relative humidity can control fungal spores to germinate the conidial sporulation and dispersal process of pathogens [31,33]. Thus, the high temperature and low dew point and humidity probably had some functions to control pathogens in the present study.

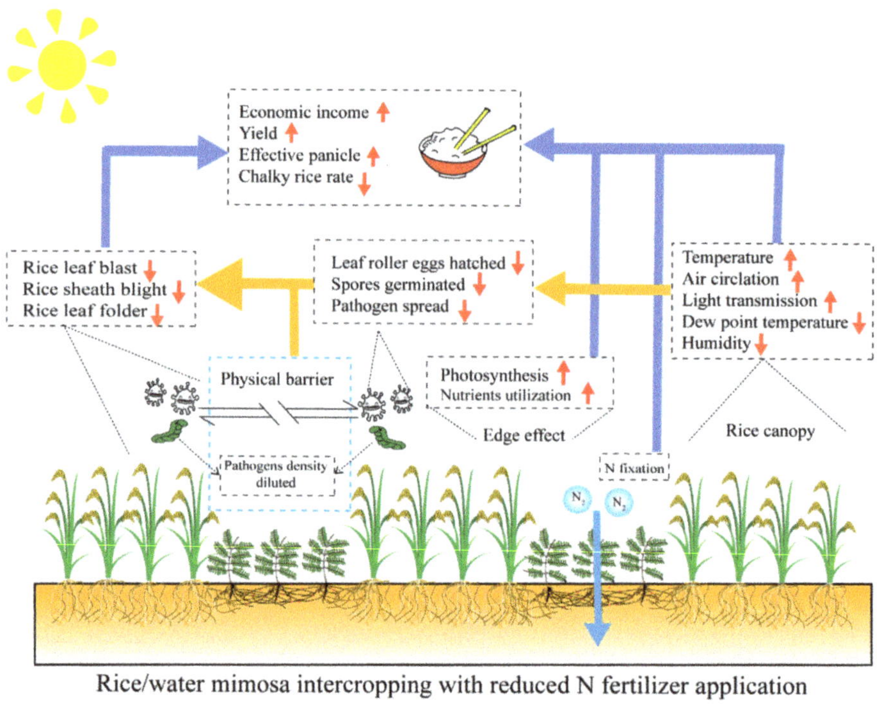

Figure 3. Mechanisms of rice/water mimosa intercropping with reduced N application to modify canopy microclimate of rice, control pest and disease, and improve rice yield, grain quality, and economic income.

According to the characteristics of the rice pest and disease, we speculated and summarized that the causes of rice/water mimosa intercropping reduced the incidences of pathogens would be as follows. (1) The strip distribution in the intercropping system established a physical barrier that blocked the horizontal spread of the pathogens [1]. (2) Microclimatic factors such as high temperature, low relative humidity and dew point temperature, stronger solar radiation, and better air circulation in the intercropping system had some functions in resisting pests' and diseases' activity, such as insect migration and reproduction [4,34]. (3) Water mimosa was neither the food nor the host of the pathogens; hence, the density of pathogens was relatively diluted. Thus, the incidences of pests and diseases reduced in the rice/water mimosa intercropping system (Figure 3).

Moreover, previous studies revealed that N fertilizer usually increases the disease susceptibility of rice [35–37]. However, in our study, there were no significant differences between N fertilizer levels in the incidences of rice leaf blast, sheath blight, and leaf folders. The reason for this phenomenon was probably that the application of N fertilizer was insufficient to increase the disease susceptibility of rice in the present study because 180 kg ha^{-1} N was the optimal N fertilizer for rice growth in paddy fields [1,4,35–37]. We also found that the pests and diseases in the monocropping with conventional N treatments were significantly higher than those in the intercropping with zero, reduced, and conventional N treatments (Figure 2, Tables S1, S3 and S5). Therefore, the reduction in rice pest and disease rates in rice/water mimosa intercropping could mainly be attributed to the border effect.

4.3. Yield and Yield Components Advantages

Environmental conditions and management methods play a crucial role in yield and yield components [2]. The increase in rice per unit yield in the early season of 2019 might be attributed to the higher rainfall compared with that in the late season of 2018 (Figure S1 and Table S8). This is consistent with a previous study that showed that sufficient irrigation could increase crop yield [38]. In the present study, we observed that the interaction between cultivation pattern and N fertilizer significantly affected the rice per unit yield and effective panicle numbers in the two seasons (Table S7). These results are in line with previous research that showed that the interactive effect between the intercropping pattern and N fertilization significantly affected plant growth in a groundnut (*Arachis hypogaea* Linn.) and sesame (*Sesamum indicum* Linn.) intercropping system [39]. In the present research, compared with zero N treatment, the reduced and conventional N treatments substantially increased the yield of rice and water mimosa (Table 1). As a comparison, previous research found that crop yield increased with N fertilizer levels [40,41]. It is acknowledged that higher grain yield was obtained due to higher panicle numbers [40]. Similar to rice yield, reduced and conventional N treatments also had higher rice effective panicle numbers than zero N treatments (Table 1). Ju et al. (2019) [24] also reported that N fertilizer application increased the panicle numbers of rice. Consequently, both reduced and conventional N treatments were beneficial to rice yield and effective panicle numbers.

Intercropping generally increased the crop yield. In the present study, intercropping treatments increased rice per unit yield by 43.00%~53.10% in the late season of 2018 and 21.40%~26.18% in the early season of 2019. Meanwhile, per unit effective panicle numbers were higher in rice/water mimosa intercropping than in rice monocropping (Table 1). Our results were consistent with previous studies. Ning et al. (2017) [1] and Liang et al. (2016) [4] both found that rice/water spinach intercropping increased the rice yield and effective panicle numbers. In addition, intercropping treatments significantly increased the yield, while there was no significant difference in grain components between cultivation patterns during most of the time (Table 1). The reason probably was that rice had higher per unit effective panicle numbers among the intercropping treatments, and the effective panicle number was one of the yield components. The increase in rice per unit yield in intercropping treatments could also be attributed to the promoting effect of water mimosa N fixation [23,42]. Moreover, the rice per unit yield and effective panicle numbers in the intercropping with reduced N treatments were higher than those in the monocropping with conventional N treatments (Table 1). These findings suggested that rice/water mimosa intercropping combined with reduced N fertilizer application would be a good practice choice to increase rice yield in paddy fields.

4.4. Grain Quality and Economic Incomes Improvement

Grain quality comprises grain appearance, milling, cooking, eating, and nutritional qualities and depends on genetic, environmental, and crop management factors [27,43]. Our results indicated that N fertilizer improved rice grain quality. For instance, the brown rice rate, protein content, length/width, and alkali value in the reduced and conventional N treatments were higher than those in the zero N treatments (Table 2). These results are consistent with previous studies which found that crop quality indexes improved with N fertilizer rates [25,44,45].

Moreover, intercropping had a positive and significant effect on brown and head rice rates (Table 2). An increase in crop quality in intercropping systems was also reported by Yusef et al. (2014) [37]. Low chalkiness is often associated with more translucent rice grain and represents a higher value and price of rice in markets [46,47]. In the present study, rice/water mimosa intercropping resulted in a significant reduction in the chalky rice rate and chalkiness degree (Table 2). The lower chalky rate and chalkiness degree of rice probably bring more profit to farmers and producers. In the rice/water mimosa intercropping system, the reason that led to better grain quality could be attributed to the following aspects: (1) Organic fertilizer and biofertilizer supplied by water mimosa could improve grain quality by improving photosynthesis and nutrient uptake by cereals, and these nutrients were eventually transported to the seed and contributed to improving grain quality (Figure 3) [48]. (2) Rice grain quality was associated with microclimatic parameters (Figure S3). Similarly, intercropping had higher temperature and good ventilation and light transmission, which could lead to less inhomogeneous substances, such as white opacity and chalkiness in grains [49]. (3) Our study found that the incidences of pathogens had negative correlations with the head rice rate and had positive correlations with the chalky rice rate and chalkiness degree (Figure S3). These results are consistent with a previous study which showed that lower pathogens in rice resulted in better grain quality (Figure 3) [50,51].

Notably, compared with monocropping with conventional N treatments, intercropping with zero N treatments resulted in lower length/width and alkali values. However, intercropping with reduced N treatment decreased the chalky rice rate relative to monocropping with conventional N treatments (Table 2). Therefore, we could infer that intercropping with reduced N treatment is an optimal choice for rice grain quality.

For economic analysis, we found that the intercropping system with reduced N treatments had the highest economic income (Table 3). Higher net income has also been reported under potato/legume and sorghum/legume intercropping systems [28,52]. Rice/water mimosa intercropping with reduced N treatments achieved the maximum economic income which would be beneficial for farmers and producers. However, there were only two seasons of performance assessment in this study, and further long-term experiments are needed in the future.

5. Conclusions

Combining rice/water mimosa intercropping and N fertilizer reduction could not only decrease N fertilizer application by approximately 40 kg ha^{-1}, which would mitigate nonpoint source pollution, but also form a canopy microclimate of rice with higher temperature, lower relative humidity and dew point temperature, and better control of rice pests and diseases to a certain extent, which would alleviate pesticide usage, finally increasing rice productivity, grain quality, and economic income, and that can be extended for green rice production to increase income for farmers and producers. In a word, our study provided an environmentally friendly eco-farming technique in rice production.

Supplementary Materials: The following are available online at https://www.mdpi.com/article/10.3390/agriculture12010013/s1, Figure S1: Rainfall (RF) and average temperature (AT) per month during the two seasons in Zengcheng, Guangzhou, Guangdong, China. Figure S2: Air temperature, relative humidity (RH) and dew point temperature of rice canopy in the monocropping and intercropping under the reduced N fertilizer application level. Figure S3: Pearson correlations among daily air temperature (AT), dew point temperature (DPT) and relative humidity (RH) of rice canopy at 50th ~ 75th after transplanting, rice leaf blast (RLB), rice leaf folders (RLF) and rice sheath blight (RSB) at 75th day after transplanting, rice yield (RY), per unit effective panicle number (EPN), head rice rate (HRR), chalky rice rate (CRR) and chalkiness degree (CD) at maturity stage. Table S1: Incidence (%) of rice leaf blast in monocropping and intercropping treatments with the three different N fertilizer application levels. Table S2: Variance in incidence of rice leaf blast between the 2018 late season and 2019 early season. Table S3: Incidence (%) of rice leaf folders in monocropping and intercropping treatments with the three different N fertilizer application levels. Table S4: Variance in incidence of rice leaf folders between the 2018 late season and 2019 early season. Table S5: Disease index (%) of rice sheath blight in monocropping and intercropping treatments with the three different N fertilizer application levels. Table S6: Variance in incidence of rice sheath blight between the 2018 late season and 2019 early season. Table S7: Interaction effects between the two factors (cultivation pattern and N fertilizer) on rice yield and yield components by two-way ANOVA. Table S8: Variance in rice yield and yield components between the 2018 late season and 2019 early season. Table S9: Interaction effects between the two factors (cultivation pattern and N fertilizer) on rice grain quality determined by two-way ANOVA. Table S10: Variance in rice grain quality between the 2018 late season and 2019 early season.

Author Contributions: Z.H. carried out the experiment, performed data analysis, and drafted the manuscript. H.X. and J.Z. (Jiaen Zhang) were responsible for conceptualization, review, and editing, supervision. K.L., J.Z. (Jiawen Zhong), M.L. and X.R. participated in data collection and interpretation. All authors have read and agreed to the published version of the manuscript.

Funding: This study was supported by grants from the Guangdong Provincial Key Laboratory of Eco-Circular Agriculture (2019B030301007), the Innovation Team Construction Project of Modern Agricultural Industry Technology System of Guangdong Province (2018LM1100, 2019KJ105, 2020KJ105), the Science and Technology Project of Guangdong Province (2015A020224040, 2016A020210094, 2017A090905030), and the Science and Technology Project of Guangzhou (201604020062).

Institutional Review Board Statement: Not applicable.

Informed Consent Statement: Not applicable.

Data Availability Statement: Not applicable.

Conflicts of Interest: We declare that we have no financial or personal relationships with other people for this work. We declare that we do not have any associative interest that represents a conflict of interest in connection with the work submitted.

References

1. Ning, C.; Qu, J.; He, L.; Yang, R.; Chen, Q.; Luo, S.; Cai, K. Improvement of yield, pest control and Si nutrition of rice by rice-water spinach intercropping. *Field Crops Res.* **2017**, *208*, 34–43. [CrossRef]
2. Li, R.; Li, M.; Ashraf, U.; Liu, S.; Zhang, J. Exploring the relationships between yield and yield- related traits for rice varieties released in China from 1978 to 2017. *Front. Plant Sci.* **2019**, *10*, 543. [CrossRef]
3. Qin, J.; He, H.; Luo, S.; Li, H. Effects of rice-water chestnut intercropping on rice sheath blight and rice blast diseases. *Crop Prot.* **2013**, *43*, 89–93. [CrossRef]
4. Liang, K.; Yang, T.; Zhang, S.; Zhang, J.-E.; Luo, M.; Fu, L.; Zhao, B. Effects of intercropping rice and water spinach on net yields and pest control: An experiment in southern China. *Int. J. Agric. Sustain.* **2016**, *14*, 448–465. [CrossRef]
5. Wang, J.; Lu, X.; Zhang, J.; Ouyang, Y.; Wei, G.; Xiong, Y. Rice intercropping with alligator flag (*Thalia dealbata*): A novel model to produce safe cereal grains while remediating cadmium contaminated paddy soil. *J. Hazard. Mater.* **2020**, *394*, 122505. [CrossRef] [PubMed]
6. Cissé, M.; Vlek, P.L.G. Influence of urea on biological N_2 fixation and N transfer from Azolla intercropped with rice. *Plant Soil* **2003**, *250*, 105–112. [CrossRef]
7. Frison, E.A.; Cherfas, J.; Hodgkin, T. Agricultural biodiversity is essential for a sustainable improvement in food and nutrition security. *Sustainability* **2011**, *3*, 238–253. [CrossRef]
8. Jacobsen, S.-E.; Sørensen, M.; Pedersen, S.M.; Weiner, J. Feeding the world: Genetically modified crops versus agricultural biodiversity. *Agron. Sustain. Dev.* **2013**, *33*, 651–662. [CrossRef]
9. Bedoussac, L.; Journet, E.-P.; Hauggaard-Nielsen, H.; Naudin, C.; Corre-Hellou, G.; Jensen, E.S.; Prieur, L.; Justes, E. Ecological principles underlying the increase of productivity achieved by cereal-grain legume intercrops in organic farming. A review. *Agron. Sustain. Dev.* **2015**, *35*, 911–935. [CrossRef]
10. Hassen, A.; Talore, D.G.; Tesfamariam, E.H.; Friend, M.A.; Mpanza, T.D.E. Potential use of forage-legume intercropping technologies to adapt to climate-change impacts on mixed crop-livestock systems in Africa: A review. *Reg. Environ. Chang.* **2017**, *17*, 1713–1724. [CrossRef]
11. Ashraf, M.; Sanusi, R.; Zulkifli, R.; Tohiran, K.A.; Moslim, R.; Ashton-Butt, A.; Azhar, B. Alley-cropping system increases vegetation heterogeneity and moderates extreme microclimates in oil palm plantations. *Agric. For. Meteorol.* **2019**, *276*, 107632. [CrossRef]
12. Ehret, M.; Graß, R.; Wachendorf, M. Productivity at the tree-crop interface of a young willow-grassland alley cropping system. *Agrofor. Syst.* **2018**, *92*, 71–83. [CrossRef]
13. Moreira, S.L.; Pires, C.V.; Marcatti, G.E.; Santos, R.H.; Imbuzeiro, H.M.; Fernandes, R.B. Intercropping of coffee with the palm tree, macauba, can mitigate climate change effects. *Agric. For. Meteorol.* **2018**, *256*, 379–390. [CrossRef]
14. Peng, X.; Thevathasan, N.V.; Gordon, A.M.; Mohammed, I.; Gao, P. Photosynthetic response of soybean to microclimate in 26-year-old tree-based intercropping systems in Southern Ontario, Canada. *PLoS ONE* **2015**, *10*, e0129467. [CrossRef]
15. Huang, S.; Wang, L.; Liu, L.; Fu, Q.; Zhu, D. Nonchemical pest control in China rice: A review. *Agron. Sustain. Dev.* **2014**, *34*, 275–291. [CrossRef]
16. Zhang, W.; Jiang, F.; Ou, J. Global pesticide consumption and pollution: With China as a focus. *Proc. Int. Acad. Ecol. Environ. Sci.* **2011**, *2*, 125–144.
17. Xiao, Y.; Li, L.; Zhang, F. Effect of root contact on interspecific competition and N transfer between wheat and fababean using direct and indirect ^{15}N techniques. *Plant Soil* **2004**, *262*, 45–54. [CrossRef]

18. Ahmad, W.; Khan, F.; Shah, Z.; Khan, M.J. Quality and crop yield potential of moderately degraded Alfisols under different nutrient inputs and cropping patterns. *Pedosphere* **2019**, *29*, 235–247. [CrossRef]
19. Mthembu, B.E.; Everson, T.M.; Everson, C.S. Intercropping maize (*Zea mays* L.) with lablab (*Lablab purpureus* L.) for sustainable fodder production and quality in smallholder rural farming systems in South Africa. *Agroecol. Sustain. Food Syst.* **2017**, *42*, 362–382. [CrossRef]
20. Rahman, M.; Khatun, S.; Ali, S.; Yasmin, S.; Kamruzzaman, M.; Rashid, M. Morpho-physiological diversity of root nodule rhizobia from mimosa (*Mimosa pudica* L.) and water mimosa (*Neptunia oleracea* L.). *J. Bacteriol. Mycol.* **2018**, *5*, 1061.
21. Wahab, A.; Ismail, S.S.; Abidin, E.Z.; Praveena, S. *Neptunia oleracea* (water mimosa) as phytoremediation plant and the risk to human health: A review. *Adv. Environ. Biol.* **2014**, *8*, 187–194.
22. Zhang, S.; Liang, K.; Zhang, Y.; Li, M.; Zhang, J. Effects of rice and water mimosa intercropping on crop yield, nitrogen uptake and soil nitrogen content. *Ecol. Environ. Sci.* **2016**, *25*, 1856–1864. (In Chinese)
23. Hei, Z.; Xiang, H.; Zhang, J.; Liang, K.; Zhong, J.; Li, M.; Lu, Y. Rice intercropping with water mimosa (*Neptunia oleracea* Lour.) can facilitate soil N utilization and alleviate apparent N loss. *Agric. Ecosyst. Environ.* **2021**, *313*, 107378. [CrossRef]
24. Ju, J.; Cai, Y.; Zuo, W.; Hai-Tao, Z.; Yang, H.; Mao, W.; Yu-Hua, S.; Ke, F. Effects of nitrogen management on soil nitrogen content and rice grain yield in double cropping rice production area with continuous full amount of straw returning. *Commun. Soil Sci. Plant Anal.* **2019**, *50*, 2655–2668. [CrossRef]
25. Yuan, L.; Zhang, Z.; Cao, X.; Zhu, S.; Zhang, X.; Wu, L. Responses of rice production, milled rice quality and soil properties to various nitrogen inputs and rice straw incorporation under continuous plastic film mulching cultivation. *Field Crops Res.* **2014**, *155*, 164–171. [CrossRef]
26. Wu, W.; Huang, J.; Cui, K.; Nie, L.; Wang, Q.; Yang, F.; Shah, F.; Yao, F.; Peng, S. Sheath blight reduces stem breaking resistance and increases lodging susceptibility of rice plants. *Field Crops Res.* **2012**, *128*, 101–108. [CrossRef]
27. Li, M.; Li, R.; Liu, S.; Zhang, J.; Luo, H.; Qiu, S. Rice-duck co-culture benefits grain 2-acetyl-1-pyrroline accumulation and quality and yield enhancement of fragrant rice. *Crop J.* **2019**, *7*, 419–430. [CrossRef]
28. Gitari, H.I.; Gachene, C.K.; Karanja, N.N.; Kamau, S.; Nyawade, S.; Sharma, K.; Schulte-Geldermann, E. Optimizing yield and economic returns of rain-fed potato (*Solanum tuberosum* L.) through water conservation under potato-legume intercropping systems. *Agric. Water Manag.* **2018**, *208*, 59–66. [CrossRef]
29. Simko, T.W.A.V. R Package 'Corrplot': Visualization of a Correlation Matrix; Version 0.92. 2021. Available online: https://cran.r-project.org/web/packages/corrplot/index.html (accessed on 4 December 2021).
30. Kresnatita, S.; Ariffin, D.; Hariyono, D. Sitawati Micro climate behavior on cauliflower plant canopy in intercropping system with sweet corn in central Kalimantan. *Int. J. Sci. Res. Publ.* **2018**, *8*, 76–83. [CrossRef]
31. Jambhulkar, P.P.; Jambhulkar, N.; Meghwal, M.; Ameta, G.S. Altering conidial dispersal of Alternaria solani by modifying microclimate in tomato crop canopy. *Plant Pathol. J.* **2016**, *32*, 508–518. [CrossRef] [PubMed]
32. Partelli, F.L.; Araújo, A.V.; Vieira, H.D.; Dias, J.R.M.; De Menezes, L.F.T.; Ramalho, J. Microclimate and development of 'Conilon' coffee intercropped with rubber trees. *Pesquisa Agropecuária Bras.* **2014**, *49*, 872–881. [CrossRef]
33. Calonnec, A.; Burie, J.-B.; Langlais, M.; Guyader, S.; Saint-Jean, S.; Sache, I.; Tivoli, B. Impacts of plant growth and architecture on pathogen processes and their consequences for epidemic behaviour. *Eur. J. Plant Pathol.* **2012**, *135*, 479–497. [CrossRef]
34. Ndiso, J.B.; Chemining, G.N.; Olubayo, F.M.; Saha, H.M. Effect of cropping system on soil moisture content, canopy temperature, growth and yield performance of maize and cowpea. *Int. J. Agric. Sci.* **2017**, *3*, 1271–1281.
35. Dusserre, J.; Raveloson, H.; Michellon, R.; Gozé, E.; Auzoux, S.; Sester, M. Conservation agriculture cropping systems reduce blast disease in upland rice by affecting plant nitrogen nutrition. *Field Crops Res.* **2017**, *204*, 208–221. [CrossRef]
36. Ballini, E.; Nguyen, T.T.; Morel, J.-B. Diversity and genetics of nitrogen-induced susceptibility to the blast fungus in rice and wheat. *Rice* **2013**, *6*, 32. [CrossRef] [PubMed]
37. Long, D.H.; Lee, F.N.; Tebeest, D.O. Effect of nitrogen fertilization on disease progress of rice blast on susceptible and resistant cultivars. *Plant Dis.* **2000**, *84*, 403–409. [CrossRef]
38. Yusef, G.S.; Khalil, J.; Mohammad, R.A.M. Evaluation of quality and quantity of corn and soybean grain yield in intercropping under deficit irrigation. *J. Biol. Agric. Healthc.* **2014**, *4*, 133–139.
39. Abdel-Galil, A.M.; Abdel-Ghany, R.E.A. Effect of groundnut—Sesame intercropping and nitrogen fertilizer on yield, yield components and infection of root-rot and wilt diseases. *Int. J. Plant Soil Sci.* **2014**, *3*, 623–643. [CrossRef]
40. Crusciol, C.A.C.; de Moraes Costa, A.; Borghi, É.; Castro, G.S.A.; Fernandes, D.M. Fertilizer distribution mechanisms and side dress nitrogen fertilization in upland rice under no-tillage system. *Sci. Agric.* **2010**, *67*, 562–569. [CrossRef]
41. Shah, Z.; Shah, S.; Peoples, M.; Schwenke, G.; Herridge, D. Crop residue and fertiliser N effects on nitrogen fixation and yields of legume–cereal rotations and soil organic fertility. *Field Crops Res.* **2003**, *83*, 1–11. [CrossRef]
42. Hei, Z.; Xiang, H.; Zhang, J.; Liang, K.; Ren, X.; Sun, Y.; Wu, R. Water mimosa (*Neptunia oleracea* Lour.) can fix and transfer nitrogen to rice in their intercropping system. *J. Sci. Food Agric.* **2021**, *1*, 156–166. [CrossRef]
43. Bienvenido, O.J. Rice grain quality: Problems and challenges. *Cereal Food. World* **1990**, *35*, 245–253. [CrossRef]
44. Hao, H.L.; Wei, Y.Z.; Yang, X.E.; Feng, Y.; Wu, C.Y. Effects of different nitrogen fertilizer levels on Fe, Mn, Cu and Zn concentrations in shoot and grain quality in Rice (*Oryza sativa*). *Rice Sci.* **2007**, *14*, 289–294. [CrossRef]
45. Tamer, F.M. Impact of organic materials combined with mineral nitrogen on rice growth, yield, grain quality and soil organic matter. *Int. J. Chemtech. Res.* **2015**, *8*, 1533–1542.

46. Zhou, L.; Liang, S.; Ponce, K.; Marundon, S.; Ye, G.; Zhao, X. Factors affecting head rice yield and chalkiness in indica rice. *Field Crops Res.* **2015**, *172*, 1–10. [CrossRef]
47. Chun, A.; Song, J.; Kim, K.J.; Lee, H.J. Quality of head and chalky rice and deterioration of eating quality by chalky rice. *J. Crop Sci. Biotechnol.* **2009**, *12*, 239–244. [CrossRef]
48. Gao, C.; El-Sawah, A.M.; Ali, D.F.I.; Hamoud, Y.A.; Shaghaleh, H.; Sheteiwy, M.S. The integration of bio and organic fertilizers improve plant growth, grain yield, quality and metabolism of hybrid maize (*Zea mays* L.). *Agronomy* **2020**, *10*, 319. [CrossRef]
49. Hu, Q.; Jiang, W.Q.; Qiu, S.; Xing, Z.P.; Hu, Y.J.; Guo, B.W.; Liu, G.D.; Gao, H.; Zhang, H.C.; Wei, H.Y. Effect of wide-narrow row arrangement in mechanical pot-seedling transplanting and plant density on yield formation and grain quality of japonica rice. *J. Integr. Agric.* **2020**, *19*, 1197–1214. [CrossRef]
50. Ogoshi, C.; Carlos, F.S.; Waldow, D.; Miranda, F.F.; Reginato, J.L.; Ulguim, A. Influence of blast on the nutrition and yield of irrigated rice in Southern Brazil. *J. Soil Sci. Plant Nutr.* **2020**, *20*, 1378–1386. [CrossRef]
51. Alves, N.B.; Balestre, M.; Pennacchi, J.P.; Fernandes, M.C.N.; Castro, D.G.; Botelho, F.B.S. Genetic progress of upland rice (*Oryza sativa* L.) lines for disease resistance. *Plant Breed.* **2020**, *139*, 853–861. [CrossRef]
52. Iqbal, M.A.; Hamid, A.; Ahmad, T.; Siddiqui, M.H.; Hussain, I.; Ali, S.; Ali, A.; Ahmad, Z. Forage sorghum-legumes intercropping: Effect on growth, yields, nutritional quality and economic returns. *Bragantia* **2019**, *78*, 82–95. [CrossRef]

agriculture

MDPI

Article

Effects of Long-Term Straw Management and Potassium Fertilization on Crop Yield, Soil Properties, and Microbial Community in a Rice–Oilseed Rape Rotation

Jifu Li [1,2,*], Guoyu Gan [1,2], Xi Chen [1] and Jialong Zou [3]

[1] Engineering Research Center of Ecology and Agricultural Use of Wetland, Ministry of Education/ College of Agriculture, Yangtze University, No.266, Jingmi Road, Jingzhou District, Jingzhou 434025, China; 202071646@yangtzeu.edu.cn (G.G.); 201804323@yangtzeu.edu.cn (X.C.)

[2] Key Laboratory of Waste and Fertilization Utilization, Ministry of Agriculture and Rural Areas, Wuhan 430070, China

[3] Agricultural Science and Technology Service Center, Bureau of Agriculture and Rural Areas, Jingzhou District, Jingzhou 434025, China; jialongzou@126.com

* Correspondence: jifuli@yangtzeu.edu.cn; Tel./Fax: +86-0716-8066314

Citation: Li, J.; Gan, G.; Chen, X.; Zou, J. Effects of Long-Term Straw Management and Potassium Fertilization on Crop Yield, Soil Properties, and Microbial Community in a Rice–Oilseed Rape Rotation. *Agriculture* **2021**, *11*, 1233. https://doi.org/10.3390/agriculture11121233

Academic Editors: Chengfang Li and Lijin Guo

Received: 25 October 2021
Accepted: 4 December 2021
Published: 7 December 2021

Abstract: The present study aims to assess the influences of long-term crop straw returning and recommended potassium fertilization on the dynamic change in rice and oilseed rape yield, soil properties, bacterial and fungal alpha diversity, and community composition in a rice–oilseed rape system. A long-term (2011–2020) field experiment was carried out in a selected paddy soil farmland in Jianghan Plain, central China. There were four treatments with three replications: NP, NPK, NPS, and NPKS, where nitrogen (N), phosphate (P), potassium (K), and (S) denote N fertilizer, P fertilizer, K fertilizer, and crop straw, respectively. Results showed that long-term K fertilization and crop straw returning could increase the crop yield at varying degrees for ten years. Compared with the NP treatment, the long-term crop straw incorporation with K fertilizer (NPKS treatment) was found to have the best effect, and the yield rates increased by 23.0% and 20.5% for rice and oilseed rape, respectively. The application of NPK fertilizer for ten years decreased the bacterial and fungal alpha diversity and the relative abundance of dominant bacterial and fungal taxa, whereas continuous straw incorporation had a contradictory effect. NPKS treatment significantly increased the relative abundance of some copiotrophic bacteria (Firmicutes, Gemmatimonadetes, and Proteobacteria) and fungi (Ascomycota). Available K, soil organic matter, dissolved organic carbon, and easily oxidized organic carbon were closely related to alterations in the composition of the dominant bacterial community; easily oxidized organic carbon, dissolved organic carbon, and slowly available K were significantly correlated with the fungal community. We conclude that long-term crop straw returning to the field accompanied with K fertilizer should be employed in rice-growing regions to achieve not only higher crop yield but also the increase in soil active organic carbon and available K content and the improvement of the biological quality of farmland.

Keywords: straw management; potassium fertilizer; rice–oilseed rape rotation; yield; bacterial community; fungal community

1. Introduction

Crop residue is a considerable renewable resource with abundant organic carbon (C) and mineral nutrients [1,2]. As the country with the largest agricultural production in the world, China produces more than 800 million tons of crop straw per year, which amounts to 3.64, 0.73, and 14.78 million tons of nitrogen (N), phosphorus (P), and potassium (K), respectively [3,4]. Straw incorporation serves as the most effective way of comprehensive straw utilization compared with other ways (as burning, compost, or cooking) at present. Many research studies have confirmed that crop residue recycling could increase crop yield and maintain soil fertility [5–7]. With the increased awareness of environmental protection and ban on burning straw, directly returning straw to the field is being accepted by more and more farmers in China.

Potassium is one of the most essential mineral nutrients for plant growth and metabolism [8]. Adequate soil K supply is beneficial for agricultural production [4]. Because of the promotion of high-yield varieties and high inputs of N and P fertilizer, K deficiency or soil K imbalance has become more widespread and critical in China, especially in the southern multi-cropping region. Paddy-upland rotation, as the main crop rotation system in the southern part of China, is mainly distributed in the rice cropping areas of the Yangtze River Basin and Huang-Huai River Basin [9,10]. However, long-term intensive cultivation removes 210–360 kg ha^{-1} of K_2O per year by crop harvest and has resulted in a substantial decrease in soil available K content [11]. In addition, potash reserves are penurious and expensive in East and South Asia [12]. As a result, farmers have employed less K fertilizer in production. Therefore, the current input of K fertilizer falls short of maintaining the soil K balance, and straw returning is indispensable to improve the K status of cropland [4,11].

Previous studies have showed that crop straw returning could improve soil available K and slowly available K content. Moreover, crop straw incorporation with K fertilizer significantly improved crop yield and maintained soil health [3,10,13]. However, the decomposition rate of straw returning to the field was significantly affected by soil water content. Compared with the upland cropping rotation, the paddy-upland rotation was found to lead to seasonal dry–wet alternation in the farmland system [14,15]. The strong conversion of hydrothermal conditions is bound to affect the decomposition rate of returned straw and the release of straw nutrients, affecting the growth of crops and their absorption and utilization of soil mineral nutrients [16]. Therefore, it is unclear how upland and paddy crops respond to the long-term combinations of K fertilizer with crop residue incorporation.

In addition, crop straw contains abundant organic ingredients, and has been widely applied in fields to promote soil C sequestration [17,18]. Soil microorganisms, such as bacteria and fungi, are the basis of soil fertility and have a great influence on plant health and growth [19,20]. Previous studies have documented the positive and significant impact of straw utilization on soil bacterial and fungal community structure under short-term or long-term straw returning [18,21,22]. However, other research has indicated that straw incorporation decreased the richness and diversity of bacterial and fungal composition. Ling et al. [23] found that the increase in the amount of soil microorganisms after long-term wheat straw returning was mainly due to the increase in the multiplication of bacteria. Additionally, some studies have shown that fertilization could alter the nutrient content of soil (i.e., total N, available P, and available K) and directly drive the evolution of soil microbial communities [24–26]. Wasaki et al. [27] found that the decrease in soil pH, caused by the application of N, was the primary cause of bacterial community changes, and soil C:P and N:P changes determine the composition of the soil microbial communities. Long-term P fertilization increased soil microbial P immobilization by decreasing the relative abundance of the P-starvation response gene and increasing that of the low-affinity inorganic-P transporter gene [28]. In black soil, the alpha diversity and the relative abundance of Acidobacteria significantly decreased with the increased rate of K fertilizer in short-term treatment [29]. Compared with the application of N and P fertilization, K fertilizer has not gained enough attention in soil microbial diversity and composition; especially, the long-term effect of straw incorporation and K fertilizers on bacterial and fungal communities in paddy soils has not been addressed. A recent study showed that the keystone taxa had higher gene copies of oxidoreductase and 71 essential functional genes associated with C, N, P, and sulfur cycling in controlling soil function and wheat production. Meanwhile, the microbial community was highly responsive to K fertilization, which was associated with lower crop production and higher abundance of potential fungal pathogens [30]. In short-term experiments, the yield-increasing effect of K fertilizer was higher than that of straw management [13,16,21]. However, in long-term experiments, the impact of bacterial and fungal community and structure on crop yield is unclear [10,16]. We hypothesized that long-term K fertilizer application would reduce soil microbial diversity and species composition, and straw returning could mitigate the toxic effects of K fertilizer on microorganisms. Therefore, this study aimed to characterize the dynamics of paddy-upland rotation yield, soil properties, microbial diversity, and community composition in the same site treated with different field management and their associated soil properties, which will provide scientific data of the long-term potential effects to compare against those of short-term field experiments.

2. Materials and Methods

2.1. Experimental Site Description

The field experimental site established in 2011 was located in the town of Chuandian, Jingzhou (part of the Jianghan Plain), Hubei Province, central China (30°33′25″ N, 112°4′53″ E, altitude 80 m). A rice–oilseed rape rotation system was implemented in 1999. The average annual rainfall was 1140 mm, and the air temperature was 15 °C. Soil type was classified as silty clay loam using the World Soil Classification of the Food and Agriculture Organization (sand, 3.5%; silt, 61.0%; clay, 35.5%). At the beginning of the experiment (June 2011), the selected soil basial properties at a depth of 0–20 cm were as follows: pH, 5.97; organic matter, 26.9 g kg^{-1}; total N, 0.61 g kg^{-1}; Olsen-P, 8.1 mg kg^{-1}; available K, 164.8 mg kg^{-1}; slowly available K, 405.4 mg kg^{-1}.

2.2. Experimental Design

A complete randomized block design was conducted with four treatments and three replications. The treatments were (1) NP, chemical fertilizer N, and P application; (2) NPK, balanced chemical fertilizer N, P, and K application; (3) NPS, application of chemical fertilizer N, P plus straw returning, where S represents crop straw; and (4) NPKS, application of chemical fertilizer N, P, K plus straw returning. The dimensions of the plot were 20 m^2, with a length of 5.0 m and a width of 4.0 m. The cropping sequence was rice followed by winter oilseed rape. Rice was transplanted at the age of five leaves at a density of 22 hills m^{-2} (row spacing: 25 cm × 18 cm) and two plants per hill in mid-June and harvested in mid-September. Winter oilseed rape was directly seeded onto the soil surface at a rate of roughly 7.5 kg ha^{-1} in early October, and the crop was harvested in early May. The water regimes were early flooding-mid season drainage intermittent irrigation for the rice season and a rain-fed agricultural regime for the oilseed rape season.

The amounts of N, P, and K fertilizer application for rice and oilseed rape under different treatments are described in Table 1. During the rice season, N (urea, 46% N) was applied in three splits: 60% as basal fertilizer before rice transplanting, 20% at tillering stage, and 20% at the booting stage. K (as potassium chloride, 60% K_2O) was applied as 60% basal fertilizer and 40% booting fertilizer. P (as superphosphate, 12% P_2O_5) was applied manually as basal fertilizers. During the winter oilseed rape season, N was applied in three splits: 60% as basal fertilizer, 20% during the overwintering stage, and 20% at the beginning of stem elongation. P, K, and B (as sodium borate, 11% B) fertilizers were applied manually as basal fertilizers.

In order to ensure the consistency of the experiment, the straw amount of the first crop season (rice) returned to field in 2011 was 2250 kg ha^{-1} of winter rape stalks and shells. The K contents of the stalk and shell were 1.82% and 2.56%, respectively. All straw should be protected from rainfall before returning to prevent K^+ leaching. In the rice season, the oilseed rape straw was crushed by a machine (to a length of 10 cm) and incorporated into the plough layer together with basal fertilizer. In the oilseed rape season, the rice straw was placed as mulch onto the soil with no tillage.

Table 1. The application rates of N, P, K, and B for treatments in both seasons per year from 2011 to 2020 (kg ha^{-1}).

Treatment	Rice Season		Oilseed Rape Season		
	Chemical Fertilizer ($N-P_2O_5-K_2O$)	Crop Straw ($N-P_2O_5-K_2O$)	Chemical Fertilizer ($N-P_2O_5-K_2O$)	Crop Straw ($N-P_2O_5-K_2O$)	Boron ($Na_2B_4O_7·5H_2O$)
NP	180-60-0	0-0-0	180-60-0	0-0-0	15.0
NPK	180-60-90	0-0-0	180-60-90	0-0-0	15.0
NPS	180-60-0	19.4-3.1-142.6	180-60-0	53.4-7.8-164.3	15.0
NPKS	180-60-90	20.2-3.5-151.5	180-60-90	61.6-8.5-179.1	15.0

Note: the nutrient (N, P, and K) apparent input of crop straw was the average value from 2011 to 2020. The conversion coefficients of N, P, and K to N, P_2O_5, and K_2O were 1, 2.3, and 1.2, respectively.

2.3. *Sample Sampling and Determination*

2.3.1. Grain Yield

For each crop season, the mature oilseed rape and rice plants were harvested and thrashed in each plot, and the grains were dried to determine the grain yield. The crop straw was moved out or fully returned to the field according to each treatment. Before the harvest, five plants of rice and oilseed rape were randomly collected for element analysis of N, P, and K. The sampled plants were partitioned into straw and grain. The dry matter was digested in 70% concentrated H_2SO_4 and 30% H_2O_2 to determine the N, P, and K content in grain and plants.

2.3.2. Soil Samples

On 10 September 2020 (after rice harvest), soil samples were randomly collected from four points in each plot at a depth of 0–20 cm using an auger with a diameter of 5.0 cm. Soil from the four core samples of a plot was mixed to obtain one composite sample. After removing stones, roots, and plant residue using a 2 mm mesh, each sample was divided in half: one half was air-dried for physicochemical analyses, and the other half was immediately stored at −80 °C for soil DNA extraction [22].

2.3.3. Determination of Soil Physicochemical Indexes

The air-dried soil samples were used to determine physicochemical properties. Soil pH was measured in water (1:2.5 w/v) by a pH meter (PHS-3C, INESA Scientific Instrument Co. Ltd., Shanghai, China). Olsen-P was extracted with 50 mL of 0.5 mol L^{-1} $NaHCO_3$ (pH 8.5) and determined using an injection pump analyzer (AA3, Bran+ Luebbe, Norderstedt, Germany). Available K and slowly available K were extracted with 1 mol L^{-1} NH_4OAc and 1 mol L^{-1} HNO_3 solution, respectively, and measured by a photoelectric flame photometer [24]. The SOM was determined using a wet oxidation procedure with potassium dichromate ($K_2Cr_2O_7$)-sulfuric acid (H_2SO_4). EOC content was measured in 15 mg of each soil sample to which 25 mL of 333 mM $KMnO_4$ was added. Afterwards, the samples were shaken at 200 rpm for 1 h and then centrifuged at 4000 rpm for 5 min. Then, the supernatant was removed and diluted 1:250 with distilled water. The absorbance of the diluted solution was measured at 565 nm. DOC was measured by adding 60 mL of distilled water to 20 g of fresh soil (3:1, v/w) in a 150 mL polypropylene bottle. The samples were shaken on a shaker for 30 min at 250 rpm and then centrifuged for 10 min at 10,000 rpm. The upper suspension was filtered through a 0.45 μm filter into a bottle, and the C content in the filtered solution was determined using a C/N element analyzer (Velp, Usmate Velate, Italy) [10]. Furthermore, soil available N content was determined by the alkaline hydrolysis diffusion method.

2.3.4. Soil DNA Extraction and High-Throughput Sequencing Analysis

DNA was extracted from the soil samples (0.5 g) using a Fast DNA Spin Kit for Soil (MP Biomedicals, Santa Ana, CA, USA) in accordance with the protocol of the manufacturer. The quantity and quality of the DNA extracts were determined using a NanoDrop

2000 spectrophotometer (Thermo Fisher Scientific, Waltham, MA, USA). The extracted DNA was stored at −20 °C for further analysis.

An aliquot of the extracted DNA from each sample was used as the template for amplification. The V3–V4 hypervariable regions of the bacterial 16S rRNA gene sequences and the ITS region of the fungal rRNA gene sequences were amplified [31]. Amplicon libraries were prepared using tagged bacterial and fungal universal primers, i.e., 338F and 806R for bacteria and ITS1F and ITS2R for fungi. The DNA samples were amplified individually using the fusion primer pairs 338F (5′-ACTCCTACGGGAGGCAGCAG-3′) and 806R (5′-GGACTACHVGGGTWTCTAAT-3′) for bacteria and ITS1F (5′-CTTGGTCATTTAGGAAGTAA-3′) and ITS2R (5 ′-GCTGCGTTCTTCATCGATGC-3′) for fungi to generate polymerase chain reaction (PCR) fragments [32]. The following thermal program was used for amplification: initial denaturation 98 °C for 2 min, followed by 27 cycles of denaturation at 98 °C for 15 s, annealing at 55 °C for 30 s, extension at 72 °C for 30 s, and a final extension at 72 °C for 5 min. The PCR reactions were performed in a 25 μL mixture containing 5 μL of 5× reaction buffer, 5 μL of 5×GC buffer, 2 μL of dNTP (2.5 mM), 1 μL of forward primer (10 uM), 1 μL of reverse primer (10 uM), 2 μL of DNA template, 8.75 μL of ddH2O, and 0.25 μL of Q5 DNA polymerase [24]. The PCR products were purified using the AxyPrep DNA Gel Extraction Kit (Axygen Biosciences, San Francisco, CA, USA) and quantified using a Quantus Fluorometer (Promega, Madison, WI, USA). The target sequences were performed on an Illumina MiSeq 250 sequencing platform by Shanghai Personal Biotechnology Co., Ltd. (Shanghai, China).

2.3.5. Sequence Processing

In sequencing the original data to remove the primer adapter sequence, the processed low-quality bases (maximum expected error higher than 1 for bacteria and 0.5 for fungi, shorter than 370 bp for bacteria and 200 bp for fungi) were removed from downstream analysis [31]. Then, the remaining data were spliced to obtain valid sequence data for each sample. Finally, using 97% as the threshold, the 16S and ITS sequences were divided into operational taxonomic units (OTUs). Using QIIME 2 software, the UCLUST sequence comparison tool was used to cluster with 97% sequence similarity. Each sequence with the highest OTU degree was selected as the representative sequence of the OTU [33]. For bacterial 16S rRNA and fungal ITS genes, both the Greengenes database and the Silva database were used as template sequences for OTU classification status identification [34]. After quality filtering and removal of chimeric sequences, 257,956 and 333,433 high-quality sequences were clustered into 14,906 and 1627 OTUs, respectively, for each bacterial and fungal sample.

2.4. Statistical Analysis

The analysis of variance procedure in SPSS 18.0 (SPSS Inc., Chicago, IL, USA) was used to perform data analysis on soil biogeochemical properties and alpha diversity. Before statistical analysis, we tested the normality of the data using the Shapiro–Wilk test. The Shannon and Simpson indexes, abundance-based coverage estimator (ACE), and Chao1 were calculated to estimate alpha diversity of each treatment using MOTHUR [35]. The yield, soil properties, and alpha diversity were tested by one-way analysis of variance (ANOVA), with Duncan's test, at a p value < 0.05. A two-way analysis of variance (ANOVA) was also used to examine the contribution of treatment (T) and year (Y) to crop yield. To determine the structural differences between bacterial and fungal communities at different treatments, an analysis of similarities was also conducted using QIIME2 based on Bray–Curtis distance measurements and abundance data. To determine which taxa were significantly affected, the linear discriminant analysis effect size (LEfSe) algorithm was implemented [17]. The "vegan" package in R language was used to perform similarity analysis. The clustering analysis was constructed using the "heatmap" package based on the Spearman correlation matrix. Each column in the heatmap represents one treatment, and each row represents a genus. The color from red to blue indicates that the abundance is from high to low. Redundancy analysis was used to access the effects of soil environmental factors on bacterial and fungal communities. To reveal how the potential pathways (soil properties, alpha diversity, and microbial community) influence rice yield and oilseed rape yield, partial least squares path models (PLS_PM) were evaluated using the Goodness of Fit (GOF) statistic [36], and assembled by the "inner plot" function using the "plspm" package of R 4.1.0.

3. Results

3.1. Grain Yield

Over the ten-year study period, grain yield was affected by straw returning, K fertilizer, and planting duration for rice and oilseed rape (Figure 1). The grain yields without K fertilizer (NP treatment) were 9.4 t ha^{-1} and 1.75 t ha^{-1} annual average for rice and oilseed rape, respectively. In the first crop rotation, when the crop straw returned to the field or K fertilizer was applied, the rice yield did not show a significant increase compared with that of NP treatment, but after two rotation cycles, a significant increase could be seen in Figure 1. For the subsequent crop oilseed rape, a yield increase effect appeared in the first rotation. Compared with the NP treatment, the average annual increments of rice and oilseed rape by NPK treatment were 1.5 t ha^{-1} and 0.13 t ha^{-1}, and the average increase rates were 15.8% and 7.4%, respectively. In straw returning (NPS treatment) compared with NPK treatment, the yield increase rate of oilseed rape was higher, while the yield increase rate of rice was the opposite. The yields of rice and oilseed rape for NPKS treatment were the highest, reaching an annual average of 11.6 t ha^{-1} and 2.11 t ha^{-1}, respectively, and the corresponding yield increase rates were 23.0% and 20.5%, respectively.

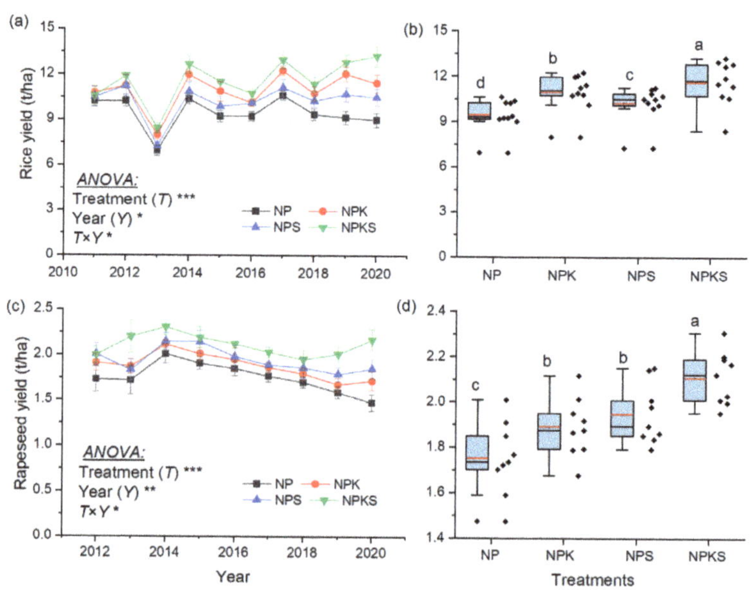

Figure 1. Variation and distribution of grain yields for rice (**a**,**b**) and oilseed rape (**c**,**d**) under different fertilization treatments NP, NPK, NPS, and NPKS over 10 years. *, **, and *** indicate significant differences at the $p < 0.05$, $p < 0.01$, and $p < 0.001$ level, respectively in (**a**,**c**). The upper, middle, and lower limits of each box represent the 75th, 50th, and 25th percentage for crop yield, respectively. Red lines indicate the mean value, and different lower-case letters indicate significant differences for the mean crop yield between treatments at $p < 0.05$ in (**b**,**d**).

3.2. Soil Properties

Table 2 shows the effects of straw incorporation and K fertilizer on the soil properties. From the results, the SOM, available N, available K, slowly available K, EOC, and DOC differed significantly among the treatments ($p < 0.05$). SOM in NPKS treatments was significantly greater ($p = 0.036$) than in NP treatment and ranked as NPKS, NPS > NPK, NP. There was no significant difference in soil Olsen-P and pH value among the treatments ($p > 0.05$). Soil available K ranged from 169.5 to 254.7 g kg^{-1}, with the highest concentration of that obtained in the NPKS treatment. Slowly available K was lowest with the NPKS treatment, and EOC, and DOC with the NPKS treatment was significantly greater ($p = 0.024$) than those with the NPK and NPS treatments.

Table 2. Effects of straw incorporation and K fertilizer on soil properties in the bulk soils.

Soil Properties	Treatments			
	NP	NPK	NPS	NPKS
pH	5.82 ± 0.03 a	5.76 ± 0.04 a	5.75 ± 0.03 a	5.73 ± 0.02 a
SOM (g kg^{-1})	29.8 ± 2.9 b	31.5 ± 3.4 b	32.8 ± 2.6 ab	35.7 ± 1.4 a
Available N (mg kg^{-1})	77.3 ± 8.9 b	71.8 ± 8.3 b	89.0 ± 6.3 a	85.7 ± 7.4 a
Olsen-P (mg kg^{-1})	9.6 ± 0.5 b	10.1 ± 0.6 b	11.2 ± 1.1 a	10.4 ± 0.8 b
Available K (mg kg^{-1})	169.5 ± 8.8 d	183.0 ± 7.6 c	208.0 ± 11.4 b	254.7 ± 16.2 a
Slowly available K (mg kg^{-1})	601.5 ± 10.3 b	572.9 ± 14.5 c	632.0 ± 16.7 a	529.0 ± 14.8 d
EOC (g kg^{-1})	5.9 ± 0.5 c	5.1 ± 0.2 c	7.6 ± 0.4 b	9.5 ± 0.3 a
DOC (mg kg^{-1})	20.8 ± 0.4 c	22.0 ± 1.0 c	27.0 ± 0.9 b	30.6 ± 1.2 a

Note: SOM, soil organic matter; EOC, easily oxidized organic carbon; DOC, dissolved organic carbon. Within a row, data (mean ± SD, $n = 3$) followed by different letters are significantly different ($p < 0.05$).

3.3. Alpha Diversity of Bacterial and Fungal Communities

A total of 364,587 and 336,310 filtered sequences remained after quality control, and 257,956 and 333,433 reads (high-quality sequence) were generated for further bioinformatic analysis (Table S1). All these sequences were subsequently clustered into 14,906 and 1627 OTUs based on 97% similarity. The number of observed OTUs detected in each sample ranged from 2890 to 4426 and 298 to 535 for bacterial and fungal groups, respectively. Good's coverage index of each sample was >0.990. The rarefaction curves (Figure S1) were close to the saturation phase, indicating that sufficient sequencing coverage was achieved and that the OTUs were representative of the overall microbial community libraries.

There were significant differences among the treatments in the alpha diversity, except the Simpson index, of bacterial and fungal populations, as shown by richness and diversity indexes (Table 3). Among the treatments, NP treatment had the highest value of Chao1 and ACE, suggesting that long-term non-K fertilizer application resulted in greater richness of bacterial populations than the other treatments. The Shannon index was significantly higher for the NP and NPS treatments than for the NPK and NPKS treatments, but no significant difference was observed between the treatments on the Simpson index. Meanwhile, this tendency was shown in the fungal group that NPK and NPKS treatments had lower richness and diversity indexes than those of NP and NPS treatments.

Table 3. Alpha diversity of bacterial and fungal gene sequences in the soil samples.

Microbe Type	Treatment	Richness Index		Diversity Index		Coverage
		Chao1	ACE	Simpson	Shannon	
Bacteria	NP	4463 ± 142 a	4422 ± 135 a	0.999 ± 0.031 a	10.98 ± 0.06 a	0.994 ± 0.035 a
	NPK	3664 ± 165 b	3470 ± 124 b	0.998 ± 0.045 a	10.27 ± 0.05 b	0.990 ± 0.051 a
	NPS	4298 ± 151 a	4128 ± 128 a	0.998 ± 0.037 a	10.65 ± 0.08 ab	0.998 ± 0.048 a
	NPKS	3045 ± 123 c	2897 ± 135 c	0.993 ± 0.033 a	9.33 ± 0.06 c	0.991 ± 0.044 a
Fungi	NP	535 ± 21 a	518 ± 19 a	0.952 ± 0.031 a	6.45 ± 0.04 a	1.000 ± 0.045 a
	NPK	298 ± 11 c	264 ± 16 d	0.958 ± 0.043 a	5.82 ± 0.03 b	1.000 ± 0.053 a
	NPS	477 ± 18 b	453 ± 21 b	0.953 ± 0.036 a	5.98 ± 0.04 b	1.000 ± 0.039 a
	NPKS	317 ± 16 c	311 ± 14 c	0.966 ± 0.045 a	6.23 ± 0.04 ab	1.000 ± 0.044 a

Note: Different letters for the same item indicate $p < 0.05$ (significant differences).

3.4. Composition of Bacterial and Fungal Communities

Long-term straw returning and K fertilizer altered the relative abundance of bacterial and fungal phyla in soil (Figure 2). Proteobacteria and Acidobacteria had the highest relative abundance in each treatment, belonging to the predominant bacterial community (relative abundance > 15.0%), with averages of 32.0% and 16.9%, respectively. The relative abundances of Actinobacteria, Chloroflexi, Nitrospirae, Rokubacteria, Bacteroidetes, and Verrucomicrobia were higher, with averages of 4.3%, 2.5%, 2.1%, 2.0%, 1.4%, and 1.1%, respectively (Figure 2a). In the fungi phylum (Figure 2b), Ascomycota was the dominant species, with an average relative abundance of 48.1%, followed by Basidiomycota and Mortierellomycota with average relative abundances of 24.8% and 4.4%, respectively.

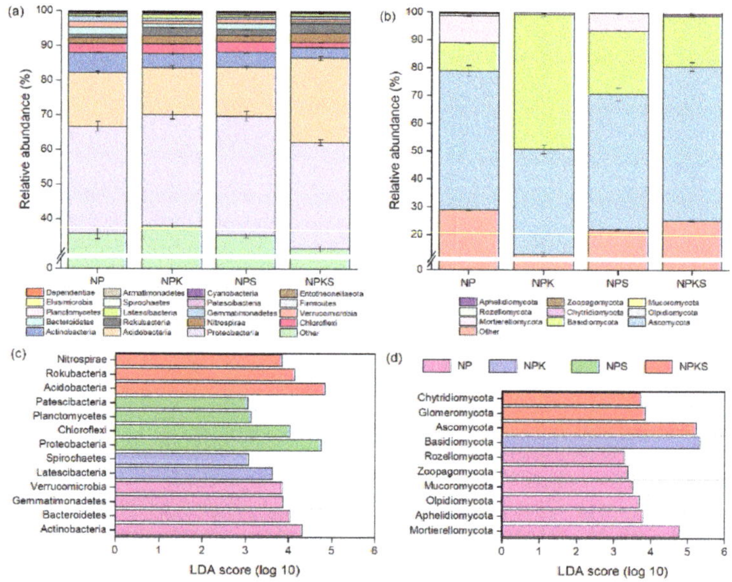

Figure 2. The relative abundance of major taxonomic groups for (**a**) bacteria and (**b**) fungi at the phylum level and the error bars are standard deviation and significant changes in bacterial (**c**) and fungal (**d**) key phylotypes identified using linear discrimination analysis effect size.

At the level of bacterial phylum (Figure 2c), Actinobacteria, Bacteroidetes, Gemmatimonadetes, and Verrucomicrobia were significantly altered taxa in the NP treatment. Latescibacteria and Spirochaetes were found to be sensitive to NPK treatment. Intriguingly, the predominant Proteobacteria, Chloroflexi, Planctomycetes, and Patescibacteria were significantly altered in the NPS treatment, and only Acidobacteria, Rokubacteria, and Nitrospirae were positively altered in the NPKS treatment. At the level of fungal phylum (Figure 2d), Mortierellomycota, Aphelidiomycota, Olpidiomycota, Mucoromycota, Zoopagomycota, and Rozellomycota were the significantly altered taxa in the NP treatment. Basidiomycota was sensitive to NPK treatment. Conversely, there were no taxa changed in the NPS treatment. Ascomycota, Glomeromycota, and Chytridiomycota were clearly altered in the NPKS treatment. These results indicated that long-term application of N and P fertilizer without K fertilizer stimulated an increase in the species and relative abundance of oligotrophic bacteria and fungi. At the same time, the application of straw with K fertilizer contributed to the increase in existing eutrophic microorganisms.

The species composition of bacteria and fungi at the genus level is shown in Figure 3. The results indicate that the relative abundances of dominant species in the fungal community for four treatments were more significant than those of the bacterial community.

At the bacterial genus level (Figure 3a), Anaeromyxobacter was the dominant genus, with an average relative abundance of 1.56%, followed by Haliangium, Nitrospira, Geobacter, Candidatus Solibacter, Candidatus, Udaeobacter, and Sh765B-TzT-35, with average relative abundances of 0.90%, 0.83%, 0.73%, 0.67%, 0.57%, and 0.54%, respectively. Among them, Haliangium, Candidatus Solibacter, and Candidatus Udaeobacter had the highest relative abundance in NP and NPS treatments, and Nitrospira and Sh765B-TzT-35 had the highest relative abundance in NPK and NPKS treatments.

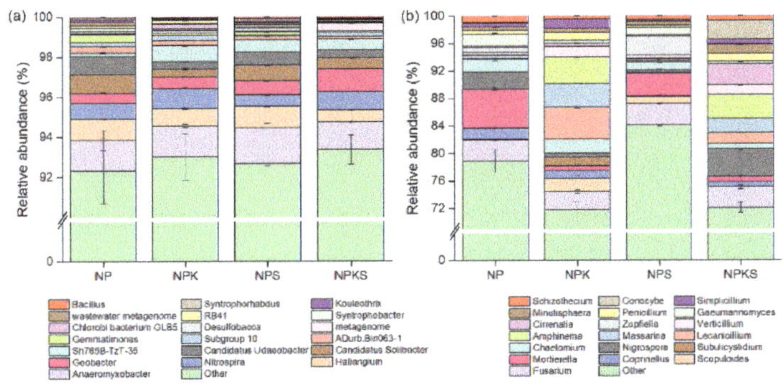

Figure 3. The relative abundance of dominant taxonomic groups for (**a**) bacteria and (**b**) fungi at the genus level. The data represent the mean values of the three replications. Values are means (n = 3); error bars are standard deviation.

At the genus level of fungi (Figure 3b), the dominant species were Fusarium and Mortierella, having average relative abundances of 2.95% and 2.61%, respectively, followed by Nigrospora, Chaetomium, Lecanicillium, Massarina, and Amphinema, with averages of 1.85%, 1.45%, 1.66%, 1.55%, and 1.89%, respectively. The relative abundance of Mortierella was increased significantly in the NP and NPS treatments than in the NPK and NPKS treatments. Among all treatments, Nigrospora (3.94%) in the NPKS treatment had the highest relative abundance, while Chaetomium, Lecanicillium, Massarina, and Amphinema had higher relative abundance in NPK treatment.

3.5. Beta Diversity of Bacterial and Fungal Communities

The Venn diagram (Figure 4a,b) shows that the microbial population had both shared components and unique parts. NPK, NPS, and NPKS treatments shared 8.75%, 11.24%, and 6.31% of the bacterial OTUs with NP treatment, while the unique OTUs of NPK, NPS, NPKS, and NP were 16.36%, 20.88%, 14.09%, and 26.23%, respectively. NPK, NPS, and NPKS shared 5.77%, 9.74%, and 5.68% of fungal OTUs with CK, while the unique OTUs of NPK, NPS, NPKS, and NP treatments were 14.58%, 23.77%, 15.2%, and 27.59%, respectively. These indicate that long-term straw returning and K fertilizer application caused differences in soil microbial communities, thereby affecting the diversity of bacteria and fungi groups among treatments.

A PCoA plot showed that bacterial communities in soils treated with K fertilizer were distinct from those in soils treated with non-K fertilizers along the x-axis (Figure 4c), and the first principal component (x-axis) accounted for 52.6% of the total variation. Still, the straw returning also regulated the communities along the y-axis, but the second principal component (y-axis) only contributed 25.0% of the variation in communities. Similarly, the first two principal coordinates represented 74.7% of the variation in fungi (Figure 4d) communities according to the PCoA, in which the first principal component (x-axis) accounted for 38.1% of the total variation.

From the heatmap (Figure 5), the distribution of dominant bacteria in each treatment was well-marked; specifically, in the NP treatment, up to 18 species could be identified in the number of abundant bacteria. On the contrary, the abundant bacteria in the NPK, NPS, and NPKS treatment were 12, 14, and 6 types of species, respectively. Furthermore, the cluster analysis results reflected that NP and NPS treatments were similar, and NPK and NPKS were similar. The results of the fungal genus also showed that the distribution and relative abundance of the dominant fungal groups in each treatment were significantly different. Additionally, the cluster analysis indicated that the NP and NPS treatments were homogeneous.

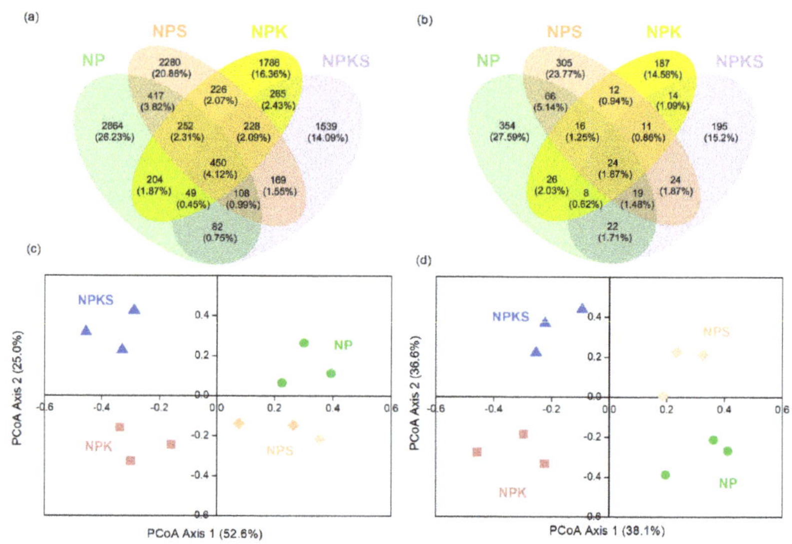

Figure 4. Venn diagram of (**a**) bacterial and (**b**) fungal operational taxonomic units from soil samples and the principal coordinates analysis (PCoA) plot of the dissimilarity between the treatments for (**c**) bacteria and (**d**) fungi based on Bray–Curtis differences.

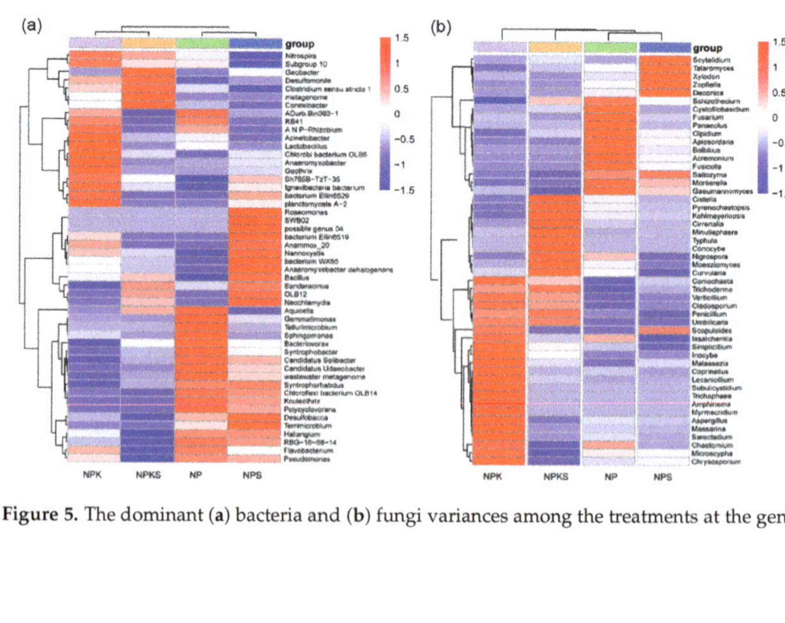

Figure 5. The dominant (**a**) bacteria and (**b**) fungi variances among the treatments at the genus level.

3.6. Correlation of Dominant Microbial Communities with Soil Properties

The redundancy analysis (Pseudo-F = 388, p = 0.002 **) showed that axis 1 and axis 2 explained 80.6% and 14.8% of the total variance in soil bacterial community composition, respectively. The phyla Rokubacteria, Nitrospirae, and Acidobacteria were clustered together to the edge of soil DOC, SOM, available K, and EOC. In contrast, the phyla Chloroflexi, Bacteroidetes, Actinobacteria, and Verrucomicrobia were highly correlated with slowly available K. The available K, SOM, DOC, and EOC had a noteworthy impact on the bacterial community, which explained the variation by 51.3%, 19.8%, 18.8%, and 9.1%, respectively (Figure 6a). The redundancy analysis (Pseudo-F = 518, p = 0.002 **) showed that axis 1 and axis 2 explained 84.6% and 13.9% of the total variance in soil fungal community composition, respectively (Figure 6b). The Ascomycota had a positive correlation with soil available K, DOC, and EOC; the Basidiomycota was highly corrected with slowly available K; and the phylum Mortierellomycota was negatively correlated with slowly available K, available N, pH, and EOC. The EOC, DOC, and slowly available K explained the variation by 27.8%, 57.9%, and 12.5%, respectively.

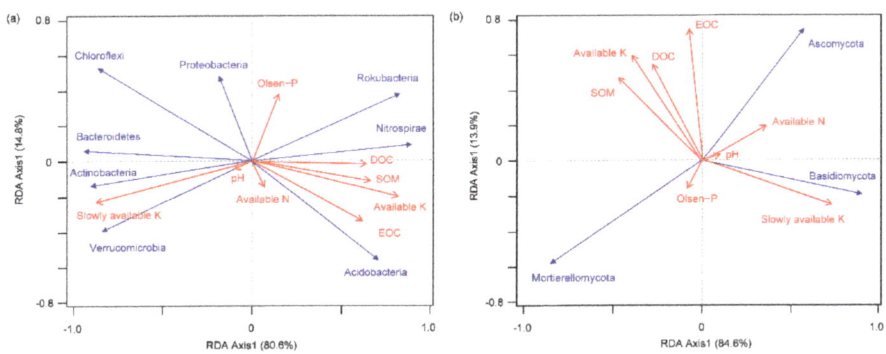

Figure 6. Redundancy analysis of soil properties and main (**a**) bacterial and (**b**) fungal communities at the phylum level in soils. Red lines represent soil properties; blue lines represent the bacterial and fungal phylum−level taxonomy.

3.7. Potential Pathways Influencing Crop Yield

The PLS_PM analysis showed that the final model had GOF of 0.81 and 0.54 for bacterial group and fungal group in 2020, respectively. The pathways of soil properties, alpha diversity, and bacterial community composition together explained 82.3% of the total variance in crop yield, while those represented 45.2% of the variation in fungal group (Figure 7b). The direct effect of soil properties (path coefficient = 0.85) on the crop yield was greater than the direct effect of bacterial community composition (path coefficient = −0.50) and alpha diversity (path coefficient = 0.47) as well as fungal community composition (path coefficient = −0.37) and alpha diversity (path coefficient = 0.31). Moreover, the PLS_PM analysis suggested that soil properties indirectly affected the crop yield by changing bacterial alpha diversity (path coefficient = 0.92) and community composition (path coefficient = −0.88) as well as fungal alpha diversity (path coefficient = 0.66) and community composition (path coefficient = −0.70) in 2020.

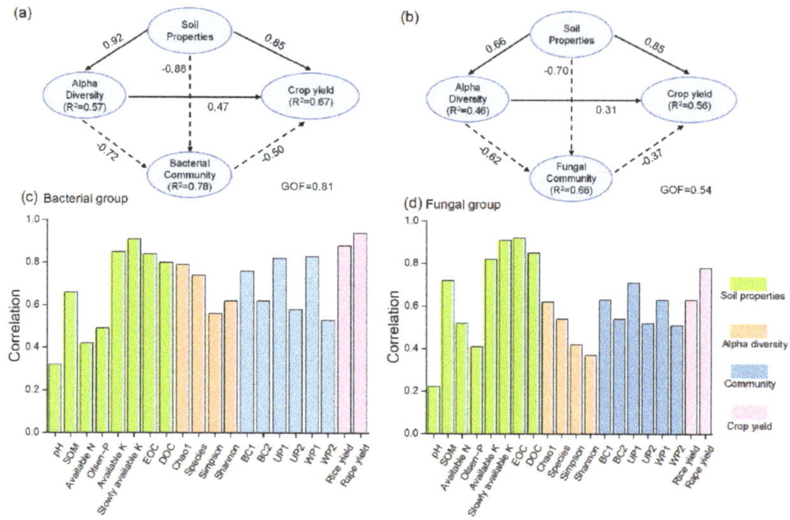

Figure 7. Partial least squares path models (PLS_PM) for the rice and oilseed rape yield in 2020. A line with an arrow indicates a causal relationship, supplemented by a path coefficient, and continuous and dashed lines indicate positive and negative relationships for (**a**,**b**), respectively; the amount of the variability explained by the variables for (**c**,**d**). Path coefficients are calculated after 1000 bootstraps. BC1: Bray−Curtis PCoA1; BC2: Bray−Curtis PCoA2; UP1: Unweighted Unifrac PCoA1; UP2: Unweighted Unifrac PCoA2; WP1: Weighted Unifrac PCoA1; WP2: Weighted Unifrac PCoA2.

4. Discussion

4.1. Effects of Long-Term of Straw Returning and K Fertilizer on Crop Yield

Many studies have reported that straw returning increased crop yields and nutrient uptake [22,37,38]. Our study showed similar results during ten-year field experiments where the same fertilizer inputs were applied among the four treatments, especially for oilseed rape (Figure 1). Through the investigation of yield structure components, the main reason for the increase in crop yield is that straw return significantly improved the number of productive ear and spike granules of rice and wheat, and the number of siliques per plant and the number of seed per pod of oilseed rape [39,40]. In paddy-upland rotation, the yield-increasing effect of the upland season (wheat, oilseed rape) was greater than the rice season. The phenomenon could also be seen in Figure 7, which shows that the soil properties, microbial alpha diversity, and community composition had higher relationships with oilseed rape yield than rice yield and the direct effect of soil properties as SOM, EOC, SOC, and available K content on yield increase were greater than those of the bacterial and fungal groups. Overall, compared with no straw returning, the increase rate of rice yield with straw returning was 5.2%, whereas the yield increase rates of oilseed rape and wheat with straw returning were 10.5% and 12.4%, respectively, in southern China, higher than that of rice [41]. The result of Figure 1 confirmed that the increase rate of oilseed rape in NPS treatment was higher than that of rice. Moreover, as the experiment progressed, the increase rate of yield in field under straw management was more noticeable compared with those with no straw returning. Wang et al. [42] showed that the yield-increasing effect of straw returning was influenced by the annual average temperature, soil nutrient status, returning period, and fertilization. In the study, the increased rate of yield in NPK and NPS treatments did not reach a significant level in the first and second year of rice season, but the third year of rice shows a significant difference compared to that of NP treatment. However, the oilseed rape season showed a significant increase in production during the first crop rotation. This is because the contribution rates of K fertilizer to the rice and oilseed rape were 8.2% and 11.5%, and the dependent rates of soil K status for rice and oilseed rape were 83.0% and 75.2%, respectively [43]. Therefore, the K element absorbed by the whole rice plant mainly comes from the soil supply, while oilseed rape and wheat are more dependent on exogenous K fertilizer supply.

From the dynamic changes of crop yield over ten years, the increase rate of NPK treatment was slightly better than that of NPS treatment in the rice season compared to NP treatment, and the result was the opposite in the oilseed rape season. Our previous multi-field results also indicated that the yield-increasing rate of K fertilizer application was better than that of straw returning in the rice season. Straw returning involved in not only the process of the release of N, P, and K mineral elements but also the process of straw self-decomposition [44]. High temperature, waterlogging, and straw rot accelerated in rice seasons, which caused a certain toxic effect on root by a higher concentration of phenols and organic acids from straw decomposition, thereby weakening the yield increasing rate of straw returning [45,46]. The decomposition rate of straw in the oilseed rape season was relatively slower, the nutrient release cycle was longer than those of the rice season, and the poison caused by straw rot was relatively weak. In addition to the biochemical effects, the straw mulching in the oilseed rape season has physical effects, such as enhancing crops to resist the resistance of adversity (low temperature and drought) and relieving temperature changes in winter, which was beneficial to crop straw returning [47–49].

4.2. Effects of Long-Term Straw Returning and K Fertilizer on the Soil Properties

In the present study, NP treatment led to the decline tendency on crop yield and lowest chemical properties after 10 years of experimentation. However, straw returning and K fertilizer application significantly improved the soil chemical properties, including the SOM, available K, slowly available K, EOC, and DOC, compared with those of NP treatment; among these treatments, NPKS treatment performed the best (Table 2). However, NPK treatment reduced soil available N content, consistent with the findings of Liu et al. [18] who reported a decline in soil total N with chemical P or K fertilizer. The reason might be that NPK treatment produces more grain than NP treatment resulting in more N uptake and soil N consumption, particularly available N under the same input of chemical N fertilizer. Moreover, the contents of available K, EOC, and DOC were significantly higher in treatments NPKS and NPS than those in NPK and NP treatments, while the slowly available K content showed the opposite trend (Table 2). These results confirmed that the accurate application of K fertilizer rate could not maintain the soil K balance without regard to straw returning in the rice–oilseed rape rotation system [50].

K fertilizer in combination with straw returning could improve the soil available K content [4]. On the one hand, straw returning brings in a large amount of straw K into the farmland; on the other hand, the root secretion of crops and the humification in the process of residual straw rot would weaken the fixation of K^+ on the clay mineral, promoting transformation of slowly available K to water-soluble K and available K, and maintain a new dynamic balance of various soil K forms [51,52]. Notably, the soil EOC and DOC had significant differences between the straw incorporation and without straw. Previous studies demonstrated that straw decomposition facilitated the accumulation of active organic carbon and the release of nutrients in the soil [10]. Therefore, NPKS treatment has been proven to significantly improve soil K and OM content synchronously compared with no straw or K fertilizer input, thereby enhancing crop yield and soil fertility via microbial activities (Figure 7).

4.3. Effects of Long-Term Straw Returning and K Fertilizer on the Soil Microbial Alpha Diversity

As the results show, the microbial community structure and composition are closely related to the soil properties (Figure 7). Different fertilization treatments changed the physical and chemical soil properties of farmland, affecting the community diversity of soil microorganisms. Consistent with the previous results reported in the paddy-upland rotation system [53,54], alpha diversity analysis showed different effects of long-term straw incorporation and fertilization on the richness and diversity of the microbial community. The PCoA plot showed significant differences ($p < 0.05$) in bacterial and fungal community structure under different treatments (Figure 4), which was consistent with the results of Bai et al. [11], who documented clear separation of bacterial and fungal community composition between the straw utilization with no straw utilization and K fertilizer application with no K fertilizer. In this experiment, compared with NP treatment, the application of chemical K fertilizer (NPK treatment) for ten years caused bacterial and fungal richness index and diversity index to decrease (Table 3). Guo et al. [25] had shown that long-term single fertilizer significantly reduces the richness and diversity of bacteria, which was consistent with our result. However, the addition of crop straw (NPS and NPKS treatment) could increase the richness index and diversity index compared with those of NPK treatment. This was because straw incorporation provided exogenous organic carbon resources for bacteria and fungi living, which was conducive to their breeding growth, reduced competition between them, and enhanced diversity of soil bacterial and fungal communities [17,38]. Therefore, straw incorporation directly affects the microbial alpha diversity by promoting or inhibiting the change of soil bacterial and fungal composition, further impacting the soil biological fertility.

Furthermore, the correlation analysis of soil properties and diversity index (Tables S2 and S3) implied that the Simpson and Shannon indexes of bacteria and fungi were significantly positively correlated with SOM, EOC, and DOC and that they had a negative correlation with pH and available K. Prior studies all have shown that the adverse impact of soil acidification in paddy soil caused by the superfluous application of N or NPK fertilizer on soil microbial diversity far exceeded the positive effect of fertilization [55,56]. Therefore, applying organic fertilizer in combination with NPK chemical fertilizer was reported to be more effective than applying NPK fertilizer alone in the future for microbial diversity.

4.4. Effects of Long-Term Straw Returning and K Fertilizer on the Soil Microbial Community

At present, there has been no more attention as to the effect of long-term straw returning with K fertilizer on soil microbial community composition in paddy soil, and the reported results were not consistent [57]. In this study, the predominant bacterial phyla (Figure 2) in four treatments were Proteobacteria and Acidobacteria, at an average of 32.0% and 16.9%, which is consistent with those reported by Wang et al. [31] and Guo et al. [25] based on agricultural soils. Although the 10-year fertilization results in differences in soil nutrient content, the effect of that on the category of predominant bacteria in each treatment was not noteworthy. Proteobacteria, Nitrospirae, Firmicute, and Actinobacteria (i.e., R-strategist) are considered as copiotrophic groups, while Acidobacteria, Actinobacteria, Chloroflexi, and Planctomycetes were typical oligotrophic (i.e., K-strategist) bacteria [58]. Sun et al. [59] found that the relative abundance of Proteobacteria was significantly positively correlated with soil C and N content. Our study also confirmed that NPS treatment had a higher relative abundance of Proteobacteria than other treatments, while the application of K fertilizer had no significant impact on the dominant bacterial groups (Figure 2). Both long-term and short-term studies have found that the abundance of Acidobacteria decreased significantly with the increase in NPK fertilizer, which was closely associated with soil pH value [56,60], whereas, in our study, the relative abundance of Acidobacteria, serving as typically oligotrophic bacteria in NP (15.4%) and NPKS (24.3%) treatments, was higher than that in NPK (13.6%) and NPS (14.8%) treatments, in contradiction with previous research. This may be related to the increase in the special subgroup function of Acidobacteria, which, in terms of the genus, were unclassified Subgroup 3, unclassified Subgroup 6, unclassified Subgroup 17, and uncultured Desulfovirga sp. (Figure 3). Additionally, the content of available K in NP was not enough for plant growth, and thereby insufficient K application could stimulate the propagation of Acidobacteria phylum community, activating the insoluble mineral ions in the soil.

In the fungi groups, Ascomycota (eutrophic) and Basidiomycota (oligotrophic) were the dominant phyla [11,31] communities in the four treatments, at an average of 48.1% and 24.8%, respectively; they are important decomposers based on organic substrates, such as wood, fallen leaves, and feces [17]. Figure 2 shows that compared with NP treatment, NPS and NPKS treatments significantly increased the abundance of Ascomycota and significantly decreased the relative abundance of Basidiomycota and Mortierellomycota. In contrast, the NPK treatment had the contrary result, which is in accordance with that reported by Wang et al. [31]. From the result of Table 2, we observed that the NPK treatment had a lower SOM and active carbon content in contrast to the NPS and NPKS treatments. As a class of fungi that can decompose cellulose into Ascomycota, Chaetomium can decompose cellulase and xylanase, which play important roles in the carbon cycle of the natural ecosystem and can result in soil improvement [61,62]. This again proves that abundant organic carbon leads to an increase in the relative abundance of eutrophic fungi and a decrease in the relative abundance of oligotrophic fungi. Moreover, the relative abundance of Mortierellomycota and Olpidiomycota in the NP treatment was higher than that in others. Some of those species belonged to pathogenic fungi generating polyketides, terpenoids, and nonribosomal peptides to cause plant disease [31]; hence, insufficient K content may induce the growth and reproduction of harmful fungi in paddy soil [63]. Particularly, according to the results of the cluster analysis at the genus level, there were significant differences in the distribution of fungal species among nutrient deficiency treatments (Figure 5), which may be closely related to soil nutrient status. This phenomenon was also found in the results of redundancy analysis, i.e., EOC, DOC, and slowly available K had the maximal influence on the fungal community (Figure 6b).

As can be seen from Table 2, after 10 years of fertilization management, the soil organic C resource and the available K content changed significantly among treatments; in particular, the available K content in the NP and NPK treatments decreased significantly, which was mainly because crop harvest took away a large amount of K, leading to the imbalance of K in farmlands [53]. However, straw incorporation could clearly increase active organic C, as EOC and DOC. Therefore, the content of EOC, DOC, available K, and slowly available K became the most important index affecting the relative abundance of microbial community. Fan et al. [63] combined ecological network theory with the ecological resistance index to evaluate the responses of microbial community to wheat production under the condition of long-term fertilization. Their results suggest that the microbial resistance indirectly drives the effects of nutrient fertilization on plant production. Two mechanisms may explain the role of microbial resistance to nutrient fertilization in the promotion of plant production: (1) resistant microbial community with organic fertilizer addition could facilitate plants acquiring more nutrients and less competition from microbial species in soil; (2) low responsive microbial community may lead to lower relative abundance of potential fungal plant pathogens. Our research also indicated that the microbial diversity and community influenced rice yield and oilseed rape yield; moreover, the bacterial community had a higher impact than fungal community on the crop yield. Therefore, crop straw residue returning could alleviate the toxic effect of long-term potash fertilization on microbial population and stimulate crop yield as well as soil C sequestration.

5. Conclusions

In the present study, we found that ten years of continuous crop residue management and K fertilizer application in the rice–oilseed rape rotation significantly improved the crop yield; altered soil physiochemical properties such as SOC, EOC, DOC, available K, and slowly available K; and, therefore, modified the diversity and composition of soil bacterial and fungal communities. The long-term application of K fertilizer and straw returning had a significant increase rate in yield after one crop rotation, and NPKS treatment resulted in the best effect. The application of K fertilizer significantly decreased the richness and diversity index of soil bacterial and fungal populations compared to NP treatment. At the same time, continuous straw incorporation could alleviate the negative effect of K fertilizer on microbial composition. The NPKS treatment increased the relative abundance of the copiotrophic bacteria, such as the Firmicutes, Gemmatimonadetes, and Proteobacteria phyla, and the relative abundance of Ascomycota. Available K, SOM, DOC, and EOC were closely related to alterations in the composition of the soil bacterial community; EOC, DOC, and slowly available K were significantly correlated with fungal community. These findings provide deep insights into the role of straw incorporation coordinated with K fertilizer on the dynamic change of yield of rice and oilseed rape, and shape soil bacterial and fungal communities and their relationship with soil properties and crop yield.

Supplementary Materials: The following are available online at https://www.mdpi.com/article/10.3390/agriculture11121233/s1, Figure S1: Rarefaction curves for (a) bacteria and (b) fungi in the four treatments. Table S1: Detailed sequencing depth results of soil samples under different treatments. Table S2: The correction of soil properties with bacterial community structures. Table S3: The correction of soil properties with fungal community structures.

Author Contributions: Conceptualization, J.L. and J.Z.; methodology, J.L.; software, G.G.; formal analysis, X.C.; investigation, G.G. and X.C.; writing—original draft preparation, J.L.; writing—review and editing, J.L. and J.Z.; funding acquisition, J.L. and G.G. All authors have read and agreed to the published version of the manuscript.

Funding: This work was supported in part by the Open Fund of Key Laboratory of Fertilizer Utilization, Ministry of Agriculture and Rural Affairs, China (KLFAW201901) and the Open-end Fund of Engineering Research Center of Ecology and Agricultural Use of Wetland, Ministry of Education (KF2020015).

Conflicts of Interest: The authors declare no conflict of interest.

References

1. Bentsen, N.S.; Felby, C.; Thorsen, B.J. Agricultural residue production and potentials for energy and materials services. *Prog. Energy Combust. Sci.* **2014**, *40*, 59–73. [CrossRef]
2. Thiagarajan, A.; Fan, J.; McConkey, B.G.; Janzen, H.H.; Campbell, C.A. Dry matter partitioning and residue N content for 11 major field crops in Canada adjusted for rooting depth and yield. *Can. J. Soil Sci.* **2018**, *98*, 574–579. [CrossRef]
3. Li, X.; Li, Y.; Wu, T.; Qu, C.; Ning, P.; Shi, J.; Tian, X. Potassium fertilization combined with crop straw incorporation alters soil potassium fractions and availability in northwest China: An incubation study. *PLoS ONE* **2020**, *15*, e0236634. [CrossRef]
4. Lv, X.; Wang, Z.; Ma, L.; Cao, N.; Meng, Y.; Zhou, Z. Crop residue incorporation combined with potassium fertilizer increased cotton canopy apparent photosynthesis and seed cotton yield in barley-cotton rotation system. *Arch. Agron. Soil Sci.* **2021**, *67*, 300–312. [CrossRef]
5. Karami, A.; Homaee, M.; Afzalinia, S.; Ruhipour, H.; Basirat, S. Organic resource management: Impacts on soil aggregate stability and other soil physico-chemical properties. *Agric. Ecosyst. Environ.* **2012**, *148*, 22–28. [CrossRef]
6. Li, J.; Lu, J.; Li, X.; Ren, T.; Cong, R.; Zhou, L. Dynamics of potassium release and adsorption on rice straw residue. *PLoS ONE* **2014**, *9*, e90440. [CrossRef] [PubMed]
7. Liu, Z.; Liu, J.; Yu, Z.; Yao, Q.; Li, Y.; Liang, A.; Zhang, W.; Mi, G.; Jin, J.; Liu, X.; et al. Long-term continuous cropping of soybean is comparable to crop rotation in mediating microbial abundance, diversity and community composition. *Soil Tillage Res.* **2020**, *197*, 104503. [CrossRef]
8. Kumar, P.; Kumar, T.; Singh, S.; Tuteja, N.; Prasad, R.; Singh, J. Potassium: A key modulator for cell homeostasis. *J. Biotechnol.* **2020**, *324*, 198–210. [CrossRef] [PubMed]
9. Li, X.K.; Lu, J.W.; Wu, L.S.; Chen, F. The difference of potassium dynamics between yellowish red soil and yellow cinnamon soil under rapeseed (Brassica napus L.)–rice (Oryza sativa L.) rotation. *Plant Soil* **2009**, *320*, 141–151. [CrossRef]
10. Yuan, G.; Huan, W.; Song, H.; Lu, D.; Chen, X.; Wang, H.; Zhou, J. Effects of straw incorporation and potassium fertilizer on crop yields, soil organic carbon, and active carbon in the rice-wheat system. *Soil Tillage Res.* **2021**, *209*, 104958. [CrossRef]
11. Bai, N.; Zhang, H.; Li, S.; Zheng, X.; Zhang, J.; Zhang, H.; Zhou, S.; Sun, H.; Lv, W. Long-term effects of straw and straw-derived biochar on soil aggregation and fungal community in a rice-wheat rotation system. *PeerJ* **2019**, *6*, e6171. [CrossRef] [PubMed]
12. Pathak, H.; Mohanty, S.; Jain, N.; Bhatia, A. Nitrogen, phosphorus, and potassium budgets in Indian agriculture. *Nutr. Cycl. Agroecosystems* **2009**, *86*, 287–299. [CrossRef]
13. Song, X.D.; Liu, F.; Wu, H.Y.; Cao, Q.; Zhong, C.; Yang, J.L.; Li, D.C.; Zhao, Y.G.; Zhang, G.L. Effects of long-term K fertilization on soil available potassium in East China. *Catena* **2020**, *188*, 104412. [CrossRef]
14. Xu, X.; Mao, X.; Van Zwieten, L.; Niazi, N.K.; Lu, K.; Bolan, N.S.; Wang, H. Wetting-drying cycles during a rice-wheat crop rotation rapidly (im)mobilize recalcitrant soil phosphorus. *J. Soils Sediments* **2020**, *20*, 3921–3930. [CrossRef]
15. Chen, H.; Jarosch, K.A.; Mészáros, É.; Frossard, E.; Zhao, X.; Oberson, A. Repeated drying and rewetting differently affect abiotic and biotic soil phosphorus (P) dynamics in a sandy soil: A 33P soil incubation study. *Soil Biol. Biochem.* **2021**, *153*, 108079. [CrossRef]
16. Erinle, K.O.; Marschner, P. Soil water availability influences P pools in the detritusphere of crop residues with different C/P ratios. *J. Soil Sci. Plant Nutr.* **2019**, *19*, 771–779. [CrossRef]

17. Hiel, M.P.; Barbieux, S.; Pierreux, J.; Olivier, C.; Lobet, G.; Roisin, C.; Garre, S.; Colinet, G.; Bodson, B.; Dumont, B. Impact of crop residue management on crop production and soil chemistry after seven years of crop rotation in temperate climate, loamy soils. *PeerJ* **2018**, *6*, e4836. [CrossRef]
18. Liu, Z.; Zhou, H.; Xie, W.; Yang, Z.; Lv, Q. Long-term effects of maize straw return and manure on the microbial community in cinnamon soil in Northern China using 16S rRNA sequencing. *PLoS ONE* **2021**, *16*, e0249884. [CrossRef]
19. Yu, C.; Li, Y.; Mo, R.; Deng, W.; Zhu, Z.; Liu, D.; Hu, X. Effects of long-term straw retention on soil microorganisms under a rice-wheat cropping system. *Arch. Microbiol.* **2020**, *202*, 1915–1927. [CrossRef]
20. Ma, Z.; Xie, Y.; Zhu, L.; Cheng, L.; Xiao, X.; Zhou, C.; Wang, J. Which of soil microbes is in positive correlation to yields of maize (Zea mays L.)? *Plant Soil Environ.* **2017**, *63*, 574–580. [CrossRef]
21. Chen, Y.; Xin, L.; Liu, J.; Yuan, M.; Liu, S.; Jiang, W.; Chen, J. Changes in bacterial community of soil induced by long-term straw returning. *Sci. Agric.* **2017**, *74*, 349–356. [CrossRef]
22. Srour, A.Y.; Ammar, H.A.; Subedi, A.; Pimentel, M.; Cook, R.L.; Bond, J.; Fakhoury, A.M. Microbial communities associated with long-term tillage and fertility treatments in a corn-soybean cropping system. *Front. Microbiol.* **2020**, *11*, 18. [CrossRef] [PubMed]
23. Ling, N.; Chen, D.; Guo, H.; Wei, J.; Bai, Y.; Shen, Q.; Hu, S. Differential responses of soil bacterial communities to long-term N and P inputs in a semi-arid steppe. *Geoderma* **2017**, *292*, 25–33. [CrossRef]
24. Tian, H.; Wang, H.; Hui, X.; Wang, Z.; Drijber, R.A.; Liu, J. Changes in soil microbial communities after 10 years of winter wheat cultivation versus fallow in an organic-poor soil in the Loess Plateau of China. *PLoS ONE* **2017**, *12*, e0184223. [CrossRef]
25. Guo, Z.; Liu, H.; Wan, S.; Hua, K.; Wang, D.; Guo, X.; He, C. Fertilization practice changes rhizosphere microbial community structure in the agroecosystem. *Ann. Appl. Biol.* **2019**, *174*, 123–132. [CrossRef]
26. Ullah, S.; He, P.; Ai, C.; Zhao, S.; Ding, W.; Song, D.; Zhang, J.; Huang, S.; Abbas, T.; Zhou, W. How do soil bacterial diversity and community composition respond under recommended and conventional nitrogen fertilization regimes? *Microorganisms* **2020**, *8*, 1193. [CrossRef] [PubMed]
27. Wasaki, J.; Sakaguchi, J.; Yamamura, T.; Ito, S.; Shinano, T.; Osaki, M.; Kandeler, E. P and N deficiency change the relative abundance and function of rhizosphere microorganisms during cluster root development of white lupin (Lupinus albus L.). *Soil Sci. Plant Nutr.* **2018**, *64*, 686–696. [CrossRef]
28. Dai, Z.; Liu, G.; Chen, H.; Chen, C.; Wang, J.; Ai, S.; Wei, D.; Li, D.; Ma, B.; Tang, C.; et al. Long-term nutrient inputs shift soil microbial functional profiles of phosphorus cycling in diverse agroecosystems. *ISME J.* **2020**, *14*, 757–770. [CrossRef] [PubMed]
29. Yu, Z.; Hu, X.; Wei, D.; Liu, J.; Zhou, B.; Jin, J.; Liu, X.; Wang, G. Long-term inorganic fertilizer use influences bacterial communities in Mollisols of Northeast China based on high-throughput sequencing and network analyses. *Arch. Agron. Soil Sci.* **2019**, *65*, 1331–1340. [CrossRef]
30. Fan, K.K.; Delgado-Baquerizo, M.; Guo, X.S.; Wang, D.Z.; Zhu, Y.G.; Chu, H.Y. Biodiversity of key-stone phylotypes determines crop production in a 4-decade fertilization experiment. *ISME J.* **2021**, *15*, 550–561. [CrossRef]
31. Wang, J.; Song, Y.; Ma, T.; Raza, W.; Li, J.; Howland, J.G.; Huang, Q.; Shen, Q. Impacts of inorganic and organic fertilization treatments on bacterial and fungal communities in a paddy soil. *Appl. Soil Ecol.* **2017**, *112*, 42–50. [CrossRef]
32. Mcbain, A.J.; Bartolo, R.G.; Catrenich, C.E.; Charbonneau, D.; Ledder, R.G.; Price, B.B.; Gilbert, P. Exposure of sink drain microcosms to triclosan: Population dynamics and antimicrobial susceptibility. *Appl. Environ. Microbiol.* **2003**, *69*, 5433–54442. [CrossRef] [PubMed]
33. Caporaso, J.G.; Kuczynski, J.; Stombaugh, J.; Bittinger, K.; Bushman, F.D.; Costello, E.K.; Fierer, N.; Pena, A.G.; Goodrich, J.K.; Gordon, J.I.; et al. QIIME allows analysis of high-throughput community sequencing data. *Nat. Methods* **2010**, *7*, 335–336. [CrossRef]
34. Radhakrishnan, S.; Varadharajan, M. Status of microbial diversity in agroforestry systems in Tamil Nadu, India. *J. Basic Microbiol.* **2016**, *56*, 662–669. [CrossRef]
35. Schloss, P.D.; Westcott, S.L.; Ryabin, T.; Hall, J.R.; Hartmann, M.; Hollister, E.B.; Lesniewski, R.A.; Oakley, B.B.; Parks, D.H.; Robinson, C.J.; et al. Introducing mothur: Open-source, platform-independent, community-supported software for describing and comparing microbial communities. *Appl. Environ. Microbiol.* **2009**, *75*, 7537–7541. [CrossRef] [PubMed]
36. Barberan, A.; Ramirez, K.S.; Leff, J.W.; Bradford, M.A.; Wall, D.H.; Fierer, N. Why are some microbes more ubiquitous than others? Predicting the habitat breadth of soil bacteria. *Ecol. Lett.* **2014**, *17*, 794–802. [CrossRef] [PubMed]
37. Yan, F.; Sun, Y.; Hui, X.; Jiang, M.; Xiang, K.; Wu, Y.; Zhang, Q.; Tang, Y.; Yang, Z.; Sun, Y.; et al. The effect of straw mulch on nitrogen, phosphorus and potassium uptake and use in hybrid rice. *Paddy Water Environ.* **2019**, *17*, 23–33. [CrossRef]
38. Huang, W.; Wu, J.F.; Pan, X.H.; Tan, X.M.; Zeng, Y.J.; Shi, Q.H.; Liu, T.J.; Zeng, Y.H. Effects of long-term straw return on soil organic carbon fractions and enzyme activities in a double-cropped rice paddy in South China. *J. Integr. Agric.* **2021**, *20*, 236–247. [CrossRef]
39. Wu, W.; Ma, B.; Uphoff, N. A review of the system of rice intensification in China. *Plant Soil* **2015**, *393*, 361–381. [CrossRef]
40. Tian, Z.; Ji, Y.; Sun, L.; Xu, X.; Fan, D.; Zhong, H.; Liang, Z.; Ficsher, G. Changes in production potentials of rapeseed in the Yangtze River basin of China under climate change: A multi-model ensemble approach. *J. Geogr. Sci.* **2018**, *28*, 1700–1714. [CrossRef]
41. Li, H.; Lu, J.; Ren, T.; Li, X.; Cong, R. Nutrient efficiency of winter oilseed rape in an intensive cropping system: A regional analysis. *Pedosphere* **2017**, *27*, 364–370. [CrossRef]

42. Wang, X.; Sun, B.; Mao, J.; Sui, Y.; Cao, X. Structural convergence of maize and wheat straw during two-year decomposition under different climate conditions. *Environ. Sci. Tech.* **2012**, *46*, 7159–7165. [CrossRef] [PubMed]
43. Bai, Y.; Wang, L.; Lu, Y.; Yang, L.; Zhou, L.; Ni, L.; Cheng, M. Effects of long-term full straw return on yield and potassium response in wheat-maize rotation. *J. Integr. Agric.* **2015**, *14*, 2467–2476. [CrossRef]
44. Li, J.F.; Zhong, F.F. Nitrogen release and re-adsorption dynamics on crop straw residue during straw decomposition in an Alfisol. *J. Integr. Agric.* **2021**, *20*, 248–259. [CrossRef]
45. Segnini, A.; Nunes Carvalho, J.L.; Bolonhezi, D.; Bastos Pereira Milori, D.M.; Lopes da Silva, W.T.; Simoes, M.L.; Cantarella, H.; de Maria, I.C.; Martin-Neto, L. Carbon stock and humification index of organic matter affected by sugarcane straw and soil management. *Sci. Agric.* **2013**, *70*, 321–326. [CrossRef]
46. Wang, Y.; Zheng, J.; Boyd, S.E.; Xu, Z.; Zhou, Q. Effects of litter quality and quantity on chemical changes during eucalyptus litter decomposition in subtropical Australia. *Plant Soil* **2019**, *442*, 65–78. [CrossRef]
47. Zribi, W.; Faci, J.M.; Araguees, R. Mulching effects on moisture, temperature, structure and salinity of agricultural soils. *ITEA Inf. Tec. Econ. Agrar.* **2011**, *107*, 148–162. [CrossRef]
48. Akhtar, K.; Wang, W.; Ren, G.; Khan, A.; Feng, Y.; Yang, G. Changes in soil enzymes, soil properties, and maize crop productivity under wheat straw mulching in Guanzhong, China. *Soil Tillage Res.* **2018**, *182*, 94–102. [CrossRef]
49. Zhang, Z.; Cong, R.H.; Ren, T.; Li, H.; Zhu, Y.; Lu, J.W. Optimizing agronomic practices for closing rapeseed yield gaps under intensive cropping systems in China. *J. Integr. Agric.* **2020**, *19*, 1241–1249. [CrossRef]
50. Zhu, D.; Zhang, J.; Lu, J.; Cong, R.; Ren, T.; Li, X. Optimal potassium management strategy to enhance crop yield and soil potassium fertility under paddy-upland rotation. *J. Sci. Food Agric.* **2020**, *101*, 3404–3412. [CrossRef]
51. Madar, R.; Singh, Y.V.; Meena, M.C.; Das, T.K.; Gaind, S.; Verma, R.K. Potassium and residue management options to enhance productivity and soil quality in zero till maize-wheat rotation. *Clean Soil Air Water* **2020**, *48*, 1900316. [CrossRef]
52. Xu, Q.; Fu, H.; Zhu, B.; Hussain, H.A.; Zhang, K.; Tian, X.; Duan, M.; Xie, X.; Wang, L. Potassium improves drought stress tolerance in plants by affecting root morphology, root exudates and microbial diversity. *Metabolites* **2021**, *11*, 131. [CrossRef] [PubMed]
53. Han, Y.; Ma, W.; Zhou, B.; Salah, A.; Geng, M.; Cao, C.; Zhan, M.; Zhao, M. Straw return increases crop grain yields and K-use efficiency under a maize-rice cropping system. *Crop. J.* **2021**, *9*, 168–180. [CrossRef]
54. Zhao, Y.; Chen, Y.; Dai, H.; Cui, J.; Wang, L.; Sui, P. Effects of organic amendments on the improvement of soil nutrients and crop yield in sandy soils during a 4-year field experiment in Huang-Huai-Hai plain, northern China. *Agronomy* **2021**, *11*, 157. [CrossRef]
55. Kallenbach, C.M.; Frey, S.D.; Grandy, A.S. Direct evidence for microbial-derived soil organic matter formation and its ecophysiological controls. *Nat. Commun.* **2016**, *7*, 13630. [CrossRef]
56. Zeng, J.; Liu, X.; Song, L.; Lin, X.; Zhang, H.; Shen, C.; Chu, H. Nitrogen fertilization directly affects soil bacterial diversity and indirectly affects bacterial community composition. *Soil Biol. Biochem.* **2016**, *92*, 41–49. [CrossRef]
57. Cai, C.; Liu, H.; Dai, X.; Whalen, J.K. Multiple selection of resistance genes in arable soil amended with cephalosporin fermentation residue. *Soil Biol. Biochem.* **2019**, *136*, 107538. [CrossRef]
58. Fierer, N.; Lauber, C.L.; Ramirez, K.S.; Zaneveld, J.; Bradford, M.A.; Knight, R. Comparative metagenomic, phylogenetic and physiological analyses of soil microbial communities across nitrogen gradients. *ISME J.* **2012**, *6*, 1007–1017. [CrossRef] [PubMed]
59. Sun, R.; Zhang, X.X.; Guo, X.; Wang, D.; Chu, H. Bacterial diversity in soils subjected to long-term chemical fertilization can be more stably maintained with the addition of livestock manure than wheat straw. *Soil Biol. Biochem.* **2015**, *88*, 9–18. [CrossRef]
60. Ma, M.; Jiang, X.; Wang, Q.; Ongena, M.; Wei, D.; Ding, J.; Guan, D.; Cao, F.; Zhao, B.; Li, J. Responses of fungal community composition to long-term chemical and organic fertilization strategies in Chinese Mollisols. *Microbiologyopen* **2018**, *7*, e00597. [CrossRef]
61. Fan, K.K.; Weisenhorn, P.; Gilbert, J.A.; Chu, H.Y. Wheat rhizosphere harbors a less complex and more stable microbial co-occurrence pattern than bulk soil. *Soil Biol. Biochem.* **2018**, *125*, 251–260. [CrossRef]
62. Gu, Y.; Wang, Y.; Wang, P.; Wang, C.; Ma, J.; Yang, X.; Ma, D.; Li, M. Study on the diversity of fungal and bacterial communities in continuous cropping fields of Chinese chives. *BioMed Res. Int.* **2020**, *5*, 1–14. [CrossRef] [PubMed]
63. Fan, K.K.; Delgado-Baquerizo, M.; Guo, X.S.; Wang, D.Z.; Zhu, Y.G.; Chu, H.Y. Microbial resistance promotes plant production in a four-decade nutrient fertilization experiment. *Soil Biol. Biochem.* **2020**, *141*, 107679. [CrossRef]

MDPI
St. Alban-Anlage 66
4052 Basel
Switzerland
Tel. +41 61 683 77 34
Fax +41 61 302 89 18
www.mdpi.com

Agriculture Editorial Office
E-mail: agriculture@mdpi.com
www.mdpi.com/journal/agriculture

www.ingramcontent.com/pod-product-compliance
Lightning Source LLC
Chambersburg PA
CBHW061816210926
51850CB00006B/722
9783036567761